World Sustainability Series

Series Editor

Walter Leal Filho, European School of Sustainability Science and Research, Research and Transfer Centre "Sustainable Development and Climate Change Management", Hamburg University of Applied Sciences, Hamburg, Germany

Due to its scope and nature, sustainable development is a matter which is very interdisciplinary, and draws from knowledge and inputs from the social sciences and environmental sciences on the one hand, but also from physical sciences and arts on the other. As such, there is a perceived need to foster integrative approaches, whereby the combination of inputs from various fields may contribute to a better understanding of what sustainability is, and means to people. But despite the need for and the relevance of integrative approaches towards sustainable development, there is a paucity of literature which address matters related to sustainability in an integrated way.

Notes on the quality assurance and peer review of this publication

Prior to publication, the works published in this book are initially assessed and reviewed by an in-house editor. If suitable for publication, manuscripts are sent for further review, which includes a combined effort by the editorial board and appointed subject experts, who provide independent peer-review. The feedback obtained in this way was communicated to authors, and with manuscripts checked upon return before finally accepted. The peer-reviewed nature of the books in the "World Sustainability Series" means that contributions to them have, over many years, been officially accepted for tenure and promotion purposes.

Firdous Ahmad Malik · Silky Vigg Kushwah ·
Muhammad Shahzad Aslam · Reyed M. Reyed ·
A. K. Haghi
Editors

Aligning ESG Goals with the UNs Sustainable Development Goals Planning

Strategies for Equitable Financing and World Sustainability

Springer

Editors
Firdous Ahmad Malik
Department of Management
University of People
Pasadena, CA, USA

Silky Vigg Kushwah
Department of Finance
New Delhi Institute of Management
New Delhi, Delhi, India

Muhammad Shahzad Aslam
Xiamen University
Sepang, Malaysia

Reyed M. Reyed
City of Scientific Research and Technology
Alexandria, Egypt

A. K. Haghi
University of Coimbra
Coimbra, Portugal

ISSN 2199-7373 ISSN 2199-7381 (electronic)
World Sustainability Series
ISBN 978-3-032-01714-7 ISBN 978-3-032-01715-4 (eBook)
https://doi.org/10.1007/978-3-032-01715-4

This Springer imprint is published by the registered company Springer Nature Switzerland AG
The registered company address is: Gewerbestrasse 11, 6330 Cham, Switzerland

Contents

Managing Global Public Goods: A Sustainable Governance Approach

Shahnawaz Qadri and Muzammil Ahad Dar

Abstract This chapter provides a comprehensive analysis of the roles of public goods, global commons, and environmental management in achieving sustainable development. There are critical environmental commons like atmospheric systems, oceanic resources, and biodiversity reserves. The significance of global commons, is examined in detail, emphasizing their vulnerability and mitigation(s) essential for maintaining various forms of capital stocknatural, human, and social to ensure economic stability and ecological resilience against global challenges. It explores the critical roles of institutional and governance challenges in managing global public goods (GPGs) and common-pool environmental resources. Through a systematic review of contemporary governance paradigms/models, we evaluate four primary intervention strategies: (1) regulatory policy instruments, (2) market-based economic incentives (Eco-economics), (3) community-centered participatory approaches, including technological innovation pathways. The study makes three contributions to the literature by identifying key critical success factors. First, that sustainable management requires interconnected policy approaches combining real-time monitoring systems, adaptive evaluation frameworks, and iterative policy refinement for effectiveness governance. Second, related to it is critical institutional design principles that enable successful long-term management of resources characterized by non-excludability and subtractability. Third, specific policy mechanisms: (a) establishment of clear property rights regimes-like tradable permit systems, payment for ecosystem services (PES), and community-based monitoring (b) Internalization of environmental externalities through innovative economic instruments, and (c) development of scalable governance models that bridge local and global conservation efforts. In conclusion, the chapter reaffirms that a sustainable governance approach, centered on cooperation, innovation, and inclusiveness, is imperative for the effective management of global public goods and the promotion of sustainable development.

S. Qadri (✉)
Faculty of Law, SGT University, Gurugram Badli Road, Gurugram 122505, Haryana, India
e-mail: shah.nawaz_fol@sgtuniversity.org

M. A. Dar
Political Science, Kumarguru College of Liberal Arts and Science, Coimbatore, India

F. A. Malik et al. (eds.), *Aligning ESG Goals with the UNs Sustainable Development Goals Planning*, World Sustainability Series,
https://doi.org/10.1007/978-3-032-01715-4_1

Keywords Sustainable management · Governance · Environmental resources · Eco-system services · Internalizing-externalities · Capital · Income-growth · Tradable permits · Total economic value

1 Introduction

Global public goods (GPGs) and global commons like clean air, oceans, atmosphere, face overuse, over-exploitation at global scale is a major cause of concern for world community. Equally important is the effective environment management as (mis-) management concerns do more harm to global commons, than its overuse, (e.g. the case of efficiency) necessitating global efforts for cooperative governance. Effective governance model for global public goods (GPGs) and integrated capital management of these resources (natural, human, social) supplement sustainable use, enhance conservation prospects for public goods and resources. Preserving or stabilizing environmental eco-systems proportionally secures, various forms of capitals like natural, human, and social, through such planned strategies. A comprehensive approach, integrated one, to include policy regulation, economic incentives, community engagement, education, and technological innovation (The Wetland Book 2018). Managing global public goods is simply an attempt to overcome governance ignorance to manage environmental sustainability, resources safeguards, and achieve sustainable development goals.

Sustainable development is a *multifaceted concept* (Whitehead et al. 2018) that aims to balance economic growth with environmental preservation and social well-being. At its core, it involves maintaining or increasing *'real disposable income'* of an economy to supplement long-term sustainability of natural, human, and social capital (Potschin et al. 2016). Public goods and global commons, such as clean air, oceans, and biodiversity, play crucial role in this process (Potschin et al. 2016). Effective environmental management sustainable use and resource conservation can be secured. Sustainable development implies that the *'real disposable income'*, measured usually in terms of *'real Net Domestic Product (NDP)'*, should increase over a period of time or either remain constant. For the *real income* (Ahmed et al. 1989) to be constant, the *'total capital stock'* of an economy, which is the total productive assets need to be maintained properly on a long-term basis (Schepelmann et al. 2010). The total capital stock consists of man-made capital (e.g. built infrastructure, machines, technology, etc.), natural capital (e.g. forests, biodiversity, water resources, etc.), human capital (e.g. education, skills, knowledge, etc.) and social capital (e.g. culture) (OECD 2004).

The natural capital, similar to other forms of capital, to generates services or benefits –called, ecosystem benefits -which can be treated as 'income' to the economy. In a developing economy like India, income or the ecosystem services from natural capital play crucial role to reduce poverty and enhance livelihoods of a large number of poor people (Patil et al. 2021). The natural capital is akin to a 'fixed deposit' in the bank and the income originating from these natural sources is equivalent to the 'interest' earned from the deposit. When the size of the deposit increases, the income originating from it will also increase. Similarly, if the entire deposit is eroded, then there will be no income. It means rather than '*income growth*', improvement of natural capital (e.g. GDP vs. NAC) drives people's welfare on long-term basis (Saha and Agarwalla 2021).

If a country's rate of harvest of its natural resources is equivalent to the rate of regeneration of the resources, then the income will be constant over a period of time. If the rate of harvest is lesser than that of regeneration, then income can be further maximized and sustained in future as well. If the rate of harvest is greater than that of regeneration, then the country will lead to eventual impoverishment. Therefore, a country's sustainable income and economic welfare depends largely on how that country manages its natural environment. However, to manage environment on a sustainable basis requires understanding of the types of environmental resources a country owns. Understanding classification of the resources is a pre-requisite for addressing an important normative issue, namely, which institution is most efficient in managing a particular type of environmental resource. For example, local community managing a village forest cannot manage a much larger sized forest, say, forest in the entire Western Ghats region; the latter may require combined institutional management from government, community and markets (Saha and Agarwalla 2021).

2 Typology of Environmental Resources

Generally, environmental resources are classified into four major categories: (a) Public Goods; (b) Private Goods; (c) Common Pool Resources; and (d) Common Property Resources (CourseHero available at www.coursehero.com). There are two criteria which are used to classify the resources into above categories. They are: *excludability* and *subtractability*. The degree of *excludability* and *subtractability* fundamentally determines the type of the resources.

Public Goods: Non-Excludable, Non-Subtractable (E.G., Carbon Sequestration).

The 'public goods' are those which possess 'non-excludability' and 'non-subtractability' characteristics. What does non-excludability mean? Once a resource or good is provided, no potential consumer can be prevented from consuming the benefits of the good. In other words, the marginal cost of excluding a consumer is exorbitantly high. Also, the marginal cost of providing the good or service to one more consumer is zero (Peterson et al. 2019). Let us take, national defense. National defense protects every citizen in the country, irrespective of their identity. So, once it is provided, everyone in the society enjoys same amount of the good and a new consumer (a citizen) also enjoys it without any additional cost being incurred. Suppose, a baby is born in the country. Does the government incur any additional cost for providing national defense to that baby? No. With the existing army, the service can be provided to the new born baby as well. Can the baby be excluded from consuming the benefits of national defense? No. Because, it is extremely difficult to do so. If at all to be excluded, the baby will have to be shifted to another country which is extremely a costly affair. So, non-excludability condition applies here in the case of national defense.

What is meant by non-subtractability. As far as public goods are concerned, consumption of the good or service by one consumer does not reduce the consumption of the same by another consumer. The benefits of national defense in India are consumed by around 1300 million citizens. One citizen's consumption does not affect another citizen's consumption and everyone consumes equal amount of benefit from national defense. Similarly, consumption by a new born baby does not reduce the consumption of the defense service by the existing 1300 million citizens in the country. There are many environmental goods and services which are public goods in nature. For example, the carbon sequestration service provided by the forests is a public good. All individuals have equal access, without being excluded, to carbon sequestration-making it non-excludable—which occurs to any consumer at any part of the world. Similarly, there is no rival-ness in consumption of this service across individuals since one person's consumption does not reduce the consumption of others, i.e., non-subtractable.

Private Goods: Private goods are just the opposite of public goods. The excludability and subtractability conditions are applicable to private goods. When we say private goods are excludable, it means that one can prevent a person from consuming a private good through a mechanism. The cost of excluding is also very negligible. For example, a shop owner can prevent a person from consuming a bottle of soft drink that she possesses—unless the consumer pays for the soft drink. Once the consumer pays for it, it becomes the consumer's own property and now the shop keeper cannot claim ownership of that soft drink. Similarly, subtractability condition is also in operation here. Consumption of a unit of soft drink by one consumer will reduce the consumption of the same unit by a potential consumer. If one soft drink in a case of 12 bottles has been consumed by somebody, only 11 bottles are available for future consumption and consumption of the 12th person in the line it will be denied. Many environmental goods are private goods in nature. For example, agricultural land, groundwater sold a by a farmer for commercial purpose and rainwater harvested by individuals are some of the examples.

3 Literature Review

Issues of environment degradation and economics undeniably are *not separate* from each other. However, whether *'growth or protection'* which one finds *'priority'* among people at large formulates a trade-off between economic gains and environmental goals (Čábelková et al. 2023). The inventory of global warming, environmental degradation took off with a parabola of *exploitation* of resources paraphrasing to more emphasis on resource protection over the loss of economy. Over the years sustainable development concept reconfigured the issue championing balancing of economic gains not against but with resources sustenance (Green Saving 2004). The

'World Development Report-2023' mentioned there is an *'interchange'* of 'environmental sustainability, 'natural asset depreciation', and level of governance that bears resonance in management of EMDEs' and attaining Sustainable Development Goals (SDGs) (World Bank 2023). Critical policy analysts have in recent researches tried to analyze *'impacts'*; how issues of environmental sustainability, natural capital preservation, with governance frameworks encourage 'market and developing economies (EMDEs)' and achieve (Barbier and Burgees 2023). Existing literature on the subject of global public goods (GPGs) and global commons depreciation offer a scenario subscribing environmental issues are interlinked to economics not with reference to the equation of eco-economics. This recent shift has occurred where it is been argued that not protection, conservation, but managing natural resources is actually *economic gains* multiplying or improving national income.

Multilateral efforts like '*UNEP-1992*' report highlighting that global environmental problems have not only multiplied but worsened and continuing. *'Paris Agreement-2015'* a roadmap regulation model to limit or reduce the greenhouse gas emissions was wrought in political wrangling. Literature on the subject has also focused how governments, markets have helped management of public goods with reference to consumerism practices, public goods as services and benefits eventually adding to real disposal income of a country. From Sierra Club's law suit 1960s in US discussing nature's legal rights (Challe 2021) (?) to issue of economic valuation global public goods and global commons have acquired center stage with respect to their management and governance. In 2013 environmental damage by global industries reached $7.3 trillion equal to 13% of world GDP. It is estimated that by 2050 the *'unchecked externalities'* would come close to 18% of global GDP seriously effecting economic stability (TruCost and TEEB 2014).

'Classifying' global public goods (GPGs) and global commons by their characteristics called 'criteria' (non-excludable, non-subtractable, excludability and subtractability) gives ample scope for research and studies on global common goods and their management. Since sustainable development of which environmental management is a part explores possible roadmaps, solutions to achieve SDGs. Many environmental goods are private goods in nature (e.g., agricultural land, rainwater harvested). Apart from typology discussed, common pool resources or global commons, since they are 'commons' and with *subtractability* character governance and management become extremely urgent as over-consumption inhabits potential danger of extinction.

Studies have clearly demonstrated that in developing countries, it is not *income growth* but improving the quantity and quality of natural capital that acts as a key factor for improving the welfare of the people on a long-term basis. Some empirical studies have demonstrated that people in forest dependent states like Madhya Pradesh earn a significant amount of their household income from forest ecosystem services (Patil et al. 2021). Further it is observed that natural capitals are analogous to Bank fixed deposit generating interest-incomes. Examples are *'Norway's REDD + programs'* (McNeilli 2015) of carbon credits, Finland's '*responsible logging*' or Ecotourism Revenue. However, to manage environment on a sustainable basis requires understanding of the types of environmental resources a country owns.

Understanding classification of the resources is a pre-requisite for addressing an important normative issue, namely, which institution is most efficient in managing a particular type of environmental resource. For example, local community managing a village forest cannot manage a much larger sized forest, say, forest in the entire Western Ghats region; the latter may require combined institutional management from government, community and markets (McNeilli 2015). Further arguments can also be raised like institutions (e.g. good governance and vibrant policy) grow due to economic growth or vice versa, however, is another debate in the domain of international political economy. Here how? sustainable development with the help of sustainable governance can help manage global public goods and global commons is where this discourse leads us to think.

Therefore, firstly, it is argued that resource-management or governance may perpetuate sustainable growth leading to welfare-maximizing especially forest harvest dependent-community centered population. Our second point of departure is that providing rights, economic valuation of productive activities-externalities (positive and negative) are core to managing global public goods and commons. Thirdly, the 'non cooperative human nature/behaviour' of Garret Hardin's, causing tragedy of the commons:

(a) on the possibility of extinction due to over-consumption added with Garret Hardin's *selfish maximizing individual's* non-cooperative behaviour a dominant strategy in the management of global public goods and global commons. (b) Has nature legal rights? Nobody owns a well-defined property right over the resource and therefore, it becomes everybody's property. Since the users maximizing objective, itself brings welfare loss to themselves, Hardin described this phenomenon 'tragedy of the commons' problem. If the resource use is not regulated adequately, the global commons become too vulnerable to tragedy of the commons problem. Therefore, the chapter explores possible roadmaps/solutions for the management of global commons with sustainable governance approach as follows:

3.1 Objectives

1. 'Optimal' degree of protection of GPGs cost–benefit analysis
2. Analyze GPG management challenges (e.g., externality internalization).
3. Evaluate policy tools (Pigouvian taxes, tradable permits) for sustainability.
4. Propose frameworks for international cooperation (example: "EU's Common Fisheries Policy" (CFP)).

Hypothesis

Efficient level of environmental conservation and governance of GPGs and global commons leads to welfare-maximizing.

Research Methodology: A mixed-methods approach combining theoretical frameworks (e.g., Hardin's Tragedy of the Commons, Ostrom's CPR theory) with policy analysis (Pigouvian taxes, Coasian solutions) and case studies (e.g., fisheries, wetlands).

4 Discussion-I

Commons Pool Resources (Sterner and Coria 2012)*(CPRs): Subtractability risks (e.g., fisheries).*

While public goods and private goods fall at two extremes in the continuum, 'common pool resources' and 'common property resources' fall in between the two extremes. Common pool resources, usually called, commons, possess 'non-excludability' but, 'subtractability' characteristics. These resources largely include, 'global commons. Deep sea fishery, atmosphere, ozone layer and biodiversity are some of the examples of global commons. Since it is extremely difficult to exclude the potential users or consumers from accessing these resources, these commons experience subtractability due to over-consumption. Garret Hardin predicted that the selfish maximizing individuals using these commons will compete with each other to over- exploit the limited resources, ultimately leading to extinction of these resources which in turn will make these individuals worse-off (Hardin 1968). If everyone in the group cooperates with each other by utilizing the resource in a limited way, then the resource may be available for all of them in the long run. But the problem here is that everyone thinks that if he or she cooperates and others do not cooperate, then the person who cooperates will loose and non-co-operators will benefit more. As a result, non-cooperative behaviour becomes a dominant strategy of the users. Nobody owns a well-defined property right over the resource and therefore, it becomes everybody's property. Since the users maximizing objective, itself brings welfare loss to themselves, Hardin described this phenomenon 'tragedy of the commons' problem. If the resource use is not regulated adequately, the global commons become too vulnerable to tragedy of the commons problem. For example, every country emits 'green house gases' (GHGs) causing global warming problem. The atmosphere as a global common is subjected to vulnerability and as a result, climate change has come into being a major challenge to the humankind. Increased draughts, frequent cyclones, increased floods, sea-level rise and loss of biodiversity are some of the serious consequences of the global warming problem. This is a major tragedy of the commons problem faced by the world now.

(i) *Policy Discussion Through Prisoner's Dilemma and Tragedy of the Commons Approach*

Tragedy of the commons can be explained through the story of Prisoner's dilemma (*illustrated in* Box 1). A murder took place in a city and the police have got some clue that individuals X and Y have committed the crime but the police have no strong evidence to punish them. They caught the suspects −X and Y− and put them in two separate cells so that there will be no communication between the two. The police tell each one of them the following: If both of them confess that they have committed the crime, then they will get 6 years jail term each (−6X, −6Y); If X confesses (that both of them have committed the crime) and Y does not confess, then X will be released immediately and Y will get 10 year's jail term for lying (0X, −10Y). If Y confesses and X does not confess, Y will be released immediately and X will get 10 year's jail term (−10X, 0Y). If both of them do not confess (i.e. have not committed the crime), each will get 1 year's jail term because the police have still got some clue that they have committed the crime (−1X, −1Y). The pay-offs are given in the respective rows and columns of the matrix below. Box 1 (Source: Hydra Digital Repository (University of Hull). hydra.hull.ac.uk).

Box 1 Confession as 'choice' explaining Tragedy of the commons through Prisoner's dilemma

Individual X → Individual Y ↓	Confess	Does not confess
Confess	(−6, −6)	(0, −10)
Does not confess	(−10, 0)	(−1, −1)

What would be the best strategy for the individuals X and Y to minimize the negative utility due to jail term? If both of them do not confess, then the punishment will be the minimum (−1, −1). However, if one does not confess but the other confesses then the person who does not confess will get maximum of 10 year's punishment and the other will be set free! Since there is complete uncertainty about other's behavior, the best strategy for both of them would be to 'confess'. Tragedy of the commons arises from Prisoner's dilemma only! If everybody cooperates, groundwater will not get over-exploited; river system will not get polluted; biodiversity will not vanish; over-fishing will not take place; CO_2 emission will not increase and global warming will not be a problem. But, each one of the polluter or resource user thinks that if she cooperates and if others do not cooperate, then she will be the net losers and non-cooperators would be the beneficiaries; in this case, the best strategy for everyone is to non-cooperate –i.e. to pollute or over-use! This ultimately would lead to tragedy of the commons!

(ii) *Common Property Resources: Community-Centered Participatory Approaches*

Common property resources, popularly known as, CPRs, possess 'excludability' and 'non- subtractability' characteristics. Village ponds, village forests and irrigation canals that are usually managed by community groups at the local level belong to CPRs. The property rights over the CPRs are usually with the community managing them and as a result, there is a possibility for the community to exclude the potential consumers. The community devices rules regarding benefit sharing and the benefits are usually shared equitably among its members. There are stringent rules against violations and violators are severely punished by the community. This avoids subtractability and paves way for sustainable use of the resource. Elinor Ostrom, the Nobel Prize Winner in Economics for the year 2009, documented various cases to demonstrate how communities across the world have been and are still managing the CPRs over centuries together (Ostrom 1998).

5 Discussion-II: Eco-economics

(i) *Market-based Mechanisms: Ecosystem Services and incentives* (Wittmer and Gundimeda 2012)

Managing GPGs through market-based mechanism is a service valuation not only of these resources on economic terms but of bringing value to eco-friendly productive activities, say *positive externalities* (elaborated further next section). The environmental resources, irrespective of their type, generate different types of ecosystem services. They are classified (Shown in Table 1) into four major categories: (a) Provisioning services; (b) Regulating Services; (c) Cultural Services; and d) Supporting Services (Venghaus and Selbmann 2014). For example, the four types of ecosystem services generated by wetlands are given in Table 1. Depending on the nature of the environmental resource, the nature and type of ecosystem services may differ. For example, a larger water body may generate all types of ecosystem services while a small water body may generate only a few of them. Some of the ecosystem services are quantifiable (e.g. volume of drinking water diverted) and some of them are non-quantifiable (e.g. biodiversity values); some of them are objective (e.g. volume of water used for industrial purpose) and some of them are subjective (e.g. religious services); some of them are visible (e.g. quantum of fish caught) and some of them are non-visible (e.g. microclimatic stabilization service); some of them are local (e.g. groundwater used for irrigation purpose locally) and some of them are global (e.g. carbon sequestration leading to reduced global warming) (TEEB 2010).

Alternatively, the wetland ecosystem benefits are classified as: input service (e.g. water for irrigation and industrial use), final demand service (e.g. water for consumption in households) and waste assimilative service (e.g. waste absorbing service). Apart from benefits utilized by various economic entities, the wetlands create ‘ecological benefits’ too as the entire ecology of wetlands depends on the primary resource, namely, water in the wetlands. Environmental economists define the ecosystem services in a different way and such definition includes values/benefits that are not captured in the conventional classification (Hassan et al. 2005). The ecosystem services framework encourages coordinated efforts between and among diverse sectors ranging from community, region, institutional/government level. These local level cooperations facilitate intra and inter level collaboration between international organizational farmwork and local government involving government institutions and agencies. These collaborations may help achieve different interests and visions into a shared interests by ‘provisioning’, ‘regulating’, ‘supporting’ and ‘cultural’ services (Wittmer and Gundimeda 2012).

Environmental economists have introduced the concept of ‘Total Economic Value’ (TEV), which encompasses both use values and non-use (or passive use) values (*elaborated in* Chart 1). Use value is derived from the direct physical interaction with a resource, such as using water from a wetland for drinking purposes (Wittmer and Gundimeda 2012). Use values are further subdivided into: (a) direct use value (OECD 2004) (e.g., water for irrigation and final consumption) and (b) indirect use value (OECD 2004, 68) (e.g., supporting biodiversity, groundwater recharge, and flood

Table 1 Wetland ecosystem services with their examples

Ecosystem services	Examples
I-Provisioning services (Video on Provisioning Services, https://www.youtube.com/watch?v=JDP67G2up3s)	Examples of provisioning services
Food Fresh water fiber and fuel biochemical Genetic Materials	Production of fish, wild game, fruits, and grains Storage and retention of water for domestic, industrial, and agricultural use Production of logs, fuel-wood, peat, fodder Extraction of medicines and other materials from biota Genes for resistance to plant pathogens, ornamental species, and so on
II-Regulating services (Video on Regulating Services, https://www.youtube.com/watch?v=On4-c_JfDAw)	Examples of regulating services
Climate regulation Water regulation (hydrological flows) Water purification and waste treatment Erosion regulation	Source of and sink for greenhouse gases; influence local and regional temperature, precipitation, and other climatic processes Groundwater recharge/discharge Retention, recovery, and removal of excess Nutrients and other pollutants Retention of soils and sediments
Natural hazard Pollination	Regulation flood control, storm protection Habitat for pollinators
III-Cultural services (Video on Cultural Services, https://www.youtube.com/watch?v=rHLeMFZ8aCA)	Examples of cultural services
Spiritual and inspirational Recreational Aesthetic educational	Source of inspiration; many religions attach spiritual and religious values to aspects of wetland ecosystems Opportunities for recreational activities Many people find beauty or aesthetic value in aspects of wetland ecosystems Opportunities for formal and informal education and training
IV-Supporting services (Video on Supporting Services, https://www.youtube.com/watch?v=pGTsR3mmGxU)	Examples of supporting services
Soil formation Nutrient cycling	Sediment retention and accumulation of organic matter Storage, recycling, processing, and acquisition of nutrients

Source MEA, 2005; TEEB, 2010, DocPlayer.net, Nature Uganda: natureuganda.org (MEA 2005)

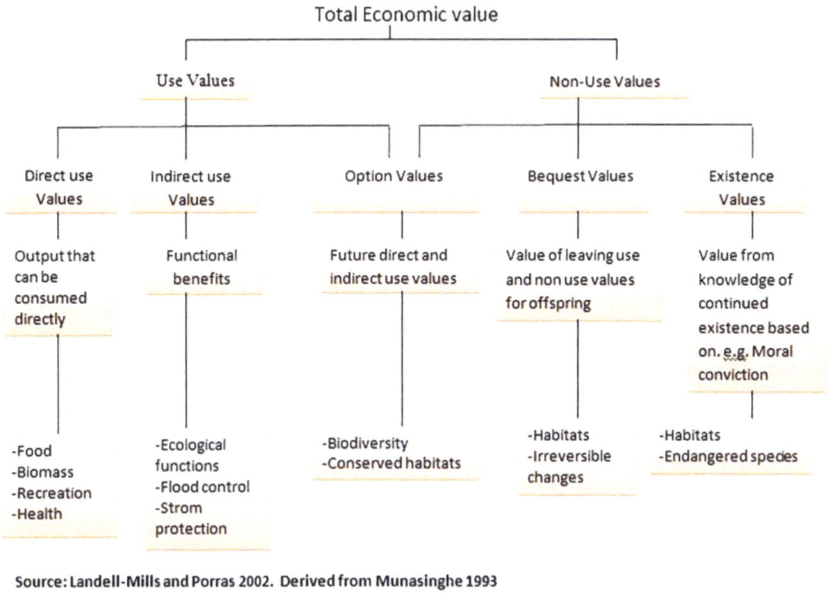

Chart 1 Sub-types of two major types of TEV-use-value and non-use-value

control) (McGuire 2017). Additionally, use values are classified into consumptive use values (e.g., drinking water, which reduces the physical availability of water) and non-consumptive use values (e.g., water used for swimming, which does not reduce its physical quantity) (McGuire 2017, Chapter 3).

Non-use or passive use values include: (a) option value—reflecting individuals' or households' willingness to pay to preserve the wetland for potential future use; (b) bequest value—willingness to pay to conserve the wetland for future generations; and (c) existence value—the satisfaction individuals derive from merely knowing that a wetland exists, regardless of direct interaction (McGuire 2017, Chapter 3). Non-use values form a distinct category, and when combined with use values, they substantially enhance the TEV of the environment (McGuire 2017, Chapter 3).

While the direct and indirect use values capture all the ecosystem services we discussed above, the option, bequest and existence values fall outside the ecosystem services and they are subjective in nature. Most of these values are 'non-marketed' in nature. That is, unlike market goods like soft drinks, most of the ecosystem values are not traded in the market (Kant and Alavalapti 2014). For example, do we buy cultural services provided by environment in a market place? No. It comes at free of cost. Even if some of the ecosystem services are traded in the market, the price does not reflect the true value of that ecosystem service (Kant and Alavalapti 2014). More precisely, the environmental values are highly valuable but there exists no market price on them! As a result, the importance of ecosystem services in enhancing and sustaining human welfare is not understood by the policymakers who always look at market prices for estimating the values of goods and services (Kant and Alavalapti

2014). Neglect of the resources makes them more vulnerable to negative externalities that destroy the resources in the long-run.

6 Result-I

Market trade-offs in GPGs: Case of Externalities (Haris and Roach 2017)-*positive and negative.*

There is a trade-off involved between different ecosystem values. If one ecosystem is used for consumption purpose, it leads to reduce the availability of other services. For example, if the disposal service of a wetland is used by an industrialist for dumping pollution, it reduces water quality thereby reducing almost all other ecosystem services, such as, drinking water, fish, recreational service, etc. The trade-off is caused mainly by externalities. Except private goods, different forms of public goods experience the problem of externality. Studying 'externalities' is important in managing environmental resources. Externality is defined as to arise when production or consumption activity of one person affecting the consumption and production activity of another person. Since most of the ecosystem services are public goods in nature, one person's production or consumption activity depends on another person's production and consumption activity.

The externalities are of two types: positive externality and negative externality. When one person's activity positively affects the activity of another person, it is called 'positive externality'. Suppose there are two neighbouring farmers. One is cultivating sunflower and the other one is doing bee keeping. The honey output goes us because the bees could collect honey from the sunflowers in the neighbourhood; similarly, the sunflower output will also go up because of the pollination service provided by the bees. If the two farms are too far away, then the output in each activity will be lesser than what they are now. The increased output is due to positive externality–positive externality of sunflowers on honey output and positive externality of bees on sunflower output.

The negative externalities are most prominent ones and are widespread in nature. Negative externality can be said to exist when one person's activity harmfully affects the activity of another person. For example, groundwater use by one farmer reduces the use of the same by another farmer. Water pollution by an industry reduces fish catch in the downstream of the river system. Fishing in the deep sea by one fisherman negatively affects the catch of other fishermen. Consumption of fossil fuels emitting greenhouse gases affects the welfare of people in another country due to change in the climate.

The ozone depletion and global warming problem caused by greenhouse gases are the result of negative externality caused by externalities –GHGs emitted by human beings due to consumption of fossil fuels; due to methane emission caused by paddy cultivation and livestock related activities; due to slash and burn cultivation and post-harvest stubble burning in agricultural fields. Negative externalities, if not controlled, will convert a public good into a common pool resource, then to a common-property resource and then finally, to a private good. Therefore, internalizing the externalities is important for sustaining the natural resource base in an economy.

7 Result-II

Market failure in GPGs: Internalizing Externalities (Haris and Roach 2017).

Externalities arise due to 'market failure'. Let us take a commodity–say, leather handbag-whose production causes water pollution and the pollution are let into a nearby wetland. Let us say that the market price of a leather bag produced and sold is Rs. 3000 per unit; the price reflects only the marginal cost of all inputs–land, labour and capital- used in the production process and an amount towards profit for the producer. As we have already seen, wetlands generate verities of ecosystem services. These services are utilized by various producers (e.g. farmers, livestock owners and other industries) and consumers (e.g. households consuming drinking water, recreationists and washer men). As we have already discussed, when disposal service is utilized by the industry by way of polluting, it reduces the water quality adversely affecting the use of other ecosystem services. Let us assume that the economic value of loss of ecosystem services due to pollution caused by the last unit of leather bag produced by the industry is equivalent to Rs.4000.00. This is called 'external cost' of producing the last unit of leather bag (Tremblay and Tremblay 2012). Since the market price (Rs. 3000) reflects only the factor cost plus a profit margin, it does not reflect the external cost (Rs. 4000). This is what is called 'market failure'. That means that if both cost of production and external costs were included in the price, the market price would have been fixed at Rs. 7000 per unit of leather bag. An increase in the price would have reduced the consumption of the leather bag to the optimum level and pollution and the resulting damage would also have been reduced to the socially desirable level. If the price includes external cost, it is said to 'internalize' the negative externality! Since market does not automatically internalize the externality, how do we ensure that the externalities are internalized?

8 Regulatory Policy Instruments

(i) *Pigouvian Solution: Internalizing Costs*

What will happen if the externalities are not internalized? The market failure theory suggests that if the externalities are NOT internalized, then the goods generating positive externalities will be under-provided and goods generating negative externalities will be provided excessively in the society. Let us first discuss positive externality. Let us take a farmer who owns 30 acres of land but cultivates teak tree in 20 acres of her land, leaving the remaining land barren. Her decision to cultivate tree crops only up to 20 acres is based on her marginal private cost and marginal private benefits. As the tree crop generate more ecosystem services (e.g. groundwater recharge, soil moisture conservation, carbon sequestration and pollination) benefiting a larger society, increasing the land area under tree crop from 20 acres up to say 25 acres will increase the ecosystem services and social benefits. But the farmers expected private benefits are lesser than the expected private costs of increasing the land under tree crop from 20 to 25 acres, the farmer will not have any incentive to increase the social benefits by increasing the land under tree crop. Here is a marker failure. What is the best way to internalize the positive externality here? This requires estimating the marginal benefits (to the society) and marginal costs (at least the private cost) of increasing the land area under tree crop. If the marginal benefits are greater than the marginal costs, then the farmers should be compensated for the marginal cost that she incurs so that she could increase the land area up to that level which maximizes the benefits to the society. So, adequate subsidy has to be provided to the farmer so that she increases the social benefits and accepting compensation to the tune of additional cost that she has to incur for growing trees on the additional 5 acres of land.

Suppose, the farmer cultivates eucalyptus tree which consumes lot of groundwater. She cultivates it in the entire 30 acres of her land. This causes negative externality to other farmers who utilize groundwater for irrigation purpose in that region. Because of cultivation of bad tree, water table declines in that region and the farmers could not cultivate crops and have to forgo their own private benefits. In this case, the eucalyptus farmer will have to reduce the land under eucalyptus tree from say, 25 acres to 20 acres, so that the external costs imposed on other farmers can be reduced. This is possible if a 'tax (Wanless 2018)' is imposed on the farmer cultivating eucalyptus tree until she reduced the land area from 25 to 20 acres. This will make the benefits (damage avoided) and costs equal. This is what is called internalization of negative externality. Pigouvian tax is being suggested for issues ranging from minimizing the global warming problems to local level pollution problems.

(ii) *Coasian Approaches: Tradable Permits*

Ronald Coase, a Nobel Prize winner in economics in 1991, argued that most of the environmental problems are due to ill-defined property rights over the resources. As we have already seen, the tragedy of the commons problem is a predominant one in atmospheric pollution because the atmosphere is not owned by anybody. So, Coase suggested that assigning property rights to the users will facilitate them to negotiate with each other, leading to an amicable solution to resolve the tragedy of the commons problem Based on the Coasian approach, many countries have devised 'tradable pollution permits' to control air and water pollution. Carbon trading is another form of Coasian approach that mitigates green house has emissions. Assigning *property rights* to local people has improved biodiversity conservation significantly (Crase and Gawne 2012).

9 Conclusion: *Managing Global Commons*

Tragedy of the commons occurring to fishery sector- especially, deep-sea fishing- is being addressed by using 'tradable permits (Sterner and Coria 2012)' that we discussed above. Countries such as the Netherlands follow this approach. Based on maximum sustainable yield, maximum amount of fish (which is called Total Allowable Catch (TAC) that can be caught from deep sea is determined by the government. For example, '*Common Fisheries Policy*' (CFP) or '*revised EU Fisheries Control Regulation*' for a sustainable use of marine resources. These regulations determined 'fishing practices and each EU country was responsible for enforcing them within their own national control systems (EUR-Lex, https://eur-lex.europa.eu/legal-content/EN/TXT/?uri=CELEX:52023DC0103). Depending on the number of fishermen and TAC for a given period, permits—called, Individual Transferable Quotas (ITC)-will be issued to the individual fishermen. The fishermen can buy or sell these quotas and a fisherman possessing ITC implies that she is entitled to catch a particular quantity of fish for a given period. The tradable permit system assigns property rights to the fishermen and the fishermen with ITC is assured that she will secure her own share of fish. In the absence of permit system, the fishermen will tend to overfish because of uncertainty over fish catch that is caused by competition from other fishermen. Under permit system, property rights allow only limited number of fishermen and assured fish catch in future provides incentives for the fishermen to conserve the fish stock for future use. Over-fishing is avoided and yield becomes sustainable. In the absence of permits, there is always a trade-off –i.e. one person's catch comes at the loss of another person's catch. However, when a fisherman buys an ITC from another fisherman the buyer values the fish more than the seller; since buyer compensates the seller by way of paying for the ITC, both the seller and the buyers are better-off under the permit system. Tragedy of the commons in the groundwater sector can be addressed by way of taxing over-extraction of groundwater or putting a Pigouvian tax on electricity being used for extracting groundwater. Permit system is also viable but monitoring groundwater use is an extremely costly affair. In many cases, strict regulations by the government that includes banning borewells in over-exploited areas are considered to curb over-exploitation (Sterner and Coria 2012).

Recent case studies on *'sustainable cities*' concept have called for innovative plans or frameworks like *'comprehensive environmental accounting of cities*' *environmental performance* for advancing sustainable development goals of sustainable living and urban development (Satterthwaite 2021). 'The Sustainable Seattle Forum', 'daily triangle' a model of integrating framework have been of recent possible solutions. *Expansionist paradigm* which posits that environmental issues have some source-model and can be remedied through global shift toward the ecological steady state (Rees). These initiatives align with while globally interconnected, can be mitigated through localized transitions to ecological steady-state economies (Satterthwaite 2021).

Addressing tragedy of global commons such as, global warming problem, requires multiple approaches. Since Prisoner's dilemma outcome becomes a dominant strategy of all the countries emitting CO_2, international organizations such as UNO and IPCC will have to play a major role in making all the countries to cooperate! The ecosystem services framework, as pointed elsewhere, are an excellent example of international cooperation fostering into multi-scale collaboration from local communities to global institutional level through conversion principle of self-interest into shared interests or goals through its standardized categories (provisioning, regulating, supporting, cultural). The United Nations Statistics Division (UNSO) in collaboration with international organizations in 1993, presented *SEEA* (Ahmed et al. 1989) (System of Environmental-Economic Accounting) as a framework. It provided a mechanism to measure environmental degradation to encourage coordination for conservation.

References

Ahmed YJ, El Serafy S, Lutz E (1989) Environmental accounting for sustainable development. The World Bank. https://documents1.worldbank.org/curated/en/156521468767069279/pdf/Environmental-accounting-for-sustainable-development.pdf

Ahmed YJ, El Serafy S, Lutz E (1989). Environmental accounting for sustainable development. The World Bank. Introduction. https://documents1.worldbank.org/curated/en/156521468767069279/pdf/Environmental-accounting-for-sustainable-development.pdf

Barbier EB, Burgess JC (2023) Natural capital, institutional quality and SDG progress in emerging market and developing economies. Sustainability 15(4):3055

Čábelková I, Smutka L, Mareš D, Ortikov A, Kontsevaya S (2023) Environmental protection or economic growth? The effects of preferences for individual freedoms. Front Environ Sci 11, Article 1129236. https://doi.org/10.3389/fenvs.2023.1129236

Challe T (2021) The rights of nature—Can an ecosystem bear legal rights? Ecology. https://www.ecology.com/rights-of-nature/

CourseHero. www.coursehero.com

Crase L, Gawne B (2012) Why rights matter. In Crase L, O'Keefe S (eds) Water policy, tourism, and recreation: Lessons from Australia (chapter 5). RFF Press

EUR-Lex, Access to European Union law. https://eur-lex.europa.eu/legal-content/EN/TXT/?uri=CELEX:52023DC0103.

IV. Green National Accounting and the Measurement of Genuine (Extended) Saving (2004). https://www.cbd.int/financial/values/several-accounting.pdf

Hardin G (1968) The tragedy of the commons. Science 162(3859):1243–1248. https://doi.org/10.1126/science.162.3859.1243

Harris JM, Roach B (2017) Environmental and natural resource economics: a contemporary approach (4th ed). Routledge. Chapter 3. https://doi.org/10.4324/9781315620190

Hassan R, Scholes R, Ash N (eds) (2005) Ecosystems and human well-being: current stateand trends, vol 1. Millennium Ecosystem Assessment. Island Press, Washington, 913. (Madras Institute of Development Studies. "Research publications". Retrieved June 10, 2024. docplayer.net. http://www.mids.ac.in

Kant S, Alavalapati JRR (eds) (2014) Handbook of forest resource economics. Routledge

McGuire CJ (2017) Environmental decision-making in context: a toolbox. Routledge

McNeill D (2015) Norway and REDD+ in Indonesia: the art of not governing? Forum Dev Stud 42(1):113–132. https://doi.org/10.1080/08039410.2014.997791

MEA (2005) TEEB, 2010, DocPlayer.net, Nature Uganda. natureuganda.org
Organisation for Economic Co-operation and Development OECD (2004). Measuring sustainable development: Integrated economic, environmental and social frameworks. OECD Publishing, p 68. https://www.oecd.org/en/publications/measuring-sustainable-development_9789264020139-en.html
Organization for Economic Co-operation and Development OECD (2004) Measuring sustainable development: Integrated economic, environmental and social frameworks. OECD Publishing, pp 53–56 https://www.oecd.org/en/publications/measuring-sustainable-development_9789264020139-en.html
Ostrom E (1998) Coping with tragedies of the commons. Annu Rev Polit Sci 1(1):493–535. https://doi.org/10.1146/annurev.polisci.1.1.493
Patil AK, Kumar V, Chaudhry P (2021) Economic dependence of forest fringe communities on threatened and near-threatened medicinal trees of Madhya Pradesh—the largest forest cover state of central India. Ecol Quest 32(2), Article 029. https://doi.org/10.12775/EQ.2021.029
Peterson GL, Swanson CS, McCollum DW, Thomas MH (2019) Valuing wildlife resources in Alaska. Routledge
Potschin M, Haines-Young R, Fish R, Turner RK (eds) (2016) Routledge handbook of ecosystem services. Routledge
Saha D, Agarwalla S (2021) People's forest dependence: a case study of Similipal biosphere reserve, India. Rev Dev Chang 26(1):63–82. https://doi.org/10.1177/0972266121998269
Satterthwaite D (2021) The Earthscan reader in sustainable cities. Routledge
Schepelmann P, Goossens Y, Makipaa A (eds) (2010) Wuppertal institute for climate, environment and energy. Towards sustainable development: alternatives to GDP for measuring progress. https://epub.wupperinst.org/frontdoor/deliver/index/docId/3486/file/WS42.pdf
Source: Hydra Digital Repository (University of Hull). hydra.hull.ac.uk
Sterner T, Coria J (2012) Policy instruments for environmental and natural resource management, 2nd ed. (Chapter 8). Routledge, pp 116–137
Sterner T, Coria J (2012) Policy instruments for environmental and natural resource management, 2nd ed. (Chapter 6 Tradable Permits). Routledge, pp 83–107
Sterner T, Coria J (2012) Policy instruments for environmental and natural resource management 2nd ed. (Chapter 25 Fisheries). Routledge, pp 460–473
TEEB (2010) Mainstreaming the economics of nature: a synthesis of the approach
The Wetland Book (2018) Springer. https://doi.org/10.1007/978-94-007-4001-3
Tremblay VJ, Tremblay CH (2012) New perspectives on industrial organization. Springer
TruCost and TEEB (2014) Natural capital at risk: The top 100 externalities of business (Report No. 978-3-9813410-9-9). TEEB for Business Coalition. http://www.teebforbusiness.org
Venghaus S, Selbmann K (2014) Biofuel as social fuel: Introducing socio-environmental services as a means to reduce global inequity? Ecol Econ 108:49–56. https://doi.org/10.1016/j.ecolecon.2014.09.013
Video on Cultural Services. https://www.youtube.com/watch?v=rHLeMFZ8aCA
Video on Provisioning Services. https://www.youtube.com/watch?v=JDP67G2up3s
Video on Regulating Services. https://www.youtube.com/watch?v=On4-c_JfDAw
Video on Supporting Services. https://www.youtube.com/watch?v=pGTsR3mmGxU
Wanless PT (2018) Taxation in centrally planned economies. Routledge, Chapter, p 5
Whitehead M (2018) Sustainable development. In Castree N, Hulme M, Proctor JD (eds) Companion to environmental studies, 1st ed. Routledge, pp 110–114
Wittmer H, Gundimeda H (2012) The economics of ecosystems and biodiversity in local and regional policy and management. Routledge. Conclusion
Wittmer H, Gundimeda H (2012a) The economics of ecosystems and biodiversity in local and regional policy and management. Routledge, Chapter, p 3
World Bank (2023) p 47

Capitalism Versus Sustainability: Finding Common Ground in Modern Markets

Renuka Mahajan

Abstract As an economic system, capitalism is fundamentally structured around perpetual growth, measured primarily through GDP expansion, corporate profits and market competitiveness. However, this growth-driven model increasingly conflicts with global sustainability objectives, emphasizing ecological balance, social equity, and long-term resource preservation. The tension between these paradigms is exacerbated by political instability, corporate short-termism, and regulatory inconsistencies. The chapter's main objective is to analyse how stakeholders define, implement, and conflict over sustainability practices, and to develop a policy-business-technology framework for authentic ESG alignment. This paper examines the structural barriers to sustainability within capitalist frameworks, analysing businesses, governments, and NGOs' roles in advancing or obstructing sustainable transitions.

Keywords Sustainable capitalism · Circular economy · Greenwashing · Artificial intelligence for resource efficienty · Extended producer responsibility (EPR) · Corporate social responsibility (CSR)

1 Introduction

As an economic system, capitalism prioritizes profit maximization, private ownership, and market competition (Friedman 1970). Capitalism relies on continuous economic expansion, which often leads to resource depletion, environmental degradation, and social inequality. Sustainable buying behaviour refers to consumers' purchasing decisions that consider environmental, social, and ethical factors, often aiming to reduce their ecological footprint (Kals et al. 2025). The inherent tension arises because capitalism thrives on endless growth, while sustainability demands resource conservation and equitable distribution (Jackson 2009).

R. Mahajan (✉)
Jaipuria Institute of Management, Noida, India
e-mail: renuka.mahajan@jaipuria.ac.in

F. A. Malik et al. (eds.), *Aligning ESG Goals with the UNs Sustainable Development Goals Planning*, World Sustainability Series,
https://doi.org/10.1007/978-3-032-01715-4_2

Conscious consumerism (Sharma and Singh 2025; Horneber et al. 2025), growing climate change concerns, zero-waste influencers (Tavman and Yalçın Doğan 2025) and increased corporate social responsibility awareness are increasing with increased consumer awareness and knowledge. Digital advancements, particularly social media and e-commerce platforms have further facilitated sustainable consumption by providing consumers with more information and access to green products (Gao et al. 2022). The market dynamics in capitalism do not actively promote environmental preservation. The capitalist model, characterized by mass production and overconsumption, leads to excessive waste and resource depletion (Næss 2006). Although consumption is the lifeblood of the late-stage capitalist economy, planetary boundaries necessitate a drastic reduction. This is putting the relationship between capitalism and sustainability under intense scrutiny. This ongoing traditional conflict to reduce costs and maximize profits is a continuous threat to business, especially when sustainability has emerged as a popular buzzword. As such, the significance of environmental stewardship is increasingly being recognized in international discussions (Falkner and Buzan 2019). These discussions are not only confined to international summits or corporate boardrooms. The battle against climate change is now taking place in consumers' restrooms, kitchens, social media and shopping carts rather than in policy rooms or activist marches. However, greenwashing—where companies falsely claim environmental benefits—challenges consumer trust and decision-making (Leonidou and Skarmeas 2017; Sinha 2023). Capitalism's pursuit of endless growth poses a profound challenge to genuine sustainability, demanding new frameworks beyond profit-driven logic. As a result, both governments and businesses are under immense pressure to balance the interests of ecological health and profit.

The increased involvement in Environmental, Social, and Governance (ESG) activities has led to a growth in the reporting and integration of ESG within the capitalist mode of production. Groups such as the International Integrated Reporting Committee (IIRC), the Carbon Disclosure Project (CDP), Global Reporting Initiatives (GRI) and Sustainability Accounting Standard Board (SASB) are essential to encourage greater accountability and transparency (Goswami et al. 2023). Recent trends have revealed that modern consumers are increasingly environmentally conscious, causing them to prioritise brands which are committed to sustainability practices. Consumer behaviour has changed immensely to the extent that a recent study by McKinsey and NielsenIQ revealed that products linked to ESG claims have experienced a 28% cumulative growth compared to compared to 20% growth of products without such claims over the past five years (McKinsey and Company 2023). During the same period, products with ESG-related claims accounted for 56% of the consumer-packaged goods (CPG) sector's growth, which accounts for over $14 trillion in annual spending in the U.S., thereby highlighting the strong correlation between consumer spending and sustainability claims.

Within the Indian context, post-pandemic trends show a change in public mindset towards issues arising from the development versus environment change debate. With events like pollution and smog not being accepted as byproducts of industrial development, there has been a rise in judicial intervention arising out of public resistance. With the increased public awareness, companies are now required to adhere to stricter regulations, resulting in the country emerging as a hub for sustainable innovation.

Swachh Bharat Abhiyan has made significant progress (Dash and Dash 2021), including prohibiting BS-III vehicles (Borkhade et al. 2022), naming the city the cleanest (Datta and Kapoor 2023), cleaning the river (Srinivas et al. 2020), banning single-use plastic (Herberz et al. 2020), and implementing pollution control measures (Petraru and Gavrilescu 2010). The election agenda focuses on cleaner air and rivers, crackdowns on crackers, recycling city waste, and Extended Producer Responsibility (EPR) regulations requiring brands (Amazon and Flipkart now recycle packaging to manage waste) (NDTV 2020). Startups like Phool.co (flower waste recycling) (Upadhyay et al. 2022) and Banyan Nation (plastic upcycling) are also gaining traction Shanker et al. 2023), while greener alternatives like solar-powered appliances and biodegradable packaging are being explored.

Corporate India's understanding of ESG imperatives is demonstrated by Reliance Industries' (Singhania and Saini 2022) $10 billion investment in green energy and the Tata Group's commitment to becoming carbon neutral by 2045 (Hasan 2024). On the other hand, third-party audits have confirmed the quantifiable impact of ITC's long-standing carbon–neutral and water-positive activities (ITC Sustainability Report 2023). The majority of Indian consumers prefer eco-friendly brands (Deloitte 2025), and hence, brands are resorting to green advertising.

Green advertising (Kong and Zhang 2014) is a trend that has been used for a long time to make products appear eco-friendly and all-natural. When customer read product descriptions physically or online, these tactics influence their perceptions by making things seem more environmentally friendly than they are. It tricks consumers by resorting to eco-friendly claims and product descriptions which are vague and 78% of which are never verified (Green Services Direct 2022). False imagery is used to entice the consumer while harming the environment. Their companies do not provide any transparency about sourcing and production and how the product is sustainable. A product that is advertised as environment-friendly and sustainable should make it obvious how its components, production processes, or lifecycle benefit the environment.

Digital influencers, particularly those with sustainability focuses, are increasingly exposing and rejecting misleading green advertising despite potential consumer backlash on social media. They are becoming watchdogs, publicly rejecting collaborations with companies accused of greenwashing (Akter 2024). The green influencers are promoting short-form videos("reels") to counter greenwashing (Bansah et al. 2024). They have been rejecting low-grade cookware, rejecting chemical-based products, rejecting unscientific claims and driving green adoption. Hence, to ensure transparency, such commitments need to be closely examined. The dangers of greenwashing were highlighted by Patanjali's legal repercussions for making false "chemical-free" claims (Vijayalakshmi 2024). Bans on Maggie (due to the presence of lead), Arvallis illegal mining show regulatory backlash, Telengana large scale deforestation, fast fashion and textile brands polluting showing that the ESG violations are increasingly viewed as unacceptable by consumers, regulators, and investors, leading to boycotts, legal penalties, and investor confidence loss for companies. The paradox between prioritizing planet or profit is more apparent than ever before. Sustainable issues are becoming crucial for investors and shareholders, leading businesses to implement policies to achieve sustainable objectives. The correlation between financial performance and ESG engagement remains unclear, with some studies supporting and others contradicting this. The standardization of sustainability reporting makes it difficult to compare businesses and their outcomes. The companies are experimenting with innovative strategies to meet the growing consumer demand for sustainability. However, implementing sustainable practices is complex and fraught with contradictions. Though 78% of U.S. consumers were found to prioritize sustainable lifestyles, CPG executives struggle to generate sufficient demand for such products. Joshi and Rahman (2019) argue that informed consumers are more likely to adopt sustainable consumption habits. However, attitude-behavior gaps often exist, where positive attitudes towards sustainability do not always translate into actual purchase behavior due to barriers such as price sensitivity, availability, and convenience (Vermeir and Verbeke 2006). Studies have examined the influence of brand sizes on ESG claims and their contribution to differentiated growth. The analysis revealed that larger companies generated 50% of the profit in all examined categories, while the smallest brands experienced a significant increase in claims in 59% of the most categories. Notably, growth rates were higher for products with less frequent ESG claims, such as "vegan" or "carbon zero," than for those with more common claims. This raises the central question of whether companies can prioritize ecological stewardship without compromising economic competitiveness. Therefore, it is pertinent to explore strategic customer-centric ESG approaches that businesses may adopt to accomplish both ESG goals and financial growth. It is also important to investigate if ESG claims confirm its genuineness and impact when positioned to achieve significant growth (Lee and Suh 2022). Businesses could implement a variety of strategies to enhance their waste management performance (Kurdve et al. 2015). The Circular Economy (CE) Model, which prioritises material reuse, recycling, and regeneration (3R), is the first step (Shmelev and Gilardi 2025).

2 Literature Review

The dynamic between capitalism and sustainability has long sparked critical discourse, particularly as the ecological crisis intensifies and the limits of current economic models become increasingly apparent. Historical context reveals that capitalist expansion has consistently been accompanied by large-scale environmental degradation. The Industrial Revolution (1760–1840) marked a decisive turning point, as economies transitioned from an agrarian system to fossil fuel-based industrial production. The widespread adoption of coal-powered steam engines, such as those improved by James Watt in 1775, enabled significant increases in productivity while dramatically raising carbon dioxide emissions (Clark and Jacks 2007). These shifts were not merely technological but systemic, embedding carbon-intensive practices into the foundation of capitalist growth.

Alongside this, escalating demand for timber to fuel construction and infrastructure, particularly for railroad, drove large-scale deforestation and habitat destruction (Williams 2003). Simultaneously, industrialisation led to rapid urbanisation, resulting in toxic smog, contaminated waterways, and deteriorating living conditions in cities like London and Manchester (Mosley 2014). These early developments illustrate how economic expansion under capitalism often comes at direct ecological cost.

In the late twentieth century, the rise of neoliberal economic policies further entrenched the commodification of nature and deregulation of environmental protections. Reaganomics and Thatcherism, which emphasised privatisation, free markets, and minimal state intervention, significantly reduced environmental oversight (Clapp and Dauvergne 2011). This shift enabled corporations in the Global North to relocate pollution-heavy manufacturing to developing countries with weaker environmental regulations, externalising environmental costs while maintaining access to cheap labour and natural resources. Concurrently, fossil fuel companies such as Exxon and Shell engaged in aggressive promotion of single-use plastics. To deflect away from corporate accountability and shield capitalism from environmental critique, these efforts were then paired with aggressive campaigns that placed the responsibility of environmental harm on consumers (Freinkel 2011).

Efforts towards the proliferation of green capitalism have accelerated in recent decades, fuelled by international agreements like the Paris Agreement of 2015, which was a landmark achievement for generating global consensus on environmental issues, but lacked enforcement mechanisms because of its structure. Cracks within such international agreements become visible when subordinate to nationalistic economic agendas, exemplified by the withdrawal of the US from the Paris Agreement under President Trump. Increasing efforts to roll back environmental regulations in the US have arisen side by side with the ESG movement. Critics argue that this movement often functions as a veneer for business-as-usual practices, with many firms receiving high ESG scores while continuing to pollute or exploit natural resources.

Within this context, greenwashing is a widely used tool by capitalist entities to superficially commit to sustainable development goals (Lyon and Montgomery 2015). Companies like Zara and H&M claim to focus on recycling clothes (Seo and Suh 2019) while continuing their unethically sourced, mass production of cheap fabrics, which are harmful to the environment. Oil giants like Shell and BP commit to 'net zero emissions' while making expansions to their drills (Niinimäki et al. 2020). There have been several instances where companies like Volkswagen, Goldman Sachs, BNY Mellon, etc. have been fined for their greenwashing tactics. While such instances create ripples in the market in the initial months, they are often forgotten and go unadmitted, thereby detracting from the long-term goal of ecological sustainability in favour of corporate profitability.

The term green capitalism emerged in the late twentieth century, gaining traction with works like Hawken et al.'s Natural Capitalism (2004), which argued that eco-efficiency and green innovation could align profit with planetary health. However, critics have pointed out how some practices, such as carbon trading, prevalent within this new initiative, often permit wealthier corporations to sustain their exploitation of the environment while negatively affecting businesses making a sincere attempt to switch to a sustainable model of production. Furthermore, 'green grabbing' is a mechanism wherein corporations appropriate land and resources from native communities under the guise of sustainability (Fairhead et al. 2012), directly goes against the principle of environmental conservation and reveals a deeper layer of neo-colonialism prevalent under capitalist economies (Scheidel and Work 2018).

So far, traditional branding suggested that consumers were perceiving significant differences between brands, suggesting that brands should prioritize visual and voice identifiers like logos or catch phrases. But modern consumerism highlights that brands must align with the triple bottom line of people, planet, and profit, prioritizing sustainable development alongside profitability. Several frameworks have emerged in contrast to the idea of green capitalism as a method of achieving sustainability. TBL—or 'Triple Bottom Line' by John Elkington is one such approach (Arowoshegbe et al. 2016). By focusing on 3 dimensions—profit, people and the planet, TBL aims to calculate a company's performance across social, environmental and economic sectors. This means that the idea of corporate accountability must move beyond marketing tactics of green capitalism. Advocates argue that the emphasis on measurable impacts in TBL would allow for a greater degree of transparency on capitalist activities. However, critics have pointed out the lack of uniform metrics to actually measure corporate impact on people and the planet. Furthermore, TBL oversimplifies the idea of trade-offs between the 3Ps since profit will always be prioritised in capitalist models of production (Shams et al. 2025).

The second approach is that of eco-modernism, which is a philosophy that posits the simultaneous growth of humanity coupled with environmental suitability. It hinges on the usage of technology for minimising corporate exploitation of the environment. Advocates of this philosophy have often pointed out how eco-modernism is suitable for capitalist economies because it does not detract from material well-being but also prioritises the reduction of human footprint on the environment (Mahadevan and Asafu-Adjaye et al. 2015). However, critics have pointed out the rather flawed understanding of 'modernisation' of this philosophy. Eco-modernists fail to consider the environmental footprint of large-scale usage of technology (Grunwald 2018).

Within the context of this critique, another idea—that of degrowth emerges. This movement propagates the reduction of production as a method of achieving ecological sustainability. Drawing on ecological economics and critical development theory, degrowth scholars argue that affluent nations must move beyond GDP as a primary measure of success and embrace post-consumerist values focused on care, sufficiency and community well-being (Kallis et al. 2020). Unlike green capitalism, degrowth challenges the core tenet of perpetual economic expansion and calls for redistributive policies, shorter working hours and democratic control over resource use. While critics dismiss degrowth as politically unfeasible or economically regressive, advocates argue that it offers one of the few frameworks that confronts the root causes of environmental degradation rather than its symptoms (Latouche 2009). Pilot initiatives in cities like Barcelona and Amsterdam have begun experimenting with post-growth indicators and circular models integrated withs social equity goals, suggesting that alternative economic imaginaries may gain traction amid growing ecological anxiety, Ultimately, degrowth represents a radical, yet increasingly influential critique of market-driven sustainability, pressing the debate towards structural transformation rather than incremental reform.

Both eco-modernism and degrowth present themselves as an alternative to green capitalism, although neither has gained traction within capitalist economies, thereby leaving space for exploration.

3 Objectives

In the contemporary landscape of capitalism, the intersection of business practices and environmental sustainability has become increasingly complex. The main objective of the article is to delve into the conflicts between opposing paradigms, such as ecological modernisation and degrowth, particularly, how governments, businesses, and NGOs co-construct sustainability in practice. This article explores how these players define and redefine sustainability in practice in the quest for a waste-free future. The chapter also explores how Corporate Social Responsibility (CSR), circular economy models, and greenwashing practices all reveal capitalism's underlying sustainability paradox, where profit interests clash with planetary limits. The chapter reports challenges and conflicts and recommends actionable steps through innovations in technology, policy and business models to develop products that adhere to ESG claims. The authors recommend the inclusion of a generative AI toolkit to automate design creation, optimize the supply chain for reduced carbon emissions and recommend sustainable product material and workflows. From the consumer perspective, Media advocacy and technology should be used to educate them on the claims and empower diverse communities. The future of environmental stewardship hinges on circular economy, technological innovation and corporate accountability. Further government regulations and consumer demand will push companies towards genuine sustainability efforts and impact. With an inclusive approach, businesses can strike a careful balance between economic prosperity and environmental responsibility. The findings of the study have implications for businesses, governments, NGOs, policy makers and marketers. It brings forth a way forward for businesses to integrate sustainable practices into the core of capitalism. The findings of the study offer a unique perspective by challenging the traditional view of capitalism and sustainability as opposing forces. It suggests a common ground where both can coexist and thrive in a reimagined economic landscape.

4 Research Methodology

This study synthesises information from several documents rather than carrying out primary research (like data collection from surveys or interviews). Like the study by Dryzek (2013), it uses descriptive and analytical methodology in the form of a critical literature review of sustainability practices defined, debated or executed. Peer-reviewed studies, policy reports, sustainable business reports, and NGO reports were considered. Emphasis was placed on recent high-impact and varied viewpoints (business, government, NGO perspectives). Thus, the study considers several stakeholders' perspectives. Since ESG transitions require multi-stakeholder alignment, synthesizing diverse sources helps propose policy recommendations. However, the study may have the limitation of researcher bias, as no primary data validation was considered (Whetten 1989). Overall, the study presents an integrated overview of sustainability issues free of primary data limitations.

5 Discussions

Based on the stated objectives, the chapter reviews the roles of different stakeholders like business, government, non-governmental organizations (NGOs) and the intricate relationship between business strategies, marketing efforts and corporate social responsibility (CSR) initiatives in how they foster sustainable practices. According to a 2025 report, the sustainable products market size was valued at approximately USD 355.3 billion in 2024 and is expected to reach USD 692 billion by 2033, growing at a compound annual growth rate (CAGR) of about 7.7%, thereby underscoring the growing demand for green offerings (Business Research Insights, 2025) Within this context, green marketing has emerged as a pivotal strategy for businesses aiming to appeal to eco-conscious consumers. From a marketing perspective, this approach involves promoting products and services based on their environmental benefits, such as reduced carbon footprints, sustainable sourcing, and eco-friendly packaging. A report by PWC found that products featuring sustainability attributes can achieve a revenue increase of 6% to 25% compared to products without such emphasis (PWC 2025). How should companies operate within a consumer culture that is driven by advertising that links being "green" with "buying more?" The issue lies in the intersection of ethical consumerism and green marketing. Since corporates are primarily profit-oriented, consumers are generally sceptical about the authenticity of claims such as 'environmentally friendly', 'sustainable models of development', etc., wherein companies try to ride on the increasing wave of environmental concerns to generate profit without actually attempting to change their business models. Recently, a YouGov survey showed that 55% of global consumers are skeptical about brands' sustainability claims, highlighting the challenge of greenwashing.

Corporate social responsibility (CSR) is the cornerstone of modern business strategy, enabling businesses to address sustainability issues. It reflects a company's long-term commitment to environmental stewardship, community engagement and ethical and sustainable practices. There is a significant relationship between branding and CSR, modern consumers prefer brands that demonstrate an authentic commitment to environmental responsibility. Thus, CSR strengthens consumers' trust and is a key differentiator in markets where products are quite similar. An important approach in this direction is that the companies report their resource use and ESG performance. Companies that exaggerate or fabricate their environmental claims pose significant risks to the integrity of CSR initiatives. Several companies have been found to promote a product as 'eco-friendly' while failing to provide evidence, deceiving consumers. Thus, it fails to address the overall environmental impact, risking the legitimacy of CSR efforts. Delmas and Burbano (2011) highlight that greenwashing can lead to customer distrust, making it challenging for sustainable businesses to differentiate themselves in the market. This ESG greenwashing is a critical issue (Kim and Lyon 2015) that erodes the entire trust. It is a practice of misleading consumers about a company's environmentally friendly green policies.

The business must adopt a more transparent approach to CSR. The company can ensure better transparency and verifiability in their sustainability practices through clear and verifiable information (Mol 2015). Several companies have successfully integrated sustainability into their business models. IKEA designs furniture that emphasizes circularity, providing a distinctive offering of repairable furniture (Mahalakshmi et al. 2024). They integrate sustainable practices across all their operations—from procurement to production to logistics. IKEA has successfully balanced environmental concerns with financial success, setting benchmarks for similar companies in the sector. Another Indian brand, TATA, has successfully integrated ESG into its operations across consumer goods, automobiles, steel, and IT, aiming to become carbon–neutral by 2045 and establish itself as a global sustainability leader (Jain et al. 2024). Among its multiple projects, Tata Nexon Electric Vehicle (EV) is spearheading the EV movement. Another globally recognized, environmentally conscious luxury brand, Moncler, incorporates eco-friendly practices into its operations (Gazzola et al. 2019). The brand uses recycled materials like polyester and nylon and adheres to the Responsible Down Standard, ensuring ethical down production. Further, beyond CSR's stakeholder-focused model, sustainable business strategies often incorporate innovative principles like the circular economy, which minimizes waste and resource consumption. Firms like H&M are designing items for reuse and recycling. By aligning their operations with environmental goals, businesses can enhance their brand reputation and expand into new markets and consumer segments that value ethical consumption. While these models present idealized frameworks for sustainable capitalism, businesses must be cautious about greenwashing risks when implementing CSR initiatives, green marketing, and sustainable business models (Kurpierz and Smith 2020). They must be transparent, accountable, and environmentally stewarding to adapt to a sustainable future. Green capitalism's success relies on companies' balancing financial interests with genuine environmental and societal concerns.

Hence, the path to achieve sustainable capitalism hinges on three players-businesses, government and non-governmental organizations (NGOs), and all of them have collective role in driving the change. Businesses are increasingly recognized as key players in advancing sustainable development. A 2023 study by PwC found that companies prioritizing sustainability in their business models experienced an increase in profitability compared to their less sustainable counterparts (Ostin and Maršíková 2023). For instance, IKEA's commitment to sustainable sourcing has led to a 30% reduction in carbon emissions across its supply chain since 2018. Furthermore, businesses should adopt a triple bottom line approach, combining economic, social, and environmental impacts, ensuring financial gains, social responsibilities, and environmental stewardship. First, for economic impact, businesses can foster sustainable growth by devising and executing strategies that promote social and environmental progress. By ethically sourcing environmentally friendly resources, businesses can drive sustainable job growth and innovation. Companies like IKEA have proven that economic performance can coexist with environmental sustainability. By relying on renewable energy and eco-friendly designs, IKEA uses a circular economic model that allows the company to maintain a loyal consumer base and thrive in an ecologically sustainable manner. Simultaneously, social responsibility demands CSR as a crucial strategy for businesses to give back to society and align their operations with ethical standards. Research shows that businesses that engage with social issues are more likely to retain consumer loyalty and brand reputation in the long run. The TATA group is a prime example of this, as running initiatives in healthcare, education and rural development, the group contributes to social development by ensuring both affordability and availability. Finally, environmental stewardship necessitates Businesses are embracing environmental stewardship by adopting sustainable practices, reducing their carbon footprint, and promoting biodiversity to minimize their environmental impact and conserve resources. For instance, Brazilian cosmetics company Natura is working with over 30 native Amazonian communities, including 7,000 families, to ethically source biodiversity ingredients. Since 1999, the company has conserved 2 million hectares of forest, valuing sustainable practices arising from traditional knowledge, thereby supporting local communities and preserving the Amazon ecosystem. Further, IKEA's IWAY Standards for suppliers that extend beyond first-tier suppliers to include second and third-tier suppliers, too, reflect a comprehensive approach to environmental stewardship.

The governments have a crucial role in encouraging businesses to invest in sustainable technologies and practices. For instance, the European Union's Green Deal aims to make Europe the first climate-neutral continent by 2050, with over €1 trillion invested in green initiatives and proposals to reduce net greenhouse gas emissions by at least 55% by 2030, compared to 1990 levels. (European Commission 2021). The government can introduce policies which promote waste reduction, renewable energy, and sustainable agriculture, such as subsidies, regulations and support plans, which can incentivise businesses to invest in sustainable technologies and practices. Previous research findings in Sub-Saharan Africa have found a link between effective energy efficiency improvements and bank profitability, signifying economic benefits, particularly in centralised political regimes (Adom et al. 2020).

NGOs complement this effort through multiple interventions. They often serve as watchdogs, raising awareness about corporate practices and advocating for social and environmental justice. Thus, they have the potential for systemic change when businesses, governments, and NGOs work together toward common sustainability goals. Corporate collaborations with NGOs can address pressing social and environmental challenges through innovative solutions. For instance, with Coca-Cola partnering up with the World Wildlife Fund, both agencies are better able to address water scarcity through the protection of freshwater ecosystems (Brownlee and Elias 2014), thereby promoting sustainable water management programs. This partnership has not only resulted in increased investment in watershed conservation projects but has also redirected focus on improving water efficiency in Coca-Cola's production processes. One key initiative within this context is the "Water Stewardship" program, via which Coca-Cola works with local communities and stakeholders to replenish the water used in the production of its beverages in order to restore natural habitats and improve water quality. This has helped foster goodwill among environmentally conscious consumers while enhancing Coca-Cola's sustainability credentials (Fundsforngos 2025).

Sustainable development necessitates collaboration among businesses, governments, and NGOs. These partnerships can help businesses find common ground, address industry-wide challenges, and drive systemic change by amplifying individual efforts and sharing best practices.

Innovations in sustainable capitalism reveal both progress and limitations. They enable businesses to create new products, services, and processes that minimize environmental impact while maximizing social benefits. For instance, a green supply chain is characterized by ethical labour practices, sustainable sourcing and reduced production waste. The increasing adoption of sustainable practices—from raw material sourcing to product delivery—by companies throughout their supply chain management is accompanied by a greater focus on energy-efficient processes to ensure competitive advantage while maintaining brand reputation. Green supply chains not only mitigate the risk of dependency on scarce resources and volatile markets but also allow companies to adhere to the regulatory standards of the government. Energy efficiency breakthroughs are adopted by businesses harnessing technologies and sources like solar and wind power, 3D printers, and green building certifications. There is greater incentive for them to step into the world of green financing and carbon pricing. This not only increases accountability on the part of corporations but also ensures resource productivity and market differentiation, which is linked to improved corporate reputation. Waste reduction strategies increasingly incorporate the circular economy model, which emphasize regenerative natural systems, elimination of waste and extension of resource productivity, and is fast gaining traction among corporations. This opens the avenue of innovation in product design and material for businesses, such as the usage of mono-materials and C2C certification, while also introducing new revenue streams. So, for instance, Apple's refurbished iPhone sales generate approximately 5 billion USD annually (Chagi 2025). Not only does this minimize the environmental impact of the corporation, but it ensures consumer retention.

However, challenges persist in implementing these solutions. Finding a common ground between sustainability and capitalism involves numerous challenges that constrain the progress towards a sustainable future. Technological limitations hinder small businesses. Biodegradable packaging costs are very high compared to conventional alternatives. Governments can facilitate environment-friendly practices, subsidies and support programs. Policies promoting renewable energy, waste reduction, and sustainable agriculture can make businesses more inclined to invest in sustainable technology and practices. Evidence from sub-Saharan Africa indicates that enhancing energy efficiency significantly boosts bank profitability, particularly in centralized political regimes. Governments prioritize sustainability agendas, but short-term political cycles can hinder long-term sustainability goals. This suggests that government needs to shatter bureaucratic silos and design environments favorable to energy efficiency programs. (Adom et al. 2020). Further, profit motives contradict long-term sustainability goals. The term "green capitalism" encourages companies to adopt sustainability practices just for profit and not through investing or demonstrating real interest in pro-environmental behaviour. Firms resort to greenwashing by falsely claiming themselves as green when, in fact, they continue to pursue unsustainable methods of operation. Hence, businesses must be encouraged to sift through real sustainability initiatives from mere superficial profit-oriented ad campaigns. Cultural diversity, language barriers, and stigma can discourage participation from diverse demographic groups like immigrants and marginalized communities.

To overcome obstacles, it is crucial to implement targeted outreach and involvement strategies that promote community ownership. The government must address externalities and regulatory gaps. However, the regulatory environment can present serious hurdles to sustainability endeavors. Regulatory mechanisms, at times, fail to penalise corporations for harming the environment. The commodification of nature and the privatization of resources lead to more responsibility in individuals towards sustainability, whereas companies continue to extract resources without facing serious repercussions. The regulatory gap in environmental degradation costs, incurred by corporations but not paid by them, hinders actual sustainability. Firms tend to shift the costs of pollution or draining natural resources to others, impacting communities and ecosystems without facing any monetary penalties. The reality puts firms engaging in sustainability activities at a competitive disadvantage, as these companies may not be able to compare favourably with those that are not using sustainability activities. It is equally important to bridge the corporate consumer divide. Sustainable "citizenship" focuses on individuals' responsibility for environmental issues, claiming that individuals' decisions can drive both social injustice and environmental protection. It creates moral division among individuals as they are judged based on their consumption choices, potentially alienating those who lack the privilege to make sustainable choices. This perspective neglects the role of firms as well as government agencies in unsustainable consumption patterns. The increasing demand for green products is being addressed by firms that have more interest in making a profit than in actual sustainability, creating a gap between what consumers want and what firms deliver. Sustainable solutions often cater to affluent consumers, and this socioeconomic disparity affects individual decision-making capacity. Addressing privileged access to sustainable choices is crucial for sustainability citizenship. Focusing on individual responsibility risks excluding those lacking access to opportunities, knowledge, or means, establishing a dichotomy where individuals who can become "sustainable" exist in contrast to those who are not, questioning equity and fairness in sustainability practice. Hence, the development of a shared economy model that depends on the collaboration of different stakeholders is crucial for its success.

6 Policy Recommendations

The analysis in this chapter highlights the urgent need for coordinated policy efforts by all key stakeholders that align capitalist growth with sustainable outcomes. An eco-capitalism system enables businesses to adopt sustainable practices, reducing costs and increasing efficiency. However, relying solely on self-regulation is insufficient; policies should guide and sometimes require businesses to act more responsibly. An important policy implication is that governments may need to rethink the basic rules of how markets function so that they encourage responsible and fair growth, not just profit-making. Governments can implement laws, regulations and incentives like carbon pricing and renewable energy subsidies to encourage sustainable practices. The government should subsidize green alternatives while making unsustainable practices costlier and more difficult to procure, making sustainability the path of least resistance for capital. Governments can invest in sustainable infrastructure such as public transportation and renewable energy sources. Financial institutions can back sustainable businesses and projects through investments and loans, while green bonds can finance these initiatives. Governments can adopt policies and regulations to promote sustainability by enacting environmental protection laws as well as developmental goals. Sustainability can be further achieved by partnerships and collaboration between businesses, governments, and NGOs. Future policy thinking should involve more partnerships with NGO and local community groups to ensure more inclusive and effective policies. Consumers can support businesses prioritizing sustainability by leveraging their purchasing power and utilizing taxes and subsidies. Businesses should prioritize longevity and switch to circular models.

In today's times of unparalleled technological advancements, advances in artificial intelligence (AI) present new opportunities for various stakeholders in designing and employing alternative resources. The domain of AI has shown potential applications of using generative AI (Gen AI) and intelligent applications that have the potential to change sustainable development and practices. GenAI can be employed by third-party audits to verify claims and identify instances of non-compliance, if any. By ensuring regulatory compliance through procedures, companies can restore consumer faith and show consistency in their sustainability culture. Technology can also help brands track sentiments on social media about their sustainability claims, respond to negative comments effectively, and resolve complaints promptly. Greenwashing is easily caught by AI, but consumers need to be empowered to boycott greenwashers. Tighter greenwashing rules can bring more transparency and accountability. The government needs to provide climate tech startup funding, provide loans for reforms in sustainability, and send capital incentives to green ventures, along with empowering consumers to boycott greenwashers. NGOs could use modern digital tools to ensure businesses are held accountable and guided towards responsible practices.

7 Conclusion

Capitalism, by its definition, defies the idea of sustainability. Profit-making has historically existed in opposition to sustainability. This chapter reveals that capitalism's sustainability crisis stems not from a lack of solutions, such as CSR, CE, and policy tools that exist, but from systemic unwillingness to prioritize long-term survival over short-term growth. However, with increasing visibility of the impacts of climate change and global warming combined with the importance attached by governments to international climate forums, there has been a shift in public mindset. This shift has fuelled companies to attempt to change their production processes. The question today is not whether corporations can adapt to consumer needs of ethicality and sustainability, but rather how fast they can make this transition.

With consumers showcasing a willingness to spend more on ecologically friendly products, corporations today are incentivised to explore innovative business practices that build on ESG claims and sustainable foundations. The future of capitalism thus hinges on systematic change that requires both government and corporations to collaborate on the idea of green growth. Future research should establish a measurable threshold where greenwashing penalties outweigh benefits, allowing activists and regulators to make sustainable business choices. Longitudinal studies of firms experiencing reputational or financial losses should be conducted. Together, these research streams could empirically demonstrate when and how capitalism's incentive structures shift from resisting to embracing genuine sustainability.

References

Adom PK, Amuakwa-Mensah F, Amuakwa-Mensah S (2020) Degree of financialization and energy efficiency in Sub-Saharan Africa: do institutions matter? Financ Innov 6(1):33

Akter M (2024) The impact of social media backlash toward a brand's sustainability initiatives on pro-environmental consumers' purchase intentions, (Master's thesis, Trent University (Canada))

Arowoshegbe AO, Emmanuel U, Gina A (2016) Sustainability and triple bottom line: an overview of two interrelated concepts. Igbinedion Univ J Account 2(16):88–126

Bansah PF, Gaffar V, Disman D, Yeliawati AK (2024) The power of green influencers: leveraging social media for environmental sustainability in fast-moving consumer goods. In: 8th global conference on business, management, and entrepreneurship (GCBME 2023). Atlantis Press, pp 697–703

Borkhade R, Bhat KS, Mahesha GT (2022) Implementation of sustainable reforms in the Indian automotive industry: from vehicle emissions perspective. Cogent Eng 9(1):2014024

Brownlee R, Elias A (2014) World wildlife fund and the coca-cola company: a global partnership for freshwater conservation
Business Research Insights (2025). https://www.businessresearchinsights.com/market-reports/sustainable-products-market-117731
Chagi (2025) The second-hand smartphone surge: refurbished sales outpace new. https://foreveryena.co.za/technology/second-hand-smartphone-surge-refurbished-sales-outpace-new/
Clapp J, Dauvergne P (2011) Paths to a green world: the political economy of the global environment. MIT press
Clark G, Jacks D (2007) Coal and the industrial revolution, 1700–1869. Eur Rev Econ Hist 11(1):39–72
Dash AK, Dash RK (2021) Environmental and sustainability campaigns: a case study of India's Swachh Bharat Abhiyan (2014–2019). J Commun Manag 25(4):385–400
Datta S, Kapoor R (2023) Clean India mission: building nation through sustainable waste management practices. Vision 27(1):135–141
Delmas MA, Burbano VC (2011) The drivers of greenwashing. Calif Manag Rev 54(1):64–87
Deloitte (2025) Sustainability has staying power. https://www2.deloitte.com/us/en/insights/environmental-social-governance/sustainable-consumption-trends.html
Dryzek JS, Norgaard RB, Schlosberg D (2013) Climate-challenged society. OUP Oxford
European Commission (2021) The European green deal. Striving to be the first climate-neutral continent. https://commission.europa.eu/strategy-and-policy/priorities-2019-2024/european-green-deal_en#:~:text=Climate%20change%20and%20environmental%20degradation,of%20greenhouse%20gases%20by%202050
Fairhead J, Leach M, Scoones I (2012) Green grabbing: a new appropriation of nature? J Peasant Stud 39(2):237–261
Falkner R, Buzan B (2019) The emergence of environmental stewardship as a primary institution of global international society. Eur J Int Rel 25(1):131–155
Freinkel S (2011) Plastic: a toxic love story. HMH
Friedman M (1970) Capitalism and Freedom, University of Chicago Press, Chicago (1972) Milton Friedman Responds, Business and Society Review (1):5–16. Friedman M (1970) The social responsibility of business is to increase its profits. New York Times Magazine 13:122–126
Fundsforngos (2025). https://www.fundsforngos.org/all-questions-answered/what-are-some-examples-of-successful-csr-ngo-partnerships/
Gao B, Li Z, Yan J (2022) The influence of social commerce on eco-friendly consumer behavior: technological and social roles. J Consum Behav 21(4):653–672
Gazzola P, Pavione E, Grechi D, Ossola P (2019) Corporate social responsibility and sustainability as market opportunities in the luxury sector. STRATEGICA 533
Goswami K, Islam MKS, Evers W (2023) A case study on the blended reporting phenomenon: a comparative analysis of voluntary reporting frameworks and standards—GRI, IR, SASB, and CDP. Int J Sustain Policy Pract 19(2):35
Green Services Direct (2022) How can I tell if a product is eco-friendly? https://www.greenservicesdirect.co.uk/how-can-i-tell-if-product-eco-friendly
Grunwald U (2018) The impact of third-generation biofuels versus fossil fuels on interstate and intrastate conflicts
Hasan J (2024) Sustainable business practices in automobile companies, india—a comprehensive study. In: Diversity, equity and inclusion. Routledge, pp 381–394
Hawken P, Lovins A, Lovins LH (2004) Natural capitalism. Sustain Urban Dev Reader 162
Herberz T, Barlow CY, Finkbeiner M (2020) Sustainability assessment of a single-use plastics ban. Sustainability 12(9):3746
Horneber D, Kaiser C, Unfried M, Schallner R, Laumer S (2025) Conscious consumerism: on the interaction between different design elements in e-commerce to promote green purchasing behaviors
ITC Sustainability Report (2023). https://www.itcportal.com/sustainability/sustainability-integrated-report-2023/ITC-Sustainability-Integrated-Report-2023.pdf

Jackson T (2009) Prosperity without growth: economics for a finite planet. Routledge
Jain V, Jain K, Vedant V (2024) Sustainable practices of emerging market multinational enterprises: a study of select Indian automotive firms. FOCUS: J Int Bus 11(1):66–83
Joshi Y, Rahman Z (2019) Consumers' sustainable purchase behaviour: modeling the impact of psychological factors. Ecol Econ 159:235–243
Kallis G, Paulson S, D'Alisa G, Demaria F (2020) The case for degrowth. Wiley
Kals E, Strubel IT, Hellbrück J (2025) Sustainable consumption. In: Environmental psychology. Springer, Berlin, Heidelberg, pp 101–116
Kim EH, Lyon TP (2015) Greenwash versus brownwash: exaggeration and undue modesty in corporate sustainability disclosure. Organ Sci 26(3):705–723
Kong Y, Zhang L (2014) When does green advertising work? The moderating role of product type. J Mark Commun 20(3):197–213
Kurdve M, Shahbazi S, Wendin M, Bengtsson C, Wiktorsson M (2015) Waste flow mapping to improve sustainability of waste management: a case study approach. J Clean Prod 98:304–315
Kurpierz JR, Smith K (2020) The greenwashing triangle: adapting tools from fraud to improve CSR reporting. Sustain Account Manag Policy J 11(6):1075–1093
Latouche S (2009) Farewell to growth. Polity
Lee MT, Suh I (2022) Understanding the effects of environment, social, and governance conduct on financial performance: arguments for a process and integrated modelling approach. Sustain Technol Entrep 1(1):100004
Leonidou CN, Skarmeas D (2017) Gray shades of green: causes and consequences of green skepticism. J Bus Ethics 144:401–415
Lyon TP, Montgomery AW (2015) The means and end of greenwash. Organ Environ 28(2):223–249
Mahadevan R, Asafu-Adjaye J (2015) Exploring the potential for green revolution: a choice experiment on maize farmers in Northern Ghana. Afr J Agric Resour Econ 10(3):207–221
Mahalakshmi S, Nallasivam A, Kumar H, Kautish S, Madan S (2024) From assembly to reassembly: Ikea's circular design for a sustainable future. In: Utilizing technology for sustainable resource management solutions. IGI Global, pp 261–280
McKinsey and Company (2023) Consumers care about sustainability—and back it up with their wallets. https://www.mckinsey.com/industries/consumer-packaged-goods/our-insights/consumers-care-about-sustainability-and-back-it-up-with-their-wallets
Mol AP (2015) Transparency and value chain sustainability. J Clean Prod 107:154–161
Mosley S (2014) Coastal cities and environmental change. Environ Hist 20(4):517–533
Næss P (2006) Unsustainable growth, unsustainable capitalism. J Crit Rism 5(2):197–227
NDTV (2020) Amazon, Flipkart need to collect plastic packaging waste generated by them: central pollution control board. https://swachhindia.ndtv.com/amazon-flipkart-need-to-collect-plastic-packaging-waste-generated-by-them-central-pollution-control-board-41286/#:~:text=New%20Delhi%3A%20E%2Dcommerce%20giants,Central%20Pollution%20Control%20Board%20(CPCB)
Niinimäki K, Peters G, Dahlbo H, Perry P, Rissanen T, Gwilt A (2020) The environmental price of fast fashion. Nat Rev Earth Environ 1(4):189–200
Ostin V, Maršíková K (2023) Fostering employee sustainability: competencies formation for a resilient business environment. In: Liberec Economic Forum, p 557
Petraru M, Gavrilescu M (2010) Pollution prevention, a key to economic and environmental sustainability. Environ Eng Manag J 9(4):597–614
PWC (2025) PwC's second annual state of decarbonization report. Essential ingredients for achieving your climate ambition. https://www.pwc.com/us/en/services/esg/library/decarbonization-strategic-plan.html#:~:text=83%25%20of%20companies%20report%20R%26D,over%20products%20without%20such%20emphasis
Scheidel A, Work C (2018) Forest plantations and climate change discourses: new powers of 'green' grabbing in Cambodia. Land Use Policy 77:9–18
Seo K, Suh S (2019) A study on the characteristics and social values of vegan fashion in H&M and Zara. J Fash Bus 23(6):86–100

Shams SR, Brown DM, Hardcastle K (2025) Brand management, marketing communication, and promotion management for people, planet, and profit. In: Sustainable marketing: strategic marketing for people, planet and profit. Springer Nature Switzerland, Cham, pp 227–298

Shanker R, Khan D, Hossain R, Islam MT, Locock K, Ghose A, Dhodapkar R (2023) Plastic waste recycling: existing Indian scenario and future opportunities. Int J Environ Sci Technol 20(5):5895–5912

Sharma R, Singh A (2025) Integrating AI, green marketing, and influencer strategies: shaping sustainable consumer behavior in the digital age. In: Strategic Brand management in the age of ai and disruption. IGI Global Scientific Publishing, pp 325–342

Shmelev SE, Gilardi E (2025) Corporate environmental, social, and governance performance: the impacts on financial returns, business model innovation, and social transformation. Sustainability 17(3):1286

Singhania M, Saini N (2022) Systems approach to environment, social and governance (ESG): case of reliance industries. Sustain Oper Comput 3:103–117

Sinha G (2023) Sustainable consumption and production versus green capitalism: What really counts? South Asia J

Srinivas R, Singh AP, Shankar D (2020) Understanding the threats and challenges concerning Ganges River basin for effective policy recommendations towards sustainable development. Environ Dev Sustain 22:3655–3690

Tavman EB, Yalçın Doğan D (2025) Greenfluencers as promoters of sustainable consumption: a cross-country analysis. In Sustainability marketing in emerging economies: conceptual and empirical perspectives. Springer Nature Switzerland, Cham, pp 135–170

Tol RS (2006) The Stern review of the economics of climate change: a comment. Energy Environ 17(6):977–981

Upadhyay D, Tyagi P, Grima S, Balusamy B (2022) PHOOL–sustainability of the Ganges through empowering women: a brighter future for the next generations of Indian children; a case study with teaching notes. In: Being a child in a global World. Emerald Publishing Limited, pp 219–260

Vermeir I, Verbeke W (2006) Sustainable food consumption: exploring the consumer "attitude–behavioral intention" gap. J Agric Environ Ethics 19:169–194

Vijayalakshmi S (2024) A case study approach to uncover greenwashing and ensure consumer protection. Presidency College (Autonomous), 128

Whetten DA (1989) What constitutes a theoretical contribution? Acad Manag Rev 14(4):490–495

Williams M (2003) Deforesting the earth: from prehistory to global crisis. University of Chicago Press

Understanding the Socio-Economic Dimensions of Net Zero Emissions in MENA Nations

Aamir Ahmad Teeli, Saima Nazir, and Aaqif Rashid Lone

Abstract The Middle East and North Africa (MENA) area faces numerous opportunities and challenges in its transition to net zero emissions. This article explores the complex socioeconomic aspects of this transformation. The urgent need to reduce greenhouse gas emissions has led to the audacious target of net zero emissions by 2050 as global climate action steps up. The shift to a low-carbon economy will need extensive reorganization in a number of sectors, with ramifications for equity, employment, energy affordability, access to technology, and infrastructure development. In order to enable a fair and inclusive transition, the study emphasizes the significance of integrated policy frameworks, financial investments, international cooperation, and community engagement. MENA nations may promote sustainable development, social justice, and economic resilience in addition to reducing the dangers associated with climate change by tackling these socioeconomic variables. The study advocates for a comprehensive strategy that makes use of cooperation, ingenuity, and group efforts in order to provide the area with a low-carbon, sustainable future.

Keywords Net zero emissions · MENA region · Climate change · Sustainable development

A. A. Teeli (✉)
Department of Economics, Central University of Tamil Nadu, Thiruvarur, India
e-mail: imamireconomics@gmail.com

S. Nazir
Department of Commerce, University of Kashmir, Srinagar, Jammu and Kashmir, India

A. R. Lone
Department of Economic Sciences, Indian Institute of Science Education and Research Bhopal, Bhopal, MP, India

F. A. Malik et al. (eds.), *Aligning ESG Goals with the UNs Sustainable Development Goals Planning*, World Sustainability Series,
https://doi.org/10.1007/978-3-032-01715-4_3

1 Introduction

The urgency of addressing climate change has grown in recent years, owing to an accumulating body of scientific evidence emphasizing the catastrophic and far-reaching effects of greenhouse gas emissions on the planet's climate system. As countries wrestle with the need to offset these effects, the concept of achieving net zero emissions has evolved as a critical pillar of global climate action (Armstrong and McLaren 2022). Net zero emissions, roughly defined as the difference between the amount of greenhouse gases discharged into the atmosphere and the amount removed or offset, are an important step toward a sustainable, low-carbon future. According to the PwC's Net Zero Economy Index 2023, the world must reach an ambitious annual decarbonization rate of 17.2% until 2050 in order to keep global warming to 1.5 °C over pre-industrial levels.

While the technical capability of achieving net zero emissions is becoming more widely known, the socioeconomic implications of this shift are as complicated and multifaceted. At its core, the transition to net zero emissions requires a fundamental restructuring of economies, industries, and social institutions, with far-reaching consequences for employment, equity, infrastructure, governance, and international collaboration (Fankhauser et al. 2022). Understanding and resolving these socioeconomic factors is critical to creating a fair, inclusive, and sustainable transition that benefits all parts of society.

The Middle East and North Africa (MENA) region, with its diverse range of nations and differing political, social, and economic contexts, faces formidable obstacles in the fight against climate change and maintaining environmental sustainability. These nations are facing more urgent problems relating to sustainable development in terms of soil degradation, broadening social and economic divides, declining fossil fuel reserves, climate change, widespread extinction of species (Göll et al. 2019). The MENA region is particularly vulnerable to the negative effects of climate change because its temperatures are rising at a rate twice as fast as the rest of the world (World Economic Forum 2023). This quick warming makes already-existing problems like water scarcity worse and poses significant challenges to agriculture in the area (Alboghdady and El-Hendawy 2016). The average global temperature increase between 1980 and 2022 was 0.18 °C per decade; by contrast, the MENA region saw a temperature rise of 0.46 °C per decade. Significant changes in precipitation patterns also occurred during this time, aggravating the already serious problems with water scarcity in a number of MENA nations (International Energy Agency 2023). For example, in 2022, Morocco experienced severe droughts, and in 2023, Tunisia experienced similar circumstances. On the other hand, severe flooding in 2022 exposed the region's susceptibility to harsh weather in the United Arab Emirates, Iran, Saudi Arabia, Qatar, Oman, and Yemen.

Because of its significant oil and gas reserves, the region is heavily dependent on fossil fuels for both domestic energy needs and as a major driver of the economy (Ferroukhi et al. 2016). Due to this dependence, even though the region contributes very little to global emissions, per capita greenhouse gas (GHG) emissions are high

(Lienard 2022). The aggravating effects of climate change on pre-existing issues like water scarcity, energy security, and political stability highlight the urgency for MENA countries to adopt sustainable development practices.

Global efforts, like the Paris Agreement, have established ambitious objectives for cutting carbon emissions in order to lessen the effects of global warming. However, because of their economic structure and reliance on fossil fuel revenues, MENA countries face significant challenges in making the transition to a low-carbon economy (World Bank 2022).

According to policy brief report by Lienard (2022), MENA's greenhouse gas (GHG) emissions in 2019 accounted for 5% of global emissions, a negligible portion when contrasted with Eastern Asia's 27% share. On the other hand, when looking at carbon emissions per person, MENA is pretty big. With an annual per capita GHG emission of 13 tons of CO_2, the region has one of the highest rates in the world. This amount is comparable to the per capita emissions in Japan, Australia, and New Zealand, but it is higher than the 19 tons in North America and less than the 7.8 tons in Europe. Interestingly, the Gulf States were the world's top emitters of greenhouse gases in 2020 when compared to per capita emissions, with Qatar leading the way. In addition to the production of fossil fuels, urbanization also has an impact on this trend; between 2010 and 2015, the Middle East saw a 30% increase in urban GHG emissions per person. In MENA countries, consumption of energy and carbon dioxide emissions have more than doubled; therefore, decelerating urbanization may aid in cutting down on pollution and energy usage (Al-Mulali et al. 2013).

Due to their significant reliance on fossil fuels, particularly oil and gas reserves, the MENA region faces numerous obstacles in their efforts to achieve net-zero emissions (IRENA 2019). The risk this reliance presents to national revenues and economic stability makes it an impediment to the reduction of carbon emissions (World Bank 2022). Furthermore, the region has some of the highest per capita greenhouse gas emissions in the world, which are a result of industrial activity and energy-intensive lifestyles that are made worse by energy subsidies that prioritize fossil fuels over renewable energy sources (Mundaca 2017). Significant investments in renewable energy infrastructure and technologies are required to achieve net-zero emissions, but many nations lack the requisite legal framework and face technological challenges brought on by extreme weather (Sever 2021). Consistent climate policy implementation is made more difficult by social unrest and political instability, as governments frequently put short-term stability ahead of long-term environmental goals (Alves et al. 2020; Bernauer. 2013; Eisenack et al. 2021; Hosseini 2023). The MENA region's susceptibility to the effects of climate change, in particular water scarcity, necessitates a balanced strategy that takes both adaptation and mitigation measures into account. Stronger incentives and regulations are necessary because the corporate sector is not keeping up with the government's commitments to set and implement net-zero goals. Moreover, the shift necessitates substantial financial and institutional resources, highlighting the significance of international financing and support.

Majority of the nations in the Middle East and Central Asia have committed to reducing greenhouse gas emissions under the Paris Agreement. To fulfill these promises, it is essential for these countries to quickly incorporate climate policies into their national economic plans. Collectively, these countries aim to cut annual emissions by 13–21% by 2030, depending on external support, requiring a reduction in per capita emissions by up to 7% in the next 8 years (International Monetary Fund 2022). The region's response to these challenges has been mixed, with some countries making notable progress in diversifying their energy sources and implementing structural reforms, while others lag behind (Göll et al. 2019). Achieving such reductions while maintaining economic growth has only been accomplished by a few countries globally.

While majority of the national governments in the MENA region have made great progress, with 60% of emissions now covered by net-zero pledges over the past couple of years, the corporate sector is falling behind. Only 12% of major business houses have set net-zero objectives, and just 6% have created detailed plans (World Economic Forum 2023). This disparity emphasizes the necessity of greater corporate capability and commitment to accomplish transformative climate action.

Aside from domestic constraints, reaching net zero emissions necessitates unprecedented levels of international cooperation and coordination. Climate change is a global concern that crosses national borders, necessitating international cooperation and solidarity to handle successfully. International agreements, such as the Paris Agreement, provide a framework for countries to establish carbon reduction targets, share best practices, and raise funds to support climate action. However, maximizing the promise of international cooperation necessitates overcoming geopolitical barriers, correcting past emissions disparities, and ensuring that climate money reaches those most in need. To this end current article aims to throw light on the socio-economic dimensions of net zero emissions in MENA countries.

The socioeconomic dimensions of achieving net zero emissions are complex and interwoven, including many aspects of society, the economy, and governance. In this in-depth examination, we attempt to have insights at each factor to understand its importance, challenges, and consequences for the transition to a sustainable, low-carbon future.

2 Employment and Labor

Transitioning to a net zero economy necessitates extensive restructuring of sectors and labor markets. While the renewable energy and clean technology sectors provide prospects for job creation, established fossil fuel businesses face job losses (Garrett-Peltier 2017). The difficulty is to ensure a fair transition for workers, particularly those in regions that rely largely on coal, oil, and gas extraction. Strategies including as retraining programs, green skill development investment, and community engagement are critical for mitigating socioeconomic disturbances and facilitating a smooth transition to a low-carbon economy.

3 Equity and Social Justice

Achieving net zero emissions requires measures to eliminate social inequities and ensure that the transition's benefits are spread evenly. Environmental pollution and climate change have disproportionately affected historically marginalized communities, such as low-income households, communities of color, and indigenous populations (Suk et al. 2016; Ehigiamusoe et al. 2022). As a result, policies and activities aimed at reaching net zero emissions must stress equity, inclusivity, and community involvement. This involves increasing access to clean energy, combating energy poverty, and encouraging marginalized groups to participate in decision-making processes affecting their environment and livelihoods.

4 Energy Affordability

The affordability of renewable energy technologies and services is critical to achieving a fair and equitable transition to net zero emissions. While the cost of renewable energy has dramatically decreased in recent years, many people, organizations, and governments continue to face upfront investment fees (Barbolini et al. 2024; El-Katiri 2014). Addressing energy affordability necessitates a combination of legislative measures, financial incentives, and novel financing models that lower costs, improve accessibility, and ensure that vulnerable people are not excluded. This could include incentives for renewable energy deployment, energy efficiency renovations, and targeted assistance for low-income households.

5 Technology Access and Innovation

Access to sustainable energy technology and breakthroughs is critical for achieving net-zero emissions. However, technology access discrepancies occur both within and across countries, with underdeveloped countries frequently encountering barriers to accessing and implementing renewable energy solutions. Promoting technology transfer, capacity building, and knowledge exchange are crucial for closing these gaps and ensuring that all countries have access to the most advanced sustainable energy technologies. Furthermore, promoting innovation and development in renewable energy, energy storage, and other sustainable technologies is critical for lowering prices, enhancing performance, and speeding up the transition to a net zero economy (Raihan and Tuspekova 2022, 2023).

6 Infrastructure Development

To achieve net zero emissions, considerable infrastructure expenditures are required, such as renewable energy generation, energy-efficient buildings, and sustainable transportation networks (Ersoy et al. 2024; Golubchikov and Deda 2012). These infrastructure investments not only cut greenhouse gas emissions, but also create job opportunities, encourage economic growth, and strengthen communities' resilience to climate change impacts (International Labour Organization 2018). To achieve socioeconomic benefits, infrastructure development must be socially inclusive, environmentally sustainable, and robust to future climate hazards. This could include incorporating climate considerations into infrastructure planning and design, encouraging community interaction and input, and prioritizing investments in underprivileged and vulnerable communities.

7 International Cooperation and Trade

Achieving worldwide net zero emissions requires international cooperation and coordination, as climate change is a global issue that transcends state borders. International agreements, such as the Paris Agreement, provide a framework for countries to establish carbon reduction targets, share best practices, and raise funds to support climate action. However, maximizing the promise of international cooperation necessitates overcoming geopolitical barriers, correcting past emissions disparities, and ensuring that climate money reaches those most in need. Furthermore, trade laws and agreements can influence the transition to a net zero economy by easing the flow of clean energy technologies, encouraging sustainable growth, and motivating low-carbon production and consumption patterns (Volz et al. 2021).

8 Consumer Behavior and Lifestyle Changes

Shifting consumer behavior towards sustainable practices is critical to achieving net zero emissions (De Marchi et al. 2022). Individuals can be encouraged to adopt low-carbon lifestyles and consumption patterns by public awareness campaigns, education, and incentives, such as reducing energy usage, switching to renewable energy sources, and using sustainable transportation options. Furthermore, supporting circular economy principles like recycling, reuse, and resource efficiency can assist reduce waste generation and emissions from resource extraction and production operations (Meys et al. 2021). However, influencing consumer behavior is a hard and varied problem that necessitates collaboration across governments, businesses, civil society, and individuals.

9 Policy and Regulation

Strong regulatory frameworks and regulations are required to accelerate the transition to net zero emissions. This includes carbon pricing mechanisms, renewable energy targets, energy efficiency standards, and rules that phase out fossil fuel subsidies (Becattini et al. 2022; Davis et al. 2018; Mundaca 2017). Furthermore, policies that encourage innovation, investment, and collaboration in clean energy research and development are critical to accelerate the deployment of low-carbon technology. Furthermore, incorporating climate issues into broader policy areas such as urban planning, agriculture, and industry can aid in the mainstreaming of climate action across several economic sectors. However, effective policy implementation necessitates political will, stakeholder participation, and continuous monitoring and assessment to ensure that targets are fulfilled and progress is maintained over time.

10 Financial Investment and Capital Flows

Mobilizing corporate and public funding for low-carbon initiatives is critical for increasing renewable energy deployment and aiding climate adaption efforts. Financial incentives, subsidies, and green bonds can direct investment toward environmentally friendly projects including renewable energy infrastructure, energy efficiency upgrades, and climate-resilient infrastructure (Fatica and Panzica 2021; Qadir et al. 2021). Furthermore, incorporating environmental, social, and governance (ESG) considerations into investment decision-making processes can help reduce climate risks and promote sustainable development (Cheema and Langa 2022). However, unlocking money for climate action necessitates overcoming barriers like as investment risks, market uncertainties, and regulatory limits, as well as ensuring that investments are consistent with long-term climate goals and contribute to beneficial socioeconomic results.

11 Resilience and Adaptation

Extreme weather events, sea-level rise, and ecosystem degradation are all serious threats to populations and businesses as a result of climate change. Building resilience and adaptive capacity is critical for avoiding these risks and maintaining a stable socioeconomic transition. This involves investing in climate-resilient infrastructure, improving early warning systems, boosting social safety nets, and promoting natural solutions like mangrove restoration and watershed management. Furthermore, empowering vulnerable communities to adapt to climate change through participatory decision-making, knowledge sharing, and capacity building can boost their resilience and minimize their susceptibility to climate-related disasters. However, addressing climate resilience and adaptation requires a comprehensive and integrated approach that takes into account the interdependence of social, economic, and environmental elements, as well as local communities' needs and aspirations.

12 Conclusion

The socioeconomic components of achieving net zero emissions by MENA nations are enormous, multifaceted, and interwoven, affecting all aspects of society, the economy, and governance. It is clear that tackling these elements necessitates a comprehensive and integrated approach that considers the numerous requirements, objectives, and perspectives of stakeholders from many sectors and locations. Societies that embrace this comprehensive perspective and actively engage with the socioeconomic components of the transition to a net zero economy can open a variety of opportunities for inclusive growth, social fairness, and sustainable development.

The road to net zero emissions is opportunities to not only alleviate the dangers and challenges of climate change, but also to create a more resilient, prosperous, and egalitarian future for all. Furthermore, acknowledging the intricacies and interdependence inherent in these socioeconomic elements emphasizes the significance of collaboration, creativity, and collective action. We can only overcome the hurdles and exploit the opportunities afforded by the transition to a net zero economy if we work together, guided by shared values and a dedication to the common good. Essentially, by understanding and addressing the socioeconomic dimensions of achieving net zero emissions, we can pave the way for a more just, inclusive, and sustainable world, one in which prosperity is shared, ecosystems thrive, and future generations inherit a vibrant, resilient, and abundant planet.

13 Policy Implications

Governments and policymakers must implement integrated policy frameworks to effectively handle the socioeconomic aspects of achieving net zero emissions. This includes prioritizing the development and implementation of just transition measures to assist workers and communities impacted by the transition to a low-carbon economy. Equity and social justice concerns should be built into policy interventions to guarantee that underprivileged groups have access to sustainable energy technologies and benefit fairly from the transition. Furthermore, utilizing financial mechanisms like carbon pricing and climate finance is critical for attracting investment in renewable energy deployment, energy efficiency improvements, and sustainable infrastructure development. Strengthening international cooperation and collaboration is also critical to advancing global efforts to attain net zero emissions and address climate change on a global scale.

14 Limitations and Scope for Future Study

While the article provides a thorough examination of the socioeconomic implications of achieving net zero emissions, it does have several shortcomings. The analysis largely provides a broad overview rather than going further into specific case studies or empirical research to substantiate its claims. Furthermore, it may fail to completely represent the diversity of stakeholders' viewpoints and experiences across areas and sectors, as well as potential trade-offs and conflicts between socioeconomic and environmental sustainability goals. Future research could address these limitations by conducting empirical studies to assess the impacts of specific policy interventions, analyzing the intersectionality of socioeconomic factors, exploring the role of social innovation and community-based initiatives, and investigating potential trade-offs and synergies between socioeconomic and environmental goals in the transition to a net zero emissions.

References

Alboghdady, El-Hendawy (2016) Economic impacts of climate change and variability on agricultural production in the Middle East and North Africa region. Int J Clim Chang Strat Manag 8(3):463–472

Al-Mulali et al (2013) Exploring the relationship between urbanization, energy consumption, and CO_2 emission in MENA countries. Renew Sust Energ Rev 23:107–112

Alves F et al (2020) Climate change policies and agendas: facing implementation challenges and guiding responses. Environ Sci Pol 104:190–198

Armstrong, McLaren (2022) Which net zero? Climate justice and net zero emissions. Ethics Int Aff 36(4):505–526

Barbolini et al (2024) Strategies to tackle energy poverty in post-conflict setting: the case study of rural Damascus. In: Sustainability in energy and buildings 2023. Springer Nature Singapore, Singapore, pp 685–695

Becattini et al (2022) A two-step carbon pricing scheme enabling a net-zero and net-negative CO 2-emissions world. Clim Chang 171(1–2):18

Bernauer T (2013) Climate change politics. Annu Rev Polit Sci 16:421–448

Cheema, Langa (2022) Environment, social, and governance (ESG) and sustainability. In: In: a director's guide to governance in the boardroom. Routledge, pp 135–171

Davis et al (2018) Net-zero emissions energy systems. Science 360(6396):eaas9793

Ehigiamuso et al (2022) The nexus between poverty, inequality and environmental pollution: evidence across different income groups of countries. J Clean Prod 341:130863

Eisenack et al (2021) Politics, profits and climate policies: how much is at stake for fossil fuel producers? Energy Res Soc Sci 77:102092

El-Katiri L (2014) Energy poverty in the Middle East and North Africa. Energy Poverty Glob Chall Local solut:273–297

Ersoy et al (2024) Industrial and infrastructural conditions for production and export of green hydrogen and synthetic fuels in the MENA region: insights from Jordan, Morocco, and Oman. Sustain Sci 19(1):207–222

Fankhauser et al (2022) The meaning of net zero and how to get it right. Nat Clim Chang 12(1):15–21

Fatica, Panzica (2021) Green bonds as a tool against climate change? Bus Strat Environ 30(5):2688–2701

Ferroukhi et al (2016) Renewable energy market analysis—the GCC region. Int Renew Energy Agency, Abu Dhabi, UAE
Garrett-Peltier H (2017) Green versus brown: comparing the employment impacts of energy efficiency, renewable energy, and fossil fuels using an input-output model. Econ Model 61:439–447
Göll Z (2019) Sustainable development in the Mena region. MENARA Futur Notes 20
Golubchikov, Deda (2012) Governance, technology, and equity: an integrated policy framework for energy efficient housing. Energy Policy 41:733–741
Hosseini SA (2023) sustainable development goal 16 and the global governance of violence: critical reflections on the uncertain future of peace. In: 13 trapped between corruption and cruelty, vol 57
ILO (2018) The employment impact of climate change adaptation
IMF (2022) How fiscal policy can help Middle East Central Asia reduce emissions. https://www.imf.org/en/Blogs/Articles/2022/11/06/how-fiscal-policy-can-help-middle-east-central-asia-reduce-emissions
International Energy Agency (2023) Climate resilience is key to energy transitions in the Middle East and North Africa
IRENA I (2019) Renewable energy market analysis: GCC 2019. Abu Dhabi
Lienard C (2022) Mitigating climate change in the MENA: shifting to a new paradigm. Brussels International Center
Marchi D et al (2022) Climate change and consumer behavior. In: Interdisciplinary approaches to climate change for sustainable growth. Springer International Publishing, Cham, pp 315–331
Meys et al (2021) Achieving net-zero greenhouse gas emission plastics by a circular carbon economy. Science 374(6563):71–76
Mundaca G (2017) How much can CO_2 emissions be reduced if fossil fuel subsidies are removed? Energy Econ 64:91–104
PricewaterhouseCoopers (2023) Net Zero Economy Index 2023. https://www.pwc.co.uk/services/sustainability-climate-change/insights/net-zero-economy-index.html
Qadir et al (2021) Incentives and strategies for financing the renewable energy transition: a review. Energy Rep 7:3590–3606
Raihan, Tuspekova (2022) Role of economic growth, renewable energy, and technological innovation to achieve environmental sustainability in Kazakhstan. Curr Res Environ Sustain 4:100165
Raihan, Tuspekova (2023) Towards net zero emissions by 2050: the role of renewable energy, technological innovations, and forests in New Zealand. J Environ Sci Econ 2(1):1–16
Sever (2021) Climate change and the energy transition in the MENA region. In: Climate change law and policy in the Middle East and North Africa region. Routledge, pp 82–106
Suk et al (2016) Environmental pollution: an under-recognized threat to children's health, especially in low-and middle-income countries. Environ Health Perspect 124(3):A41–A45
Volz (2021) 9th IMF statistical forum. In: Transboundary climate-related risks: analysing the impacts of a decarbonisation of the global economy on international trade, finance, and money, pp 17–18
World Bank (2022) Middle East and North Africa climate roadmap: Driving transformational climate action and green recovery in MENA FY 2021–2025
World Economic Forum (2023) How to cut direct emissions in MENA by 2030. https://www.weforum.org/agenda/2023/12/cut-direct-emissions-mena-2030/

Economic Advancement and the Indian Economy Under the Aegis of Sustainable Development Goals (SDGs)

Arshad Bhat, Abid Sultan, M. M. Kachroo, Parvaiz Rashid, and M. Lateef

Abstract The last two decades have seen impressive levels of Indian economic growth, with valuable increases in GDP growth, poverty reduction, and industrial development. However, the journey to achieve the Sustainable Development Goals is especially full of twists and turns, vis-à-vis trends on income inequality, regional disparities, and environmental sustainability. Complications arising from the informal sector, high carbon emissions, and industrial pollution have complicated the country's ability to combine economic growth with more sustainable development. This paper reviews progress in India toward SDG-linked economic growth, where it has placed greater emphasis on key sectors such as renewable energy, electric vehicles, sustainable agriculture, and the digital economy. The growth of green industries and expansion of employment in these sectors are promising as far as the alignment of economic growth with environmental and social objectives is concerned. The policy initiatives such as Make in India, Start-up India, and Skill India have been helpful in fostering innovation and employment. The biggest challenges remain those of formal employment, low wages, and environmental degradation. In addition, the paper argues on the relevance of green financing, skill development, public-private partnerships (PPPs), and digital transformation as the main levers through Sustainable development goals (SDG)-driven economic growth. These policy recommendations are critical to promote sustainability, fostering more inclusive development, and promoting decent work. While much has been achieved, structural inefficiencies

A. Bhat (✉)
Amity Institute of Liberal Arts, Amity University Mumbai, Mumbai, Maharashtra, India
e-mail: bhatarshad09@gmail.com

A. Sultan · M. M. Kachroo
Sher-e-Kashmir University of Agricultural Sciences and Technology of Kashmir, Institute of Business and Policy Research, Shalimar, Jammu and Kashmir, India
e-mail: abidsultan@skuastkashmir.ac.in

P. Rashid
Government Degree College Ganderbal, Ganderbal, Jammu and Kashmir, India

M. Lateef
Department of Commerce, Gandhi Memorial College Srinagar, Srinagar, Jammu and Kashmir, India

F. A. Malik et al. (eds.), *Aligning ESG Goals with the UNs Sustainable Development Goals Planning*, World Sustainability Series,
https://doi.org/10.1007/978-3-032-01715-4_4

in public finance, infrastructure development, and governance stand in the way of tapping the full potential of the SDGs in India. India will need a rethink on economic growth to ensure a balance where sustainability and social equity are prioritised. The governmental, private sector, and civil society sphere all have their respective roles to play to attain self-driven inclusive and green development. This transformation is integral to building a resilient economy sustained over time for the betterment of the coming ages.

Keywords Progress · Technology · Challenges · SDG'S · FDI · GDP

1 Introduction

In fact, during the last two decades, India has been characterized by high, consistent growth of its gross domestic product (GDP), tremendous industrialization, and general social-economic development. From the beginning of liberalization reforms in 1991, structural adjustments have increased foreign direct investment (FDI) into the country while turning toward a market-oriented economy that promoted a strong boom. From 2000 to 2010, the GDP growth rate was approximately 7–8% on an annual basis, marking this economy amongst the globe's most rapidly growing economies (World Bank 2022). There was much infrastructural development in this period and in the field of technology as well as the financial sector.

Between 2011 and 2020, India's economy continued to grow despite the global challenges of the lingering effects of the 2008 financial crisis and trade disruptions. The country made strategic policy interventions, such as Make in India, Startup India, and Digital India, aimed at strengthening domestic industries and fostering entrepreneurship (Government of India 2023). One of the significant sectors that have been contributing to the economy and added value to more than 50% of India's GDP is the service sector, mainly information technology (IT), Business Process Outsourcing (BPOs), and financial services, as mentioned in NITI Aayog (2022).

However, the COVID-19 pandemic severely affected India's economic growth and contracted its GDP by −7.3% in 2020–21 (IMF 2023). Stimulus packages came in the form of Atmanirbhar Bharat to make the economy self-reliant, more digital, and sustainable (Ministry of Finance 2023). During the recovery stage with a growth rate of 7.2% in 2023, the focus had been on high economic momentum that is sustainable and inclusive (Reserve Bank of India 2022).

The United Nations Sustainable Development Goals adopted in 2015 provide a broad framework for the world to be sustainably developed. It covered all the issues with 17 goals and 169 targets, from poverty and economic growth to gender equality, environmental sustainability, and innovation. India has integrated the SDGs into national policies aligned to long-term economic and social planning (NITI Aayog 2021).

Some of the SDGs directly reflect the economic strategies that India has proposed. Goal 8-Decent Work and Economic Growth lies at the heart of the ambition to attain a

$5 trillion economy and focuses on increasing productivity, developing labor markets, and promoting more inclusive financial growth (Government of India 2023). SDG 9-Industry, Innovation, and Infrastructure focuses on the country's industrial modernization initiatives, the rollout of green technologies, and Indian investments in digital infrastructure. Furthermore, SDG 12 – responsible consumption and production has become more important in the country as it adopted circular economy models and sustainable business practices (UNDP India 2022).

The government of India developed the SDG India Index, which tracks progress on various development indicators. It is prepared by NITI Aayog and uses data from ministries, national surveys, and independent agencies to assess the performance of states and union territories in achieving the targets of the SDGs (NITI Aayog 2022). As of 2023, three top-performing states are the states of Kerala, Himachal Pradesh and Tamil Nadu whereas the biggest laggards Bihar, Jharkhand and Uttar Pradesh represent a challenge about realization of SDGs (NITI Aayog 2023).

India has also made efforts to harmonize the budgetary and policy frameworks of the country with the SDGs. The Union Budget 2023–24 has invested in renewable energy, social infrastructure, and employment generation, while owing to its sustainability targets (Ministry of Finance 2023). PM-KUSUM, National Hydrogen Mission and Smart Cities Mission for sustainable urbanization for India, reduce carbon emission, further the initiative. This marks a transition from purely economic growth to a more comprehensive model of development, balanced between economic growth and environment as well as social considerations.

The relationship between economic growth and sustainability is much more complicated for the case of the developing economy of India. Though in consequence, a high growth rate has helped in increasing levels of living standards, infrastructural facilities, and employment opportunities, it has also created challenges in the form of degradation of environmental elements, depletion of natural resources, and socio-economic disparities (World Economic Forum 2022). That is why this country faces the dual challenge of maintaining sustainable high economic growth.

Climate change and environmental degradation are significant issues. India ranks third among carbon emitters in the world, with huge carbon emissions from coal-based electricity generation, industrial processes, and transportation (IEA 2023). Its growth has been historically fossil-fuel based, although it has committed to achieving net-zero carbon emissions by 2070 and raising renewable energy capacity to 500 GW by 2030 (Ministry of Power 2022). The shift towards green energy is reflected in investments in solar, wind, and hydroelectric power, with India ranking among the top five countries in renewable energy deployment (IRENA 2023).

Inclusive growth and poverty alleviation are considered another indispensable dimension for sustainable development. Despite India's impressive GDP growth, there still prevails a wide income gap that deems rural areas more economically less mobile than city centers (UNDP India 2022). With schemes like PM-KISAN, an income support scheme for farmers; MGNREGA, an employment guarantee program; and Aspirational Districts Programme, which the government looks towards bridging the gap between the developed urban centers and less fortunate rural districts (NITI Aayog 2023), there are other schemes like financial inclusion schemes of opening more than 400 million accounts for the unbanked, Jan Dhan Yojana, and similar schemes (Reserve Bank of India 2022).

Technological advancement and a digital economy are tremendous growth opportunities for the country in terms of economic sustainability. Digitalization in India has made tremendous strides in enhancing governance, financial inclusion, and entrepreneurship with more than 1.2 billion mobile phone users and 900 million internet users (MeitY 2024). Increasing fintech solutions, digital payments, or UPI and e-commerce have transformed the fabric of the economy, giving the country more business models to create a low-carbon economic scenario (World Bank 2022). This is also as a result of the growth of artificial intelligence and automation, which will enhance efficiency in industries and reduce resource wastage while increasing productivity (NASSCOM 2024).

On the other hand, economic growth, however, carries governance and policy issues along with sustainability. India needs to prepare itself to absorb the burden of policy integration, institutional reform, and cross-sectoral interaction to tap SDG-based economic growth (UNDP India 2022). Public Private Partnership initiatives shall require enhancement, clean technologies innovation, and green finance mechanism shall be in demand to trigger sustainable growth sans the degradation of the environment and social well-being over the long run (Ministry of Finance 2023).

The Indian economy, in the last two decades, has been one of the fastest-growing major economies in the world. Its growth trajectory has been notable and characterized by substantial contributions from the services sector, with some major developments around the agricultural and manufacturing sectors. India is aligned with the United Nations' Sustainable Development Goals (SDGs), mainly SDG 8 Decent Work and Economic Growth; SDG 9 Industry, Innovation, and Infrastructure; and SDG 12 Responsible Consumption and Production, that fall in tandem with global frameworks on sustainability.

Impressive GDP growth rates characterize the growth of India's economy. The average annual GDP growth in the country was at about 7.5% between 2012 and 2023, with growth that is primarily led by the services sector, which saw its share rise from 48 – 50% during that period (Roland Berger 2025). Meanwhile, during the same period, the industrial sector's share in GDP declined to 25% from 29%. The immediate conclusion here is issues in growth of manufacturing. The Indian government has taken several initiatives to address these disparities and promote sustainable development.

The 2025–26 budget brought in significant cuts in personal income tax to boost spending by the middle class and improve domestic demand (Reuters 2025). On the other hand, investments in agriculture, manufacturing, and infrastructure were stressed for balanced economic growth. Such provisions befit the general strategy of achieving sustainable and inclusive development.

The services sector forms the base of Indian economic growth and contributes nearly 50% of GDP as per the data for the year 2023 (Roland Berger 2025). These sectors include information technology, telecommunication, finance services, and tourism. For example, tourism once again touched 5% to the GDP level in FY23 (Press Information Bureau 2025).

Agriculture and allied activities remain unavoidable, accounting for about 16% of the GDP in FY24 (Press Information Bureau 2025). Horticulture, livestock, and fisheries have emerged as those high-value areas that contribute to agricultural growth. An increase in Kharif food grain production to 1647.05 lakh metric tonnes in 2024 is also a testimony to its resilience and potential.

Although the share of manufacturing in GDP has declined, it still stands as the sector with the greatest employment and exports. It has plans to reinvigorate the sector through reforms in the business environment, logistics infrastructure improvements, and tax rate rationalization (World Bank 2025). Buoyant services, with a growth rate of 9.9%, were matched by the subsector which, however had continued to remain buoyant and hence balanced out the weakness in agriculture. Growth was driven by public investment in infrastructure and household investment in real estate.

The commitment to the SDGs has been in India's economic policies and planning, where there is a special focus on SDG 8, SDG 9, and SDG 12. SDG 8 focuses on the promotion of sustained, inclusive, and sustainable economic growth, full and productive employment, and decent work for all (United Nations, n.d.). Government efforts in achieving this are improved labor market conditions, unemployment levels, and a fair wage scale. Formalizing the gig economy and providing health care and social welfare for its workers is government action toward fulfilling these goals (Associated Press 2025). SDG 9 focuses on resilient infrastructure, promoting inclusive and sustainable industrialization, and fostering innovation. The vision of the Indian economy is embedded in this in the form of infrastructure creation and development, enabling startup culture, and encouraging the use of renewable energy sources (Reuters 2025). Nuclear Energy Mission-100 GW nuclear power by 2047-the country would be more industrialized but sustainable. SDG 12 relates to sustainable consumption and production patterns. In India, the programs include high-yielding crops, sustainable agriculture, and the development of the food processing industry (Press Information Bureau 2025). These are designed to improve resource efficiency and reduce environmental degradation.

The economic performance of India from 2015 to 2023 reflects a dynamic yet irregular growth trajectory that is conditioned by global economic conditions, policy reforms in the domestic arena, and sustainability initiatives. It has shown resilience and recovery, especially after the situation of COVID-19; however, much remains to be achieved with respect to employment generation, controlling inflation, and sustainable development.

The Make in India and Skill India initiatives had been boosting manufacturing, FDI, and workforce training. Even though investment inflows were boosted through these policies, India was still a service-dominated economy; manufacturing contributed only 16% to GDP-indicating the slow pace of industrialization (NITI Aayog 2022). The growth declined after demonetization in 2016 and GST in 2017, which had short-term shocks to the economy. Startup India promoted entrepreneurship but was restricted due to bureaucratic red tape and its limited rural reach. The growth slowdown signaled that structural reforms require better phasing to avoid economic shocks (World Bank 2019). Global economic conditions had weakened by 2019, and domestic consumption had slowed. Stress in the financial sector, the crisis of the NBFCs constrained further credit. Sustainable Infrastructure has now become the theme for the long-term resilience of the economy, but private sectors involvement has been limited owing to the regulatory bottlenecks and risks in investments. This was stated by IMF 2022. 2021 Lockdown Recovery Growth rate 8.9%.

Given the aftermath of the contraction of 2020, stimulus packages, growth in the digital economy, and fiscal interventions ensured that India's economy rapidly recovered from all setbacks. MSMEs and informal workers were still on shaky grounds. Atmanirbhar Bharat is a boost to self-reliance but could not change supply chain dependencies of the world at large (ILO 2022).

With India undertaking Green Energy Initiatives in a commitment to the implementation of SDG 12 (Responsible Consumption and Production), the growth became flat for the economy. Renewable energy was invested in further, but higher transition costs coupled with infrastructure gaps made the longer term uncertain. Growth was good but with rising inflation and job disorder, this economic gain had not been reasonably shared (IEA2023).

Inflation was moderate but volatile, influenced by commodity price volatility, policy actions, and broader global economic trends. Controlled Stability Inflation eased back due to monetary policy tightening and low oil prices. The government's focus on controlling food price volatility and improving supply chain efficiency helped maintain stability. Supply Chain Pressures Inflation rose due to the food and fuel prices that were going up while the structural inefficiencies in agriculture and logistics remained unabated. The sustainable infrastructure efforts tried to fill these gaps, but investment was slow-moving. Pandemic Aftermath Inflationary pressures resurfaced due to demand from stimulus-driven, supply chain disruptions, and global price shocks. Critical goods experienced a rise in price, impacting the most the poor households. RBI, 2022. 2023 Cost of Green Transition Inflation- 6.2%. As the Indian government increased investment in renewable energy sources, green technologies became more costly with rising inflation. Although energy transition will be continued by policymakers, its affordability will be compromised at the cost of sustainable ambition (IMF 2023).

As the Indian economy grew, unemployment worsened, which pointed out the disconnect NITI Aayog.

GDP expansion and job opportunities. The growth in the economy and employment in the informal sector helped keep unemployment at a relatively low level. However, there was a significant issue with job quality, where most workers were not covered under social security in 2015 (5.0% Unemployment was found. Challenges of Formalization Startup India gave a boost to entrepreneurship but not enough formal employment. Besides, demonetization and GST compliance resulted in losses of jobs (6.1% Unemployment in 2017) (Table 1) in informal and small businesses that increased unemployment (ILO 2018). 2019 (7.6% Unemployment): Structural Job Crisis Despite growth-oriented policies, employment generation lagged, particularly in manufacturing. Automation and slow industrial expansion limited job opportunities, raising concerns about jobless growth. In 2021 7.5% Unemployment was recorded due to COVID-19 Shock. Pandemic-induced mass job loss struck the worst through tourism, retail, and construction sectors. Government has launched

Table 1 India's economic growth trends (2015–2024)

Year	GDP Growth rate (%)	Inflation rate (%)	Unemployment rate (%)	SDG-Related policy implemented
2015	8.0	4.9	5.0	Make in India, skill India
2017	7.1	3.3	6.1	Start-up India
2019	4.2	4.8	7.6	Sustainable infrastructure
2021	8.9	5.1	7.5	Post-COVID recovery measures
2023	7.2	6.2	7.1	Green energy initiatives

Source World Bank, 2025 and Reserve Bank of India 2022

relief measures in response, but informal workers lack social protection. In 2023 7.1% Unemployment was recorded. Green Economy Opportunities and Gaps Green Energy Initiatives created high-skilled jobs in renewable sectors but failed to absorb large numbers of low-skilled workers. Rural employment challenges persisted, which required better skill development programs (NITI Aayog 2023).

2 Review of Literature

The implementation of the Sustainable Development Goals (SDGs) by India has placed an indelible mark on its economic policy agenda since 2015. SDGs form a plan for making social development and environmental protection march alongside economic development (Statista 2025, United Nations 2015). The flagship government initiatives such as "Make in India," "Startup India," and "Digital India" complement SDG 8 (Decent Work and Economic Growth) and SDG 9 (Industry, Innovation, and Infrastructure) and reflect an inclination to promoting industrialization, entrepreneurship, and employment creation (Chakraborty and Pradhan 2021). Kumar et al. (2020) posit that the schemes not only promote economic productivity but also foster sustainability in the form of innovation and skill upgradation. Other than these, the NITI Aayog's SDG Index has also provided a data-driven framework for monitoring progress, inducing cooperative and competitive federalism among Indian states (NITI Aayog 2020).

Nevertheless, despite these, India still has very serious problems in attaining inclusive and equitable growth across regions. Studies have suggested that while urban and industrialized states rank fairly well on the majority of economic SDG indicators, rural and emerging areas lag behind in poverty alleviation, education, and infrastructure (Kundu and Chakraborty 2022). Furthermore, environmental concerns like carbon emission and unsustainable land use are still issues to economic long-term sustainability (Financial Times 2025, BBC News 2024, Raghunandan and Nambiar

2021). Experts are of the opinion that filling these gaps requires policy reforms but also improved governance, innovative financing, and people's engagement to localize the SDG agenda (Bhattacharya and Mitra 2021). Thus, although the SDGs offer a strategic roadmap to India's economic advancement, their realization in practice calls for an integrated, inclusive, and locally responsive effort.

3 Objectives and Hypothesis

The basic objectives of the chapter are

- To examine the contribution of SDG-based policies to India's economic development and growth.
- To examine the level of integration of SDGs into national and state economic plans.
- To consider the regional imbalances of economic progress in the context of SDG deployment.
- To measure the contribution of governance, institutions, and public-private partnerships to sustainable growth.
- To determine challenges and opportunities to align India's economic path to the 2030 SDG agenda.

In consonance with the objectives of the chapter the following hypotheses have been made.

- H_1: There is a strong positive correlation between the passage of SDG-congruent policies and Indian economic growth overall.
- H_2: States with higher SDG Index values reflect improved economic performance compared to states with low SDG values.
- H_3: Public-private partnerships have a statistically significant impact on the success of SDG-driven economic activities in India.
- H_4: Regional differences exert significant impact on the uniform application of SDGs in India, resulting in disparate economic progress.
- H_5: Institutional quality of governance positively moderates SDG implementation—economic development outcomes relationship.

4 Methodology

This research utilizes a mixed-methods approach, which combines qualitative and quantitative methods to examine the intersection of economic growth with SDG adoption in India. Secondary data will be accessed from valid national and global sources like NITI Aayog SDG India Index, Ministry of Statistics and Programme Implementation (MoSPI), Reserve Bank of India (RBI), United Nations Development

Programme (UNDP), and World Bank. The sources will provide statistical information for economic performance measurements and SDG achievement at states and national levels. For quantitative analysis, regression analysis, descriptive statistics, and correlation will numerically measure the association of SDG implementation with economic development among states in India. For the qualitative component, questionnaires and structured interviews administered to the stakeholder groups of policymakers, economists, regional planning authorities, and development practitioners will be used as tools to identify experts engaged in SDG planning and grassroots implementation through purposive sampling.

Thematic analysis will be conducted on the qualitative data to gain insights into challenges, opportunities, and policy efficiency. The convergence of empirical evidence and stakeholders' views will try to provide a detailed explanation of how sustainable development policies influence the trajectory of India's economy under the SDG plan.

5 Discussions

India's economic policies aligned with SDG targets, yet their effectiveness varied Make in India, Skill India-Encouraged manufacturing growth, but regulatory barriers and lack of infrastructure limited success (UNDP 2022). 2017 (Startup India)-Boosted the startup ecosystem, but remained urban-focused, benefiting tech-based firms over rural enterprises (World Bank 2020). 2019 (Sustainable Infrastructure)-Investments in transport and energy targeted long-term growth, but faced funding delays (Talent Advisory Group 2023, IMF 2023). 2021 (Post-COVID Recovery Measures) Economic stimulus helped mitigate the shock; on the other hand, income inequality worsened (ILO 2022)). 2023 (Green Energy Initiatives) -Aligned with SDG 12, yet faced costly energy transition challenges (IEA 2023).

India's industrial growth has been quite a transformational journey. The focus has shifted towards sustainability and employment generation. Green industries, SDG-driven initiatives, and the role of MSMEs have all positively contributed to economic growth while taking care of environmental and social challenges. While policy progress has been witnessed, structural barriers, financial constraints, and slow technological adoption remain major factors that impede full realization of sustainable industrial development.

Expansion of green industries like renewable energy, electric mobility, sustainable manufacturing, and waste management has been a crucial component of India's industrial strategy owing to its linkages with the SDGs. The government for example, with the support of SDG 9, Industry, Innovation and Infrastructure, and SDG 12, Responsible Consumption and Production, encouraged green industrialisation through schemes like the National Solar Mission, Faster Adoption and Manufacturing of Hybrid and Electric Vehicles, and the Green Hydrogen Mission. India is also the third-largest producer of renewable energy globally and has set its target for 2030 at achieving a non-fossil fuel energy capacity of 500 GW (IEA 2023). Rise of electric vehicles space, including subsidy under the FAME scheme that supports the local EV manufacturing. Circular economy initiatives like waste-to-energy plants, sustainable packaging, etc. One major deterrent for SMEs to move towards sustainable practices is the huge upfront investment in adopting green technology. Bottlenecks in the regulatory and infrastructure space include charging infrastructure for electric vehicles and delayed land acquisition for renewable projects. The reliance of critical raw materials, such as lithium and rare earth elements, on imports endangers self-sufficiency in green industrialization.

Sustainable development policies of India have been significant to employment generation and are doing exceptionally well in areas like renewable energy, eco-tourism, waste management, and sustainable agriculture. In the initiatives of Skill India, Start-up India, and Green Skill Development Programme (GSDP), workforce skills are provided for the green economy. More than 1 million people are involved in the renewable energy sector, and this number is likely to increase with the promotion of solar and wind energy in India (ILO 2022). Employment has been generated in the rural sector through agroforestry and organic farming. This has resulted in reduced migration to cities. Eco-tourism has created new jobs in hospitality, conservation, and environmental management. Although green jobs require higher technical expertise, India's skilling programs still lag industry needs, causing a skills gap. The informal sector, employing a large chunk of India's workforce, lacks the ease to formalize and is deprived of green job opportunities. As newer jobs emerge, older ones from coal mining and heavy industries are diminishing. So, the essence is developing a just transition strategy that should minimize the likely large-scale unemployment in the long term.

MSMEs are the backbone of India's industrial economy and contribute almost 30% to GDP while providing employment for over 110 million people (Ministry of MSMEs 2023). The large presence of MSMEs in manufacturing, agriculture, and services makes it important for MSMEs in driving SDG 8, Decent Work and Economic Growth, and SDG 12, Responsible Consumption and Production. MSMEs have been using cleaner production techniques and promoting sustainable packaging. MSMEs contribute to non-agricultural employment and avoid overcrowding in cities. Small-scale enterprises are involved in recycling, upcycling, and waste reduction activities to contribute to sustainable consumption. Capital investments are required in green business transitions; however, there are limitations of credit availability and high interest rates in the MSME sector. Most of the MSMEs are denied access to new energy-saving technologies that restrain their chances of using eco-friendly production techniques. Strict environmental regulations often increase costs for small enterprises, discouraging them from transitioning to green practices (Table 2).

Data has emerged with dramatic shifts in employment dynamics in major industries of India and especially focused on sustainable development goal sectors. Growth of employment in areas like renewable energy, electric vehicles, sustainable agriculture, and the digital economy reflect encouraging momentum; however, several crucial factors are worth discussion for understanding the consequences of such employment change in the overall economy stability, sustainability, and social equity.

Table 2 Employment growth in key SDG-aligned sectors (2015–2024)

Sector	Jobs in 2015 (Million)	Jobs in 2024 (Million)	Growth (%)	SDG impacted
Renewable energy	0.3	1.2	300	SDG 7, SDG 8
Electric vehicles (EV)	0.1	0.9	800	SDG 9, SDG 13
Sustainable agriculture	5.2	6.5	25	SDG 2, SDG 12
Digital economy	2.0	4.8	140	SDG 8, SDG 9

Source World Bank, 2020 and Reserve Bank of India 2022

The renewable energy sector has experienced high growth in India, with the number of jobs expected to rise by 300%, from 0.3 million jobs in 2015 to 1.2 million jobs in 2024. This growth is attributable to India's ambitious renewable energy targets, such as their commitment to achieve 500 GW of non-fossil fuel capacity by 2030 (Ministry of New and Renewable Energy 2022). The sector aligns strongly with SDG 7 (Affordable and Clean Energy) and SDG 8 (Decent Work and Economic Growth), contributing to clean energy access and employment generation. However, despite rapid growth, job creation in this sector is still a fraction of what is required for sustained, equitable development. The jobs created are highly technical in nature, and therefore, not always available in rural and underserved areas. There is, therefore, a need for skills development programs to be expanded across the sectors involved, particularly among low-income groups. The local transition for green jobs also must encompass the people displaced by the exiting traditional energy industries.

The electric vehicle (EV) sector demonstrates an impressive 800% growth in jobs, from 0.1 million in 2015 to 0.9 million in 2024. This surge reflects India's transition toward greener transportation and its push for reducing emissions as part of its climate change commitments. This sector contributes to SDG 9 (Industry, Innovation and Infrastructure), focusing on innovation in green technology and clean transportation, and SDG 13 (Climate Action) by reducing carbon emissions from the transport sector. It should be noted, though, that with all that's happening within the EV industry, structural factors to adopt-whether in charging infrastructure, very high purchase prices, or dependency on raw materials (such as lithium and cobalt) in making the EV batteries-do bear much significance. Such growth in the sector will result in employment displacements for automotive manufacturing in case of inadequate retraining efforts. The jobs created in EV manufacturing are largely in urban centers, so regional imbalances in employment may persist unless there is rural-focused growth as well.

Sustainable agriculture is expanding at a relatively slower rate and creates 25% more jobs, growing from 5.2 million jobs in 2015 to 6.5 million jobs in 2024.

Even though agriculture forms the core of the Indian economy, growth in agricultural employment has been slow due to climate vulnerabilities and low adoption of sustainable practices in rural areas. The sector is very much aligned with SDG 2 (Zero Hunger) through food security and SDG 12 (Responsible Consumption and Production) through environmentally friendly agriculture. Slow growth in agricultural employment indicates that the challenges are deep-rooted: low productivity, fragmented land holdings, and a lack of access to modern and sustainable farming techniques. While focus on organic farming and water-efficient irrigation has been increasing, adoption of sustainable agriculture is not even. Moreover, youth migration from agriculture to urban areas in search of better employment opportunities continues to drain this sector of young workers. Therefore, investments in rural infrastructure, technology transfer, and policy reforms are necessary to foster more robust growth in sustainable agricultural jobs.

The sector of digital economy has been the prime employment stimulant in India, with job growth as high as 140% in this sector-from 2 million in 2015 to 4.8 million in 2024. With rapid expansion, the IT services, e-commerce, and the digital platform- together, they helped generate large-scale employment, primarily in urban regions. The digital economy impacts SDG 8 (Decent Work and Economic Growth) by creating a vast number of digital jobs, and SDG 9 (Industry, Innovation and Infrastructure) by fostering technological innovation. While the digital economy offers immense potential for job creation, its growth has been primarily urban-centric, leaving rural areas with limited access to digital jobs. The skills gap remains a significant issue, as low-income and rural workers often lack access to digital literacy training and technology infrastructure. Furthermore, job quality in the digital economy varies significantly, with many digital jobs in e-commerce and gig economy platforms characterized by precarious working conditions, lack of job security, and no social protection. This sector requires a policy focus on inclusive digital education, skills training, and protection policies for the workers to ensure good growth in the sector. The efforts by India towards SDGs are undoubtedly excellent; however, before the country can realize these goals to their fullest extent, challenges still prevail. Income inequality, regional disparities, and environmental degradation, among others, are deep-seated issues affecting the pursuit of SDG-linked economic growth. Structural inefficiencies in public finance and infrastructure are also related to these challenges. In other words, the impact of global disruptions like the COVID-19 pandemic and geopolitical tensions also generally affect the pace and scope of progress toward sustainable and inclusive economic growth.

One of the most pressing challenges to achieving SDG-linked economic growth in India is growing income inequality and the significant regional disparities in wealth, resources, and opportunities. Indian economic growth has been lopsided, with some states and regions forging ahead in rapid development, and others also are still behind. According to the World Inequality Report (2022), India ranks among the most unequal countries in terms of income distribution, with the top 1% of the population holding a disproportionate share of national wealth. Economic inequality not only hinders social cohesion but also limits the capacity of disadvantaged groups to access quality healthcare, education, and decent work (SDG 3, SDG 4, SDG 8).

This is partly due to the increasing gap between the rural poor and the urban elites that widens the inequality gap further, with the growth in the urban areas being more pronounced than in the rural areas. State-level disparities, however, worsen this condition, as states like Tamil Nadu and Maharashtra showed advanced SDG progress over Bihar and Uttar Pradesh (NITI Aayog 2023). Such disparities hinder the objective of inclusion in growth and resource utilization. Redistributive policies, progressive taxation, and investments in social safety nets would be required to address these inequalities, along with a focus on regional development to bridge the gaps between states and rural-urban areas.

Rapid industrialization in India has led to serious environmental challenges that directly challenge the country's ability to achieve several of the SDGs, particularly SDG 13 (Climate Action), SDG 12 (Responsible Consumption and Production), and SDG 6 (Clean Water and Sanitation). Unchecked expansion of industries has resulted in severe industrial pollution, deforestation, and water scarcity, all of which continue to undermine sustainable development. India is one of the world's largest polluters, with air and water pollution reaching alarming levels, particularly in major industrial and urban centers. The Central Pollution Control Board (CPCB) reports that air quality in cities like Delhi, Mumbai, and Kolkata is often classified as hazardous, posing serious health risks to the population. Industrial emissions, particularly from the coal-based power sector, contribute significantly to air pollution (WHO 2022). India continues to lose forest cover due to agriculture expansion, urbanization, and illegal logging. While efforts have been made to curb deforestation, including the implementation of afforestation programs and environmental regulations, these efforts remain insufficient to halt the overall trend. According to the Forest Survey of India (2021), India has lost around 3 million hectares of forest cover in the last two decades. Acute water shortages characterize many regions within India, making it increasingly dependent on a more severe water crisis. According to the NITI Aayog's Composite Water Management Index (2020), 21 major cities across India, from Delhi to Bangalore and Chennai, are under extremely high levels of water stress threatening public health along with agricultural productivity (NITI Aayog 2020). For instance, stronger regulatory frameworks, better waste management practices, as well as sustainable industrial practices that respect the environment can be achieved. Practices related to renewable energy as well as sustainable agriculture should take precedence in reducing the pressure on natural resources.

India faces structural inefficiencies in public finance and infrastructure development, which limit the government's ability to effectively implement SDG-driven policies. Public finance is often constrained by fiscal deficits, poor tax collection efficiency, and ineffective allocation of resources across sectors. This, in turn, affects infrastructure development, which is crucial for achieving SDG-related economic growth. Despite the growth of the Indian economy, the country struggles with underfunding in key sectors like education, healthcare, and infrastructure. The Indian government's expenditure on social sectors (as a percentage of GDP) remains low compared to other emerging economies. According to the World Bank (2022), India's fiscal deficit was projected to be over 6% of GDP in 2022–2023, constraining the government's ability to invest in key public services and infrastructure. Inadequate infrastructure in transportation, energy, and sanitation continues to hinder economic growth and development. The World Economic Forum (2022) ranks India 91st out of 140 countries in terms of infrastructure quality. A lack of investment in rural infrastructure, including roads, water supply systems, and electricity grids, impedes inclusive growth and the achievement of SDG 9 (Industry, Innovation and Infrastructure). India needs to prioritize infrastructure development through public-private partnerships (PPPs) and focus on improving public financial management to ensure efficient allocation of resources to critical sectors.

The COVID-19 pandemic and other global disturbances have caused India to experience a slowdown in the progress of its SDGs. These disturbances have exposed the weaknesses in India's economic systems and increased the demand for resilience in the country's sustainable development strategies. The COVID-19 pandemic brought forth a sharp halt in activities across all economic sectors, exacerbated inequalities, and reversed the improvement achieved in the reduction of poverty, access to healthcare, and employment. The World Bank (2021) estimates that India's GDP contracted by 7.3% in 2020, pushing millions back into poverty and significantly impacting SDG 1 (No Poverty) and SDG 3 (Good Health and Well-being). In addition, the pandemic severely affected supply chains, particularly in manufacturing and agriculture, delaying progress in achieving SDG 8 (Decent Work and Economic Growth). Geopolitical issues, such as border tensions with neighboring countries and global trade disruptions, have also impacted India's SDG progress. Disruptions in global supply chains, coupled with increased costs of goods and services, have slowed India's recovery post-pandemic and raised the cost of implementing SDG-related projects. India needs to focus on creating resilience in the economy and thereby ensure that some social safety net is available that can be developed to protect sensitive populations during emergencies. It could also explore an alternative supply chain, technological innovations, and build local production capabilities to reduce vulnerability to global conditions. The table captures a snapshot of India's progress toward three important Sustainable Development Goals: No Poverty (SDG 1), Decent Work and Economic Growth (SDG 8), and Climate Action (SDG 13). Though it portrays some of India's successes in these areas, critical challenges require immediate attention. Below is a critical analysis of each of these SDGs, focusing on the underlying challenges, status, and the implications for India's sustainable development.

India's economic growth has been good over the last decades, but poverty is still persistent. Poverty in the country still occurs mainly in rural areas due to a lack of resources, education, health care, and economic opportunities. It is reported that India's poverty rate stands at 10%, which is relatively lower than that of past decades but leaves millions of people still in poverty. Many in the rural belts, mostly from Bihar, Uttar Pradesh, and Madhya Pradesh, continue to live under poverty lines due to lack of availability of minimum amenities. India ranks at 120 from 193 nations on the index of poverty reduction, meaning though it has marked some progress on the scale of poverty reduction the country is a long way to achieve the targets of poverty removal by 2030. It has a rate of 10% in this regard, that may seem so good, but the scenario has more complexity; however, there are factors increasing the problems regarding income inequality and regional disparities. Rural areas and sometimes particular communities are experiencing the worst in terms of poverty, including Dalits and Adivasis. All the initiatives the government is undertaking to relieve the poverty line, such as PMAY or MGNREGA, have succeeded in some measure in changing conditions, but all these have cases of inefficiency and improper implementation. To fill these gaps, India should have targeted interventions in the spheres of economic empowerment, education, health care, and inclusive growth in underdeveloped regions.

The informal sector continues to dominate India's labor market, where young workers often experience precarious employment conditions including low wages, no social protection, and inadequate working conditions. This is an expression of general issues pertaining to the structure of India's labor market, which fails to provide ample formal and stable employment for the growing workforce. As such, India boasts one of the highest rates of informal employment globally, with an estimated 80 percent of its workforce employed in the informal sector. This number therefore indicates the level of informality along with job insecurity among most Indian workers. The government has initiated several schemes to regularize employment and protect labor rights; for example, the Code on Wages 2019 has been introduced, but enforcement still lags, and labor rights violations and horrible working conditions are witnessed. India is ranked 102 out of 193 countries. It has a huge advancement in the global contribution to the workforce but still has a lack of significant proportion in decent work and inclusive economic growth. This has the tendency to keep most of its labor force outside the formal economy; in other words, excluding from it, which impacts economic stability, social protection, and a decent wage. Dominance of informal employment is not only a structural inefficiency in the labor market but also inadequacy of job creation in the formal sector of India. Formalization of labor is very important for improving working conditions, ensuring social security, and providing equal opportunities for all workers. Efforts like Skill India and Start-up India are steps in the right direction, but their effectiveness will depend on broader systemic reforms in labor laws, inclusive economic policies, and support for MSMEs (Micro, Small, and Medium Enterprises).

India is one of the largest carbon emitters in the world, mainly due to its reliance on coal for energy production, rapid industrialization, and population growth. India is the third-largest emitter of greenhouse gases in the world and has significant challenges in balancing economic growth with climate action. Currently, India is the third largest emitter of carbon dioxide (CO_2), after China and the United States. This high carbon footprint is due to energy-intensive industries and the growing transportation sector, coupled with a lack of proliferation of clean energy technologies across the nation. In India, good intentions are made in terms of climate action with ambitious targets for achieving net-zero emissions by 2070 and shifting to renewable energy sources, but the changeover is quite slow because it has run into challenges such as fossil fuel dependence, access to energy, and unavailability of available technologies. India is ranked 108th out of 193 countries for its climate action efforts. The country has demonstrated leadership in international climate negotiations, such as COP26, but domestic actions do not meet the Paris Agreement's targets. India's carbon emissions are a challenge not only for SDG 13 (Climate Action) but also for the country's economic stability and health outcomes. Air pollution in major cities like Delhi and Mumbai incurs huge public health costs and affects millions of people. Though India has upped its efforts in shifting towards renewable energy sources such as solar and wind power, the speed of transition is still slow because of financial, technical, and infrastructural barriers. The major dependency on coal remains a major barrier, besides international financing and technology transfer for an accelerated shift toward a greener economy. India must focus on investing in cleaner energy, improving energy efficiency, and building climate resilience to mitigate the adverse effects of climate change while promoting sustainable growth (Table 3).

India's ambitious Sustainable Development Goals (SDGs) call for a comprehensive strategy that integrates economic growth with social equity and environmental sustainability. The country indeed witnessed visible progress in most of the SDGs, but the challenges to achieve them in a holistic and inclusive manner call for focused policy interventions. Some of the key policy recommendations that can strengthen SDG-driven economic growth in India are as follows:

Green finance is going to play a very crucial role in ensuring economic growth is on the right side of sustainability and climate action. Sustainable investments can contribute to the growth of industries that become vital to the achievement of SDG 7 (Affordable and Clean Energy), SDG 9 (Industry, Innovation, and Infrastructure), and SDG 13 (Climate Action). Therefore, India requires considerable investments in renewable energy, energy efficiency, and sustainable infrastructure for the transition into a green economy. The Indian government needs to incentivize private sector investment in clean energy through green bonds, impact investing, and environmental, social, and governance (ESG) criteria. Green financing mechanisms can also be scaled up further through public-private partnerships (PPPs) to support climate-smart projects and green businesses. According to the report UNEP (2020), United Nations Environment Programme, green financing has been established as the keystone to unlocking capital for sustainable development; and for India, for instance, schemes such as National Action Plan on Climate Change, or NAPCC, and the creation of an India Green Bond Market. International climate finance and further development of domestic green financial tools will bring needed capital for scaling clean energy projects, energy-efficient infrastructure, and development resilient to climate change. However, the green financing market in India is constrained by low

Table 3 Key SDG challenges and India's performance (2023 Report)

SDG Goal	Challenge for India	Current status (2023)	Global rank
SDG 1 (no poverty)	High rural poverty	10% poverty rate	120/193
SDG 8 (decent work)	High informal employment	80% of the workforce in informal sector	102/193
SDG 13 (climate action)	High carbon emissions	3rd largest emitter	108/193

Source World Bank, 2019 and Reserve Bank of India 2022

investor awareness, weak regulatory frameworks, and a lack of standardized ESG metrics. All these requirements are to be addressed by ensuring the development of proper guidelines in green projects as well as channelizing the availability of capital flow through innovative instruments such as a sustainable bond, and tax benefits for green industry sectors (Macrotrends 2025, Sachs et al. 2021).

Thus, in India, the demographic dividend for achieving SDG-driven economic growth can be realized only if the labor force is adequately skilled to match this highly changing global economy. The workforce of the future will need to be resilient, digitally literate, and ready and equipped with skills for emerging industries like renewable energy, digital technologies, and sustainable manufacturing. The Indian government needs to expand and align skill development programs to the needs that the economy is projected to face in the future. The skills development programs must target renewable energy technologies, green jobs, AI, data science, and sustainable agriculture. These programs must also include youth employment opportunities, especially targeting rural and underserved areas, by collaborating with industry stakeholders, educational institutions, and training centers. Initiatives such as Skill India and NSDM have so far taken skill development for Indian workers significantly further, but green industries demand focus on skills preparing workers to not only succeed today but, in a future, tomorrow as well. According to the India Skills Report 2023, the skills of the young people and their employability show improvement; however, a mismatch between industry demands and youth skill sets has become a cause for concern in emergent sectors, such as clean energy and digital technologies. While India has been able to successfully launch many skill development initiatives, there is a need to ensure that these programs should be not just in line with the industry's demand but also accessible to the marginalized sections of society. So, there should be emphasis on ensuring that the skill development process is an inclusive one that will provide gender equality and equal opportunities for all communities.

Public-private partnerships (PPPs) have the potential to spur innovation in sustainable development, particularly in clean energy, infrastructure, and agriculture sectors. PPPs can foster innovation and scalability through the combination of public policy strength and private sector efficiency in making the transition towards a green economy. India needs to strengthen its PPP frameworks for private sector participation in sustainable infrastructure projects. These include solar power, EV infrastructure, waste management, and green urban planning. The government may provide financial incentives such as subsidies, tax benefits, and soft loans to entice private investment and entrepreneurship towards green projects. PPPs can be a game-changer, as evidenced by the success of the Make in India and Smart Cities Mission. According to the Asian Infrastructure Investment Bank (AIIB 2021), PPPs can provide the necessary capital investment and expertise needed to implement large-scale sustainable infrastructure projects. For example, the Narmada Basin Solar Power Project was developed through a PPP model, significantly contributing to India's renewable energy capacity. Despite the success stories, PPPs in India have faced challenges related to regulatory hurdles, lack of transparency, and delays in project execution. It will require the provision of more substantial legal and regulatory frameworks through which private sector partners are challenged to invest in long-term and sustainable development projects. Public resources must be put to a much better effect toward public goods.

Little more can be said about how the digital transformation of the economy has a bearing on governance, but it is most crucial when relating to inclusive economic growth. The use of technology is bridging the gaps within public service delivery, financial inclusion, and education. Digital platforms can be harnessed in order to create efficiency and inclusiveness in SDG-linked economic growth. The Indian government should accelerate the digital transformation of governance by investing in e-Government initiatives that streamline processes, improve service delivery, and increase transparency. The adoption of digital technologies in areas like education, healthcare, agriculture, and financial services will not only enhance public sector efficiency but also increase access to resources for marginalized populations. Improvement in digital literacy and ensuring access to e-governance services for the rural population are facilitated by programs like Digital India and the Pradhan Mantri Gramin Digital Saksharta Abhiyan. Financial inclusion and digital transactions are on the rise in India, due to online banking and mobile payment systems, under the National e-Governance Plan (NeGP) and have facilitated millions of unbanked citizens in Indian society (Government of India 2022). While digital transformation holds immense potential, India faces challenges such as digital divide, internet connectivity issues, and cybersecurity risks. Addressing these barriers will require investments in digital infrastructure, cybersecurity frameworks, and digital literacy programs to ensure that all citizens benefit from the digital revolution and that no one is left behind.

6 Policy Implications

The results of the study carry significant policy implications for the alignment of India's economic development trajectory with the Sustainable Development Goals (SDGs). Plans for economic development must be ensured not only to increase GDP but also to ensure that they place greater emphasis on inclusiveness, eco-sustainability, and regional equity. These initiatives like "Make in India," "Startup India," and "Digital India"have to be yet more tailored to match SDG indicators-more specifically those that concern decent work, innovation, and industrialization that is sustainable. NITI Aayog's SDG India Index also has to be used more aggressively as a tool for planning and monitoring in order to better identify gaps, rank, and target resources at the state and district levels.

Also, the research suggests more inter-sectoral coordination and decentralized leadership to effectively localize the SDG agenda. Enhancing institutional capacity, investment in green infrastructure, and encouraging public-private partnerships can ensure economic development as well as sustainable development. Backward districts and rural towns need to be particularly prioritized for economic inequality reduction as well as ensuring balanced progress. The integration of sustainability principles into budgeting, policymaking, and performance measurement can enable India's long-term economic sustainability as well as the fulfillment of its commitments to the 2030 Agenda.

7 Conclusion

India's story of journey towards economic sustainability in the form of SDGs is characterized by glorious success as well as the persistence of challenges. The economic growth curve of India over the last two decades presents a nation that has achieved incredible strides in terms of GDP growth, poverty alleviation, and industrial expansion. Even though growth indicators and significant exports and high values of GDP would be markers of good times, it is found that a few factors are a hurdle between India and the accomplishment of its set SDG targets. To begin with, this would involve critical income inequality that reflects on huge differences in various states of India and, therefore, affects environmental aspects that affect development. Though promising initiatives have been undertaken through schemes such as Skill India and Start-up India, still a very large section of the population remains excluded from the domain of formal employment and opportunities for decent work. Simultaneously, carbon footprint and environmental degradation, ranking the country at number three in the list of CO_2 emitters in the world, are manifestations of challenges towards sustainable industrial growth with continued economic growth. This is what green industries look like, and the increased scale of renewable energy; it reflects India's commitment to balancing economic advancement and climate action. While the development of sectors like EV, sustainable agriculture, and the digital economy is commendable, these sectors still need stronger policy support as well as significant investment in green technologies before they can reach their full potential. Even though green financing is an up-and-coming phenomenon, it still has a long way to go to ensure that such investments reach their optimum channels into sustainable projects. Overall, the prospects for the future of economic growth dovetailing SDG objectives look bright but depend on the collective effort of many sectors. Green financing can further be augmented by developing certain additional skill programs and digital transformation towards governance for a sustainable economy. However, actualizing the SDGs demands focusing on the most basic problems, like income inequality, lack of formal job opportunities, and environmental degradation. It would demand multi-stakeholder collaboration in this regard. Winning the goals is not the government's work alone but active participation by private sector actors, civil society, and international organizations. India, with the strength of these stakeholders, can push innovative solutions that ensure sustainable economic transformation. Public-private partnerships, along with green financing and innovative policy frameworks, can create the synergies needed to overcome structural inefficiencies in sectors such as infrastructure, agriculture, and renewable energy. The path to economic sustainability for India lies in its ability to align its growth strategies with the principles of sustainability and inclusive development. By concentrating on sustainable industrial growth, decent work, and environmentally considerate elements, India can create a resilient, green, and inclusive economy, allowing it to fulfill its current and future needs. Therefore, India's future will be governed by all elements of society working together to ensure that economic growth does not compromise social equity or environmental health. This transformation requires a shared commitment to creating a prosperous, sustainable, and equitable future for all citizens, ensuring that the goals of economic growth and sustainable development are harmoniously realized.

References

Asian Infrastructure Investment Bank (2021) Public-private partnerships in sustainable infrastructure

Associated Press (2025) Making climate-friendly lifestyle choices isn't always easy. India learned the hard way. https://apnews.com/article/ec1700fb6eea9b730c63b4394b627623

BBC News (2024) India's economy: the good, bad and ugly in six charts. https://www.bbc.com/news/world-asia-india-68823827

Bhattacharya S, Mitra A (2021) India's progress towards SDGs: challenges and policy imperatives. Econ Polit Wkly 56(34):21–27

Chakraborty L, Pradhan BK (2021) SDGs and the Indian economy: policy perspectives. J Econ Policy Res 16(2):45–60

Chancel L, Piketty T, Saez E, Zucman G (eds.) (2022) World Inequality Report 2022. World Inequality Lab. https://wir2022.wid.world

Financial Times (2025) Narendra Modi's government rolls out tax cuts to stimulate slowing growth.https://www.ft.com/content/1d76cc7f-0f84-4dbb-9c17-0ecd29eb6340

Forest Survey of India (2021) India state of forest report 2021. Ministry of environment, forest and climate change, Government of India. https://fsi.nic.in/isfr2021

Government of India (2022) National e-Governance Plan

International Energy Agency (IEA) (2023) Tracking SDG7: the energy progress report 2023. IEA. https://www.iea.org/reports/trackingsdg7- the-energy-progress-report-2023

International Labour Organization (ILO) (2018) World Employment and Social Outlook 2018: greening with jobs. https://www.ilo.org/sites/default/files/wcmsp5/groups/public/%40dgreports/%40dcomm/%40publ/documents/publication/wcms_628654.pdf

International Labour Organization (ILO) (2022) World employment and social outlook 2022: the role of digital labour platforms in transforming the world of work. https://www.ilo.org/global/research/global-reports/weso/2022/WCMS_836622/lang--en/index.htm

International Renewable Energy Agency (IRENA) (2023) Tracking SDG7: the energy progress report 2023. International Renewable Energy Agency. https://www.irena.org/Publications/2023/Jun/Tracking-SDG7-2023

International Monetary Fund (2023) India—IMF DataMapper. https://www.imf.org/external/datamapper/profile/IND

Kumar P, Rao KS, Singh R (2020) Sustainable development goals and economic transformation in India. Indian J Sustain Dev 8(1):12–25

Kundu A, Chakraborty M (2022) Regional inequality and SDG implementation in India. South Asian Econ J 23(1):56–74. https://doi.org/10.1177/13915614221089654

Macrotrends (2025) India GDP growth rate 1961–2025. https://www.macrotrends.net/global-metrics/countries/ind/india/gdp-growth-rate

Ministry of Electronics and Information Technology (2024, February 2) Annual report 2022–23. Government of India. https://www.meity.gov.in/sites/default/files/2024-02/AR_2022-23_English_24-04-23-1.pdf
Ministry of Finance, Government of India (2023) Sustainable development goals. Press information bureau. https://pib.gov.in/Pressreleaseshare.aspx?PRID=1911844
Ministry of Micro, Small and Medium Enterprises (MSMEs) (2024) Annual Report 2023–24. Government of India. https://msme.gov.in/sites/default/files/FINALMSMEANNUALREPORT2023-24ENGLISH.pdf
Ministry of New and Renewable Energy (2023) Annual report 2022–23. Government of India. https://mnre.gov.in/en/annual-reports- 2022-23
Ministry of Power, Government of India (2022) National energy data management report. https://beeindia.gov.in/sites/default/files/Energy%20Data%20Management%20Report%20_pdf.pdf
Ministry of Statistics and Programme Implementation Government of India (2023) Sustainable development goals (SDGs). https://www.mospi.gov.in/sustainable-development-goalssdgs
NASSCOM (2024, August 28) Driving social impact: can tech companies help India attain its SDGs faster? https://community.nasscom.in/communities/tech-good/driving-social-impact-can-tech-companies-help-india-attain-its-sdgs-faster
NITI Aayog (2020) SDG India index and dashboard 2019–20. https://www.niti.gov.in/sdg-india-index-dashboard-2019–20
NITI Aayog (2021) SDG India Index 2023–24: towards viksit bharat, sustainable progress, inclusive growth. https://www.niti.gov.in/sites/default/files/2024-07/SDG_India_Index_2023-24.pdf
NITI Aayog (2022) An overview of SDGs. https://www.niti.gov.in/overview-sustainable-development-goals
NITI Aayog (2023) Release of SDG India Index 2023–24 Press Information Bureau. https://pib.gov.in/PressReleasePage.aspx?PRID=2032857
Press Information Bureau (2025) Highlights of economic survey 2024–25. https://pib.gov.in/PressReleasePage.aspx?PRID=2097919
Press Information Bureau Government of India (2025) India: world's fastest-growing major economy. https://pib.gov.in/PressNoteDetails.aspx?ModuleId=3&NoteId=153669
Raghunandan M, Nambiar S (2021) India's renewable energy policy and SDG alignment energy. Pol Rev 15(3):34–47
Reserve Bank of India (2022, June 22) Monetary policy statement 2022-23: Sixth bi-monthly review. https://www.rbi.org.in/commonman/english/scripts/BS_PressReleaseDisplay.aspx?head=Monetary%20Policy%20Statement%202022-23%3A%20Sixth%20Bi-Monthly%20Review
Reuters (2025) Comment: India's renewables push holds lessons for the world. https://www.reuters.com/sustainability/climate-energy/comment-indias-renewables-push-holds-lessons-world-2025-02-03/
Reuters (2025) Reactions to India's 2025–26 budget.https://www.reuters.com/world/india/reactions-indias-2025-26-budget2025-02-01/
Roland Berger (2025) India's remarkable economic ascent: a distinct story of growth. https://www.rolandberger.com/en/Insights/Publications/India-s-remarkable-economic-ascent-A-distinct-story-of-growth.html
Sachs J, Schmidt-Traub G, Kroll C (2021) Sustainable development SolutionsNetwork: SDG financing
Statista (2025) GDP of India 1987–2029. https://www.statista.com/statistics/263771/gross-domestic-product-gdp-in-india/
Talent Advisory Group (2023) India skills report 2023
United Nations (2015) Transforming our world: the 2030 agenda for sustainable development. https://sdgs.un.org/2030agenda
United Nations (2015) Transforming our world: the 2030 Agenda for Sustainable Development. https://sustainabledevelopment.un.org/post2015/transformingourworld

United Nations Development Programme (2022, August 30) Independent country programme evaluation: India. https://www.undp.org/evaluation/publications/icpe-india
United Nations Environment Programme (2020) Green financing for sustainable development
World Bank Group (2019) World bank group partnership fund for the sustainable development goals: annual report 2019. World Bank Group. https://documents.worldbank.org/en/publication/documents-reports/documentdetail/106391567056944729/world-bankgroup-partnership-fund-for-the-sustainable-development-goals-annual-report-2019
World Bank Group (2020) Sustainable development goals and targets. World Bank Group. https://datatopics.worldbank.org/sdgatlas/archive/2020/targets
World Bank (2022) GDP growth (annual %)—India. https://data.worldbank.org/indicator/NY.GDP.MKTP.KD.ZG?locations=IN
World Bank (2024) India overview: development news, research, data. https://www.worldbank.org/en/country/india/overview
World Bank Group (2025) World development report 2025: standards for development. World Bank Group. https://www.worldbank.org/en/publication/wdr2025 United Nations Environment Programme (2020) Green financing for sustainable development
World Economic Forum (2022, November 10) Has the UN reached its sustainable development goals? Yes and no. https://www.weforum.org/stories/2022/11/the-un-sustainable-development-goals-a-positive-perspective/
World Health Organization (WHO) (2022) Monitoring health for the SDGs, sustainable development goals health-related targets. https://www.who.int/publications/i/item/9789240051157

Digitalization, Sustainability and Green Banking: A Case Study of Turkey

Işil Demirtaş and Seda Yildirim

Abstract Climate change issues necessitate adaptation for environmental sustainability across all sectors. The banking sector also plays a crucial role in ensuring sustainability and reducing environmental risks by engaging in environmentally friendly activities. The provision of green services, products, and strategies by banks, and the continuation of their activities in line with sustainable development policies, is referred to as green banking. On one hand, government regulations on environmental sustainability, and on the other, practices driven by social awareness, are encouraging banks to become more environmentally conscious. The acceleration of digitalization further supports the greening of the banking sector. Online services, mobile banking applications, paperless processes, and ATMs significantly contribute to this transformation within the banking system. Green banking in Türkiye has been advancing through both institutional initiatives and government regulations. The initiatives implemented to promote sustainable banking in Türkiye encourage banks to adopt and expand their green banking strategies. In particular, digital banking practices such as e-banking, mobile banking services, and paperless transaction systems make a significant contribution to achieving the Sustainable Development Goals (SDGs). This study examines the relationship between digitalization, sustainability, and green banking in Türkiye. The main aim of the study is to analyze the impact of green banking practices on environmental sustainability and the role of digitalization in the implementation of these practices. In this context, several banks from Türkiye were selected as case examples, and the relationship between their green finance instruments, green banking activities, and the 2030 SDGs was evaluated.

Keywords Green banking · Sustainabilitiy · Digital economy

I. Demirtaş (✉)
Giresun University, Giresun, Turkey
e-mail: isil.demirtas80@gmail.com

S. Yildirim
Tekirdag Namık Kemal University, Tekirdag, Turkey

F. A. Malik et al. (eds.), *Aligning ESG Goals with the UNs Sustainable Development Goals Planning*, World Sustainability Series,
https://doi.org/10.1007/978-3-032-01715-4_5

1 Introduction

The environmental consequences of global warming and climate change, along with the acceleration of digital transformation driven by technological advancements, have led to the increasing integration of sustainability, digitalization, and the banking/finance sector, and have gained more visibility in the literature. Banks, which play a key role in economic growth by transforming collected deposits into investments, have undertaken a significant mission in shaping sustainable development. The banking sector influences sustainable development both through its institutional structure, operational workflows, and activities, and through its role in directing investments and financial decision-making processes.

The ability of banking activities and banks to channel their resources into environmentally friendly investments plays a significant role in achieving the United Nations Sustainable Development Goals (SDGs). In this context, banks are integrating eco-friendly practices into their operational processes, contributing to environmental and social awareness projects, developing green products and services, and formulating strategies that prioritize environmentally conscious investment choices in project financing. These activities also support the environmental, social, and economic dimensions of sustainable development.

The provision of green services, products, and strategies by banks, and the alignment of their operations with sustainable development policies, is referred to as green banking. Green banking encompasses the process of making banking operations and procedures environmentally friendly and climate-conscious. In this process, banks are taking on increasing responsibilities each day. On one hand, government regulations aimed at environmental sustainability, and on the other hand, initiatives driven by rising social awareness, are pushing banks to become more environmentally sensitive. The acceleration of digitalization contributes to making the banking sector greener, while also amplifying the positive impact of green banking on the 2030 Sustainable Development Goals. The conveniences offered by technological advancements have facilitated the transition to green banking (Kesebir 2022). A significant portion of environmentally friendly activities is now made possible through digitalization. Digital technologies such as Artificial Intelligence (AI), blockchain, 3D printing, and nanotechnology enable companies to develop new products, services, and strategies. These technologies enhance data processing capacities, improve data storage and accessibility, and increase the integration and performance levels of business processes. AI refers to the performance of tasks—traditionally carried out by human intelligence—by computer systems through learning. The use of AI forms that optimize systems like neural networks and deep machine learning is increasingly integrated into all areas of life (OECD 2020). This rapid digital transformation enables customers to access banking services quickly from anywhere without visiting a branch, supports the development of new products tailored to customer needs, facilitates transactions through tools like ATMs, credit cards, and debit cards, and allows for the execution of many operations we may not yet foresee but are likely to encounter in the near future. Such rapid technological advancement also provides many environmental benefits, including paperless banking, opportunities to ensure energy efficiency and savings in bank branches, improved employee productivity, and a reduction in customers' ecological footprint through digital access to banking services. In addition, banks are utilizing digital technologies as tools to support environmental strategies aligned with sustainable development goals. However, although it is widely acknowledged that emerging digital technologies promote environmentally friendly practices both in green banking and in other operations, it is also undeniable that these technologies consume large amounts of energy and create significant waste problems by quickly rendering older tools and technologies obsolete. Therefore, the impact of technology on sustainable development remains uncertain for the time being.

In Turkey, green banking has shown progress thanks to both institutional-level initiatives and government regulations and guiding policies. In particular, the rapid trends observed in the digitalization process offer significant opportunities for the implementation of green banking in the country. Advancements in digital technologies enable the development of new business models and strategies, the design of environmentally friendly investment projects, financial products and services, and more effective management of environmental risks.

In this context, the study will address the relationship between digitalization, sustainability, and green banking within a holistic framework. The role of green banking in achieving the Sustainable Development Goals will be analyzed through examples from banks in Turkey. The study will focus on how the banking sector in Turkey has adopted the green banking approach, how eco-friendly tools have been integrated into business strategies, and how banks have adapted to green and institutional transformation processes. Additionally, the role and contribution of digitalization in this process will also be thoroughly evaluated.

2 Literature Review

The financial sector has the transformative power to direct the allocation of funds in relation to all dimensions of sustainability. It has a strong influence in guiding all economic actors toward sustainable development (Zhang et al. 2011). Moreover, the financial sector has the capacity to support, promote, and steer the achievement of sustainable development goals through a wide range of strategies and objectives, such as channeling resources toward sustainable investment projects, encouraging and expanding sustainable finance products, prompting financial institutions to adopt sustainability goals and strategies, and directing companies operating within the system toward this field. Green finance practices support environmentally friendly investments and encourage the development of green products and services (International Finance Corporation 2016). The use of sustainable products and services, along with business models and processes that support environmental sustainability, are considered within the scope of green banking. Another widely adopted practice in the financial sector is Socially Responsible Investment (SRI). These investments refer to sustainability-oriented practices in which social, environmental, and ethical principles are considered in processes such as evaluating investment projects and credit allocation decisions (Renneboog et al. 2008).

Weber (2014) suggests that the relationship between the financial sector and sustainability can be examined in three dimensions. Actors who provide funds within the financial sector indirectly influence sustainability through the way these funds are used and through their activities. For example, if a company that receives funding uses these funds for investments in renewable energy, this contributes to environmental sustainability. In particular, risks associated with environmental issues are increasingly incorporated into risk assessment processes. In addition, stakeholder sensitivity and pressure related to sustainable development are pushing financial sector actors to become more responsive to sustainability and take action. As a result, the financial sector influences sustainable development through its business models and processes, the activities of stakeholders within the system, investment projects, credit products offered, and risk identification. At the same time, the sector itself is undergoing transformation through this interaction. The new tools introduced by the financial sector to support sustainable development also contribute to the overall advancement of the system.

The level at which financial institutions adopt sustainability strategies varies. Ziolo et al. (2021), in a study addressing the attitudes and behaviors of financial institution managers toward sustainability, revealed that bank finance managers are more aware of sustainability compared to capital market managers. However, they also showed that banks' intentions and thinking about sustainable financial systems are mostly limited to green investment-related activities and that there is a lack of integration of sustainability into their operational processes.

Zimerman (2019), in his study on the German banking sector, classified the extent to which banks adopt sustainability practices and proposed different motivations for the integration of sustainability into business processes. According to Zimerman, banks participate in sustainability activities driven by social, environmental, and commercial motivations, and he classifies the relationship between these activities and sustainability under five different strategies: These are expressed as narrow strategy, where sustainability is addressed only in terms of visibility and image and is not integrated into business strategies; peripheral strategy, where sustainability activities are carried out within the scope of marketing and sponsorship activities but remain separate from core business strategies; balanced strategy, where corporate and social strategies are integrated in a balanced manner; Integrated Strategy Content with a Social Focus, where sustainability is integrated into bank strategies by prioritizing social factors; and Integrated Strategy Content with an Environmental Focus, where sustainability is integrated into bank strategies and processes, especially in terms of environmental impacts.

In the literature, the adoption of sustainable development policies and strategies by banks is referred to as green banking. Green banking refers to banking activities through which economies gain environmental benefits. If a traditional bank conducts its operations in a way that does not harm the environment, or minimizes the harm done, this is considered green banking (Lalon 2015). Developing environmentally conscious business models in banking, prioritizing environmental strategies in operational processes, and conducting all banking activities in a way that considers environmental concerns fall within this scope (Khairunnessa et al. 2021).

Green banking practices encompass all economic actors, from customers to managers and employees. At the same time, they include all business processes. For instance, the development of mobile banking applications reduces paper usage and lowers the carbon footprint associated with customers traveling to and from bank branches, thus creating a positive environmental impact. Such practices not only enhance operational efficiency—by reducing the number or size of branches—but also support environmental sustainability through changes in customer behavior. Lalon (2015) classifies green banking activities into internal and external dimensions. Environmentally responsible behavior in banks' own operational activities is referred to as internal green banking practices. Examples include the use of renewable energy, energy-saving initiatives, green building practices, waste management, and the promotion of internet and mobile banking. In contrast, actions such as prioritizing environmentally friendly projects in financing decisions are considered external green banking practices. Punitha (2021) emphasizes the dual nature of green banking practices: on the one hand, conducting banking operations with attention to energy and resource efficiency and carbon footprint; and on the other hand, promoting environmentally friendly investment projects. Energy efficiency initiatives implemented by banks not only reduce operational costs but also have a positive impact on national energy consumption. Moreover, social and environmental activities carried out by banks as part of their sustainability efforts aim to raise public awareness and enhance reputation, generating benefits for all stakeholders.

Green banking activities carried out by banks do not only represent environmental sensitivity through their internal operations. The decisions made by banks regarding which projects to finance also determine the extent to which their practices align with environmentally friendly principles and how closely they conform to green banking principles. A bank's willingness to fund green investment projects, its proactive stance in this regard, or its prioritization of firms that minimize environmental risks during financing decisions can make significant contributions to aligning economic growth with sustainable development goals (Khairunnessa et al. 2021).

Mandagie et al. (2024) evaluate banks across various activities related to customers, employees, operational processes, and institutional policies. These include customers benefiting from green finance products and using digital services; employees developing environmentally conscious management practices and engaging in eco-friendly behaviors; banks improving their operations through energy savings, efficiency, digital service use, and waste management; and banks contributing to environmental policies by offering green loans, supporting green projects, and providing green financial instruments.

Burhanudin et al. (2019) examined the use of online banking in relation to sustainability. Compared to traditional branch-based banking, online banking represents a more environmentally friendly approach in terms of service delivery. The study analyzes whether the use of online banks—which are inherently more environmentally friendly than conventional banks—is driven by customer awareness of environmental sustainability or by the convenience and efficiency these services provide.

With the digitalization of banking services, the scope of banking has expanded, and interactions have become more accessible. Banks now provide services such as financial consulting via mobile devices, enabling customers to access information quickly and easily. Digitalization also brings benefits such as increased competitiveness, improved efficiency, and greater market share (Siswanti et al. 2024).

Advancements in paperless banking have increased with the development and widespread adoption of digital technologies. Many processes—such as executing transactions and managing documents electronically—have contributed to sustainability by reducing paper consumption (Kumari 2021). Mobile banking services offer various advantages, such as ease of transaction, 24/7 accessibility, and time and cost savings. These applications support sustainability by minimizing the need to visit bank branches and reducing paper use (Goel 2022).

The replacement of physical infrastructure with online platforms contributes to the reduction of both paper usage and carbon emissions. Increasing awareness of green banking encourages both banks and customers to use these platforms (Kuosuwan et al. 2024). In their study, Kuosuwan and colleagues (2024) concluded that digital banking services are effective in promoting environmental sustainability and reducing the carbon footprint. Digital banking also plays a role in raising awareness about green banking. Srivastava (2013) notes that digital banking can facilitate access to financial services and contribute to sustainability.

3 Objective and Hypothesis

The main purpose of this study is to examine green banking and green financing applications and to explore the changes in banking within the scope of digitalization and sustainability. This study employed qualitative research methodology. It is thought that qualitative research methods are proper for examining the link between green banking practices and sustainable development.

In this context, Turkey was selected as a case study. In the study, current banking indicators and data in Turkey will be examined and green banking and green financing products and services offered by banks actively operating in Turkey will be examined in detail. The data will be collected via recent reports and open access data of TBB (Turkish Banks Association) and selected bank's official web site.

As it is purposed, the main hypothesis is:

H1: Green Banking Practices Contribute 2030SDGs Positively.

4 Research Methodology

To support examination of Green Banking practices and cases, it may be useful to give a general profile of the banking market in Türkiye. There are many different types of banks operating in Türkiye. Some of these banks operate according to the conventional banking system and some according to the participation banking system. The basic indicators for both types of banking systems are shown in tables below.

The data and indicators were reviewed based on TBB's web site. Founded in 1958, the Turkish Banks Association (TBB) is an institution that operates to implement basic practices and regulations for the banking system in Türkiye with the principles of a free market economy (TBB n.d.a).

The number of ATMs, POS and employees in the banking system is important data. Average population per branch, Average population per employee and Average population per ATM in Türkiye are given in Tables 1 and 2 shows main indicators of Turkish Banking System.

When Table 1 is examined, it is evident that a digital transformation has taken place in the Turkish banking system. While the banking sector expanded through physical accessibility in the early 2000s, it has evolved into a more digital structure in the 2020s, accompanied by a decreasing need for personnel and branches.

Table 3 shows the number of call center employees for conventional banking system.

TBB and Green Transformation in Banking System: Case Study

As a result of the importance that TBB attaches to the green transformation of the banking system in Türkiye, the relationships between 2030SDGs and green banking activities are shown in detail. TBB finds green transformation important and draws a roadmap for the sustainability of the banking and finance market (TBB 2022a).

Table 1 Number of ATM, employee, branch

Years	Average population per branch	Average population per employee	Average population per ATM
2000	8.652	398	–
2005	11.536	545	4.862
2010	7.789	413	2.762
2015	7.035	391	1.800
2020	8.413	448	1.783
2021	8.648	457	1.821
2022	8.827	452	1.835
2023	8.989	451	1.800

Source Adapted from TBB (n.d.b), TBB (n.d.c)

Table 2 Main indicators of Turkish banking system

Indicators	2023		December 2024	
	Branches (number)	Employees (number)	Branches (number)	Employees (number)
Sector total	9.497	189.107	9.329	187.918
Public capital banks	3.792	65.386	3.837	65.740
Private capital banks	3.418	64.700	3.317	62.897
Assigned deposits insurance fund transfer	1	230	1	201
Foreign capital banks	2.211	52.474	2.093	52.556
Development and investment banks	75	6.317	81	6.524

Source Adapted from TBB (2025a), TBB (n.d.c)

Table 3 Number of call center employees

	December 2024	March 2025
Number of customer representatives	8.735	9.052
Number of support staff	1.107	1.122
Number of managers	1.079	1.122
Total	10.921	11.296

Source Adapted from TBB (2025b)

Table 4 shows the link between green banking practices and 2030SDGs. Essentially, TBB launched a theme for banks to guide them for sustainable transformation. Except Goal-1 and Goal-2, practices of banks can contribute to most of 2030 SDGs in the scope of three main dimensions of sustainable development (economy-social welfare-environment) (TBB 2022b).

It can be said that the green transformation and green financing practices in banking put into effect by TBB have received positive responses from various banks in Türkiye. Examining the elements related to green banking practices offered by banks actively operating in Türkiye on their websites will be more explanatory for green banking and finance.

Below is information on the green banking and financing practices of some selected banks in Turkey.

Case 1: Türkiye İş Bankası (İşbank)-Green Banking Practices

İşbank presents sectoral roadmaps and concrete action plans in line with the needs of its customers with the detailed Climate Transformation Plan it has created to achieve these targets. İşbank has announced its targets to reduce emission intensity in the cement, iron and steel and energy production sectors in 2023 (İşbank 2025).

Table 4 Link between 2030SDGs and green banking practices

Contribution of TBB for 2030 SDGs	2030SDGs														
	G3	G4	G5	G6	G7	G8	G9	G10	G11	G12	G13	G14	G15	G16	G17
Employee satisfaction-employee rights	X	X	X			X		X							
Equity issues			X					X						X	
Digital transformation						X	X								
Employee capacity development		X	X			X									
Fight against corruption, Legal compliance						X								X	
Stakeholder Satisfaction, Relationship management										X					
Stakeholder education		X	X			X									
Corporate governance						X				X					
Stakeholder data privacy						X	X			X					
Corporate social responsibility		X													
Environmental responsibility					X	X						X	X	X	X
Strong corporate collaborations						X								X	

Adapted from TBB (2022b)

When İşbank's website is examined, it is seen that it includes the following elements under the title of Responsible Banking (İşbank n.d.a). İşbank made a decision that will shape the banking sector in Türkiye in 2020, announcing that it will not allocate resources to new thermal power plant investments based on coal and natural gas in electricity generation and that it will not provide financing for new coal mine investments in 2021. Thus, İşbank announced its commitment to gradually exit the financing of coal and coal-related activity areas by 2040 in 2023 and its interim reduction targets for emission intensity in the cement, iron and steel and energy production sectors. Accordingly, it was targeted to reduce emission intensity by 7% in aluminum, 36% in real estate, 20% in road freight transportation and 15% in oil and gas by 2030 compared to the base year of 2023 (İşbank İs Bankası n.d.b).

As part of the fight against climate change, İşbank has set goals to reduce emissions from banking operations by 38% by 2025, by 65% by 2030, and to conduct carbon–neutral operations by eliminating emissions by 2035 to minimize its environmental footprint (İşbank n.d.c).

Furthermore, İş Bankası is among the first banks to sign the Turkey Sustainable Finance Declaration implemented by Global Compact Türkiye. The Bank takes care to ensure that the projects it finances include planned investments aimed at reducing environmental impacts (İş Bankası n.d.d). In this context, attention is paid to including afforestation activities in projects to reduce carbon emissions, taking measures to protect biodiversity, prioritizing contracted land purchases, and assuming social responsibilities for local people in the project impact area. In addition, projects should be structured according to stakeholder views; biological impact analyses and environmental analyses should be evaluated together. For employees who will take part in projects, work environments compatible with occupational health and safety (OHS) should be provided.

Apart from green finance instruments, İşbank has also given importance to environmental impact and has undertaken projects that will contribute to 2030SDGs. All buildings and branches belonging to İşbank are monitored with the Environmental Management System. In other words, İşbank buildings are managed in a way that ensures resource and energy savings. A collaboration was established between TEMA Foundation and İşbank to evaluate paper waste. Paper waste and paper waste are prevented with digital transformation. For example; digitalization of contracts, forms, invoices, etc. between the bank and the customer prevents paper waste. At this point, İşbank has prevented the consumption of 21.6 million pages annually.

Finally, İşbank creates both environmental and economic and social impact with its new and innovative green products. These products are shown in Fig. 1.

Case 2: Akbank-Green Banking Practices

Akbank is also one of the privately-owned banks that has adapted to the green transformation in banking. It has adopted green banking and finance practices to contribute to the 2030SDGs. Akbank has categorized some green practices to reduce its environmental footprint as Fig. 2.

AKBANK aims to reduce its environmental footprint, reduce social inequalities and ensure continuity in economic activities by adopting sustainable finance tools and

Fig. 1 Green products in İşbank. *Source* İşbank (n.d.e)

Products and Services Contributing to the Green Economy	**Products and Services That Contribute to Society**
Financing Renewable Energy Green Bond İş'te Solar Loan Urban Transformation Loan Unlicensed Electricity Production Loan Energy Efficiency Loan Electric Charging Station Installation Loan Water Value Loan Our Solar Powered ATMs Forest for the Future Maximum TEMA Card TEMA Environment Variable Fund İş'te Kadın Stock Fund İş Portfolio Renewable Energy Mixed Fund	Support for Women Entrepreneurs, Young Entrepreneurs and Technological Startups Support for the Agricultural Sector and Farmers Barrier-Free Banking
Economic impact + Environmental impact + Social impact	

Fig. 2 AKBANK's green banking theme. *Source* created by the authors based on Akbank (2022):4

practices. In other words, it serves 3 main areas of sustainable development (Akbank 2022): environmental, technological, and social transformation. Within the scope of environmental transformation, Akbank focuses on combating the climate crisis, protecting and efficiently using natural resources, transitioning to renewable energy, and supporting sustainable agriculture, animal husbandry, and forestry practices. In the area of technological transformation, the Bank emphasizes the importance of Industry 4.0 applications, circular economy applications, and business model transformation. As for social transformation, Akbank works toward reducing inequality, promoting sustainable cities, improving access to education, and supporting low-cost housing solutions to facilitate home ownership (Akbank 2022).

In terms of green financing instruments, AKBANK has investment instruments that will promote sustainable and green growth (Akbank 2022). The Bank supports green and renewable energy investment practices, including projects focused on energy efficiency, investments in renewable energy sources, and other efforts aimed at reducing carbon emissions. Additionally, AKBANK places emphasis on investments that protect natural resources, such as circular economy focused applications, projects aimed at reduce raw material consumption, Projects aimed at protecting the ecosystem, and sustainable agricultural practices.

In the area of water and wastewater management, the Bank finances projects that aim to increase water efficiency, treat wastewater, recycle water, and collect rainwater to promote sustainable water use. AKBANK also engages in social purpose investments, which include initiatives that support disadvantaged groups, aimed at increasing the female workforce, and supporting education and health services, while also supporting and generating employment.

Furthermore, in its efforts to combat pollution, the Bank provides support for electric and hybrid vehicles, promotes clean transportation solutions, contributes to environmental protection, and supports the development of green buildings. These green financing instruments demonstrate AKBANK's commitment to fostering a more sustainable and inclusive economy.

AKBANK's green financial tools can be shown as Table 5.

Case 3: Garanti BBVA Bank-Green Banking Practices

Garanti BBVA (2021) report showed, the bank's green banking practices have made significant contributions to the achievement of the 2030 Sustainable Development Goals (SDGs). Within this framework, the Agricultural Support Package offered by Garanti BBVA has positively contributed to Goal 2 (Zero Hunger) by supporting sustainable agricultural practices and indirectly aiding the fight against climate change. In relation to Goal 3 (Good Health and Well-Being), the bank's sustainable finance transactions and Healthcare sector infrastructure investments have enhanced access to health services. Through its talent development program in collaboration with ÖRAV, Garanti BBVA supports Goal 4 (Quality Education). The bank's ongoing services and support programs for women entrepreneurs significantly contribute

Table 5 AKBANK's green financial tools

Commercial products	SME products
Blue financing product package	SME eco transformation package
ROOF SPP investment loan	Agricultural credits
Low carbon economy support loan	Women SME package
Renewable energy loans	
Sustainable deposit	

Source Akbank (n.d)

to Goal 5 (Gender Equality). Financing for the treatment plant infrastructure and sustainable finance transactions, offered by Garanti BBVA are beneficial for the TURMEPA Blue Breath Project Goal 6 (Clean Water and Sanitation).

Garanti BBVA's investments in renewable energy and related sustainable finance transactions provide substantial support for Goal 7 (Affordable and Clean Energy). Moreover, its comprehensive portfolio of green finance tools and services fosters economic growth, thereby supporting Goal 8 (Decent Work and Economic Growth). The bank's services and green financial tools such as digital payment system, sustainable transaction system, supporting programs for women etc. are useful for Goal 9 (Industry, Innovation, and Infrastructure). By employing over 18.486 people and offering financial tools, Garanti BBVA contributes to Goal 10 (Reduced Inequalities). Its environmental and social credit policies and infrastructure investments further support Goal 11 (Sustainable Cities and Communities).

Furthermore, Garanti BBVA's sustainable loan programs are aligned with Goal 12 (Responsible Consumption and Production). The bank's extensive efforts—such as its Sustainable Finance Transactions, Environmental and Social Credit Policy, Sustainable Finance Declaration, communication activities centered on sustainability, and the BBVA 2025 Pledge—all reflect its dedication to Goal 13 (Climate Action). The TURMEPA Blue Breath Project, a key environmental initiative, specifically addresses Goal 14 (Life Below Water), while carbon offsetting activities contribute to Goal 15 (Life on Land).

Through the implementation of ethical programs and sustainable banking practices, Garanti BBVA also supports Goal 16 (Peace, Justice and Strong Institutions). Finally, with its active participation in the Sustainable Finance Declaration, related policy papers, and sustainability-themed sectoral working groups coordinated through the Banks Association of Turkey (TBB), the bank plays an instrumental role in advancing Goal 17 (Partnerships for the Goals).

Table 6 demonsttrates Garanti BBVA's sustainability products.

Case 4: Vakıfbank and Green Banking Practices

Vakıfbank has transformed its banking activities in line with the 2030 SDGs. Table 7 summarizes the relationship between the 2030 SDGs and Vakıfbank's banking activities.

Case 5: Ziraatbank and Green Banking Practices

Ziraat Bank attaches importance to environmental and social impact management for sustainable transformation. In this context, Ziraat Bank has stated that it will not provide financial support to sectors and businesses that are not sustainable and green (Ziraat Bank n.d.a). Specifically, Ziraat Bank excludes from financing any entities involved in the production of weapons of mass destruction, the use of forced child labor, or operations within protected areas such as RAMSAR sites or CITES areas. Furthermore, it refrains from supporting activities that involve the products with high chemical content, fishing practices using nets longer than 2.5 km, actions that damage the ozone layer, or the production of asbestos.

Ziraat Bank's sustainable loan products can be summarized as Table 8.

Table 6 Sustainable products of Garanti BBVA

For house	For transportation	For job	For investment
Building insulation loan Individual roof Solar power plant investment loan Green housing loan urban transformation loan	Electric/hybrid vehicle loan Electric bike special shopping loan Electric charging Station loan	Corporate roof Solar Power plant loan Agricultural loans (Organic farming, irrigation loans, direct planting, digital farming, solar power plant loans in agriculture) Women entrepreneur loans Gender equality loan Sustainable supply Chain financing Sustainable loans Sustainability-related loans Electric/hybrid vehicle loan	Sustainable Investment funds Sustainable Retirement fund Green IPO Sustainable bonds

Source Adapted from Garanti BBVA (n.d.a), https://www.garantibbva.com.tr/surdurulebilirlik/surdurulebilirlik-urunleri

Case 6: BIST Sustainability Share Indices

In Turkey, BIST Sustainability 25 indices are included in green finance applications. Since November 4, 2014, BIST Sustainability Index has been actively included in Türkiye. In the index calculation, it creates a new competitive area for institutions and new investment products are launched to the market. The index is based on and registers the approaches and activities of companies to sustainability issues in areas such as global warming, depletion of natural resources, health, etc. In other words, the index is an important performance tool for companies in terms of corporate sustainability (Borsa Istanbul n.d). Figure 3 shows key principles of BIST Sustainability Indices in Turkey.

The sectoral distribution for the BIST Sustainability Indices can be given as follows (Borsa Istanbul n.d.b). Banks constitute the largest share with 17.8%, followed by holding and investment companies at 14%. The share of the retail sector stands at 11.4%, while the transportation and storage sector accounts for 10.7%. 8.5% of the index consists of companies operating in the chemicals, petroleum, rubber, and plastic products sector. The telecommunications sector holds a 5.7% share. The remaining 18.7% is comprised of various other sectors.

According to the data as of May 7, 2025, the companies in the top 10 institutions among BIST Sustainability are distributed as follows: Transportation and Storage; Retail Trade; Defense; Chemicals Petroleum Rubber and Plastic Products; Telecommunications; Banks; Holding and Investment Companies (Borsa Istanbul n.d.b).

Table 7 2030 SDGs and Vakıf Bank's focus

Goals	Issue	Banking practice
Goal 8	Vakıfbank continues its banking activities to improve its financial performance	Financial and economic performance
Goal 4 Goal 8	Vakıfbank employees are the most important supporters of achieving sustainable development goals. Therefore, the work environment and employee satisfaction are important	Employees satisfaction
Goal 4 Goal 8	At Vakıfbank, importance is given to the self-development of employees	Talent management
Goal 17	With the Risk Management approach at Vakıfbank, non-financial risks such as financial risks and reputational risks are also addressed	Risk management
Goal 13	The environmental impact of banking sector activities may seem low, but Vakıf Bank is improving its own performance	Environment perofrmance
Goal 5 Goal 7 Goal 10 Goal 13 Goal 17	Vakıf Bank supports sustainability by offering a wide range of sustainable ministry tools. Financial opportunities such as supporting women entrepreneurs and farmers are important tools	Sustainable product and services
Goal 5 Goal 9 Goal 10 Goal 12 Goal 13 Goal 17	The financing provided by Vakıf Bank to projects creates environmental and social contributions. For example, supporting renewable energy projects	Responsible financing
Goal 9 Goal 12 Goal 17	Customer experience is important in keeping the bank and the customer going. It is important to establish long-term relationships with customers and communicate through various channels	Customer experience
Goal 9 Goal 12	With digital transformation, better services are provided and many banking processes have improved	Digital transformation
Goal 9 Goal 12 Goal 17	Privacy Policy and principles for protecting personal data are important for banking activities	Customer data security

Source created by the authors based on Vakıfbank (2018)

Table 8 Sustainable loan products of Ziraat bank

Personal loan products	Corporate loan products
Individual energy efficiency credit Energy efficiency management credit	Energy efficiency investment loan Energy efficiency operating loan Renewable energy investment loan Renewable energy operating loan Wastewater treatment and recycling investment loan Wastewater treatment and recycling operating loan SPP Roof SPP licensed SPP investment and operating loans

Source created by the authors based on Ziraat Bank (n.d.b)

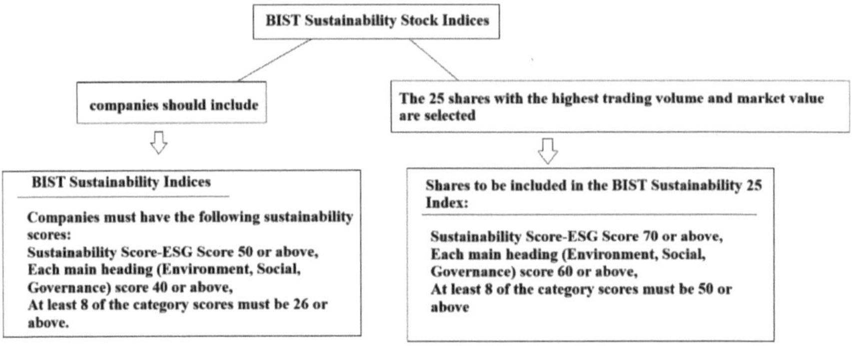

Fig. 3 BIST Sustainability Indices and Key Principles. *Source* Borsa Istanbul (n.d.a)

The sectoral distribution for the BIST Sustainability 25 Index can be given as follows (Borsa Istanbul n.d.c). Banks hold the largest share in the index with 22.6%, followed by holding and investment companies at 20.7%. The transportation and storage sector accounts for 14% of the index, while companies operating in the chemicals, petroleum, rubber, and plastic products sector make up 10.5%. The telecommunications sector has a share of 8.4%, and the sector of fabricated metal products, machinery, electrical equipment, and transportation vehicles represents 7.3%. The retail sector contributes 5.3% to the index, while the remaining 11.1% consists of other sectors.

According to the data as of May 7, 2025, the companies in the top 10 institutions among BIST Sustainability 25 are distributed as follows: Transportation and Storage(1); Chemicals Petroleum Rubber and Plastic Products(1); Abricated Metal Products Machinery Electrical Equipment And Transportation Vehicles(1); Telecommunications(1); Banks(3); Holding and Investment Companies(3) (Borsa Istanbul n.d.c).

The sectoral distribution for the BIST Participation Sustainability 25 Index is as follows (Borsa Istanbul n.d.d). The chemicals, petroleum, rubber, and plastic products sector holds the largest share at 20.3%, followed by the retail sector at 16.9%. Basic metals account for 11.5% of the index, while the defense sector represents 10.7%. The shares of non-metallic mineral products and electricity, gas, and steam sectors are 9% and 9.1%, respectively. Human health and social work activities contribute 5.3% to the index. The remaining 17.3% consists of various other sectors grouped under the others category.

According to the data as of May 7, 2025, the companies in the top 10 institutions among BIST Participation Sustainability are distributed as follows: Defense(1); Basic Metals(1); Retail Trade(2); Non-Metallic Mineral Products(1); Human Health And Social Work Activities(1); Chemicals Petroleum Rubber And Plastic Products(3); Holding And Investment Companies(1) (Borsa Istanbul n.d.d).

5 Results and Discussion

The growing awareness of sustainability in the Turkish banking sector has led stakeholders to develop more active policies. The Banking Regulation and Supervision Agency (BRSA) summarized various initiatives in its 2021 report that aim to support the development of sustainable strategies by banks. These include: the Sustainability Guide for the Banking Sector (2014), the Borsa Istanbul Sustainability Index (2014), issuance of green bonds (2016), the Global Compact Turkey Sustainable Finance Declaration (2017), the Climate Risks Report (2021), and the issuance of partnership-based green lease certificates (2021).

Awareness of sustainability in the sector has increased, and in line with this, the Sustainable Banking Strategy Document was prepared by the Banking Regulation and Supervision Agency (BRSA) within the scope of the Turkey Green Deal Action Plan, published in 2021. This document aims to establish the infrastructure for sustainable banking and to promote its adoption at the institutional level. The BRSA (2021) has also undertaken several initiatives to promote sustainable banking practices. These include membership in the IFC Sustainable Banking and Finance Network (2015), membership in the Task Force on Climate-related Financial Risks (TCFD), establishment of the Sustainable Banking Working Group (2021), and membership in the Network for Greening the Financial System (NGFS) (2021).

In line with sustainability goals, several banks from Turkey have joined the Net-Zero Banking Alliance (NZBA), which was established in 2021 by the United Nations Environment Programme Finance Initiative (UNEP FI n.d.). Within this framework, banks have committed to aligning their lending and investment portfolios with the net-zero emissions target by 2050. For example, Garanti BBVA has pledged a 71% reduction in emissions by 2035 and has announced that it will phase out coal-related investments by 2040 (Garanti BBVA n.d.b).

According to a report prepared by 350 Türkiye (2022), 8 out of 17 banks have announced that they will no longer finance coal-related projects. Additionally, Akbank, Garanti BBVA, HSBC, ING, İş Bankası, Yapı Kredi, and Halkbank have committed to achieving net-zero targets by 2050. Garanti BBVA, Halkbank, VakıfBank, and the Industrial Development Bank of Turkey (TSKB) are among the banks that have pledged carbon neutrality targets.

These developments demonstrate that sustainable banking activities have accelerated, particularly due to the operational efficiency enabled by digitalization. However, despite all these positive steps, it is evident that the Turkish banking sector still faces certain structural limitations in the field of sustainability (BRSA 2021). One major challenge is the absence of a defined green taxonomy in Turkey, which makes it difficult to conduct sustainability assessments and generate relevant data. Additionally, the calculation and reporting of carbon emission data remain inadequate. The lack of an institutional framework for carbon markets limits the development of green financing mechanisms. Moreover, gaps in the supervision and regulation regarding sustainable finance hinder consistent and effective implementation.

The initiatives implemented to promote sustainable banking in Turkey have encouraged banks to adopt and expand their green banking strategies. However, the development of the infrastructure for green banking is also closely linked to the effective use of opportunities provided by digitalization. Applications such as e-banking, mobile banking services, and the advancement of paperless banking, offered by digital banking, provide significant support for achieving the Sustainable Development Goals (SDGs). Nevertheless, to fully achieve the SDGs, strengthening the regulatory framework—particularly through the development of a green taxonomy, the establishment of carbon markets, and the enhancement of supervisory mechanisms—remains critically important.

As a result of the study, it is seen that green banking activities and green finance instruments have positive and positive contributions to 2030SDGs. Therefore, the hypothesis of the study is accepted. Banks can create positive effects for 2030SDGs by completing their own green transformation.

6 Policy Implications

To achieve the 2030 Sustainable Development Goals and reach the 2053 net-zero emission target, the banking sector in Turkey plays a pivotal role—both by implementing its own sectoral transformation at the operational level through its institutional structure and by transforming the economy through the services it offers, and the investment projects it finances.

In this context, the development of a green banking approach requires the formulation, regulation, implementation, and supervision of green and sustainable policies. Although there have been various initiatives aimed at achieving the Sustainable Development Goals and advancing green banking activities in Turkey, there remains a noticeable lack of legal and regulatory frameworks. To enable effective policymaking, it is essential to define data sets and financial instruments within the domain of green and sustainable finance. There must be a clear and transparent classification of which instruments fall within the scope of green banking activities and which products and services do not. Moreover, the development of green banking also necessitates the diversification of green financial products.

For this reason, increasing the number of green products, improving accessibility, and encouraging their use are of critical importance. Digital technologies have a positive impact on green banking practices by facilitating the dissemination of environmentally conscious services. In this regard, promoting digital technologies that support the development of environmentally sensitive banking services will contribute significantly to strengthening sustainable finance.

Green banking is seen as a holistic activity. In this context, it has been observed that banks have undergone both institutional and activity scope changes with sustainable transformation. It has been observed that sample banks examined as cases have planned their own resource and waste management processes sustainably within the scope of their institutional changes. It has been observed that they have prevented paper waste and excess waste in document management with the economical use of energy and water resources and digitalization. On the other hand, banks provide support to sustainable sectors and entrepreneurs with green financing instruments. The green transformation of banks will be an important supporting factor in the realization of the SDGs in the long term.

7 Conclusion

The new normal of the twenty-first century world is to be sustainable and digital (Yıldırım et al. 2025a, 2025b; Bostancı and Yıldırım 2025a, 2025b; Yildirim et al. 2022). In this context, both digital transformation and green transformation are taking place in the banking sector. In this study, green banking, digitalization and sustainable development relationships were examined. For the study, some banks from Turkey were selected as cases and the relationships between green financing instruments, green banking activities and 2030 SDGs were analyzed.

The banks examined in the study were observed to have undergone a green transformation within the scope of the 2030 SDGs. At first glance, green banking practices and green financing instruments may seem to have a limited sustainable development contribution, but a more in-depth examination of green banking activities provides a different perspective. In other words, green banking practices, digital banking and green financing instruments provide significant contributions to the 2030 SDGS targets and the balance between economic-social welfare and environmental quality. When the banking activities examined are considered, it was observed that banks implemented both green and digital transformation in parallel. As a result of digital transformation, significant solutions were brought to both resource saving and waste problems. For example, banks using opportunities such as digital invoices, digital contracts and forms have made great contributions to the problem of paper waste and waste. One of the important benefits of the digital banking system is that it increases customer satisfaction. Digital channels are an important tool in improving relations with stakeholders.

When green financing instruments are examined, the effect of the expectation of sustainable benefits is seen in the sectors and business areas where credit is given. The new credit and support areas of banks can be summarized as women entrepreneurs, farmers, entrepreneurs who do sustainable business. On the other hand, it has been seen that state-owned banks have decided not to give credit to unsustainable and environmentally harmful business areas. The new support areas can now be stated as areas such as renewable energy, energy efficiency, sustainable fishing, sustainable animal husbandry, sustainable agriculture.

It has been observed that banks are not only changing their activities but also their institutional sustainability. For example, some banks pay attention to having green buildings and structures in their own branches and buildings. Areas such as work environment layouts, lighting, water resource usage, energy extraction are carefully arranged.

References

350 Türkiye (2022) Climate change approaches of Turkish banks. https://350turkiye.org/files/2022/11/climate-change-approaches-of-turkish-banks-report.pdf

Akbank (2022) TBB İyi Uygulamalar, Mayıs 2022. https://www.tbb.org.tr/arastirma-ve-yayinlar/surdurulebilirlik

Akbank (n.d). Sürdürülebilir Finansman. https://www.akbank.com/bankamiz/surdurulebilir-finansman

Aksin-Sivrikaya S, Bhattacharya CB (2017) Where digitalization meets sustainability: opportunities and challenges. In Osburg T, Lohrmann C (eds) Sustainability in a digital world: new opportunities through new technologies. Springer, pp 37–49. https://doi.org/10.1007/978-3-319-54603-2_3

Borsa Istanbul (n.d.a) BIST sustainability stock indices. https://www.borsaistanbul.com/en/sayfa/10918/bist-sustainability-stock-indices

Borsa Istanbul (n.d.b) BIST sustainability. https://www.borsaistanbul.com/en/index-detail/324/bist-sustainability

Borsa Istanbul (n.d.c) BIST sustainability 25. https://www.borsaistanbul.com/en/index-detail/521/bist-sustainability-25

Borsa Istanbul (n.d.d) BIST participation sustainability. https://www.borsaistanbul.com/en/index-detail/114/bist-participation-sustainability

Bostancı SH, Yıldırım S (2025a) Environmental justice in the himalayan region. In: Borthakur A, Singh P (eds) People and mountain environments. Springer, Cham. https://doi.org/10.1007/978-3-031-83553-7_12

Bostancı SH, Yıldırım S (2025b) Role of government and international institutions in fostering social responsibility and environmental harmony. In Singh P, Daga S, Yadav K (eds) Integrating environmental sustainability and corporate social responsibility: a move beyond profit. World sustainability series. Springer, Cham. https://doi.org/10.1007/978-3-031-81469-3_10

BRSA (2021) Sustainable banking strategic plan (2022–2025). https://www.bddk.org.tr/KurumHakkinda/EkGetir/19?ekId=143

Burhanudin B, Ronny R, Sihotang ET (2019) Connecting intention to use online banking, commitment to environmental sustainability, and happiness: The role of nature relatedness. Procedia Comput Sci 161:341–349. https://doi.org/10.1016/j.procs.2019.11.132

Garanti BBVA (2021) Sürdürülebilir Kalkınma Amaçları ve Hedeflerine Katkı. https://www.garantibbvainvestorrelations.com/tr/entegre-faaliyet-raporu-2021/surdurulebilir-kalkinma-amaclari.aspx

Garanti BBVA (n.d.a) Sürdürülebilirlik Ürünleri. https://www.garantibbva.com.tr/surdurulebilirlik/surdurulebilirlik-urunleri

Goel A (2022) Digitization: road ahead for banking industry. OJAS: expanding knowledge horizon. An Int J Res Manag 10(1):32–37. https://jsb.jaipuria.edu.in/wp-content/uploads/2023/02/ojas-journal-Jan-Jun-2022-Final.pdf

IFC—International Finance Corporation (2016) Green finance: a bottom-up approach to track existing flows. Washington, D.C.: World Bank Group

İşbank (2025) İş Bankası, tüm karbon yoğun sektörlerde İklim Dönüşüm Planı'nı açıklayan Türkiye'nin ilk bankası oldu. https://www.isbank.com.tr/bankamizi-taniyin/is-bankasi-tum-karbon-yogun-sektorlerde-iklim-donusum-planini-aciklayan-turkiyenin-ilk-bankasi-oldu

İşbank (n.d.a) Sorumlu Bankacılık. https://www.isbank.com.tr/bankamizi-taniyin/sorumlu-bankacilik

İşbank (n.d.b). Karbonsuzlaşma Hedefleri. https://www.isbank.com.tr/bankamizi-taniyin/karbonsuzlasma-hedefleri

İşbank, (n.d.c). İklim Değişikliği ile Mücadele,https://www.isbank.com.tr/bankamizi-taniyin/iklim-degisikligi-ile-mucadele

İşbank (n.d.d). https://www.isbank.com.tr/bankamizi-taniyin/cevresel-etki, https://www.isbank.com.tr/bankamizi-taniyin/cevresel-etki

İşbank (n.d.e) Sorumlu Ürün ve Hizmetlerimiz.,https://www.isbank.com.tr/bankamizi-taniyin/sorumlu-urun-ve-hizmetler

Kesebir M (2022) YEŞİL BANKACILIK, eds (Dayı, F.& Esmer, Y.) In: book: KÜRESEL SİSTEMDE YEŞİL YÖNETİM, 1. Baskı, Ocak, Ankara 2022, Gazi Kitabevi

Khairunnessa F, Vazquez-Brust DA, Yakovleva N (2021) A review of the recent developments of green banking in Bangladesh. Sustainability 13(4):1904. https://doi.org/10.3390/su13041904

Kumari DAT (2021) Customer loyalty towards paperless banking (with special reference to systemically important banks in Sri Lanka). Int J Bus Manag Inven 10(1):51–63. https://doi.org/10.35629/8028-1001025163

Kuosuwan B, Risman A, Dudukalov E, Kozlova E (2024) Digital banking and environmental impact: how Fintech supports carbon footprint reduction. BIO Web Conf 145:05017. https://doi.org/10.1051/bioconf/202414505017

Lalon RM (2015) Green banking: going green. Int J Econ, Financ Manag Sci 3(1):34–42. https://doi.org/10.11648/j.ijefm.20150301.15

Mandagie YRO, Hatta IH, Wahyoeni SI, Ahmar N (2024) Development of a green banking and green financing practice model for enhancing sustainability development goals (SDGs). SDGsReview 5:e02775. https://doi.org/10.47172/2965-730X.SDGsReview.v5.n01.pe02775

OECD (2020) The digitalisation of science, technology and innovation

Punitha K (2021) Transition of environmental sustainability towards green banking. Sambodhi 44(1):1–5

Renneboog L, Ter Horst J, Zhang C (2008) Socially responsible investments: institutional aspects, performance, and investor behavior. J Bank Finance 32(9):1723–1742. https://doi.org/10.1016/j.jbankfin.2007.12.039

Siswanti I, Riyadh HA, Cahaya YF, Prowanta E, Beshr BAHB (2024) Unlocking sustainability: exploring the nexus of green banking, digital transformation, and financial performance with foreign ownership moderation. Discov Sustain. https://doi.org/10.1007/s43621-024-00597-5

Srivastava A (2013) Mobile banking and sustainable growth. Am J Econ Bus Adm 5(3):89–94. https://doi.org/10.3844/ajebasp.2013.89.94

TBB (2022a) Türkiye Bankalar Birliği Sürdürülebilirlik Raporu, p 16. https://www.tbb.org.tr/arastirma-ve-yayinlar/surdurulebilirlik

TBB (2022b) TBB Sürdürülenilirk Çalışma grubu Toplantısı TKYB İyi Uygulamaları. https://www.tbb.org.tr/pdf/faaliyetler/116/783

TBB (2025a) Banka, Çalışan ve Şube Sayıları, Aralık 2024, Rapor Kodu: DT13. Ocak

TBB (2025b) 2025 Mart—Çağrı Merkezi İstatistikleri. https://www.tbb.org.tr/istatistiki-raporlar/2025-mart-cagri-merkezi-istatistikleri

TBB (n.d.a) TBB Hakkında. https://www.tbb.org.tr/tbb-hakkinda

TBB (n.d.b) TBB Veri Girişi. https://verisistemi.tbb.org.tr/index.php?/tbb/report_bolgeler

TBB (n.d.c) 2023 Yıllık—Türkiye'de Bankacılık Sistemi (1958'den İtibaren). https://www.tbb.org.tr/istatistiki-raporlar/2023-yillik-turkiyede-bankacilik-sistemi-1958den-itibaren

UNEP FI (n.d.) Web Page. https://www.unepfi.org/banking/bankingprinciples/

Vakıfbank (2018) Rapor ve Sertifikalar. https://www.vakifbank.com.tr/tr/bankamiz/surdurulebilirlik/rapor-ve-sertifikalar

Weber O (2014) The financial sector's impact on sustainable development. J Sustai Finance Investment, Taylor Francis J 4(1):1–8

Yildirim S, Demirtas I, Yildirim DC (2022) A review of alternative economic approaches to achieve sustainable development: the rising digitalization and degrowth post COVID-19. In Castanho R (ed) Handbook of research on sustainable development goals, climate change, and digitalization. IGI Global Scientific Publishing, pp 288–307

Yıldırım DÇ, Yıldırım S, Turan T, Tulia Gattone T, Magazzino C (2025a) Balancing green power: hydropower and biomass energy's impact on environment in OECD countries. Renew Energy 241(2025):122352. https://doi.org/10.1016/j.renene.2025.122352

Yildirim S, Yildirim DC, Erdoğan S, Aydin K, Ektiren B (2025b) The role of digitalization on the way to sustainable future: Turkey case. In Castanho R (ed) Bridging technology and development for

sustainable innovation and geopolitical dynamics. IGI Global Scientific Publishing, pp 363–380. https://doi.org/10.4018/979-8-3693-9072-6.ch015
Zhang B, Yang Y, Bi J (2011) Tracking the implementation of green credit policy in China: top-down perspective and bottom-up reform. J Environ Manag 92(4):1321–1327. https://doi.org/10.1016/j.jenvman.2010.12.019
Zimmermann S (2019) Same same but different: How and why banks approach sustainability. Sustainability 11(8):2267. https://doi.org/10.3390/su11082267
Ziolo M, Filipiak BZ, Bak I, Cheba K (2021) Sustainable financial systems toward sustainability in finance: institutional and managerial approach. Procedia Comput Sci 192:3427–3436
Ziraat Bank (n.d.a) T.C. Ziraat Bankası A.Ş. Kredilendirme Faaliyetlerinde Çevresel ve Sosal Etki Yönetimi Politikası. https://www.ziraatbank.com.tr/tr/yatirimci-iliskileri-ZB/kurumsal-yonetim/Documents/cevresel-sosyal-etki-politikasi.pdf
Ziraat Bank (n.d.b) Çevresel Kredi Ürünleri. https://www.ziraatbank.com.tr/tr/bankamiz/surdurulebilirlik/cevresel-kredi-urunleri

Innovative Finance for Sustainability: Bridging ESG and SDGs

Natia Surmanidze, Iza Gigauri, and Catalin Popescu

Abstract Integrating financial innovation into sustainable development strategies stands now a climactic priority for global economic planning. Policymakers and the corporate community want to integrate the Sustainable Development Goals (SDGs) of the UN with the principles of environmental, social, and governance (ESG). In response to pressing global issues like social inequality, climate change, and economic sustainability, the UN approved the SDGs framework in 2015. However, their successful implementation depends on sufficient financial resources, efficient capital allocation, and equitable investment strategies. While traditional financing mechanisms have played a crucial role in pursuing the Sustainable Development Goals, they are often insufficient to address the scale of the challenges presented by the SDGs. Even if traditional financing methods have been essential to achieving the Sustainable Development Goals, they frequently fall short of meeting the scope of the problems the SDGs confront. From green and sustainable finance to blockchain-based investment platforms, financial innovations have emerged as critical tools to fill financing gaps and ensure that capital is directed towards sustainable and socially responsible projects. The study aims to analyze the prospects for using financial innovations to achieve ESG objectives, especially in the context of sustainability. This provides an analytical framework for designing equitable financing strategies for global sustainability. In particular, this chapter examines how new financial instruments, namely impact investment funds, sustainability loans, and green bonds, are changing investment priorities by incentivizing sustainability-oriented decision-making. In addition, it explores the position of financial technology (FinTech), blockchain, and decentralized finance (DeFi) in increasing transparency, accountability, and access to ESG-compliant financial markets. This research identifies key

N. Surmanidze · I. Gigauri (✉)
School of Business and Administrative Studies, The University of Georgia, Tbilisi, Georgia
e-mail: i.gigauri@ug.edu.ge

N. Surmanidze
e-mail: natia.surmanidze@iliauni.edu.ge

C. Popescu
Department of Business Administration, Petroleum-Gas University of Ploiesti, Ploiesti, Romania
e-mail: cpopescu@upg-ploiesti.ro

F. A. Malik et al. (eds.), *Aligning ESG Goals with the UNs Sustainable Development Goals Planning*, World Sustainability Series,
https://doi.org/10.1007/978-3-032-01715-4_6

trends in sustainable finance by analyzing the interactions between regulatory policies, technological advances, and institutional investment strategies. It highlights best practices for integrating financial innovations into global economic frameworks.

Keywords Finance · Innovation · ESG · SDG · Sustainability · Digital · Financial technology · Developing countries

1 Introduction

Sustainability has become essential in global finance as climate risks, social inequalities, and governance challenges increasingly shape investment decisions (Fu et al. 2023). While traditional financing mechanisms have played a crucial role in pursuing the Sustainable Development Goals, they are often insufficient to address the scale of the challenges presented by the SDGs. In the modern world, profit-driven financial markets have historically turned their attention to sustainable strategies, with business and investment choices becoming more and more centred on environmental, social, and governance (ESG) considerations (Colombage et al. 2025). The authors' approach related to the chapter outline is summarized in Fig. 1.

The focus on sustainable finance is driven by the perception of environmental risks and the severity of social risks, which companies have equated with financial risks (Mlanga 2024). Climate-related disasters, resource depletion, and social unrest represent significant economic threats. This has, in turn, forced investors and financial institutions to effectively direct a substantial portion of capital towards sustainable projects (Cromwell and Pepra 2025). In response, financial innovations, namely impact investing, sustainability-linked loans, and green bonds, have emerged as financing mechanisms for sustainable development (Jesuka et al. 2025).

Decentralised finance (DeFi) and financial technologies (FinTech) have grown in importance as pillars of sustainable finance, allowing businesses to restructure sustainable finance while taking ESG factors into account. In this regard, the focus is on investment access, increasing transparency and efficiency. These innovations reduce traditional barriers to entry, providing a wider range of opportunities for more investors to establish themselves in financial markets focused on sustainability (Sahid et al. 2023). Through digital platforms and blockchain technology, FinTech facilitates fractional ownership of green assets, ultimately enabling individuals and small institutions to engage in impact investing (Panait et al. 2024). At the same time, DeFi introduces decentralized lending and peer-to-peer investment models that bypass traditional financial intermediaries, lowering costs and improving capital allocation efficiency (Mhlanga 2024).

Beyond accessibility, fintech also enhances transparency in ESG performance tracking. Blockchain technology provides an immutable record of sustainability-linked financial transactions, ensuring accountability and reducing risks associated with greenwashing. Artificial intelligence further strengthens the integrity of ESG investments by analyzing vast datasets to provide real-time sustainability insights

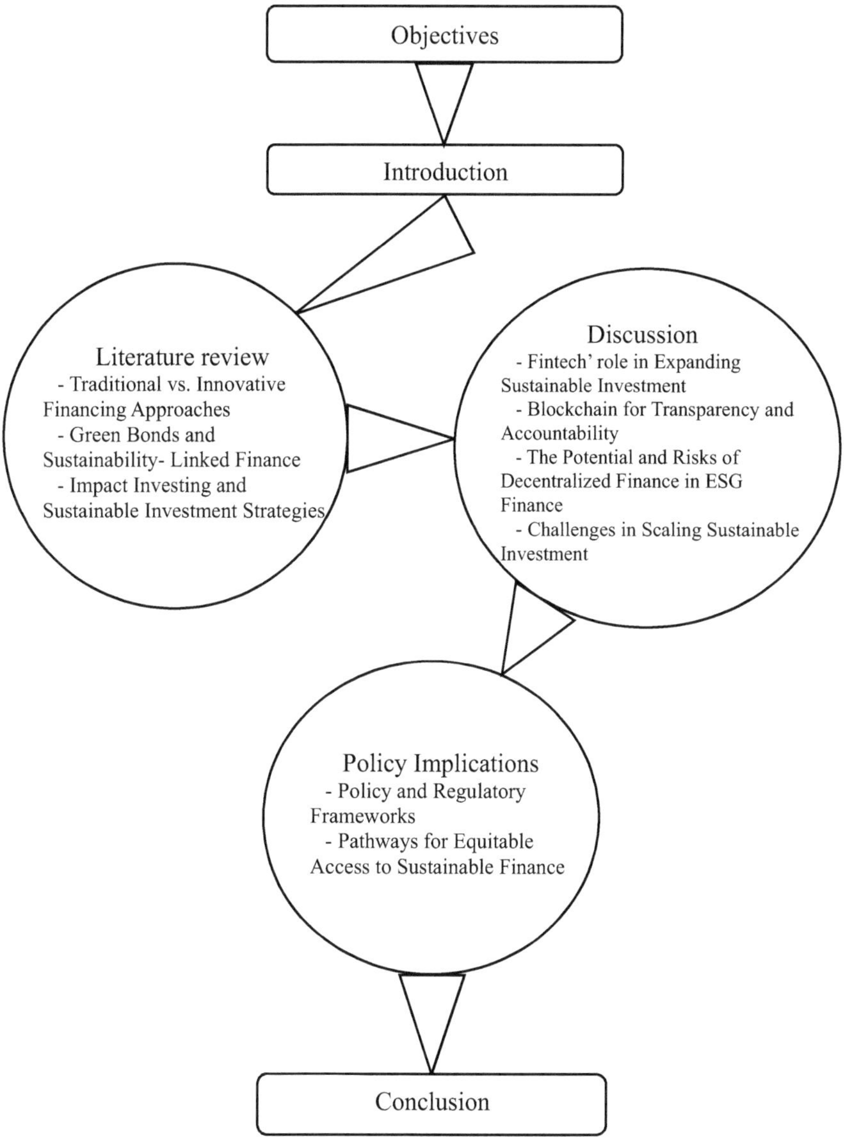

Fig. 1 The authors' approach in organizing the chapter' workflow

and automated compliance verification, helping investors make informed decisions (Pancholi and Shukla 2025). These technologies provide dynamic, data-driven financial strategies that match investments with long-term sustainability objectives and boost confidence in ESG markets. Integration with sustainable finance will be crucial as fintech and defi grow in order to close the capital gap and hasten the world's shift to a greener, more sustainable economy.

With an emphasis on analysing the effects of labour practices, governance structures, and environmental risks, Environmental, Social, and Governance (ESG) principles offer a framework and an organised process for evaluating how investments affect sustainability (Trahan and Jantz 2023; Popescu et al. 2022). It consists of three key pillars: Environmental (E), which evaluates factors such as carbon footprint, renewable energy adoption, biodiversity conservation, and waste management; Social (S), which focuses on workplace diversity, human rights policies, community engagement, and ethical labor practices; and Governance (G), which evaluates elements of corporate governance such as the structure of the governing board, the remuneration of executive directors, the rights enjoyed by shareholders, and the requirements of business ethics. These ESG criteria help investors mitigate financial risks associated with poor sustainability performance while identifying long-term opportunities in responsible business practices (Martínez-Falcó et al. 2025; Rosca et al. 2025). However, while ESG principles facilitate corporate sustainability assessments, they do not inherently guarantee alignment with broader global sustainability objectives, as their primary focus remains on risk management and financial materiality rather than systemic social and environmental transformation (Madison and Schiehll 2021).

In contrast, the UN's global framework of Sustainable Development Goals (SDGs) focuses on addressing pressing challenges such as social inequality, social well-being, and climate. It emphasizes creating measurable positive impacts (Gigauri et al. 2023). In order to ensure sustainable development, the United Nations ratified 17 Sustainable Development Goals in 2015. These goals include SDG 7 (affordable and clean energy), which encourages financing for renewable energy projects; SDG 9 (industry, innovation, and infrastructure), which pushes for investments in sustainable infrastructure; and SDG 13 (climate action), which calls for raising funds for efforts to mitigate and adapt to climate change. Although ESG investing is essential for funding sustainability projects, it is still very difficult to align ESG investment strategies with SDG-driven financial frameworks. This calls for improved data standardisation, increased transparency in ESG reporting, and the creation of financial instruments created especially to finance SDG-aligned projects (Bengo et al, 2022). Without these structural improvements, ESG investments risk failing to deliver tangible contributions toward achieving the SDGs, highlighting the urgent need for regulatory harmonization, impact measurement frameworks, and investor accountability mechanisms (Raihan et al. 2024).

The primary obstacle to accomplishing the Sustainable Development Goals (SDGs) is the financial component, which is crucial even if nations all around the world acknowledge and support these initiatives. According to the United Nations' 2030 prediction, the proper implementation of the SDGs will need an annual expenditure of $5–7 trillion USD (Kutateladze et al. 2025). However, current financial contributions from the public and private sectors fall significantly short, with developing economies facing the most severe capital constraints (Mhlanga 2024; Surmanidze et al. 2025). This shortfall is exacerbated by the limited capacity of government budgets, which are insufficient to fund large-scale sustainability projects alone. While private investment funds are crucial to serve the financing gap, many investors still hesitate to commit capital to sustainability-related projects due to uncertainty about risk-adjusted returns and ESG performance measurement (Zhang and Zhang 2025).

Investor hesitancy is further exacerbated by the absence of standardised ESG indicators and the ubiquity of "greenwashing," which damages the legitimacy of financial products that comply with ESG standards by placing an excessive focus on sustainability issues (Dempere et al. 2024). At the same time, the fragmentation of the regulatory framework and unbalanced policies hinder the efficient flow of capital into sustainable finance, potentially creating barriers for retail and institutional investors, for whom aligning their portfolios with sustainable development goals is crucial (Fu et al. 2023).

Green bonds, digital financial platforms, and blended finance may be crucial in addressing financial difficulties (Naysary and Shrestha 2024). For instance, green bonds offer a formal way to raise money just for sustainability and climate projects, while blended finance mixes public and private investments to lower risk and draw in more money for high-impact projects (Jesuka et al. 2025).

Additionally, fintech-driven solutions, including blockchain-based investment platforms and sustainability-linked digital assets, democratize access to influence finance by enabling fractional ownership of ESG investments and improving transparency in sustainability reporting. The success of digital financial mechanisms is correlated with effective regulatory frameworks, straightforward implementation of ESG standards, and market-based incentive practices, which in turn encourage the mobilization of long-term investments in projects consistent with sustainable development goals (Bengo et al. 2022). Policy harmonization and institutional cooperation are jointly responsible for motivating the introduction of financial innovations. Otherwise, it is reasonable to assume that the risk of restraint is significantly higher. Naturally, the need for a global regulatory system in the digital financial sector is evident, without which an adequate allocation of capital will not be tangible to ensure sustainable development.

In order to help raise money for sustainable objectives, the chapter examines the function of financial innovation in relation to ESG investment strategies and the execution of projects that align with SDGs. It also emphasises the kinds of financial instruments—impact investments, loans with a sustainability focus, and green bonds—that support sustainable objectives. It also examines how decentralised finance (DeFi) and financial technology (fintech) might increase access to sustainable finance via digital lending platforms, blockchain, and AI-driven ESG analytics.

This chapter also aims to identify challenges in sustainable finance expansion, including greenwashing, regulatory inconsistencies, and barriers to equitable capital distribution. Through an assessment of current trends, policy frameworks, and investment strategies, strategy recommendations are presented on expanding ESG-compliant finance, improving the regulatory environment, and increasing access to sustainable investment instruments. In the end, this chapter adds to the larger conversation on sustainable finance, technology, and policy by highlighting the necessity of cooperation between investors, financial institutions, and governments in order to create a sustainable finance ecosystem that is more inclusive, transparent, and scalable.

2 Literature Review

2.1 *Traditional versus Innovative Financing Approaches*

Sustainable development requires significant financial resources, yet conventional financing mechanisms have often failed to address the scale and complexity of global sustainability challenges. Traditional finance includes public sector financing, equity financing, and loans from financial institutions. These sources are distinguished by their lack of adaptability and their disregard for the incentive systems required for long-term, sustainable development. The main limitation of these types of financing sources is the risk-reward mismatch. Historically, investors have placed a higher priority on immediate financial gains, which has resulted in a lack of funding for long-term sustainability projects including infrastructure for climate resilience, renewable energy, and circular economy efforts (Mhlanga 2024; Popescu et al. 2024a, b). Moreover, the predicted USD 5–7 trillion yearly finance gap required to accomplish the United Nations Sustainable Development Goals (SDGs) by 2030 cannot be filled by public spending alone (The Sustainable Development Report 2023).

To address these limitations, financial innovations have emerged as crucial tools for mobilizing capital toward ESG-aligned investments. This type of innovation includes fintech-based mechanisms, efficient investment strategies, sustainable loans, and green bonds. Traditional finance is based on statistical analysis of risks and returns. At the same time, innovative financial instruments are oriented towards performance-based models, which are aimed at projects that measure sustainable development results (Gutterman 2024). For instance, sustainability-linked finance modifies interest rates according to a company's capacity to satisfy ESG performance goals, whereas green bonds raise money only for environmental projects. Increasing sustainability and ensuring transparency in financial markets is achieved through fintech-based solutions, such as blockchain-enabled carbon credit trading and ESG analysis performed by artificial intelligence (Alshahmy and Sahiner 2024).

Changing regulatory landscapes also drive the shift from traditional to innovative financing. Policymakers increasingly integrate sustainability reporting requirements, carbon-pricing mechanisms, and green finance taxonomies to incentivize responsible investment practices (Bilbao-Terol et al. 2018). As a result, financial innovations are shifting the focus of investments and encouraging governments, businesses, and institutional investors to match their capital allocation plans with the objectives of sustainable development.

2.2 *Understanding Sustainability-Linked Finance and Green Bonds*

Green bonds are unavoidably brought up when talking about sustainable financing because they are fundamentally linked to environmentally beneficial enterprises. The market for green bonds was introduced by the European Investment Bank (EIB) in 2007 and has been expanding rapidly ever since. By 2024, the global amount of issuance will exceed 2 trillion USD (Singh et al. 2024). Unlike traditional corporate bonds, green bonds are explicitly designated for renewable energy, climate adaptation, sustainable transport, and pollution reduction projects. Investors are drawn to green bonds because they combine financial returns with measurable environmental impact, making them an attractive tool for ESG-focused portfolios (Hakim 2024; Popescu et al. 2024a, b).

The dynamic growth trend of green bonds demonstrates investors' growing demand for sustainable assets in the market and the support of regulatory institutions. Governments and financial institutions are making a concerted effort to incorporate the idea of green bonds into the framework of sustainable financing. For instance, China's Green Bond Guidelines and the European Union's Green Bond Standard (EUGBS) have set strict requirements for green bond eligibility, guaranteeing increased credibility and transparency in sustainable finance markets (Hu and Jin 2023; Maragopoulos 2023; Zhang 2020). In addition, sovereign green bonds issued by national governments have gained traction, with countries such as Germany, France, and Indonesia leveraging these instruments to finance large-scale sustainability initiatives (Mhlanga 2024).

The emergence of sustainability-linked finance, which includes sustainability-linked bonds (SLBs) and sustainability-linked loans (SLLs), is a noteworthy development in the field of sustainable finance. Sustainability-linked finance structures modify financial terms (such interest rates or coupon payments) in response to the borrower's capacity to satisfy predetermined ESG performance criteria, in contrast to green bonds, which allocate funding for particular environmental projects (Gutterman 2024). For instance, if a company fails to meet its carbon reduction goals, it may face higher interest rates on sustainability-linked loans, creating a direct financial incentive for sustainable performance (Alshahmy and Sahiner 2024). Companies like Enel and Danone have successfully issued SLBs, aligning corporate financing strategies with global climate and social objectives (Kandpal et al. 2024; Berrada et al. 2022).

Despite their advantages, green bonds and sustainability-linked finance face challenges, including greenwashing risks, inconsistent reporting standards, and a lack of universally accepted ESG performance metrics (Hakim 2024). To improve openness and credibility in the green finance markets, regulatory agencies and financial institutions are actively attempting to create international sustainability disclosure standards, such as the International Sustainability Standards Board (ISSB) (De Villiers and Dimes 2023).

2.3 Impact Investing and Sustainable Investment Strategies

A growing subset of sustainable finance is impact investment, which aims to provide monetary profits as well as observable benefits for the environment and society.

Impact investing proactively seeks companies and projects that support the SDGs, such as financial inclusion programs, clean energy, and affordable housing, in contrast to typical ESG investing, which concentrates on risk reduction and ethical corporate conduct (Bradford et al. 2024). Impact investing has grown significantly on a worldwide scale, with assets under management (AUM) expected to reach USD 1 trillion by 2024. This increase demonstrates how institutional and individual investors are becoming more interested in making investments that put sustainability first (Trelstad and Walske 2025).

Using performance-based investment models, such development impact bonds (DIBs) and social impact bonds (SIBs), is a key component of impact investing. These financial instruments follow a pay-for-success structure, where investors provide initial funding for social or environmental initiatives, and returns are based on meeting specific impact targets (Mhlanga 2024). As an illustration of the scalability of outcome-driven investment models, SIBs have been utilised to finance early childhood education in the US and job training programs in the UK (Tse and Warner 2020; Knoll and Fraser 2024).

The lack of standardised frameworks for assessing impact, the difficulty of quantifying social and environmental consequences, and worries about greenwashing are just a few of the major obstacles impact investment must overcome. In order to improve accountability in impact-driven financial markets, investors and financial institutions are increasingly using impact assessment tools like the Sustainable Accounting Standards Board (SASB), the UN Principles for Responsible Investment (PRI), and the Impact Reporting and Investment Standards (IRIS +) of the Global Impact Investing Network (GIIN) (Gutterman 2024).

In impact investing, a notable challenge is "greenwashing," or the fraudulent representation of their sustainability impact by business entities to gain benefits (Vollero 2024). This has led to reputational damage to impact investing when the ESG label is fraudulently applied to those less compliant with the criteria (Hakim 2024). Strong industry regulators like the European Securities and Markets Authority (ESMA) and the US Securities and Exchange Commission (SEC) have strengthened and reviewed criteria that introduce a high degree of transparency into impact investing practices and enable the detection of fraudulent sustainability claims (Alshahmy and Sahiner 2024).

Impact investing is anticipated to become more crucial in filling funding shortages for SDG-aligned projects as sustainable finance develops. To de-risk sustainable investments and speed up capital flows towards impact-driven projects, institutional investors, development finance organisations, and philanthropic groups are expanding blended finance strategies, which combine public and private funds (Agrawal and Hockerts 2021). Impact investment, which links financial incentives with quantifiable social and environmental consequences, is well-positioned to spearhead the next wave of sustainable economic transformation.

3 Objectives

In light of the Sustainable Development Goals (SDGs), this chapter analyses the literature review and looks at the connection between financial innovations and ESG investments. With a focus on financial technologies and decentralised finance (DeFi) in advancing sustainable finance, it explores the ways in which financial innovations support sustainability. The chapter explores a number of financial products, including impact investment, sustainability-linked loans, and green bonds. Furthermore, it investigates how financial innovations, such as FinTech and decentralised finance, may make sustainable finance more accessible by utilising tools like blockchain, AI-powered ESG analytics, and digital lending platforms.

In addition, the chapter identifies opportunities and challenges for expanding sustainable finance. It analyzes the vicious practices of "greenwashing," regulatory frameworks, and barriers to equitable capital distribution. Accordingly, the chapter aims to develop recommendations that identify ways to improve the formation of the type of finance and sustainable investment mechanisms needed for the ESG concept.

4 Discussion

4.1 The Role of Fintech in Expanding Sustainable Investment

By enabling sustainable finance projects and expanding access to ESG-aligned investments, the swift development of financial technology, or fintech, has drastically changed the investing environment. Traditional investment models have often excluded small-scale investors and underrepresented communities, creating barriers to participation in sustainability-focused financial markets (de Sousa Gabriel and Rodeiro-Pazos 2020; Popescu and Gheorghiu 2021). These inefficiencies are addressed by fintech solutions, including blockchain, big data, and artificial intelligence (AI), which improve transparency, lower transaction costs, and maximise investment methods. (Ojukwu et al. 2024). AI-powered robo-advisors and machine learning algorithms are now used to assess ESG performance, helping investors align their portfolios with sustainable development goals (SDGs) while mitigating risks associated with greenwashing and misleading sustainability claims (Schwinn and Teo 2018).

Blockchain technology is changing the game for ESG investing, enabling a safe and secure way to raise awareness of sustainable development indicators. However, if we want to verify data on corporate ESG performance in real-time, distributed ledger technology (DLT) is the perfect solution. It also ensures that companies are more accountable for meeting and adhering to sustainability criteria (Dissanayake et al. 2023; Kwong et al. 2023). Fintech-driven digital platforms are also revolutionizing impact investing, allowing for fractional ownership of green assets, carbon credit tokenization, and automated ESG compliance checks (Hasan et al. 2024). It becomes equally possible for retail and institutional investors to focus on ethics when implementing investment policies and to exercise long-term oversight over financial decisions (Gumel and Zunguze 2024).

The deficit observed in financing sustainable projects can be eliminated by the involvement of financial technologies, especially in emerging markets with challenges in access to finance (Gigauri 2022). Alternative financing models are offered by smart contracts based on blockchain technology and digital financing platforms, which, unlike the conservative banking system, offer fewer barriers to customers, as a result of which green enterprises and climate-oriented businesses have increased chances of attracting adequate capital (Patel 2024). These fintech innovations reshape sustainable finance by promoting financial inclusion, improving investment efficiency, and reinforcing accountability in ESG-aligned financial markets (Macchiavello and Siri 2022).

4.2 Accountability and Transparency Reach by Blockchain

ESG-compliant finance requires a high level of accountability and transparency, which blockchain technology provides, addressing challenges in terms of sustainability compliance, fraud, and "greenwashing." Prior to the standard ESG reporting framework, company frameworks often relied on self-reported data from corporations, making it difficult to verify the accuracy of sustainability claims (Ho 2023). Blockchain-based ESG reporting eliminates this challenge by ensuring that all ESG data is securely recorded, timestamped, and immutable, preventing companies from altering or misrepresenting their sustainability performance (Almadadha 2024).

Smart contracts, a core application of blockchain technology, have revolutionized sustainability-linked financial instruments by enabling automated enforcement of ESG performance targets. For instance, sustainability-linked loans (SLLs) and green bonds can be programmed with predefined ESG conditions, ensuring that borrowers receive financial incentives (such as lower interest rates) only if they achieve their sustainability targets (Auzepy et al. 2023). This mechanism reduces the risk of greenwashing, requiring transparent and verifiable ESG performance metrics to trigger contractual obligations (Bhambri and Starostka-Patyk 2025).

Blockchain's impact extends beyond financial instruments to carbon markets and supply chain transparency. The tokenization of carbon credits allows for real-time trading of verified emissions reductions, ensuring that carbon offset programs maintain credibility and traceability (Khalil et al. 2025). In addition, stakeholders may monitor the process, evaluate the sustainability of raw materials, and evaluate ethics and responsibility thanks to blockchain technology's transparent supply chain, which tackles a variety of social justice and environmental concerns (Unsal and Ozkan 2024).

Despite its benefits, blockchain implementation in ESG finance faces a number of technological and regulatory challenges. Investors find it difficult to assess sustainability performance across jurisdictions because to inconsistent data verification procedures caused by the absence of a single global standard for blockchain-based ESG reporting (Hasan et al. 2024). A move towards low-energy consensus techniques like Proof-of-Stake (PoS) and hybrid blockchain models has also been prompted by worries about the environmental impact of decentralised finance applications due to the energy consumption of certain blockchain networks (Auer et al. 2024). However, blockchain technology continues to be one of the most promising developments in sustainable finance for guaranteeing accountability, transparency, and trust (Macchiavello and Siri 2022; Rosca and Stancu 2025).

4.3 The Potential and Risks of DeFi in ESG Finance

With its decentralised, borderless, and intermediary-free financial services, Decentralised Finance (DeFi) has become a disruptive force in the worldwide financial scene. DeFi reduces reliance on traditional financial institutions by enabling peer-to-peer lending, yield farming, and decentralised asset management through the use of smart contracts, blockchain technology, and decentralised exchanges (DEXs) (Alamsyah et al. 2024). DeFi's ability to increase access to sustainable investment options is what makes it important in ESG (Environmental, Social, and Governance) finance. It enables eco-aware companies, underserved populations, and smaller investors to actively participate in financially significant sustainability-focused marketplaces (Bhambri and Starostka-Patyk 2025).

Facilitating direct investing in sustainability-linked assets is one of DeFi's primary benefits in ESG finance. Through tokenized green bonds, carbon credit trading platforms, and decentralized ESG lending pools, DeFi enables investors to allocate capital toward climate resilience projects, renewable energy initiatives, and social impact enterprises without relying on centralized financial intermediaries (Turi 2023). Additionally, liquidity pools and automated market makers (AMMs) in DeFi ecosystems enable the seamless trading process of sustainability-related digital assets, increasing market efficiency by reducing transaction costs (Ho 2023).

Resolving the stability issue and improving the regulatory environment are directly related to enhancing DeFi's potential as a sustainable finance tool. The lack of regulatory oversight in DeFi markets creates risks related to fraud, cyberattacks, and financial instability, deterring institutional investors from fully embracing decentralized sustainability solutions (Kwong et al. 2023). However, the high volatility of cryptocurrency assets associated with DeFi protocols makes it difficult for investors to prioritize long-term stability, which is fully consistent with the ESG framework (Panait et al. 2024).

Another pressing concern is the environmental impact of specific blockchain-based DeFi applications. While DeFi has the potential to finance green initiatives, its reliance on energy-intensive blockchain networks contradicts ESG principles, leading to debates over whether DeFi aligns with sustainability objectives (Hasan et al. 2024). Developers are incorporating carbon offset mechanisms into DeFi protocols and embracing eco-friendly blockchain solutions in greater numbers to address this problem and lessen their environmental impact (Patel 2024).

In parallel with the development of DeFi, those involved in blockchain development, whether financial institutions or regulatory entities are emerging with common goals. Through their combined efforts, these objectives pertain to the development of a regulatory framework for the field that will address security concerns, lower the likelihood of instability, and raise awareness of global environmental issues. By tackling these issues, DeFi may play a significant role in advancing financial inclusion, accountability, and transparency in ESG finance, which will ultimately help achieve the SDGs and create a more just financial system (Macchiavello and Siri 2022).

4.4 Scaling Sustainable Investment and It's Challenges

Because there is no established standard for assessing ESG metrics, the work of investors is becoming more difficult, as they cannot conduct a comparative analysis between companies in the direction of sustainable development. Various agencies offer ESG ratings; however, they lack objective criteria and are not uniform, primarily due to the absence of a universal assessment scheme and the taking into account different regional or similar priorities (Gutterman 2024). This fragmentation results in conflicting sustainability scores for the same company, leading to investor confusion and reduced confidence in ESG-driven financial decision-making (Nicholls 2021).

Without globally accepted sustainability reporting standards, greenwashing risks increase, allowing corporations to claim ESG compliance without implementing meaningful sustainability practices (Oliveira and Leal 2024). Initiatives like the International Sustainability Standards Board (ISSB) and the European Union's Corporate Sustainability Reporting Directive (CSRD), which seek to establish a single ESG disclosure framework that improves accountability and transparency, are among the attempts to address this issue (Bhattacharya and Sachdev 2024).

Greenwashing remains a critical credibility risk in sustainable finance as companies increasingly seek to attract ESG-conscious investors without substantiating their claims with verifiable impact data (Dempere et al. 2024). Many firms use vague sustainability statements, misleading marketing, and selective data disclosure to appear more sustainable than they are, eroding investor trust and threatening the long-term viability of ESG finance (Gutterman 2024; Kiladze et al. 2024).

By enacting stronger sustainability disclosure regulations and enforcing penalties for false ESG claims, regulatory bodies such as the European Securities and Markets Authority (ESMA) and the U.S. Securities and Exchange Commission (SEC) have increased their scrutiny of greenwashing practices (Dempere et al. 2024). Enforcement remains a significant challenge, as inconsistencies in jurisdictions reflect ESG regulations, and consequently, global accountability by companies is not tangible (Wahid and Rofiah 2024). This is why, firstly, the existence of an independent third party that specifically defines the criteria for assessing ESG metrics, and, secondly, regulatory oversight, which will ultimately focus on strengthening efforts to detect "greenwashing" and expose fake sustainability (Dimmelmeier 2024; Joiţa et al. 2023).

These issues are hindering the development of sustainable finance, as they create a whole chain of barriers due to inconsistent policies, lack of transparency standards, and geopolitical considerations, ultimately damaging the investment process for projects related to ESG (Shirai 2023). Governments and financial institutions have declared incentives as the priority in sustainable finance. However, the policy implementation is inconsistent, fragmenting the market, highlighting compliance issues, and raising doubts among institutional investors (Lei and Yu 2024). Clear rules for ESG investments are established by the EU's "green" taxonomy and the Sustainable Finance Disclosure Regulation (SFDR). Other areas, especially emerging economies, may not yet have robust regulatory structures, though, which hinders their capacity to draw in international funding for sustainable development projects Chiu (2022). Coordinating harmonized global standards for ESG investing will be crucial in overcoming regulatory barriers and accelerating the transition to a sustainable financial ecosystem (Bengo et al. 2022).

5 Policy Implications

5.1 Policy and Regulatory Frameworks

Creating a sustainable financial policy is the responsibility of government circles, as capital allocation, encouraging green initiatives, and shaping the legal framework are precisely the spheres of influence of the public sector to create incentives for the private sector to implement ESG practices (Patel 2024).

Many governments have introduced green bonds, carbon pricing mechanisms, and tax incentives for sustainability-linked investments, aiming to direct financial flows toward projects that align with climate resilience, clean energy transition, and social equity objectives (Bhattacharya and Sachdev 2024). Realising that climate change presents systemic hazards to financial markets, central banks have also started to include climate risk factors in evaluations of financial stability and monetary policy (Migliorelli 2021). To encourage banks and institutional investors to include ESG concerns into their credit allocation strategies, the European Central Bank (ECB) and the People's Bank of China (PBoC) have specifically introduced green lending programs (Wahid and Rofiah 2024).

International organisations including the United Nations, the International Monetary Fund (IMF), and the Financial Stability Board (FSB) have been at the forefront of the growing global push in recent years to create standardised rules for sustainable finance (Shirai 2023). Important frameworks for climate risk reporting have been established by significant initiatives such as the Network for Greening the

Financial System (NGFS) and the Task Force on Climate-related Financial Disclosures (TCFD), which help investors assess the financial effects of environmental and social risks (Migliorelli and Dessertine 2020; Migliorelli 2021). Simultaneously, the Sustainable Finance Action Plan and the European Green Deal established mandatory ESG (Environmental, Social, and Governance) disclosure requirements, which demand that businesses and financial institutions submit thorough reports on their risk profiles and sustainability practices (Dempere et al. 2024). These regulatory developments aim to tackle information asymmetry, enhance market efficiency, and ensure that capital is directed towards sustainable and impactful initiatives (Dimmelmeier 2024).

The lack of uniformity in global regulatory bodies remains a key challenge as countries focus on implementing sustainability goals, which will improve their economic structures, change their approaches to resources, and transform their political ecosystem (Gutterman 2024).

In developing economies, governments often struggle to balance sustainability mandates with economic growth objectives, leading to conflicts between environmental policies and financial market stability (Oliveira and Leal 2024).

Additionally, ESG regulatory fragmentation between the United States, Europe, and Asia creates compliance challenges for multinational corporations and limits cross-border investment flows into sustainable assets (Mhlanga and Adegbayibi 2024). Addressing these discrepancies requires greater international cooperation, aligning sustainability taxonomies, and coordinated efforts to integrate ESG considerations into global financial regulations (Bengo et al. 2022).

5.2 *Pathways for Equitable Access to Sustainable Finance*

In areas where traditional financing methods are inadequate or unavailable, blended finance has become a potent instrument for directing private funds towards sustainable development initiatives (Shirai 2023). Blended finance approaches can de-risk sustainable initiatives and increase their appeal to institutional investors by integrating private investment, charitable resources, and public-sector support (Gutterman 2024).

In underdeveloped nations, where governments lack the financial resources to independently finance large-scale sustainability projects, this strategy is especially pertinent (Nicholls 2021). Blended financing arrangements have been effectively used by organisations like the World Bank, the International financing Corporation (IFC), and the Green Climate Fund (GCF) to support social impact investments, climate adaption initiatives, and renewable energy projects (Oliveira and Leal 2024).

Public–private partnerships (PPPs) case studies illustrate how collaborative investment models can drive sustainable development while balancing financial returns with social impact (Wahid and Rofiah 2024). In Kenya, for example, the Kenya Off-Grid Solar Access Project (KOSAP) has leveraged blended finance to expand access to clean energy in rural communities, mobilizing investment from the private sector while receiving technical and financial support from the World Bank (Patel 2024). The religious-innovative example of Indonesia, which blended the ideas of green finance and Islamic finance, was an intriguing experience. The Green Sukuk initiative attracted investors who promised in accordance with Sharia law and simultaneously invested in renewable energy (Ali et al. 2024). The examples presented clearly show that implementing financial sustainability projects and implementing ESG practices through public–private sector cooperation is more effective (Dimmelmeier 2024).

Ensuring equitable access to sustainable finance requires addressing structural barriers that prevent small enterprises, marginalized communities, and underserved regions from participating in green investment markets (Surmanidze et al. 2023). Initiatives for financial inclusion, like impact investment funds, digital financial platforms, and microfinance programs, are crucial for opening doors for people and companies without access to traditional banking services (Mhlanga and Adegbayibi 2024; Gigauri et al. 2023). Impactful financing practices that are transparent and trustworthy ensure that financial institutions and policymakers can implement broad economic sustainability and promote social equity (Gutterman 2024).

6 Conclusion

Markets are increasingly considering Environmental, Social, and Governance (ESG) considerations in their lending processes, governance frameworks, and investment choices as the need for sustainability in the financial industry grows. This change has prompted businesses to set quantifiable environmental and social performance goals and resulted in the creation of financial products connected to sustainability objectives. This strategy aids in making sure that financial choices are made with long-term sustainability in mind rather than just immediate financial gain. A wider spectrum of participants, from small business owners to individual investors, can now participate in activities that support the Sustainable Development Goals (SDGs) thanks to the growth of digital platforms in the finance industry. There are still issues, though, such the absence of widely recognised ESG indicators, worries about deceptive environmental claims (greenwashing), and the regulatory ambiguity that erodes investor confidence.

The advantages and drawbacks of financial innovations in sustainable financing are examined in this study. By examining successful case studies of financial innovations in global sustainability efforts, the study provides valuable strategies for scaling investment in projects aligned with SDGs. The results emphasise how crucial it is for governments, financial institutions, and the private sector to work together to create policies that encourage innovation while maintaining accountability and openness. Moreover, the research emphasizes the necessity of developing financing mechanisms to ensure equitable access to sustainable investment opportunities, especially in developing countries with critical financing gaps.

Financial innovations are revolutionizing how individuals interact with money and make investment decisions. At the front of this change are technologies like decentralised finance (DeFi), blockchain, and artificial intelligence (AI). Blockchain removes the need for conventional middlemen by providing a transparent and safe method of conducting transactions. AI enhances the financial services sector by providing personalized experiences and utilizing predictive analytics to optimize lending, insurance, and investment processes. DeFi opens up financial services to a global audience, lowering entry barriers and democratizing access to finance.

Additionally, innovations such as asset tokenization and instant digital payment systems—like Central Bank Digital Currencies (CBDCs)—create new opportunities for more efficient and diverse financial transactions. Adopting these innovations is becoming indispensable for maintaining competitiveness and staying informed in a rapidly evolving financial environment. Quick adaptation and improved financial literacy are essential for thriving in this digital age.

The role of financial innovations in achieving global sustainability objectives cannot be overstated. However, their success hinges on effective regulatory oversight, collaboration among institutions, and the seamless integration of new technologies into established financial markets. This study adds to the continuing discussion on sustainable finance by bringing these innovations into line with the larger ESG and SDG frameworks. It does this by providing insights into how new financial instruments and digital technologies may propel the shift to a more sustainable and inclusive global economy.

References

Agrawal A, Hockerts K (2021) Impact investing: review and research agenda. J Small Bus Entrep 33(2):153–181

Alamsyah A, Kusuma GNW, Ramadhani DP (2024) A review on decentralized finance ecosystems. Futur Internet 16(3):76. https://doi.org/10.3390/fi16030076

Ali Q, Rusgianto S, Parveen S, Yaacob H, Zin RM (2024) An empirical study of the effects of green Sukuk spur on economic growth, social development, and financial performance in Indonesia. Environ Dev Sustain 26(8):21097–21123

Almadadha R (2024) Blockchain technology in financial accounting: enhancing transparency, security, and esg reporting. Blockchains 2(3):312–333

Alshahmy S, Sahiner M (2024) Enhancing the issuance and monitoring of sustainable finance instruments through AI. In: Artificial intelligence, finance, and sustainability: economic, ecological, and ethical implications. Springer Nature Switzerland, Cham, pp 217–241

Auer R, Haslhofer B, Kitzler S, Saggese P, Victor F (2024) The technology of decentralized finance (DeFi). Digit Financ 6(1):55–95. https://doi.org/10.1007/s42521-023-00088-8

Auzepy A, Bannier CE, Martin F (2023) Are sustainability-linked loans designed to effectively incentivize corporate sustainability? A framework for review. Financ Manag 52(4):643–675. https://doi.org/10.1111/fima.12437

Bengo I, Boni L, Sancino A (2022) EU financial regulations and social impact measurement practices: a comprehensive framework on finance for sustainable development. Corp Soc Responsib Environ Manag 29(4):809–819

Berrada T, Engelhardt L, Gibson R, Krueger P (2022) The economics of sustainability linked bonds. Swiss finance institute research paper, pp 22–26. https://doi.org/10.2139/ssrn.4059299

Bhambri P, Starostka-Patyk M (2025) Decentralized finance (DeFi) and environmental impact. In: Digital sustainability. CRC Press, pp 142–151

Bhattacharya S, Sachdev BK (2024) Regulatory challenges and opportunities for future fintech and green finance. In: Blockchain's transformative potential of financial technology for sustainable futures, pp 103–122
Bilbao-Terol A, Arenas-Parra M, Cañal-Fernández V, Obam-Eyang PN (2018) Multi-criteria analysis of the GRI sustainability reports: an application to socially responsible investment. J Oper Res Soc 69(10):1576–1598
Bradford M, Earp JB, Williams PF (2017) Understanding sustainability for socially responsible investing and reporting. J Cap Mark Stud 1(1):10–35
Chiu IH (2022) The EU sustainable finance agenda: developing governance for double materiality in sustainability metrics. Eur Bus Organ Law Rev 23(1):87–123
Colombage S, Nanayakkara M, Barua S, Colombage UN (2025) Leveraging financial technology and innovation for sustainability in the post-covid era. In: Financing of sustainable development goals (SDGs) challenges and opportunities. Springer Nature Switzerland, Cham, pp 217–242
Cromwell RS, Peprah JA (2025) Banking business models and sustainable development: challenges and opportunities. In: Strategic approaches to banking business and sustainable development goals, pp 3–22
de Sousa Gabriel VM, Rodeiro-Pazos D (2020) Environmental investment versus traditional investment: alternative or redundant pathways? Organ Environ 33(2):245–261
De Villiers C, Dimes R (2023) Will the formation of the international sustainability standards board result in the death of integrated reporting? J Account Organ Chang 19(2):279–295. https://doi.org/10.1108/JAOC-05-2022-0084
Dempere J, Alamash E, Mattos P (2024) Unveiling the truth: greenwashing in sustainable finance. Front Sustain 5:1362051. https://doi.org/10.3389/frsus.2024.1362051
Dimmelmeier A (2024) Expanding the politics of measurement in sustainable finance: reconceptualizing environmental, social and governance information as infrastructure. Environ Plan c: Polit Space 42(5):761–781. https://doi.org/10.1177/23996544231209149
Dissanayake H, Popescu C, Iddagoda A (2023) A bibliometric analysis of financial technology: unveiling the research landscape. FinTech 2(3):527–542. https://doi.org/10.3390/fintech2030030
Fu C, Lu L, Pirabi M (2023) Advancing green finance: a review of sustainable development. Digit Econ Sustain Dev 1(1):20. https://doi.org/10.1007/s44265-023-00020-3
Gigauri I, Janjua LR, Panait M, Guillen FP (2023) Exploring the nexus between financial inclusion, poverty, digitalization and sustainable development goal. Lumen Proc 19:63–81. https://doi.org/10.18662/lumproc/gekos2022/0
Gigauri I (2022) The promise of financial inclusion for developing economies. Int J Manag Sci Bus Adm 8(6):7–20. https://doi.org/10.18775/ijmsba.1849-5664-5419.2014.86.1001
Gumel BI, Zunguze L (2024) Investing in a sustainable future: how fintech and blockchain enable green asset investments for financial security. Int J Blockchain Technol Appl 2(2):100–109
Gutterman AS (2024) Sustainable finance and impact investment: a guide for sustainable entrepreneurs. SSRN. https://doi.org/10.2139/ssrn.4944162
Hakim A (2024) Bridging the gap: can Sukuk instruments foster green investments in the Islamic bond market? SSRN 4860680. https://doi.org/10.2139/ssrn.4860680
Hasan MH, Hossain MZ, Hasan L, Dewan MA (2024) FinTech and sustainable finance: how is fintech shaping the future of sustainable finance? Eur J Manag Econ Bus 1(3):100–115
Ho J (2023) Banking, blockchain, and ESG. In: Leo S, Panetta IC (eds) The role of distributed ledger technology in banking: from theory to practice. Cambridge University Press, pp 235–266
Hu Y, Jin Y (2023) Unraveling the influence of green bonds on environmental sustainability and paving the way for sustainable energy projects in green finance. Environ Sci Pollut Res 30(52):113039–113054
Jesuka D, Cunha MA, Borsatto JMLS, Araújo AA (2025) The impact of green innovation and ESG performance on firm debt. BBR Braz Bus Rev 22:e20231610
Joița D, Dobrotă CE, Popescu C (2023) "Do no significant harm" principle and current challenges for the EU taxonomy towards energy transition. In: Machado C, Paulo Davim J (eds) Corporate

governance for climate transition. Springer, Cham, pp 187–208. https://doi.org/10.1007/978-3-031-26277-7_7

Kandpal V, Jaswal A, Santibanez Gonzalez ED, Agarwal N (2024) Sustainable financing for ESG practices. In: Sustainable energy transition: circular economy and sustainable financing for environmental, social and governance (ESG) practices. Springer Nature Switzerland, Cham, pp 167–200

Khalil MA, Sinliamthong P, Khalil R (2025) From paper to pixels for a world of purpose: a systematic review of sustainable finance, blockchain, and digital assets. Sustainable financing—a contemporary guide for green finance, crowdfunding and digital currencies: sustainable financing, pp 61–78

Kiladze L, Surmanidze N, Mushkudiani Z (2024) Social entrepreneurship and corporate social responsibility driving sustainable solutions: comparative analysis. Access J 5(1):85–101

Knoll L, Fraser A (2024) Social impact bond assetization struggles: a comparative case study of the United Kingdom and Germany. Econ Soc 53(1):92–111. https://doi.org/10.1080/03085147.2024.2309802

Kutateladze R, Nozadze M, Chelidze M, Beridze T (2025) Green financing and its instruments: Foreign and Georgian experience. Eur J HumIties Soc Sci 1:58

Kwong R, Kwok MLJ, Wong HS (2023) Green FinTech innovation as a future research direction: a bibliometric analysis on green finance and FinTech. Sustainability 15(20):14683

Lei X, Yu J (2024) Striving for sustainable development: Green financial policy, institutional investors, and corporate ESG performance. Corp Soc Responsib Environ Manag 31(2):1177–1202

Macchiavello E, Siri M (2022) Sustainable finance and fintech: can technology contribute to achieving environmental goals? A preliminary assessment of 'green fintech'and 'sustainable digital finance.' Eur Co Financ Law Rev 19(1):128–174. https://doi.org/10.1515/ecfr-2022-0005

Madison N, Schiehll E (2021) The effect of financial materiality on ESG performance assessment. Sustainability 13(7):3652

Maragopoulos N (2023) Toward a European green bond standard: a European initiative to promote sustainable finance. Greening the bond market: a European perspective. Springer International Publishing, Cham, pp 21–50

Martínez-Falcó J, Sánchez-García E, Marco-Lajara B, Millán-Tudela LA (2025) Unraveling the relationship between ESG strategy and sustainable performance in the Spanish wine industry: a structural equation analysis. Soc Responsib J. ahead-of-print(ahead-of-print). https://doi.org/10.1108/SRJ-12-2023-0682

Mhlanga D, Adegbayibi AT (2024) Promoting sustainable finance for enhanced economic growth in Sub-Saharan Africa. In: Sustainable finance and business in Sub-Saharan Africa. Springer, Cham, pp 473–492. https://doi.org/10.1007/978-3-031-74050-3_23

Mhlanga D (2024) Green finance and the evolution of financial markets in Sub-Saharan Africa: a comprehensive analysis. In: Sustainable finance and business in Sub-Saharan Afric*a*. Springer, Cham, pp. 209–225

Migliorelli M (2021) What do we mean by sustainable finance? Assessing existing frameworks and policy risks. Sustainability 13(2):975

Migliorelli M, Dessertine P (2020) Sustainability and financial risks. Palgrave Macmillan, Cham. https://doi.org/10.1007/978-3-030-54530-7

Naysary B, Shrestha K (2024) Financial technology and ESG market: a wavelet-DCC GARCH approach. Res Int Bus Financ 71:102466

Nicholls A (2021) Impact measurement and management in sustainable finance. Asian development outlook: financing a green and inclusive recovery, Asian Development Bank. https://www.adb.org/sites/default/files/institutional-document/691951/ado2021bp-measurement-mgt-sustainable-finance.pdf. Last accessed 12 Mar 2025

Ojukwu PU, Cadet E, Osundare OS, Fakeyede OG, Ige AB, Uzoka A (2024) Exploring theoretical constructs of blockchain technology in banking: applications in African and US financial institutions. Int J Front Res Sci Technol 4(01):035–042

Oliveira B, Leal CC (2024) Leading the way toward a sustainable future: the role of sustainable finance and environmental, social, and governance investing. Circ Econ Manuf 53–82

Panait MC, Apostu SA, Gigauri I, Confetto MG, Palazzo M (2024) Defeating the dark sides of fintech: a regression-based analysis of digitalization's role in fostering consumers' financial inclusion in Central and Eastern Europe. Risks 12(11):178. https://doi.org/10.3390/risks12110178

Pancholi K, Shukla P (2025) Harnessing AI for sustainability: innovations, policies, and investment paradigms. In: Diversity, AI, and sustainability for financial growth. IGI Global Scientific Publishing, pp 97–124

Patel K (2024) Digitalization and the rise of green finance: opportunities, challenges, and the path forward. In: Impact of digitalization on reporting, tax avoidance, accounting, and green finance. IGI Global Scientific Publishing, pp 179–199. https://doi.org/10.4018/979-8-3693-1678-8.ch008

Popescu C, Gheorghiu SA (2021) Economic analysis and generic algorithm for optimizing the investments decision-making process in oil field development. Energies 14(19):6119. https://doi.org/10.3390/en14196119

Popescu C, Hysa E, Kruja A, Mansi E (2022) Social innovation, circularity and energy transition for environmental, social and governance (ESG) practices—a comprehensive review. Energies 15(23):9028. https://doi.org/10.3390/en15239028

Popescu C, Dissanayake H, Mansi E, Stancu A (2024a) Eco breakthroughs: sustainable materials transforming the future of our planet. Sustainability 16(23):10790. https://doi.org/10.3390/su162310790

Popescu C, Gabor MR, Stancu A (2024) Predictors for green energy versus fossil fuels: the case of industrial waste and biogases in European Union context. Agronomy 14(7):1459. https://doi.org/10.3390/agronomy14071459

Raihan A (2024) A review of the potential opportunities and challenges of the digital economy for sustainability. Innov green Dev 3(4):100174. https://doi.org/10.1016/j.igd.2024.100174

Rosca C-M, Stancu A (2025) A comprehensive review of machine learning models for optimizing wind power processes. Appl Sci 15(7):3758. https://doi.org/10.3390/app15073758

Rosca C-M, Stancu A, Tănase MR (2025) A comparative study of azure custom vision versus google vision API integrated into AI custom models using object classification for residential waste. Appl Sci 15(7):3869. https://doi.org/10.3390/app15073869

Sahid A, Maleh Y, Asemanjerdi SA, Martín-Cervantes PA (2023) A bibliometric analysis of the FinTech agility literature: evolution and review. Int J Financ Stud 11(4):123

Schwinn, R., & Teo, E. G. (2018). Inclusion or exclusion? Trends in robo-advisory for financial investment services. In: Handbook of Blockchain, digital finance, and inclusion. Academic Press 2:481–492. https://doi.org/10.1016/B978-0-12-812282-2.00021-8

Shirai S (2023) Global climate challenges, innovative finance, and green central banking. Asian Development Bank Institute. https://www.adb.org/sites/default/files/publication/894961/global-climate-challenges-innovative-finance-and-green-central-banking.pdf. Last accessed 11 Mar 2025

Singh V, Singh S, Jain S (2024) Green bond premium diagnosis: an interplay of repayment obligation structure. Int Rev Econ Financ 96:103689

Surmanidze N, Beridze M, Amashukeli M, Tskhadadze K (2023) Empowering small businesses in Georgia: access to finance, economic resilience, and sustainable growth. Agora Int J Econ Sci 17(2):158–169. https://doi.org/10.15837/aijes.v17i2.6453

Surmanidze N, Beridze M, Tskhadadze K, Mushkudiani Z, Chichinadze R (2025). Exploring the role of leaders in CSR in sustainable enterprises: a case study in Georgia. In: Navigating corporate social responsibility through leadership and sustainable entrepreneurship. IGI Global Scientific Publishing, pp. 407–434

The Sustainable Development Report (2023) The costs of achieving the SDGs. UN Trade and Development. https://unctad.org/sdg-costing/about. Last accessed 12 Mar 2025

Trahan RT, Jantz B (2023) What is ESG? Rethinking the "E" pillar. Bus Strateg Environ 32(7):4382–4391

Trelstad B, Walske JM (2025) An introduction to impact investing. In: An Introduction to Early-Stage Impact Investing. Edward Elgar Publishing, pp 1–25

Tse AE, Warner ME (2020) The razor's edge: social impact bonds and the financialization of early childhood services. J Urban Aff 42(6):816–832. https://doi.org/10.1080/07352166.2018.1465347

Turi AN (2023) Financial technologies and DeFi. a revisit to digital financial revolution. Springer, Cham. https://doi.org/10.1007/978-3-031-17998-3

Unsal O, Ozkan B (2024) Technologies and emerging trends shaping fintech. The emerald handbook of fintech, Emerald Publishing Limited, Leeds, pp 399–418. https://doi.org/10.1108/978-1-83753-608-520241046

Vollero A (2024) The impact of greenwashing on the sport industry: shades of deception. In: Integrity and sustainability in sport. Routledge, pp 52–65

Wahid SH, Rofiah K (2024) Renewing the legal and policy framework for sustainable financing: bibliometric analysis and systematic literature review. Int J Sustain Policy Pract 21(1):1

Zhang H (2020) Regulating green bond in China: definition divergence and implications for policy making. J Sustain Financ Invest 10(2):141–156. https://doi.org/10.1080/20430795.2019.1706310

Zhang Q, Zhang Z (2025) Investment and economic growth: a dilemma in China. J Asia Pac Econ 30(1):211–242. https://doi.org/10.1080/13547860.2023.2196886

Exploring the Role of Digital Financial Literacy in Promoting Sustainable Digital Financial Inclusion Among Salaried Peoples

Mansi Yadav, Akanksha Kathuria, Neeraj Chopra, and Animesh Singh

Abstract This study explores the critical relationship between Digital Financial Literacy (DFL) and Digital Financial Inclusion (DFI) among salaried individuals in the Delhi NCR region, with a particular emphasis on the influence of family and social networks. Despite increased availability of digital financial services such as mobile banking, e-wallets, and online investment platforms, disparities persist in their adoption and usage, largely due to gaps in digital financial literacy. Based on a structured survey of 399 salaried respondents, the research identifies a strong positive correlation between DFL and DFI, establishing that higher digital financial literacy leads to greater inclusion, more informed financial decision-making, and improved personal financial management. The findings suggest that social and familial influences play a pivotal role in empowering individuals to navigate digital finance confidently and securely. Older and more educated individuals are shown to be more digitally financially literate and inclusive, underscoring the need for targeted interventions aimed at younger and less-educated segments. Integrating a sustainability perspective, this study highlights how promoting digital financial literacy fosters economic resilience, social equity, and sustainable financial empowerment, which are crucial for reducing inequalities and supporting enduring digital financial ecosystems. The research underscores that sustainable digital inclusion requires not only technological access but also ongoing education and community engagement to ensure long-term financial well-being and environmental consciousness in

M. Yadav
K.R. Mangalam University, Gurugram, Haryana, India

A. Kathuria (✉) · N. Chopra · A. Singh
Manav Rachna University, Faridabad, India
e-mail: akanksha.singh123@gmail.com

N. Chopra
e-mail: neerajchopra@mru.edu.in

A. Singh
e-mail: animesh@mru.edu.in

F. A. Malik et al. (eds.), *Aligning ESG Goals with the UNs Sustainable Development Goals Planning*, World Sustainability Series,
https://doi.org/10.1007/978-3-032-01715-4_7

economic behaviors. The study advocates digital financial literacy as a key enabler of sustainable development, aligned with global sustainability goals.

Keywords Digital financial literacy · Digital financial inclusion · Social networks · Family influence · Sustainability · Economic empowerment

1 Introduction

The sudden digitalization of the financial world has revolutionized the access and use of financial products by individuals. Digital financial inclusion (DFI) is such that individuals, especially salaried, are able to use digital channels to access banking, investment, and payment. Yet, to what extent digital financial literacy (DFL) helps in achieving DFI remains less researched, particularly among salaried people who are early adopters of digital finance (Ozili 2020).

The sudden digitalization of the financial world has revolutionized access to financial products, particularly for salaried individuals. Digital financial inclusion (DFI) enables individuals to access and use digital channels for banking, investment, and payments. However, true financial inclusion goes beyond mere access; it involves the sustainable use and empowerment of individuals to participate confidently in the digital economy. This sustainable usability is essential not only for personal financial health but also for fostering resilient economic systems that support social equity and environmental stewardship.

With global agendas increasingly emphasizing sustainable development and economic equity, digital financial literacy (DFL) emerges as a critical lever to bridge gaps in digital finance access and usage, especially among salaried populations. The ability to make informed, secure, and ethical financial decisions in digital environments enhances economic participation while promoting sustainable financial behaviors that align with long-term societal well-being.

This study recognizes DFL as a catalyst for sustainable digital financial inclusion, where empowerment through knowledge can reduce economic disparities, lower dependence on environmentally taxing cash economies, and create financial ecosystems that are inclusive, secure, and enduring. The sustainable development perspective highlights that digital financial literacy contributes not only to economic outcomes but also supports responsible consumption and investment decisions that benefit both people and the planet.

This study explores the relationship between DFL and DFI among salaried individuals, recognizing that digital finance is not just a matter of access but of sustainable usability and empowerment. With increasing global emphasis on financial equity, understanding how individuals—especially those with steady incomes—can sustainably integrate into the digital economy is essential (Liu et al. 2021).

Digital Financial Inclusion (DFI): DFL extends beyond technical knowledge; it encompasses the ability to make informed financial decisions securely, ethically, and confidently in digital contexts. This capability is essential not only for individual financial well-being but also for building a resilient and sustainable financial ecosystem (Chhillar & Arora 2022). When individuals are equipped with the knowledge and tools to engage in digital finance responsibly, they contribute to economic inclusion, reduce dependence on cash economies, and help foster financial systems that are inclusive, secure, and enduring.

Digital Financial Literacy (DFL): Evolution of digital financial services translates into the opportunity to connect with previously excluded and underserved markets with more tailored financial services (Tay et al. 2022a). However, the speed of growth comes with a dark side in the form of consumer risks.

Limited financial and digital literacy hinder the appropriate adoption of digital financial services. DFL was defined by the study as acquiring knowledge, skills, confidence and abilities to use safely digitally delivered financial products and services, and to make suitable financial decisions (Mbatane & Kekana 2024). DFL has been on the policymakers' agenda for decades (Fig. 1).

Digital Financial Inclusion (DFI) is a paradigm shift in the accessibility and use of financial services, especially in a time where technological advancements dominate (Mpofu & Mhlanga 2022). Simply put, digital financial inclusion (DFI) refers to the availability, affordability, and access to digital financial products such as mobile banking, electronic wallets (e-wallets), online investment platforms, and digital credit services. Escalating mobile phone usage, online penetration, and fintech innovations have all converged to propel the growth of such services, intended to bridge the divide between incumbent banking systems and underbanked populations (Adesola Oluwatosin Adelaja et al. 2024).

Despite these improvements, the majority of the world's population continues to be out of the digital financial sphere. The World Bank's Global Findex Database (2022) puts the number of people in the world who are unbanked at about 1.7 billion,

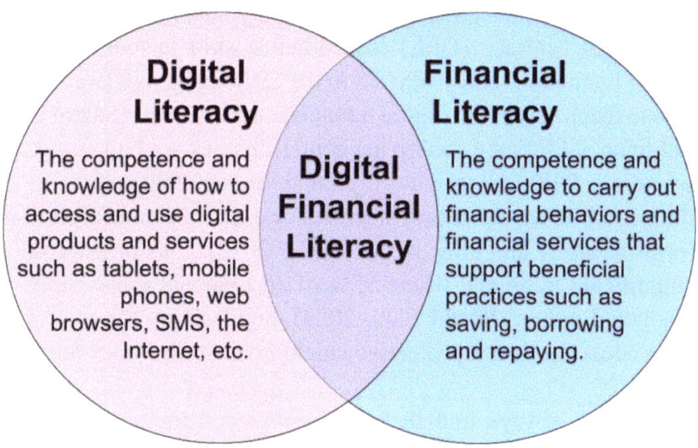

Fig. 1 Venn diagram depicting the intersection of digital literacy and financial literacy (Lyons & Kass-Hanna 2021)

with a significant percentage of these not having access to basic financial services (Nandru et al. 2021). While structural constraints (e.g., poor internet connectivity) exist, there are also significant behavioural and cognitive obstacles—namely, low levels of financial literacy, distrust of digital systems, and self-perceived sophistication—Ozili (2020). For example, in developing nations, many people prefer cash transactions owing to security concerns or a lack of experience with digital payment methods (Bahia & Muthiora 2019).

Surprisingly, even among paid people—a group traditionally associated with steady earnings and higher education levels—engagement with digital finance is inconsistent and even poor. While this group has more access to formal banking channels, research shows that many paid professionals underuse modern digital financial instruments (e.g., stock trading apps, peer-to-peer lending platforms) owing to knowledge gaps or risk aversion (Moenninghoff & Wieandt 2013). According to a poll conducted by the Reserve Bank of India in 2023, just 37% of paid workers in urban areas regularly utilise digital investing platforms, despite the fact that 89% possess bank accounts. This mismatch highlights the need to study whether digital financial literacy (DFL)—rather than simple access—is the major driver of DFI in this group (Zaimovic et al. 2024). The study positions DFL as a catalyst for sustainable digital financial inclusion, emphasizing the need for structured interventions to bridge literacy gaps and promote equitable access. By empowering individuals through DFL, we move closer to achieving a more inclusive and sustainable financial future.

In addition, the COVID-19 pandemic accelerated global digital finance use, yet it also exposed weaknesses in users' ability to use these technologies effectively (IMF 2021). Salaried workers, even though they enjoy economic security, struggled to make a transition to fully digital money management, highlighting the importance of DFL in ensuring inclusion (Mmari, n.d.). This study attempts to close some of these spaces by examining the impact of modifications in DFL on DFI among paid samples, presenting learning for financial teachers and policymakers similarly (Fig. 2).

Digital Financial Literacy (DFL) is a valuable skill in today's financial environment, taking financial literacy beyond to the technical, cognitive, and behavioral competencies to reach success in digital financial environments (Salem 2023). Unlike conventional financial literacy, where the central themes are budgeting, saving, and calculating interest, DFL incorporates digital skills such that people are able to use online banking, mobile payments, cryptocurrency exchanges, and algorithmic investment platforms securely and effectively (OECD 2022). This broader set of skills is especially significant in an age where 75% of all financial transactions worldwide are now conducted online (World Bank 2023), highlighting the role of DFL as an enabler of the economy and a protective shield against digital exclusion (Showkat et al. 2025).

Empirical studies always find that higher financial literacy is associated with improved financial decisions and economic welfare in the long run (Hastings et al. 2013). For instance, individuals with strong financial literacy are 43% more likely to maintain emergency savings and 30% less likely to engage in high-risk borrowing (CHONG et al. 2021). However, these studies predominantly measure

Fig. 2 Core principles of digital financial inclusion (DFI) (Mpofu 2024)

traditional literacy, failing to capture the unique challenges of digital finance, such as, Cybersecurity risks (e.g., phishing scams, identity theft), Algorithmic bias in fintech lending (Yadav & Banerji 2023) and Cognitive overload from complex digital interfaces (Xiao and O'Neill 2022).

The limitations of traditional frameworks become starkly apparent when examining digitally excluded demographics (Bach et al. 2013). For example, in India—a leader in digital payment adoption—68% of salaried workers use mobile banking, yet only 22% understand how to dispute fraudulent transactions (RBI 2023). This "usage-competency gap" highlights how digital proficiency mediates the relationship between access and empowerment (Ibrahim & Aldawsari 2023). Without DFL, even tech-savvy users may:

Overestimate security (e.g., sharing PINs via SMS).

Underutilize advanced tools (e.g., tax-optimized digital investments).

Fall prey to predatory fintech (e.g., algorithmic micro-loans with 300% APR).

The COVID-19 pandemic accelerated digital financial adoption by 5–7 years (McKinsey, 2021), but also exposed critical literacy deficits (Mpofu 2023).

In Brazil, for instance, 41% of new digital banking users could not interpret basic loan terms (Central Bank of Brazil, 2022), leading to widespread over-indebtedness. Such findings necessitate a paradigm shift—from measuring financial literacy in isolation to assessing its integration with digital capabilities (OECD 2023; Zieser et al. n.d.).

Theoretical Underpinnings

DFL's importance is further validated through two theoretical lenses:Capability Approach: Positions DFL as a "conversion factor" transforming financial access into tangible well-being (Martin & Hill 2015) and Technology Acceptance Model (Davis 1989): Suggests that perceived usefulness/ease-of-use of fintech hinges on literacy.

Policy implications are clear: National financial education strategies must evolve to address digital dimensions. South Korea's "MyData" initiative—which combines open banking with compulsory DFL training—increased safe fintech usage by 35% among salaried workers (Souza n.d.; Yadav & Banerji 2024). This evidences DFL's pivotal role in achieving inclusive digital financial ecosystems.

The interconnection between Digital Financial Literacy (DFL) and Digital Financial Inclusion (DFI) is central to understanding how individuals, especially salaried employees, interact with and gain from digital financial systems. Digital financial literacy can be defined as the education and expertise needed to make efficient use of digital financial services like mobile banking, digital wallets, and e-investment platforms (Maji & Laha 2023). Digital financial inclusion, on the contrary, is the availability and usage of these services among various sections of society so that people are able to access the formal financial sector (Kim et al. 2018).

Digital financial literacy increase, in turn, increases the capacity of an individual to use digital financial platforms, thus raising the degree of financial inclusion (Al-shami et al. 2024a). Research has established that individuals with higher incomes and greater DFL will be more likely to use digital payment methods, invest in digital funds, and use online banking (Lumpkin & Schich 2019). Patel and Joshi (2020) discovered, for example, that greater digital financial literacy among employees increased their use of fintech apps by 30% compared to their less knowledgeable counterparts.

Also, financial literacy in the digital age reduces distrust and perceived risk when using digital transactions (Koskelainen et al. 2023). As stated by Mehta and Saxena (2022), those who possess sufficient DFL are less vulnerable to fraud and feel more comfortable utilizing digital financial instruments, thus contributing to continued utilization of formal financial systems. Reserv Bank of India (RBI 2023) further points out that financial literacy initiatives have greatly enhanced the uptake of digital banking among urban salaried employees (Kumar 2024).

The Role of Informed Decision-Making in Managing Digital Financial Tools:

Informed decision-making is a critical outcome of digital financial literacy, enabling individuals to, compare financial products (e.g., loans, insurance, investments) effectively (Das & Sen 2020). Avoid predatory lending and scams by recognizing secure platforms (Bose & Roy 2023). Optimize financial management through budgeting apps and automated savings tools (Lusardi & Mitchell 2020).

In a study by Yadav et al. (2025), it was illustrated that salaried individuals with high DFL levels were likely to have diversified their digital financial portfolios, such as investments in mutual funds and digital gold, than those who had lower levels of literacy (Lu et al. 2021).

We see the interconnection between digital financial literacy and digital financial inclusion—greater DFL results in more DFI through enabling trust, adoption, and proper utilization of digital financial services (Zaimovic et al. 2025). Policymakers and banks need to put much emphasis on financial education programs in order to close the digital gap and provide equal financial involvement (Ozili 2021).

2 Literature Review

Uthaileang and Kiattisin (2023) investigate the pivotal contribution of DFL to improving the abilities of small entrepreneurs, especially in developing nations. Their research indicates that the explosive expansion of digital financial services, fueled by the COVID-19 pandemic, has created the need for enhanced digital financial literacy to enable enhanced access to online loans and other digital financial instruments (Ozili 2018). The research points out the inter-relationship between knowledge, motivation, and digital financial skills such that motivation is central in inspiring small entrepreneurs to enhance their digital financial skills. This inter-relationship highlights the need to enhance both the digital financial skills and attitudes required in order to access digital financial services.

Similarly et al. (2024) highlight the increasing relevance of digital financial literacy globally. They elucidate how fintech advancements have reshaped the financial scenario, offering new opportunities but presenting challenges, particularly regarding the intricacies and risks of digital financial products (Tay et al. 2022b). Authors consider digital financial literacy to be an overall concept that includes awareness of digital financial products, risk perception, and awareness to guard against digital financial risk. Authors indicate that these aspects are needed to allow people to better use digital financial services, especially in developing countries such as India. Their research invites the formulation of standardized measurement of DFL financial literacy to further facilitate cross-country comparison and the detection of financial literacy gaps particularly among vulnerable groups.

Al-Majali et al. (2024) carried out a thorough study to analyze the effects of digital financial literacy (DFL) on financial development and governance in 108 nations during the period of 2017–2021. Using a panel vector autoregressive (PVAR) model and Granger causality tests, the authors found evidence of a time-sensitive and significant relationship between the Composite Digital Financial Literacy Index (CDFLI) and the Financial Development Index (FDI). They establish that DFL improvement results in the growth of financial development through heightened access and consumption of digital financial products (Uthaileang & Kiattisin 2023).

Moreover, authors cited the necessity for support in terms of digital financial education in bridging financial inclusive governance as well as the inclusion of the finance sector with technological advancements.

Li and Liu (2023) analyzed how digital financial inclusion (DFI) impacts local income and its component arrangements in China based on panel data obtained from the China Household Finance Survey (CHFS) for 2015–2019. Recent scholarship increasingly situates digital financial literacy and inclusion within the broader framework of sustainable development. Studies (Liu et al. 2021; Tay et al. 2022a) emphasize that sustainable digital financial inclusion facilitates economic resilience, social equity, and environmental responsibility by enabling underserved populations to access and responsibly use digital financial services. This sustainable approach fosters inclusive growth by bridging the digital divide, thereby contributing to poverty alleviation and reducing systemic inequalities (Abdurrahman & Nugroho 2024).

The nexus between digital financial literacy and sustainability is evident in research advocating for financial education programs that empower individuals to make sustainable financial choices, such as investing in green technologies or ethical funds, avoiding predatory lending, and utilizing digital platforms to optimize resource use. As digital financial ecosystems evolve, literacy must encompass not only technical competence but also awareness of the environmental and social impacts of financial behaviors, aligning digital inclusion efforts with United Nations Sustainable Development Goals (SDGs), especially goals related to reduced inequalities (SDG 10), decent work and economic growth (SDG 8), and responsible consumption and production (SDG 12).

Al-shami et al. (2024) carried out reflective research on the contribution of financial literacy, digital financial literacy (DFL), and social media utilization towards financial inclusion of batik small businesses in Indonesia. Their research depicts the critical contribution of financial literacy towards the improvement of the capacity of small business owners to access financial services, especially in industries such as batik that play critical roles in the Indonesian economy (Gangani & Raval, n.d.). They contend that enhancing both financial and digital financial literacy enables entrepreneurs to make informed financial choices, hence improving the capacity to utilize formal financial systems positively. In addition, they pose the question of the moderating role of social media in proving that proper use of the web-based platform greatly enhances the positive impact of financial literacy on financial inclusion. This research contributes to the body of knowledge by demonstrating how DFL, when used in conjunction with social media, can close the financial services-smalls business gap and offer a sustainable option for expanding access to finance in developing economies.

Similarly, Kumar et al. (2024) investigated the role of digital financial literacy in fintech adoption and financial inclusion in India. Based on their findings, digital financial literacy has a significant impact on the adoption and acceptance of fintech services, which are now part of financial inclusion policies. They suggest a conceptual model that includes four essential components of DFL: financial product knowledge about digital finance, knowledge of digital financial risk, knowledge of the risk control mechanism, and rights of the consumer. Their research demonstrates that those with higher levels of digitally financial literacy are likely to embrace fintech services and, in doing so, gain exposure to otherwise inaccessibilites in finance (Mohakud & Biswal 2024). Additionally, their study connects fintech adoption to better financial inclusion and further supports the contention that digital literacy is key to enabling individuals to navigate digital financial systems.

Financial literacy in the digital world is crucial for millennials, as argued by Dube et al. (2023), since they utilize various digital financial products like mobile wallets, online payment, and investment platforms. They conclude that the more financially literate the millennials are, the more they are likely to make improved financial choices, thereby improving their capacity to engage with the formal financial system (Desy Wulan Ayuning Gumilar et al. 2024). The authors concluded notable differences in the degree of DFL based on urban vs. rural residence, emphasizing the need for focused education in attempting to close the digital divide in financial literacy.

More generally, Tony and Desai (2020) analyzed the connection between digital financial literacy and digital financial inclusion in India. What they found through their research is that DFL is the primary mover of increased access to digital financial services (Bushra & Mir 2024). They contended that although there exists huge potential for online financial inclusion, most do not have the level of basic digital literacy to be able to use the products effectively. The authors posited that increasing DFL would reduce the threats of fraud and cyber attacks due to financial transactions over the internet and consequently increase financial inclusion. Their work supports policy advocating for digital literacy initiatives, especially in the rural regions, in order that they can promote robust development.

Building on the premises of DFL, Mishra et al. (2024) investigated the role of digital financial literacy to enhance women's financial decision-making in India. They presumed that DFL empowers women by enhancing their personal finance management capability, and thus their economic decision-making capability (Vibhuti Shivam Dube et al. 2023). The research established that the higher the DFL, the more women would engage in the digital economy in terms of savings and investments, thereby increasing their financial independence. Authors also came up with a number of deterrents for women obtaining DFL, which included low exposure to digital channels and socio-cultural determinants, and indicated that it might be easier to improve women's financial decision-making and inclusion by doing away with such deterrents.

Gangani and Raval (2021) carried out research on the level of digital financial inclusion among Gujarat's urban youth. The study looked at challenges to digital financial inclusion, particularly in the context of the fact that low digital financial literacy, trust, and technological limitations are challenges to the mass use of digital financial services (Xu et al. 2025). The research aimed that although digital financial inclusion is on the rise, there is a vast demographic gap, specifically by age, gender, and educational levels. The authors clarified that elimination of digital financial literacy barriers is the key to universal financial inclusion, especially in urban areas where technology exposure is higher.

Mohakud and Biswal (2024) emphasized the level of digital financial literacy among rural working people in Odisha, Mayurbhanj district. Based on their findings, most of the respondents had low levels of digital financial literacy, which largely constrained their use of digital financial services (Santhanakrishnan, n.d.). The research also emphasized that rural people were extremely susceptible to digital financial fraud because they were unaware of digital financial risks. The authors emphasized the importance of focused education and sensitization campaigns to enhance digital financial literacy, particularly in rural poverty enclaves.

Gumilar, Sangka, and Totalia (2024) offered a thorough review of literature that explored the intersection of digital financial inclusion and digital financial literacy within the era of digital disruption. Their review of literature established that digital financial literacy is central to the optimal use of digital financial services and plays an important role in mitigating economic and social inequalities. The writers noted that although digital financial services hold the promise of greater financial inclusion, problems such as poor financial literacy levels, distrust, and poor systems for managing risk continue to hamper their widespread adoption (Tony & Desai 2020). They argued that enhancing digital financial literacy is central to making digital financial services effective, especially in developing nations, where unequal access to financial services is still prevalent.

Ozili (2018) reflected on the nexus between digital finance and financial inclusion and its advantages and disadvantages. His research shows that digital finance has the potential to greatly enhance access to financial services in emerging and developing economies, leading to economic growth and poverty alleviation. However, Ozili (2018) also indicated issues such as the digital divide, where technological and access to finance inequalities continue to limit the full potential of digital finance (Mishra et al. 2024). Bridging such gaps is vital to achieving the broader benefits of digital financial inclusion.

Tay et al. (2022) examined the role of digital financial inclusion in promoting development, especially in the period of the COVID-19 pandemic. Their study pointed out that digital financial inclusion plays a very important role in mitigating the economic and social impacts of the pandemic, mostly in developing countries (Al-shami et al. 2024b). Although the pandemic accelerated the use of digital financial services, its impact continues to be weakened by issues of the existing rural–urban, gender, and income gaps. The authors recommended stepping up digital infrastructure, simplifying banking processes, and financial education to facilitate greater financial inclusion.

Sadaf and Hussain (2025) was interested in the gender gap in digital financial literacy and women entrepreneurs' financial inclusion in Punjab, Pakistan. Her research found a large gap between male and female entrepreneurs in financial literacy, digital financial literacy, and access to formal financial services. Women entrepreneurs, particularly in rural areas, encountered more obstacles in accessing funds because of low digital financial literacy and cultural barriers (Kumar et al. 2024). Sadaf and Hussain (2025) recommended focused intervention to raise financial literacy of women in the digital world to economically empower them as well as to support their financial inclusion. This, she said, would help in filling the gender gap and enhancing regional economic growth.

Abdurrahman and Nugroho (2024) examined the connection between digital financial literacy, financial technology, financial behavior, and financial well-being. Their research showed that digital financial literacy has a significant contribution to enhancing financial confidence and better financial behavior, resulting in better financial well-being (Al-Majali et al. 2024). They established that people with greater digital financial literacy were more likely to use digital financial services, resulting in better financial resilience and well-being. The authors highlighted the importance of improving digital financial literacy in driving financial inclusion and well-being, especially in the emerging economies (Li & Liu 2023).

3 Objectives

To assess the relationship between digital financial literacy (DFL) and digital financial inclusion (DFI) among salaried professionals in Delhi NCR.

To evaluate the impact of family and social influences on digital financial literacy (DFL) and its role in digital financial inclusion (DFI).

4 Hypothesis

H1: There is a significant positive relationship between digital financial literacy (DFL) and digital financial inclusion (DFI) among salaried professionals in Delhi NCR.

H2: Family and social influences significantly impact the level of digital financial literacy (DFL) among salaried professionals in Delhi NCR, which in turn enhances their digital financial inclusion (DFI).

5 Research Methodology

Research Design

The current research attempted to identify the connection between digital financial inclusion and digital financial literacy among salaried staff. Because the research design used was qualitative in nature, it used a quantitative method. Quantitative research has been proven effective for the gathering of data on the basis of figures and its examination. This research was fundamentally looking into how digital financial literacy measures among salaried officers and whether this inspires them for their inclusion under digital financial arrangements.

Cross-sectional survey design was applied in the process of gathering relevant data from a specific sample within one moment of time. Such design was thought most appropriate because it was a picture of reality as it happens to be currently when it came to digital financial literacy and inclusion for working professionals. Quantitative approach, besides this, made possible statistical methods application for scrutiny of the associations among variables as well as to form intelligent conclusions.

Sample Selection

The population under consideration for this current study involved 399 salaried working professionals. Using a non-probability convenience sampling, whereby participants were selected keeping in view their availability as well as willing involvement in the research, facilitated covering professionals based in various professions, such as finance, information and technology, medical, as well as teaching profession, for retaining the representational aspect based on salaried working participants based on diversity of backgrounds.

The inclusion criteria for participants were, Participants must be employed full-time in a salaried role. Participants must have access to digital financial services, such as online banking, mobile wallets, or other forms of digital financial tools. Participants should have a basic understanding of financial concepts to ensure meaningful responses to the questionnaire.

Given the exploratory nature of this research, the sample size was selected to ensure sufficient statistical power for analysis while maintaining practicality in terms of time and resource constraints.

Data Collection

Data were obtained through a self-report questionnaire that was developed to capture both digital financial literacy and digital financial inclusion. The questionnaire contained numerous sections, each of which addressed various constructs with respect to research goals.

Digital Financial Literacy: This section measured participants' knowledge and competence in utilizing digital financial instruments. It asked the participants about their understanding of online banking, mobile payments, security features, and their capacity to make effective financial choices via digital means. The answers were ranked on a Likert scale between "strongly agree" and "strongly disagree."

Digital Financial Inclusion: This section was employed to measure to what extent participants were engaged with digital financial systems. The most significant indicators included the engagement in digital financial services (e.g., mobile banking, online payments, digital lending platforms) and whether they were exposed to these services.

Demographic Information: Basic demographic data like age, gender, income level, and education level were also gathered to explore their possible contribution to digital financial literacy and inclusion.

The survey was pre-tested in a pilot sample of 30 participants to establish the clarity of questions and measure reliability. Small changes were made from the feedback to enhance the readability and clarity of questions.

Data collection was carried out using web-based questionnaires, enabling rapid distribution and ease to the respondents. The internet platform was used to facilitate a broad reach and ease for working professionals. Data collection was conducted within a month, and reminders were sent to achieve a high response rate.

Data Analysis

For guaranteeing validity and reliability of information, responses were examined using various statistical tools. Data analysis procedure was carried out using SPSS (Statistical Package for the Social Sciences) and Smart PLS (Partial Least Squares Structural Equation Modeling) since both are efficient to perform basic as well as advanced statistical analysis.

Descriptive Analysis: Descriptive statistics, including frequencies, means, and standard deviations, were computed to provide a comprehensive overview of the sample characteristics and the general trends in digital financial literacy and inclusion.

Reliability Analysis: To determine the internal consistency of the survey tool, Cronbach's alpha was computed for every scale in the questionnaire. A Cronbach's alpha of more than 0.7 was deemed acceptable, which meant that the tool had high reliability.

Parametric Tests: To establish the associations among the variables, parametric tests like the ANOVA were conducted. These analyzed whether statistically significant differences in digital financial inclusion and literacy existed with respect to demographic variables like education level, income, and age.

Correlation Analysis: Pearson correlation coefficient was applied in investigating the magnitude and direction of the association between digital financial literacy and digital financial inclusion. It was pivotal in uncovering the impact that digital financial literacy has on the level of involvement individuals have with digital financial systems.

Structural Equation Modeling (SEM): To test the model and evaluate the interdependencies between the variables and test the model as proposed, Smart PLS was employed to do Partial Least Squares Structural Equation Modeling (PLS-SEM). This method is well-suited for small sample exploratory research and appropriate for modeling latent and observed variable relationships. The PLS-SEM method enabled the testing of direct and indirect effects between digital financial inclusion and digital financial literacy, considering the different factors that may affect these constructs.

Ethical Considerations

Ethical considerations were a built-in component of the research process. Before collecting data, each participant was provided with an informed consent form stating the intent of the study, the voluntary aspect of participation, and confidentiality promises. Participants were notified that responses would be kept anonymous and employed solely for the purpose of conducting research.

Additionally, participants were free to withdraw from participation in the study at any time without penalty. Data were kept safely and accessed by authorized persons only for analysis.

Limitations of the Study

Even as this research captures greater insight in the connection of digital financial literacy and inclusion, it is limited. Convenience sampling reduces the external validity of results to the salaried labor population at large. Additionally, the research involves self-report measures that are liable to social desirability or recall bias. Future research may seek to review these connections according to longitudinal information or experimental data to better evaluate causality.

6 Results and Discussion

This section speaks and reveals the main findings of the study conducted to scrutinize the effect of Digital Financial Literacy (DFL) on Digital Financial Inclusion (DFI) among the working population. Analysis is keen to investigate the degree to which digital financial products are embraced and utilized effectively, and to what degree social and family influence (FSI) plays a role in promoting digital financial literacy. The argument draws attention to substantial trends of age, education and social environment in the formation of digital financial attitudes, providing an outlook on the larger implications for digital financial empowerment and inclusion (Table 1).

Cronbach's Alpha of 0.948 indicates high internal consistency for measuring the set of items measuring 44. The high value in this case suggests that the items used for the assessment or survey are highly reliable and will most likely provide consistent results, reflecting good reliability in the measurement of the desired construct.

The findings of the study on Digital Financial Literacy (DFL) and Digital Financial Inclusion (DFI) among employed individuals. The outcomes point to major trends in the adoption of digital financial instruments, the role of family and social networks, and the effect of education and age on digital financial conduct. By examining survey data, the conversation touches on how social support, digital literacy, and access factors come together to bring about the use and proper adoption of digital financial services, which are useful considerations for policymakers, educators, and financial institutions interested in increasing financial inclusion (Table 2).

The statistics reveal the break-down of people by age group. The highest percentage (34%) is aged 55–64, and then 20% for both 35–44 and 45–54 age brackets. The 25–34 age bracket covers 17%, and the under 25 category covers the

Table 1 Reliability statistics for the survey instrument

Reliability statistics	
Cronbach's alpha	N of items
0.948	44

Table 2 Distribution of respondents by age group

Age group			
		Frequency	Percent
Valid	Under 25	9	9.0
	25–34	17	17.0
	35–44	20	20.0
	45–54	20	20.0
	55–64	34	34.0
	Total	100	100.0

lowest percentage at 9%. The statistics are generally spread evenly across the middle to older age brackets, with a clear drop-off in the younger age brackets (Table 3).

The data reveals that the highest level of education completed by most individuals is either a postgraduate degree (31%) or a doctoral degree (29%). High school graduates make up 18%, while 13% have completed an undergraduate degree. The smallest group (9%) consists of individuals with no formal education. This suggests that the majority of the population has attained higher education, with a significant concentration in advanced degrees (Table 4).

The crosstabulation shows the distribution of education levels across different age groups. The majority of individuals in the 55–64 age group (34%) have completed postgraduate or doctoral degrees, with a higher concentration of higher education in older age groups. Conversely, younger adults (Under 25 and 25–34 age groups) have fewer education levels, with greater numbers in high school and bachelor's degrees. Generally, as age goes up, the highest education level attained also increases, mirroring the pattern of higher educational attainment among older populations (Table 5).

The statistics show the means, medians, standard deviations, and Cramér-von Mises p-values for different variables, all reflecting high consistency between the measurements, as reflected by the p-values of 0, which imply statistically significant outcomes. The mean values for the majority of items, from 3.53 to 4.35, indicate a positive response trend overall across categories. The fact that the medians are close to the means also confirms this consistency. Standard deviations are fairly

Table 3 Distribution of respondents by highest level of education completed

Highest level of education completed			
		Frequency	Percent
Valid	No formal education	9	9.0
	High school	18	18.0
	Undergraduate degree	13	13.0
	Postgraduate degree	31	31.0
	Doctoral degree	29	29.0
	Total	100	100.0

Table 4 Crosstabulation of age group and highest level of education completed

Age group * highest level of education completed crosstabulation							
		Highest level of education completed					Total
		No formal education	High school	Undergraduate degree	Postgraduate degree	Doctoral degree	
Age group	Under 25	2	2	3	1	1	9
	25–34	1	3	1	6	6	17
	35–44	1	5	1	6	7	20
	45–54	3	3	3	5	6	20
	55–64	2	5	5	13	9	34
Total		9	18	13	31	29	100

Table 5 Descriptive statistics and cramér-von mises *p*-value for survey variables

Name	Mean	Median	Standard deviation	Cramér-von Mises *p* value
Age	3.53	4	1.345	0
Education	3.53	4	1.315	0
DFL1_H1_IV	3.58	4	1.226	0
DFL2	4.21	5	1.08	0
DFL3	4.23	5	1.076	0
DFL4	4.26	5	0.986	0
DFL5	4.26	5	1.006	0
DFL6	4.23	5	0.988	0
DFL7	4.02	4	1.233	0
DFL8	4.05	4	1.152	0
DFL9	4	4	1.086	0
DFL10	4.24	5	1.001	0
DFL11	4.17	4	1.049	0
DFL12	4.2	4	0.938	0
DFI1_H1_DV	4.05	4	1.178	0
DFI2	4.18	5	1.004	0
DFI3	4.17	4	1.02	0
DFI4	4.24	5	1.001	0
DFI5	4.14	4	1.068	0
DFI6	4.35	5	0.921	0
DFI7	4.21	5	1.08	0
DFI8	4.23	5	1.076	0

(continued)

Table 5 (continued)

Name	Mean	Median	Standard deviation	Cramér-von Mises p value
FSI1_H2_IV	4.26	5	0.986	0
FSI2	4.26	5	1.006	0
FSI3	4.23	5	0.988	0
FSI4	4.02	4	1.233	0
FSI5	4.05	4	1.152	0
FSI6	4	4	1.086	0
FSI7	4.24	5	1.001	0
DFL1_MV_H2	4.17	4	1.049	0
DFL_2	4.2	4	0.938	0
DFL_3	4.05	4	1.178	0
DFL_4	4.18	5	1.004	0
DFL_5	4.24	5	1.001	0
DFL_6	4.23	5	0.988	0
DFL_7	4.02	4	1.233	0
DFI1_H2_DV	4.05	4	1.152	0
DFI_2	4	4	1.086	0
DFI_3	4.24	5	1.001	0
DFI_4	4.17	4	1.049	0
DFI_5	4.2	4	0.938	0
DFI_6	4.05	4	1.178	0
DFI_7	4.18	5	1.004	0
DFI_8	4.24	5	1.001	0

small, usually 1 or less, showing that the answers are close together with the mean, showing little variability in the data. It is an indication that respondents feel similarly on most of these variables, with slight variability seen in some items. The data shows participants uniformly score the different factors, with high agreement between the means and medians, and minimal variability in answers (Fig. 3).

Hypothesis 1 Independent Variable (IV): Digital Financial Literacy (DFL).

Dependent Variable (DV): Digital Financial Inclusion (DFI).

The neural net model wherein various input variables, denoted as DFL1 through to DFL12, and other such specific inputs such as DFL_H1_IV and DFI_H1_DV, are all linked to two important output nodes: DFL and DFI. The figures next to the links are the weights or impacts of these inputs on the respective outputs. Notably, the input DFL7 has the highest weight (0.456) towards the DFL output, while the DFI output shows a prominent weight of 0.828 from the DFL node. This suggests a more significant influence of the DFL on the DFI output, with many other connections

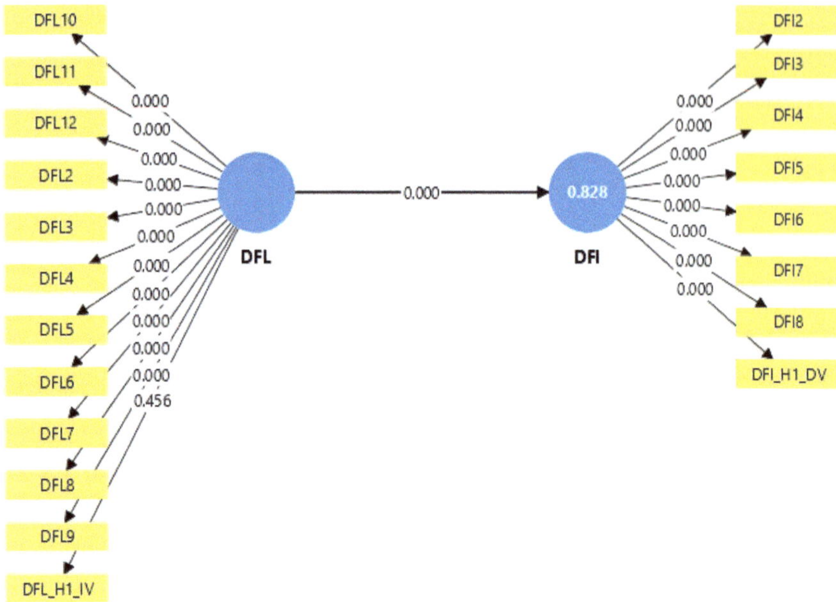

Fig. 3 Structural relationship between digital financial literacy (DFL) and digital financial inclusion (DFI) (*Source Smart PLS*)

having negligible influence, as indicated by the near-zero weights. The diagram indicates that some inputs have little to no impact on the outputs, suggesting that the model is possibly well-calibrated to focus on the most relevant inputs (Table 6).

Path coefficients

The path coefficient between DFL and DFI is 0.910, indicating a strong positive relationship between these two variables. The sample mean (M) is 0.915, suggesting consistency in the estimated effect across the sample. The standard deviation (STDEV) of 0.023 indicates low variability in the path coefficient across samples, demonstrating the stability of the relationship. The T-statistic of 39.197 is significantly large, well above the typical threshold (e.g., 1.96 for a 95% confidence level), confirming that the effect of DFL on DFI is statistically significant. The P-value of 0.000 further supports this conclusion, indicating that the observed relationship is highly unlikely to be due to chance (Table 7).

Table 6 Hypothesis testing for the relationship between DFL and DFI

	Original sample (O)	Sample mean (M)	Standard deviation (STDEV)	T statistics (|O/STDEV|)	P values
DFL -> DFI	0.910	0.915	0.023	39.197	0.000

Table 7 T-Test results for DFL to DFI relationship

	Original sample (O)	Sample mean (M)	Standard deviation (STDEV)	T statistics (|O/STDEV|)	P values
DFL -> DFI	0.910	0.915	0.023	39.197	0.000

Table 8 Outer loadings for DFL and DFI constructs

	Outer loadings
DFI2 <- DFI	0.861
DFI3 <- DFI	0.751
DFI4 <- DFI	0.802
DFI5 <- DFI	0.738
DFI6 <- DFI	0.733
DFI7 <- DFI	0.785
DFI8 <- DFI	0.820
DFI1 <- DFI	0.852
DFL10 <- DFL	0.798
DFL11 <- DFL	0.808
DFL12 <- DFL	0.807
DFL2 <- DFL	0.832
DFL3 <- DFL	0.846
DFL4 <- DFL	0.824
DFL5 <- DFL	0.803
DFL6 <- DFL	0.852
DFL7 <- DFL	0.843
DFL8 <- DFL	0.865
DFL9 <- DFL	0.817
DFL1 <- DFL	0.098

Total effects

The overall effect of DFL on DFI is 0.910, which indicates that DFL has a highly positive effect on DFI. The sample mean of 0.915 implies that the effect is uniform throughout the sample with a low standard deviation of 0.023, which means that there is little variation in the size of the effect. The T-statistic of 39.197 is extremely high, which implies that the effect is highly significant and very reliable. P-value of 0.000 asserts that the noted effect is not random, pointing towards the strength and significance of the interaction between DFI and DFL in the model (Table 8).

Outer loadings

The loadings of the outer loadings for the DFI and DFL variables demonstrate the strength of the relationship between each indicator and its corresponding latent construct. The DFI indicators (DFI1 to DFI8) have strong positive loadings with values ranging from 0.733 to 0.861, which confirm that these indicators are highly appropriate in measuring digital financial inclusion. Particularly, DFI2, DFI1, and DFI7 have the maximum loadings and thus are hypothesized to contribute most to digital financial inclusion as a construct. Likewise, indicators of DFL (DFL1 to DFL12) have predominantly good positive loadings ranging from 0.098 to 0.865. The majority of the DFL indicators have very high loadings, such as DFL8 (0.865), DFL3 (0.846), and DFL6 (0.852), pointing toward their prominent part in capturing digital financial literacy. Yet DFL1 has a quite low loading of 0.098, significantly lower than the rest and indicating this indicator is possibly not playing an important role in the construct of digital financial literacy within the model. Its low loading might need further examination to ascertain its relevance or the necessity for its removal. The findings show that the majority of the indicators are good predictors of their respective constructs except for DFL1, which might need more focus (Table 9).

Oneway

The ANOVA for DFL shows that there are statistically significant differences between age categories since the p-value of 0.029 is less than the usual significance level of 0.05. The result implies that age significantly influences digital financial literacy (DFL). The F value of 2.817 shows a moderate degree of variation between the groups, and the p-value shows that this variation is not likely to be due to chance. Thus, the findings suggest that digital financial literacy differs by age groups, and age is a significant factor determining digital financial literacy in this dataset (Table 10).

Table 9 One-way ANOVA results for digital financial literacy (DFL) across age groups

ANOVA between age						
		Sum of squares	df	Mean Square	F	Sig.
DFL	Between groups	15.942	4	3.986	2.817	0.029
	Within groups	134.418	95	1.415		
	Total	150.360	99			

Table 10 One-way ANOVA results for digital financial inclusion (DFI) across age groups

ANOVA between age						
		Sum of squares	df	Mean square	F	Sig.
DFI	Between groups	8.459	4	2.115	1.542	0.256
	Within groups	130.291	95	1.371		
	Total	138.750	99			

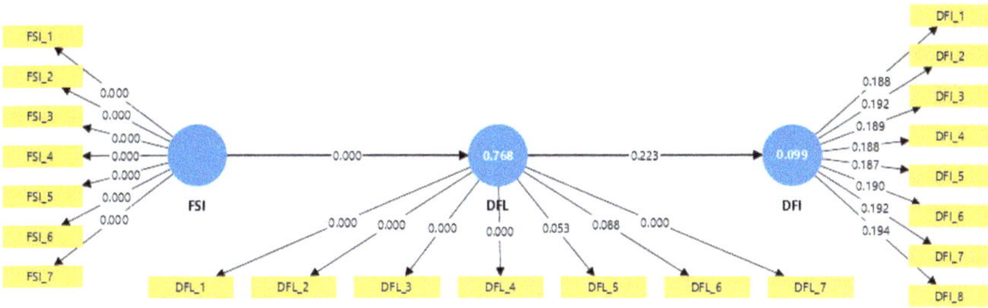

Fig. 4 Structural model of FSI, DFL, and DFI relationships (*Source Smart PLS*)

ANOVA result of DFI reveals that differences are not statistically significant between the different age groups, since p-value of 0.256 is higher than the typical value of 0.05 that serves as significance level. Thus, age cannot significantly affect digital financial inclusion (DFI) based on ANOVA findings. In spite of some variation in the data, as reflected in the F value of 1.542, the p-value suggests that the differences are not significant and may be due to random chance. Hence, age does not seem to have any significant impact on digital financial inclusion in this data (Fig. 4).

Hypothesis 2 Independent Variable (IV): Family and Social Influences (FSI).

Mediator Variable (MV): Digital Financial Literacy (DFL).

Dependent Variable (DV): Digital Financial Inclusion (DFI).

A relationship between Digital Financial Literacy (DFL) and Digital Financial Inclusion (DFI), reveals some inconsistencies that need clarification. As shown in the model, the direct path coefficient from DFL to DFI is 0.223, indicating a positive relationship. However, the total effect values, which are not fully visible in the diagram, have been noted to show negative values, contradicting the conclusion that there is a strong positive relationship (Table 11).

Total effects

The analysis of the model reveals that among the proposed hypotheses, only the relationship between Financial Social Inclusion (FSI) and Digital Financial Literacy (DFL) demonstrates a strong and positive total effect. Specifically, the total effect of FSI on DFL is 0.876, indicating a substantial and meaningful positive relationship. This suggests that as individuals experience higher levels of social inclusion in financial contexts, their level of digital financial literacy is likely to improve significantly.

Table 11 Total effects of financial self-interest (FSI) and digital financial literacy (DFL) on digital financial inclusion (DFI)

	Total effects
DFL -> DFI	−0.315
FSI -> DFI	−0.276
FSI -> DFL	0.878

Table 12 Outer loadings of indicators for financial self-interest (FSI), digital financial inclusion (DFI), and digital financial literacy (DFL) constructs

	Outer loadings
FSI1 <- FSI	0.821
FSI 2 <- FSI	0.867
FSI3 <- FSI	0.878
FSI4 <- FSI	0.878
FSI5 <- FSI	0.821
FSI6 <- FSI	0.801
FSI7 <- FSI	0.836
DFI_1 <- DFI	0.795
DFI_2 <- DFI	0.879
DFI_3 <- DFI	0.864
DFI_4 <- DFI	0.842
DFI_5 <- DFI	0.859
DFI_6 <- DFI	0.833
DFI_7 <- DFI	0.879
DFI_8 <- DFI	0.841
DFL_1 <- DFL	0.842
DFL_2 <- DFL	0.856
DFL_3 <- DFL	0.844
DFL_4 <- DFL	0.836
DFL_5 <- DFL	−0.307
DFL_6 <- DFL	−0.250
DFL7 <- DFL	0.861

Based on this strong positive effect, and assuming statistical significance, Hypothesis (FSI → DFL) is supported and considered accepted (Table 12).

Outer Loading

The structural model results indicate that among the hypothesized relationships, only the path from Financial Social Inclusion (FSI) to Digital Financial Literacy (DFL) demonstrates a strong and positive total effect. Specifically, the total effect of FSI on DFL is 0.876, confirming a substantial positive relationship. This suggests that higher levels of financial social inclusion contribute significantly to improving individuals' digital financial literacy. The outer loadings for the FSI construct range from 0.801 to 0.878, and for the DFL construct (excluding DFL_5 and DFL_6, which have negative loadings), the loadings mostly exceed 0.83, indicating strong reliability and good measurement validity for the respective indicators. Based on this evidence, Hypothesis 2 (FSI → DFL) is supported and considered accepted (Table 13).

The ANOVA results for FSI show that there are no significant differences between the age groups, as indicated by the p-value of 0.243, which is greater than the standard significance threshold of 0.05. This suggests that age does not significantly affect the

Table 13 One-way ANOVA results for financial self-interest (FSI) across age groups

ANOVA between age						
		Sum of squares	df	Mean square	F	Sig.
FSI	Between groups	5.376	4	1.344	1.390	0.243
	Within groups	91.864	95	0.967		
	Total	97.240	99			

family and social influences (FSI) variable. Although the F value (1.390) suggests some variation between the groups, the p-value indicates that these differences are not statistically significant, and any observed variations are likely due to random chance rather than a true effect of age on FSI (Table 14).

The ANOVA for DFL shows no significant variations in age groups, since the p-value of 0.234 is larger than the usual significance level of 0.05. This means that age has no statistically significant effect on the digital financial literacy (DFL) variable. Even though the F value of 2.807 shows that there is some variation across the groups, the p-value reveals that the differences seen are not statistically significant. Hence, any difference in digital financial literacy by age would be due to chance rather than an actual association between age and DFL (Table 15).

ANOVA for DFI: There are no significant variations across the age groups in terms of DFI, with a p-value of 0.690, which is much higher than the standard significance level of 0.05. This implies that age has no statistically significant impact on DFI. The F value of 0.563 also confirms this, as it shows little variation between the groups. Thus, the differences observed between age groups for DFI are not statistically significant, suggesting that age is not a factor influencing digital financial inclusion in this sample.

Table 14 One-way ANOVA results for digital financial literacy (DFL) across age groups

ANOVA between age						
		Sum of squares	df	Mean square	F	Sig.
DFL	Between groups	7.785	4	1.946	2.807	0.234
	Within groups	102.325	95	1.077		
	Total	110.110	99			

Table 15 One-way ANOVA results for digital financial inclusion (DFI) across age groups

ANOVA between age						
		Sum of squares	df	Mean square	F	Sig.
DFI	Between groups	3.076	4	0.769	0.563	0.690
	Within groups	129.674	95	1.365		
	Total	132.750	99			

This study reveals a strong and positive relationship between digital financial literacy (DFL) and digital financial inclusion (DFI) among salaried individuals, highlighting the pivotal role of family and social networks in facilitating this connection. These findings align with previous research emphasizing that beyond mere access to digital financial tools, sustainable inclusion requires individuals to be empowered with the knowledge, skills, and confidence to use these tools responsibly and effectively.

From a sustainability standpoint, digital financial literacy acts as a critical enabler of economic resilience and social equity. When individuals are digitally literate, they are better equipped to engage in financial behaviors that promote long-term stability—such as prudent saving, informed investing, and avoidance of exploitative credit—thereby reducing vulnerability to economic shocks and fostering inclusive economic growth. This responsible financial participation contributes to sustainable development by helping bridge economic disparities and promoting equitable access to resources.

The influence of family and peer groups underscores the social dimension of sustainability. Social support networks enhance digital financial literacy, encouraging knowledge sharing and fostering a culture of safe and ethical use of digital financial services. These community-based influences help embed sustainable financial habits and reinforce trust in digital platforms, which is essential for maintaining the integrity and longevity of digital financial ecosystems. Policy implications arising from this discussion suggest that sustainable digital financial inclusion efforts must go beyond infrastructure and technology deployment. There is a critical need for integrated approaches that combine digital literacy education with community engagement, inclusive policies, and security awareness. Such approaches will support the creation of resilient financial systems that empower all demographic segments—including younger and less-educated populations—thus aligning digital financial inclusion with broader sustainable development goals.

7 Policy Implications

Promote Digital Financial Literacy Across All Age Groups: The statistics indicate that older age groups are more educated and are more aware of digital financial instruments. Policymakers need to concentrate on increasing digital financial literacy programs for young people and lower-educated groups. This can narrow the knowledge gap and promote wider adoption of digital financial instruments among various age groups and educational levels. Targeted campaigns and educational materials could aid in this process.

Encourage Family and Peer Influence in Financial Decision-Making: The role played by peers and family in confidence and the use of digital money tools indicates the role of family-targeted and community-targeted programs. Families and social networks must be supported by policies to discuss digital financial services openly with the aim of increasing the adoption rates as a whole. Banks and governments can collaborate to develop policies that utilize familial and social influence to increase financial inclusion.

Accessibility and Inclusivity of Digital Financial Services: The overall sentiment that digital financial services are available to all income groups lays a strong base to facilitate financial inclusion. But the policies must further emphasize the underrepresented and the underserved so that digital financial platforms can reach everybody, be they of any income group or geographical area. This may involve enhancing infrastructure, lowering costs, and making mobile available to rural or economically backward classes.

Encouraging Safe Digital Financial Practices: With such vast faith in utilising digital finance platforms, policymakers should be concerned with making users aware of the security parameters necessary to safeguard them against cyber threats. Public awareness campaigns for digital security, like encryption and two-factor authentication, need to be highlighted to combat fraud and generate trust in digital platforms.

Leveraging social media and Peer Influence for Financial Education: The data demonstrate that social media and peer influence are significant drivers in the adoption of digital financial instruments. Policymakers could look into researching collaborations with social media influencers and sites to spread sound financial information and instruments. Promoting the use of social media for financial literacy and awareness could speed the use of safe digital financial behavior among young and technology-loving individuals.

Support for Digital Financial Tools Adoption: The positive effect of digital financial tools on the financial management of users shows that the platforms are efficient in assisting people in managing their finances. Policymakers should promote the use and creation of such tools, particularly for savings and investment. Offering incentives or subsidies for the uptake of these platforms, particularly among lower-income and less digitally advanced communities, can boost their usage and enhance financial security.

Fostering Peer and Family Encouragement: Because family and peer support exert such a critical influence on whether and how individuals utilize digital finance tools, such policies should advocate community-based interventions promoting sharing learning on digital finance. This could take the forms of workshops, training, and support groups meant to create climates where people would be more open to embracing and using digital financial tools.

8 Conclusion

The findings of this study affirm that digital financial tools—such as mobile banking, e-wallets, and online platforms—are increasingly integrated into the financial habits of salaried individuals, particularly within the Delhi NCR region. The findings of this study affirm that digital financial literacy is foundational to achieving sustainable digital financial inclusion among salaried professionals. By equipping individuals with the competencies to confidently and securely navigate digital financial services, DFL fosters economic empowerment that is inclusive, equitable, and enduring. The significant role played by family and social networks further highlights that sustainable financial inclusion is not solely a technological challenge but also a social one, requiring community-driven support and education.

Incorporating sustainability into digital financial literacy initiatives ensures that inclusion efforts contribute to reducing inequality, building economic resilience, and promoting responsible financial behaviors that support long-term societal well-being. This approach aligns with global sustainable development priorities, including the United Nations Sustainable Development Goals, particularly those targeting economic growth, reduced inequality, and strong institutions.

For digital financial inclusion to be truly sustainable, policies and programs must address the educational, social, and security dimensions of digital finance. Empowering individuals through continuous learning and community support will not only increase adoption rates but also ensure that digital financial systems are trusted, ethical, and resilient.

The empirical evidence supports the hypothesis that DFL significantly and positively influences DFI. In addition, the analysis confirms that family and social support have a meaningful impact on enhancing DFL, indicating that community-based interventions can be pivotal in building a digitally literate and financially included population. These relationships are especially crucial in the context of sustainable development. Digital financial literacy, when equitably distributed and inclusively promoted, becomes a powerful enabler of economic sustainability, reducing inequalities, encouraging responsible financial behavior, and fostering economic resilience across diverse income and education groups.

From a sustainability perspective, the digital divide is not merely a technological issue—it is a socio-economic challenge. Bridging this divide by enhancing DFL, especially among younger and less-educated segments, serves both as a strategy for economic empowerment and as a commitment to sustainable, inclusive growth. In particular, the support of peers, families, and educational institutions becomes essential in closing knowledge gaps, improving decision-making capabilities, and increasing trust in digital systems. This holistic approach promotes long-term financial well-being and resilience, aligning closely with the United Nations Sustainable Development Goals (SDGs), particularly those related to reducing inequality, building strong institutions, and promoting inclusive economic growth.

Another key implication is the need for safe and secure engagement with digital financial platforms. Participants reported greater financial control, increased savings and investment capabilities, and higher satisfaction when using digital tools. However, true sustainability in digital finance demands more than access—it requires robust knowledge of security practices, awareness of rights, and confidence in navigating digital environments. Without this, digital financial inclusion may remain superficial and unsustainable.

Ultimately, this study reinforces the idea that digital financial literacy is a cornerstone of sustainable digital financial inclusion. It goes beyond individual competence to impact systemic change—empowering individuals, supporting families, and contributing to a more inclusive and resilient financial system. Policymakers, educators, and financial service providers must continue to develop comprehensive strategies that integrate DFL into broader digital and economic inclusion efforts. Future research should further explore the intersection of DFL, sustainability, and social equity to inform programs that are not only inclusive but enduring and transformative.

References

Abdurrahman A, Nugroho DA (2024) The role of digital financial literacy on financial well-being with financial technology, financial confidence, financial behavior as intervening and sociodemography as moderation. J Ekon Dan Bisnis 27(2):191–220. https://doi.org/10.24914/jeb.v27i2.11891

Adesola Oluwatosin Adelaja, Stanley Chidozie Umeorah, Bibitayo Ebunlomo Abikoye, Michelle Chibogu Nezianya (2024) Advancing financial inclusion through fintech: Solutions for unbanked and underbanked populations. World J Adv Res Rev, 23(2), 427–438. https://doi.org/10.30574/wjarr.2024.23.2.2379

Al-Majali AA, Al-Oshaibat SD, Al-Sarayreh AA, Al-Manaseer SR (2024) The effect of digital financial literacy on financial development and governance: Using panel vector autoregressive model. J GovAnce Regul, 13(2, special issue), 465–473. https://doi.org/10.22495/jgrv13i2siart21

Al-shami SA, Damayanti R, Adil H, Farhi F, Al Mamun A (2024) Financial and digital financial literacy through social media use towards financial inclusion among batik small enterprises in Indonesia. Heliyon 10(15):e34902. https://doi.org/10.1016/j.heliyon.2024.e34902

Bach A, Shaffer G, Wolfson T (2013) Digital human capital: developing a framework for understanding the economic impact of digital exclusion in low-income communities. J Inf Policy 3:247–266. https://doi.org/10.5325/jinfopoli.3.2013.0247

Bahia K, Muthiora B (2019) The mobile money regulatory index. GSMA Intelligence, London

Bushra Mir MA (2024) Digital financial literacy and financial well-being: In Singh D, Bansal R, Gupta S, Ansari Y (Eds), Adv Financ, Account, Econ (pp. 57–73). IGI Global. https://doi.org/10.4018/979-8-3693-1750-1.ch003

Chhillar N, Arora S (2022) Personal financial management behavior using digital platforms and its domains. J Financ Manag, Mark Insts 10(02):2250009. https://doi.org/10.1142/S2282717X22500098

Chong KF, Sabri MF, Magli AS, Abd Rahim H, Mokhtar N, Othman MA (2021) The effects of financial literacy, self-efficacy and self-coping on financial behavior of emerging adults. J Asian Financ, Econ Bus, 8(3), 905–915. https://doi.org/10.13106/JAFEB.2021.VOL8.NO3.0905

Davis FD (1989) Technology acceptance model: TAM. In: Al-Suqri MN, Al-Aufi AS (eds) Information seeking behavior and technology adoption, vol 5. pp 205–219

Desy Wulan Ayuning Gumilar, Khresna Bayu Sangka, Salman Alfarisy Totalia (2024) Digital financial literacy and digital financial inclusion in the era of digital disruption: systematic literature review. Formosa J Multidiscip Res, 3(5), 1563–1576. https://doi.org/10.55927/fjmr.v3i5.9213

Gangani DJ, Raval DD (n.d.) A study on level of digital financial inclusion among the urban youth of Gujarat

Hastings JS, Madrian BC, Skimmyhorn WL (2013) Financial literacy, financial education, and economic outcomes. Annu Rev Econ 5(1):347–373. https://doi.org/10.1146/annurev-economics-082312-125807

Ibrahim RK, Aldawsari AN (2023) Relationship between digital capabilities and academic performance: The mediating effect of self-efficacy. BMC Nurs 22(1):434. https://doi.org/10.1186/s12912-023-01593-2

Kim M, Zoo H, Lee H, Kang J (2018) Mobile financial services, financial inclusion, and development: A systematic review of academic literature. Electron J Inf Syst Dev Ctries 84(5):e12044. https://doi.org/10.1002/isd2.12044

Koskelainen T, Kalmi P, Scornavacca E, Vartiainen T (2023) Financial literacy in the digital age—A research agenda. J Consum Aff 57(1):507–528. https://doi.org/10.1111/joca.12510

Kumar G, Murty A, Ratna Dr R, Ranjan Dr A (2024) Impact of digital financial literacy on financial inclusion—the role fintech services. SSRN Electron J. https://doi.org/10.2139/ssrn.4954800

Kumar I (2024) Banking services and financial inclusion in India's poorest regions. J Bank Regul 25(2):145–159. https://doi.org/10.1057/s41261-023-00224-9

Li Q, Liu Q (2023) Impact of digital financial inclusion on residents' income and income structure. Sustain 15(3):2196. https://doi.org/10.3390/su15032196

Liu S, Gao L, Latif K, Dar AA, Zia-UR-Rehman M, Baig SA (2021) The behavioral role of digital economy adaptation in sustainable financial literacy and financial inclusion. Front Psychol 12:742118. https://doi.org/10.3389/fpsyg.2021.742118

Lu X, Guo J, Zhou H (2021) Digital financial inclusion development, investment diversification, and household extreme portfolio risk. Account & Financ 61(5):6225–6261. https://doi.org/10.1111/acfi.12863

Lumpkin S, Schich S (2019) Banks, digital banking initiatives and the financial safety net: theory and analytical framework. J Econ Sci Res, 3(1). https://doi.org/10.30564/jesr.v3i1.1113

Lyons AC, Kass-Hanna J (2021) A methodological overview to defining and measuring "digital" financial literacy. Financ Plan Rev 4(2):e1113. https://doi.org/10.1002/cfp2.1113

Maji SK, Laha A (2023) Role of financial and digital literacy in determining digital transaction behaviour: Evidence from student level survey in West Bengal (India). Int J Bus Environ 14(2):183. https://doi.org/10.1504/IJBE.2023.129917

Martin KD, Hill RP (2015) Saving and Well-being at the base of the pyramid: implications for transformative financial services delivery. J Serv Res 18(3):405–421. https://doi.org/10.1177/1094670514563496

Mbatane S, Kekana K (2024) The role of digital financial literacy in the use of financial technology products and services for university students. SSRN. https://doi.org/10.2139/ssrn.4974284

Mishra D, Agarwal N, Sharahiley S, Kandpal V (2024) Digital financial literacy and its impact on financial decision-making of women: evidence from India. J Risk Financ Manag 17(10):468. https://doi.org/10.3390/jrfm17100468

Mmari PJ (n.d.) Financial Literacy as a determinant of financial inclusion in Tanzania

Moenninghoff SC, Wieandt A (2013) The future of peer-to-peer finance. Schmalenbachs Z Für BetrSwirtschaftliche Forsch 65(5):466–487. https://doi.org/10.1007/BF03372882

Mohakud M, Biswal SK (2024) An investigation into the Digital Financial Literacy of employed individuals in rural India: A case study focusing on Mayurbhanj district. In Review. https://doi.org/10.21203/rs.3.rs-3887682/v1

Mpofu FY (2023) Digital financial inclusion and digital financial literacy in Africa: the challenges connected with digital financial inclusion in Africa. In Mhlanga D, Ndhlovu E (Eds), Econ Incl Post-Indep Afr (pp. 123–147). Springer Nature Switzerland. https://doi.org/10.1007/978-3-031-31431-5_7

Mpofu FY (2024) Industry 4.0 in finance, digital financial services and digital financial inclusion in developing countries: opportunities, challenges, and possible policy responses. Int J Econ Financ Issues, 14(2), 120–135. https://doi.org/10.32479/ijefi.15081

Mpofu FY, Mhlanga D (2022) Digital financial inclusion, digital financial services tax and financial inclusion in the fourth industrial revolution Era in Africa. Econ 10(8):184. https://doi.org/10.3390/economies10080184

Nandru P, Chendragiri M, Velayutham A (2021) Determinants of digital financial inclusion in India: Evidence from the World Bank's global findex database. In Review. https://doi.org/10.21203/rs.3.rs-329541/v1

OECD (Organisation for Economic Co-operation and Development) (2022) Advancing financial inclusion through digital financial literacy: policy guidance and considerations. OECD. https://doi.org/10.1787/9c6ccf6b-en

OECD (2023) OECD/INFE 2023 international survey of adult financial literacy (OECD Business and Finance Policy Papers, No. 39). OECD Publishing. https://doi.org/10.1787/56003a32-en

Ozili PK (2018) Impact of digital finance on financial inclusion and stability. Borsa Istanbul Rev 18(4):329–340. https://doi.org/10.1016/j.bir.2017.12.003

Ozili PK (2020) Theories of financial inclusion. In: Uncertainty and challenges in contemporary economic behaviour. Emerald Publishing Limited, pp 89–115

Ozili PK (2021) Covid-19 pandemic and economic crisis: the Nigerian experience and structural causes. J Econom Adm Sci 37(4):401–418

Reserve Bank of India (2023) Annual report 2022–23. RBI

Sadaf I, Hussain B (2025) Exploring energy consumption trends: insights from developed and developing nations. Global Environ Change Adv 5:100020

Salem SS (2023) Financial Literacy role in achieving Financial Inclusion; Evidence from Egypt. *المجلة العلمية للبحوث والدراسات التجارية*,37(4), 1439–1471. https://doi.org/10.21608/sjrbs.2023.232694.1524

Santhanakrishnan DD (n.d.) Digital financial literacy

Showkat M, Nagina R, Baba MA, Yahya AT (2025) The impact of financial literacy on women's economic empowerment: Exploring the mediating role of digital financial services. Cogent Econ & Financ 13(1):2440444. https://doi.org/10.1080/23322039.2024.2440444

Smart PLS. (n.d.). [Computer software]. https://www.smartpls.com/

Souza TLD (n.d.) The perceived relationship between training programs and employee performance at fintech companies in estonia

Tay L-Y, Tai H-T, Tan G-S (2022) Digital financial inclusion: A gateway to sustainable development. Heliyon 8(6):e09766. https://doi.org/10.1016/j.heliyon.2022.e09766

Tony N, Desai K (2020) Impact of digital financial literacy on digital financial inclusion. 9(01)

Uthaileang W, Kiattisin S (2023) Developing the capability of digital financial literacy in developing countries: A Case of online loan for small entrepreneurs. Heliyon 9(12):e21961. https://doi.org/10.1016/j.heliyon.2023.e21961

Vibhuti Shivam Dube, Pandey P, Asthana PK, Abdullah H (2023). Digital financial literacy: an empirical study on millennials in India. https://doi.org/10.5281/ZENODO.8253419

World Bank (2022) The Global Findex Database 2021: financial inclusion, digital payments, and resilience in the age of COVID-19. World Bank. https://doi.org/10.1596/978-1-4648-1897-4

World Bank (2023) Financial inclusion: lessons from World Bank Group experience, fiscal years 2014–22 (Independent Evaluation Group). World Bank, Washington, DC

Xiao JJ, O'Neill B (2022) Disability type, financial capability, and risky asset holding. J Disabil Policy Stud 32(4):269–279

Xu M, Shi L, Zhao J, Zhang Y, Lei T, Shen Y (2025) Achieving agricultural sustainability: Analyzing the impact of digital financial inclusion on agricultural green total factor productivity. Front Sustain Food Syst 8:1515207. https://doi.org/10.3389/fsufs.2024.1515207

Yadav M, Banerji P (2023) A bibliometric analysis of digital financial literacy. Am J Bus 38(3):91–111

Yadav M, Banerji P (2024) Systematic literature review on digital financial literacy. SN Bus Econ 4(11):142

Yadav M, Banerji P, Garg A (2025) Digital financial literacy and retirement planning with mediating effect of saving behaviour. Am J Bus 40(3)

Zaimovic A, Meskovic MN, Dedovic L, Arnaut-Berilo A, Zaimovic T, Torlakovic A (2024) Measuring digital financial literacy. Procedia Comput Sci 236:574–581. https://doi.org/10.1016/j.procs.2024.05.068

Zaimovic A, Omanovic A, Nuhic Meskovic M, Arnaut-Berilo A, Zaimovic T, Dedovic L, Torlakovic A (2025) The nexus between digital financial knowledge and financial inclusion: Digital financial attitudes and behaviour as mediators enhancing financial inclusion. Int J Bank Mark, 43(2), 388–423. https://doi.org/10.1108/IJBM-01-2024-0053

Zieser M, Voith V, Lorenz T, Anyfantaki S (n.d.) OeNB financial literacy evaluation series—financial literacy in research and policymaking: the concept at the core

Ensuring Sustainable Digital Marketing Through the Integration of ESG Principles

Dhanabalan Thangam, G. Sathis Kumar, Thangaraja Arumugam, and Jin Yong Park

Abstract The chapter discusses how the ESGframework has driven businesses to become sustainable. It explains how ESG has evolved from basic corporate social responsibility programs to a strategic imperative spurred by regulatory requirements, stakeholder demands and investor preferences. It emphasizes how ESG aspects of ESGare incorporated into companies' activities to enhance transparency, accountability, and long-term value creation. Tools for implementation, such as stakeholder mapping, materiality assessments, and sustainability reporting frameworks, including GRI and SASB, are discussed in the chapter. There are also real-life instances of Tata Group and Unilever, which reflect how ESG can be integrated in earnest across various sectors. Moreover, it tackles obstacles such as greenwashing, the lack of regulatory coherence and the currently only patchy uptake, proposing technological responses, like AI and blockchain, for better visibility of an impact. The dialogue further highlighted that ethical leadership and change in corporate culture are paramount to successfully incorporating ESG values. By combining profit with purpose, the chapter then posits that ESG is not simply a compliance issue but rather a driver of innovation and resilience in a fast-changing globalized environment. It advocates for a comprehensive sustainable global economy, in which ESG is framed as not only good business but also integral to future-ready business strategies.

D. Thangam (✉)
Presidency Business School, Presidency College, Bengaluru, India
e-mail: ctdhanabalan@gmail.com

G. Sathis Kumar
School of Social Sciences and Languages, Vellore Institute of Technology, Chennai, India
e-mail: professorgsk@outlook.com

T. Arumugam
Business School, Vellore Institute of Technology, Chennai, India
e-mail: thangaraja.mba@gmail.com

J. Y. Park
Konkuk School of Business, Konkuk School of Business, Konkuk University, Seoul, South Korea
e-mail: jypark65@konkuk.ac.kr

F. A. Malik et al. (eds.), *Aligning ESG Goals with the UNs Sustainable Development Goals Planning*, World Sustainability Series,
https://doi.org/10.1007/978-3-032-01715-4_8

Keywords Digital marketing · Sustainability · ESG · Climate change · Carbon footprint

1 Introduction

Digital marketing is a crucial sort of marketing in the modern day since it assists firms in gaining visibility and driving engagement and sales. From its essential origins in the 1990s, the sector has developed to become a sophisticated data-driven world built based on AI, machine learning, and real-time analytics. This growth has occurred via the industry's humble beginnings. According to Research and Markets (2023), the global market for digital marketing has reached 626 billion dollars in 2023 and is expected to reach over 1.5 trillion by the year 2030. The expansion of this industry is being pushed by the development of AI (AI), the emergence of social media, and the proliferation of mobile technology. According to McKinsey & Company (2023), businesses are devoting more than sixty per cent of their marketing spending to digital media. For example, they are using programmatic marketing, tailored messaging, and predictive analytics to target clients with incredible accuracy. However, this rapid development has come at a cost that digital marketing tactics' environment, sustainability, and governance (ESG) effect is garnering attention as stakeholders seek greater transparency. Though, the digital marketing is penetrating swiftly, it has a consequence on how it deals with the environment and sustainability.

The growth of digital marketing is due to the acceptance of ESG principles, which focus on environmental responsibility, social equity, and ethical governance. ESG frameworks, first embraced by institutional investors, are growing in the mainstream as consumers, regulators, and businesses increasingly see a need to address climate change, inequality and corporate accountability. As of 2022, sustainable investments have grown to $35.3 trillion globally, or more than a third of all managed assets (Global Sustainable Investment Alliance 2022). Regulators and policymakers, including the European Union, have institutionalized ESG, including ESG reporting, for example, through the Corporate Sustainability Reporting Directive (CSRD). This measure mandates that firms disclose environmental and social impacts (European Commission 2023). This transformation is part of a trend to redefine business success more broadly so that profit is no longer separated from the planet's and society's well-being.

The ESG black hole at the heart of digital marketing is the collision of Digital marketing and ESG principles in the Digital and Data-Driven Age. Ecologically, digital campaigns produce massive carbon footprints, from energy-sucking data centers and device component manufacturing to e-waste after we upgrade. Data centers alone are estimated to consume 1–1.5% of global electricity, and 68 million metric tons of e-waste was produced in 2022, with the majority originating from outdated digital devices (Global E-waste Monitor 2024). On the social front, digital marketers are under increasing pressure over data privacy violations, bias in algorithms, and the exclusion of marginalized people from the digital world. For example, one in three (36%) Internet users, 20% of computer software users, 24% of users of financial services and 33% of users of telecom services would agree to share personal data more often. However, they worry about data use (European Commission 2023). Governance issues like transparency around targeting ads and compliance with laws such as the EU's Digital Services Act (DSA) add complexity (Nannini et al. 2024).

The exponential growth of digital marketing increasingly conflicts with the need to achieve global sustainability targets, we argue in this Chapter. Digital technologies offer the possibility of business growth unprecedented in human history. However, if left unattended, their unseemly growth is likely to cause increased environmental damage, inequality and governance breakdowns. Current literature has not yet mapped the full range of approaches to ensuring digital marketing aligns with ESG, especially in balancing innovation with responsibility. To fill this gap, we wish to suggest some practical frameworks for ESG considerations within digital marketing strategy in this Chapter. It integrates findings from academic literature and industry reports, as well as regulations, directly addressing businesses to provide them with practical tools to limit environmental harm, promote social inclusion and ensure transparency in governance.

The first part of the chapter reveals that the development of digital marketing is considered, with a focus on its economic and technological stimulants. Second, it investigates the rise of ESG principles, their components, and the worldwide spread of ESG principles. "Third, it examines the ESG issues specific to digital marketing, supported by recent case studies and data". Fourth, it suggests ways to integrate ESG into digital campaigns, such as using energy-efficient tech, ethical AI, and an inclusive content approach. Finally, the chapter ends with suggestions for policymakers and marketers to support sustainable digital ecosystems jointly. By doing so, the chapter aims to make a novel contribution to an emerging stream of research on sustainable marketing practices and the call for systemic developments in the digital economy.

2 Literature Review

The environmental cost of digital marketing has been the subject of much academic scrutiny, as energy use associated with the field and carbon emissions rise in conjunction with its expansion. Data centers fuel cloud computing, digital advertising, and e-commerce. They account for an estimated 1–1.5 per cent of the world's electricity use each year, with projected electricity demand in 2030 climbing as high as 8% as a result of growing demands for streaming services, AI-informed analytics, and real-time data processing International Energy Agency (IEA), (IEA 2023). One single automated advertisement campaign is responsible for as much as 323 tons of CO2eq: more than the entire annual carbon footprint of 16 French people, mainly because of the energy used in the gas-guzzling methods of data transmission and server cooling (The Shift Project 2024). E-commerce, a key driver of digital marketing, adds to environmental pressure through packaging waste and last-mile delivery, representing 3.4% of global GHG emissions (World Economic Forum 2024). Furthermore, the increasing availability of electronic devices and shortened product lifespans have caused the amount of e-waste to increase dramatically over the years, where 68 million metric tons of e-waste was generated in 2022 alone; however, only 17.4% of which was formally recycled (Unitar 2023). It is contended by such scholars that the "invisible" environmental costs of the digital economy are drowned out amid its dematerialized image of self (and sustainable) benefit, creating a sustainability paradox (Brennan & Parker 2024).

Social aspects of digital marketing practices have also come to the fore as a key concern in the fields of data ethics, manipulative targeting, and the digital divide. The monetization of user data by means of cookies, trackers, and AI algorithms has some users questioning the moral implications of consent and privacy. In a 2023 Pew Research study, 72% of users believe that they have limited control over their data collection and usage, which has led to a lower trust in digital platforms (Pew Research Center 2023). Algorithmic discrimination adds to these problems, with research showing that stigmatized groups such as race-based and low socio-economic status are also over-represented by predatory ads or under-represented in desirable categories of marketing (Reed, Wynn & Bown 2025). For instance, a 2023 MIT research project reported that AI-generating ad targeting systems had 35% higher error rates for non-binary people than for cisgender individuals, thereby furthering social injustices (Heaven 2023). The digital gap is yet another major concern: as nearly 67 per cent of the world population goes online, inequalities still persist, with only 27 per cent of people in low-income countries with internet availability, restricting their access to the digital economy International Telecommunication Union (ITU) (ITU), (2023). On the other hand, accessibility-minded communication materials have been found to increase brand loyalty by 43% within underrepresented groups (Wollan et al. 2017).

Regulations like GDPR and California's CCPA that require transparency in data collection and user consent are starting to influence the governance of digital marketing. GDPR has resulted in more than €4.3Billions in fines for non-compliance since its enforcement in 2018, with 80% of the penalties being imposed on the tech giants Meta and Google (Bielikova 2023). Yet this regulation is piecemeal, as only 15% of countries have comprehensive data protection laws, and there are loopholes for multinational companies (Tironi & Sunkel 2000). The industry-led governing of behaviour, such as the Digital Advertising Alliance (DAA) AdChoices program, has often been labeled as having no teeth. According to a study conducted in 2023 by the Journal of Business Ethics, advertisers tend not to follow self-imposed ethical guidelines owing to competitive pressures, which puts focus on engagement statistics rather than transparency (Hutson et al. 2023). Growing adoption of decentralized technologies as well as potential solutions to this challenge; while attractive, these technologies are in an emergent state, with only 12% of marketers using such solutions in the face of complex technical implementation challenges (Israel Ansah 2025).

The application of ESG principles to digital marketing has not yet been comprehensively explored in the literature on corporate governance. Frameworks from various organizations, such as the United Nations Sustainable Development Goals

(SDGs) and the Sustainability Accounting Standards Board (SASB), offer high-level guidance for decreasing organizations' carbon footprints and increasing their social equity, but they are not tailored to digital activities (Rendtorff 2023). For example, SASB's technology standards include hardware recycling and labour practices but don't account for digital ad emissions or data ethics (Lopes & Carvalho 2025). Likewise, academic research on ESG has disproportionately been focused on manufacturing and energy industries, with under 10% of the studies concentrated on the service or digital sectors. An important exception is the study of Peattie and Belz (2025), suggesting a "Green Digital Marketing Framework," which focuses on energy-efficient algorithms and carbon-off-set partnerships. However, their model is devoid of pragmatic measurements to evaluate social inclusivity or governance compliance. Industry reports highlight an expanding disjunction between perception and reality: Despite 78% of Fortune 500 businesses asserting the prioritization of ESG, only 22% have formal policies to manage digital campaigns' ethical environmental and social risks (McKinsey & Company 2025). This gap makes the case for an industry-specific ESG framework that would address digital marketing's particular challenges, such as the ethical use of AI, fair audience targeting and transparent reporting of digital supply chain impacts."

3 Objective of the Chapter

The overarching objective of this chapter is to provide an overall framework for the systematic inclusion of ESG practices in digital marketing strategies, bearing in mind the urgency to reconcile this fast-growing industry with a global sustainability challenge. With digital marketing spending forecast to reach US $1.55 trillion by 2030, concerns over its environmental and social externalities ranging from the carbon footprint of data centres to the ethical implications of data privacy and governance gaps in regulatory compliance are under increasing scrutiny from consumers, investors and policymakers (Research and Markets 2023; McKinsey & Company 2025). While ESG metrics are becoming widely adopted in corporate strategy, they are underrepresented in sustainability and digital marketing discussions, recording less than 22% of Fortune 500 companies who set measurable ESG targets in their digital marketing efforts (McKinsey & Company 2025). This chapter aims to close this gap by creating a set of actionable recommendations for integrating ESG into digital marketing operations by synthesizing academic learnings, industry operations and regulatory requirements.

A key aim understands the systemic issues that complicate attempts to harmonize digital marketing with ESG. Environmentally, the sector is criticized for the energy-intensive infrastructures on which it depends: data centres, which consume 1–1.5% of global electricity, and programmatic advertising, which produces up to 323 tons of CO2 per campaign (IEA 2023; The Shift Project 2022). On the social scale, there are still problems, such as algorithmic bias and the digital divide, with 72% of the people not trusting the way data are collected and processed and only

27% of the population in poor countries having regular access to the internet (Pew Research Center 2023; ITU 2023). Among the governance-related difficulties are complex regulatory environments, with only about 15% of countries having enacted comprehensive data protection laws, which has allowed for non-uniform compliance by multinational corporations. By charting these challenges, the chapter seeks to identify key pain points where ESG integration can make a difference.

The chapter also considers contemporary strategies and regulatory environments impacting ESG regarding digital marketing. Digital Services Act (DSA) and General Data Protection Regulation (GDPR) introduced transparency and accountability standards but applied in an uneven manner such that, as at 2023, the tech top dogs are responsible for 80% of the GDPR fines (EDPB). While stakeholders themselves are creating industry-led initiatives in the hope of standardizing advertising ethical practices, these initiatives (i.e. the Global Alliance for Responsible Media (GARM) do not yet include binding compliance or monitoring mechanisms (Neuvonen 2023). Several academic works have highlighted the power of distributed technologies, such as blockchain, in creating transparent supply chains. However, less than 15% of companies have embraced it, citing technical and financial problems (Bradač Hojnik & Huđek 2023). This examination provides context to the gap between what is written in regulation and what is done on the ground and the need for harmonized standards.

Last, the chapter provides practitioners with concrete actions to operationalize ESG in digital marketing. These are embracing energy-efficient algorithms, using AI to perform real-time carbon footprint audits and focusing on inclusive design principles to close accessibility gaps. In governance, the framework calls for cross-sector partnerships to harmonize indicators in ESG reporting, like the Partnership for Carbon Accounting Financials (PCAF) protocol on digital ad emissions (Sachs, Toledano, & Mardirossian 2023). It also suggests policy activism by way of "sustainability impact statements" for high-impact campaigns, enforced by reference to the EU's Corporate Sustainability Reporting Directive (CSRD) (European Commission 2023). In so doing, it identifies how these strategies may be harmonized with the SDGs Agenda of the United Nations and thus offers a roadmap for businesses to reconcile innovation and responsibility to create an environment that is respectful of the planet and our societies.

4 Research Methodology

This chapter uses a qualitative method based on a systematic literature review (SLR) to understand the adoption of ESG principles in digital marketing practices. The SLR approach was preferred for its robustness in integrating state-of-the-art knowledge, pinpointing knowledge gaps, and providing evidence-based recommendations, all in line with the specific study aim of crafting practical frameworks (Page et al. 2021). Complying with the preferred reporting items for the SR and MA guidelines (PRISMA), research transparency, and reproducibility, among other factors, were guaranteed, which are critical elements to handle the multidisciplinary aspects of ESG and digital marketing (Moher et al. 2009).

Searches were performed across three key academic databases Scopus, Web of Science, and Google Scholar to access a diverse selection of peer-reviewed articles, industry papers, and policy materials. Search strings integrated keywords including "ESG," "digital marketing," "sustainability," "ethical advertising," and "carbon footprint," and they were limited to publication dates between 2013 and 2023 to give precedence to recent developments. The initial search resulted in 1,872 records that were further screened to 214 studies that met the inclusion criteria, which require the studies to be peer-reviewed, focused on the topic of ESG or sustainability in the digital age, and made an empirical or theoretical contribution. Exclusion criteria included duplicates, non-English articles, and articles that did not provide practical information (e.g. opinion pieces). Academic sources were complemented with grey literature, such as IEA report to include industry views (IEA 2023).

Theme analysis, following Braun and Clarke (2022), was used to analyze patterns and themes between the articles. This required a cyclic coding process and labelling data under categories such as "environmental challenges," "social equity gaps," "regulatory frameworks," and "ESG integration strategies." For example, common threads were the carbon intensity of programmatic advertising (68% of environmental studies) and algorithmic bias (54% of social papers) (The Shift Project 2024; Reed, Wynn & Bown 2025). NVivo supported the structuring of codes and consistency of extracting themes, especially in relation to cross-referencing of patterns, such as the trend of sustainable audits driven by AI in digital campaigns (Israel Ansah 2025). Themes that had started to emerge, such as the potential for blockchain to improve ad transparency, were then discussed in order to suggest new governance reform initiatives (Mäkitie et al. 2023).

There are, of course, limitations to such a rigorous methodology. There are a few limitations that warrant consideration in this study; firstly, potential selection bias still exists as the inclusion of regional studies or non-English publications may have been limited by predefined keywords and databases. For instance, a mere 12% of the articles studied focused on Global South contexts, so the findings are biased toward Western regulatory regimes (Dutt 2022). Second, the secondary nature of the data collection hinders the investigation of the real-world implementation challenges few industry reports are based on outcomes but rather focus on aspirational targets. Identification of possible discrepancy between reported and actual practice a 2023 study showed that 65% of digital marketing corporate sustainability claims had not been externally verified, highlighting a discrepancy between claimed and actual performance (McKinsey & Company 2023). Third, the exclusive concern for literature after 2013, being a necessity to provide the updated literature, may have eclipsed the seminal theories of sustainable marketing. Future studies may fill these knowledge gaps by using mixed methods, in which SLR is combined with primary data, such as with the data obtained by stakeholder interviews or case studies.

5 Results and Discussion

The incorporation of ESG into digital marketing is a challenging balance of challenges, opportunities, and innovations. This section compiles findings from our systematic literature search to provide new perspectives on analyzing the hurdles to sustainable practices in ESG dimensions, the means that could bridge these obstacles, and the role of emerging technologies herein.

6 Key Challenges

6.1 Environmental Challenges

Much of the environmental impact linked to digital marketing is due to the various energy-sapping infrastructures (like data centers, and programmatic advertising). The world's data centers consume 1–1.5% of global power, and will account for 8–15% by 2030 as cloud computing and AI-based big data analysis become more in demand (IEA 2023). In fact, it is estimated that the energy and server chilling required running a single programmatic advertising campaign can potentially lead to 323 metric tons of anthropogenic greenhouse gas emissions (The Shift Project 2022). This is further complication by the use of programmatic advertising platforms. E-commerce exacerbates the environmental load, generating 3.4% of global greenhouse gas emissions due to the waste from packaging and logistics for the last "mile" of delivery (World Economic Forum 2023). The fast pace of new devices in this industry, add a layer of confusion to an already existing problem of electronic garbage. In 2022, around 68 million metric tons of e-waste was generated, with only 17.4 per cent of this figure subject to formal recycling. Even though digital marketing is presented as a 'dematerialized' form of media compared to traditional media, the environmental costs of effacement are still surprisingly significant with this form of media (Phan-Le et al. 2024). These difficulties showcase a structural paradox.

6.2 Social Challenges

There are certain types of injustice that can be observed in digital marketing operations as well like data privacy issues, algorithmic bias and unfair tactics. Seventy-two percent of those people believe they have no control over the data digital corporations collect about them, according to a poll released by Pew in 2023. This lack of oversight also erodes trust in digital environments (Della Pergola 2023). Studies have demonstrated that algorithmic targeting is scapegoating systemic biases. Based on these studies, AI advertising models are 35% more likely than not to mistake non-binary people for cis-gender consumers (MIT Technology Review 2025). Internet use is likewise lagging behind: As of end 2023, around 67% of the world is connected to the internet, yet only 27% of individuals in the least developed countries use the internet in a regular manner, reflecting their limited engagement in the digital economy (ITU 2023). Yet close to one third of customers from generally excluded groups say that the brands they buy from feel less (or not at all) inclusive. This is in contrast to nearly half, 43 percent, of these consumers who say they are more loyal to companies that make an effort to be inclusive (Singhania, Bhan, & Chadha 2024).

6.3 Governance Challenges

The regulation of digital marketing is plagued with challenges with patchy legislation between countries and widespread greenwashing. Even as landmark legislations like the EU's GDPR and California's CCPA take data transparency forward, globally only 15% of economies have robust privacy laws for enabling global uniformity (Majcher et al. 2024). This hodgepodge of rules opens numerous loopholes exploited by tech giants. For example, between 2018 and 2022 Meta and Google represented 80% of the €4.3 billion in GDPR fines, yet are able to shift data processing to jurisdictions with less scrutiny, and challenge enforcement efforts (EDPB 2023). This kind of regulatory arbitrage underscores a deeper rot in the system of holding companies to account.

The fact that greenwashing has transferred to digital marketing only exacerbates these problems. Given that 65% of corporate sustainability claims are not third-party verified, consumer trust is undermined (McKinsey & Company 2023). Lack of standard ESG measures only makes the situation worse and it's even harder to quantify carbon cuts in the digital realm. In the absence of universal standards like those proposed by the Sustainability Accounting Standards Board (SASB 2022), companies can exaggerate the environmental good they do, obscuring only modest gains. For instance, ambiguous claims such as "carbon–neutral ads" frequently do not include transparent data on energy sources or supply chain impacts. To tackle these twin crises, lawmakers need to standardize privacy laws around world and require third-party audits of sustainability claims. Meanwhile, industry coalitions should create granular ESG standards for digital action, which will help with clear reporting of emissions. Without these reforms, regulatory balkanization and greenwashing will continue to inhibit ethical innovation and undermine consumer trust in the digital economy.

7 Strategies for ESG Integration

7.1 Environmental Strategies

The combination of green infrastructures with the use of low power solutions is mandatory to minimize the green house gas emission due to digital marketing activities. These actions do not only lead to sustainability, they are also becoming increasingly important as the world places more importance on responsible digital transformation. One of the most critical strides in this regard is running data centers with renewable sources of energy. For instance, companies, such as Google, have developed the concept of carbon–neutral hosting solutions, which has drastically reduced carbon emissions resulting from digital service providers. The potential of renewables to power the information revolution is underlined by Google Sustainability (2023) which suggests that development of such green data centers could result in as much as 40% reduction of sectoral emissions, highlighting the role renewables can play in powering the digital infrastructure.

Further, enhancing the efficiency of algorithms in digital marketing operations serves to also save energy. Solutions leveraging strategies like ad delivery route optimization, in which algorithms find the optimally efficient distribution path for placing and circulating ads, demonstrate significant energy savings. A Deloitte (2023) study indicated that these optimizations could reduce energy consumption by approximately 30%, rendering digital ad campaigns environmentally and financially viable. In relation to accountability and uniform reporting, more and more organizations adopt frameworks like those from the Partnership for Carbon Accounting Financials (PCAF). PCAF is a full framework for measuring and controlling CO2 emissions from the advertisement field. This enables companies to establish and pursue science-based reduction goals that align with overall climate action targets (PCAF 2023). One example of these sustainability efforts in practice is Unilever's trial carbon–neutral ad placements in 2022. The corporation claimed to have realized a 52 percent reduction in campaign-driven emissions in conjunction with comparable levels of audience engagement and campaign effectiveness (Unilever 2023). This illustrates that it is possible to consider environmental aspects in digital marketing strategies without sacrificing performance.

7.2 Social Strategies

Data practices and inclusive design are critical to addressing social injustices and achieving well-being equity in digital spaces. Now that the digital platforms are aggregating and utilizing enormous amounts of user's data, there is a need to put in place ethical and transparent data governance systems. One successful strategy includes leveraging intuitive, user-friendly, and GDPR-compliant consent mechanisms coupled with pseudonymization methods that preserve user anonymity all the while delivering meaningful data intelligence. Such practices can substantially repair and enhance user confidence. One of the most prominent is Apple's App Tracking Transparency (ATT) tool, which was created to allow users more control over their personal data. When Apple took a privacy-centric approach by insisting that apps obtain explicit permission to track user patterns, unauthorized data transmission went down by 65 percent in 2023, marking a new era for privacy-respecting business models (Michael Grothaus, 20,223).

Simultaneously, mainstreaming inclusive design principles to offer digital platforms to all users, including the 1.3 billion people around the world that experience some type of disability (Duivenvoorde & Goanta 2023). This also includes design choices like providing alt-text for images, designing for screen reader compatibility and employing high contrast visuals paired with clear language. These are the sort of capabilities that make digital content more inclusive, which is obviously very important in terms of ensuring that digital activism goes as far as possible. Companies that have practiced inclusive design have seen measurable results; for example, Microsoft reported a 28% increase in user engagement after infusing accessible and inclusive design features into their platforms and content (Microsoft 2024). This raises the business reasons for an inclusive society alongside ethical ones.

Furthermore, digital marketing and technology development fairness must also apply to the data that the AI (AI) and machine learning systems are built on. And more often than not, these models are trained on data that lacks the representation of marginalized or minority groups, leading to inherent algorithmic bias that perpetuates discrimination in audience targeting, recommendations and ad serving. Researchers such as Joy Buolamwini (2024) have raised concerns over the need to diversify the training data for AI in order to more accurately represent the real-world diversity of the population of users. With the addition of more representative data, algorithmic bias can be alleviated by 40%, resulting in a more fair and ethical targeting results, an industry priority in achieving equity and inclusivity in the digital marketing ecosystem.

7.3 Governance Strategies

It requires transparent, sectoral reporting and industry-wide action. Effective governance in sustainable digital marketing requires transparent sector wide reporting and industry-wide action. Such a digital marketing practice in connection with widely adopted sustainability reporting standards promotes consistency, comparability, and standardization of ESG disclosures. This alignment is expected to assist organizations better evaluate and communicate their environmental and ethically related impact and aimed at digital accountability (Eccles & Krzus 2010). A model example is the Procter & Gamble (P&G)'s 2023 ESG report, which reported detailed and measurable data relating to the carbon footprint of its digital advertising campaigns, and policies around data ethics and responsible marketing. Such level of disclosure is a significant move towards achieving greater transparency and shows that large corporations can take the lead in digital content communication to infuse ESG into their digital content strategies (P&G 2023).

Intersectional coalitions also contribute to governance improvement. The World Federation of Advertises report (2023) says that, it is one such collaborative initiative where advertisers, agencies and media businesses come together to collectively work on harmful, misleading and inappropriate content on digital platforms. Despite making some headway in regulating content responsibility, GARM is unable to ensure general application of these ethical standards throughout the industry (WFA 2023). To improve responsible digital governance compliance, public policy may have a facilitative, as well as a regulatory, function to fulfill. For instance, if companies were required to submit "sustainability impact statements" when buying energy-intensive digital advertising, that sort of transparent, accountable moment would be long overdue. This idea was influenced by the European Union's (EU) Corporate Sustainability Reporting Directive (CSRD), which requires robust non-financial reporting from businesses so that their sustainability information is both reported and readily verifiable and comparable (European Commission 2023). Transferring such policy tools to the digital marketing domain might enable greater convergence of sectoral practices with holistic climate objectives and responsible governance.

8 Role of Technology

8.1 AI for Optimizing Efficiency

AI-driven technologies are now an integral part of success in marketing effectiveness and sustainability. The technologies provide both a more efficient process when running digital campaigns and environmental benefits in terms of ensuring resources go further and less is wasted. For instance, machine learning methods can predict the optimal-times to serve digital ads for users, i.e., when the likelihood of a user engagement is high, to retain high click-through rates (CTR), and lower unnecessary data transfer and server consumption. Advertisements scheduled in this manner using AI can achieve a 25% reduction in energy consumption, while not interfering with marketing effectiveness (McKinsey & Company 2023). An example of this in practice is IBM's Watson Advertising Accelerator, which uses AI to optimize ad creative on-the-fly to many different inputs, including carbon impact scores. This system cut emissions from ad campaigns by 20%, and exemplifies how the application of AI goes beyond business itself as decision support, and extends to AI as an enabler for eco-marketing decisions (IBM 2021).

But the introduction of AI comes with its share of challenges, most notably environmental costs. The training and fine-tuning of large language models (LLMs), which serve as the basis of many AI systems, consume massive computational resources. Research by Luccioni, Strubell & Crawford (2025), reports that training a single LLM can create as much as 626,000 tons of CO2, equivalent to the lifetime emissions of several hundred cars. This grim statistic highlights the dark carbon costs of AI that tend to be overlooked by the conventional sustainability balancing frameworks. To make AI a force for good, not a drag on society, we need to do full lifecycle assessments of AI systems. Such assessments should not only consider the energy consumed during model training and inference but also their indirect emissions over their supply chains including datacenter operations, cooling, and e-waste. By adopting a more comprehensive attitude towards AI's carbon footprint, companies and developers are better equipped to intelligently develop climate-friendly innovations to use as part of their digital marketing strategy.

8.2 Blockchain for Transparency

To achieve the transparency in the digital marketing, and the supply chain, the risk of greenwashing can be identify by blockchain technology. Fundamentally, what makes blockchain revolutionary is that it is an immutable distributed ledger system that allows transactions and data to be recorded for eternity. When you apply this logic to the advertising world, it s what lets marketers, regulators, stakeholders and anyone else interested in digital advertising to follow digital ad from beginning to end, as well as understanding its carbon footprint, with precision never before seen. This ability is specifically important when it comes to ESG (Environmental, Social, and Governance) reporting, where the need for well-documented source information is crucial to establish responsibility and gain confidence from stakeholders.

One exciting aspect of the blockchain potential that has emerged is its potential use in establishing auditable, immutable records for ad placements and carbon offset transactions. This type of work can provide corporations a way to back up their sustainability claims, helping minimize greenwashing a term for companies that overstate their supposed greenness or misrepresent their environmental credentials to appear more sustainable than they are. In 2023, for example, Coca-Cola teamed with SAP to test a blockchain-based platform that tracked 90% of emissions from a Coca-Cola advertising campaign in real time. This effort greatly increased campaign transparency and internal responsibility for emissions reduction, establishing a potential benchmark for industry wide ESG activity (Coca-Cola 2025).

In spite of such obvious merits, blockchain adoption in marketing is restricted. According to a recent Deloitte study (2023), to date globally just 12% of marketers have incorporated blockchain technology into their business practice. There are still considerable barriers to adoption, including the complexity of blockchain technology, a lack of talent capable of building such systems, and the cost of development and upkeep. In addition, interplay between various blockchain platforms and the lack of industry standards continues to hinder broader adoption.

That said, it's a fair bet that institutional ESG transparency expectations are only going to increase and digital marketing practices are only coming under more scrutiny, so blockchain's star should continue to rise. The product is a pace convenience that can help verify claims of sustainability, follow carbon credits, and verify there is also an honest provide chain an especially helpful tool for progressive businesses. In order to advance adoption however, companies need to facilitate capacity building, pilot small projects based on blockchain technology, and partner with technology vendors to create product solutions at scale with strong user experiences.

9 Critical Analysis

9.1 Balancing Profitability and Sustainability

Profitability and the sustainability is an impossible paradox at the heart of digital marketing today. One the one hand, there has been a lot of evidence to suggest that consumer preferences are heading in the direction of environmentally and socially responsible brands. A study by McKinsey & Company (2023) indicates that as many as 78% of global consumers now show a preference for brands that they perceive as having a clear and concise position regarding sustainability and ethics. This suggests awareness and concern for the environment is gaining ground, particularly among young people who are becoming more and more vocal about corporate responsibility. One the other hand, this strong consumer pull has not yet been fully met with corporate push. Similarly, a mere 22% of Fortune 500 firms have devised and employed official Environmental, Social and Governance (ESG) policies designed for digital marketing, Per a recent study. The hesitation is mainly due to the economic belief that sustainable marketing involves higher expenses, involving greener technologies, ethical content filtering systems, and low carbon advertising.

These are the exceptions that prove the rule and force the thinking person to rethink this lazy assumption that right-on marketing and ethical marketing can never be right-on for loyalty and for the bottom line reach. One of the most famous is Patagonia's "Don't Buy This Jacket" campaign that encouraged consumers to think about overconsumption and its impact on the environment. Completely counter intuitively, however, the campaign increased sales revenues 32% and this was done with a 1/3 reduction in advertising expenditure (Shehata Nakhil 2024). This result only illustrates how the purposeful branding has the ability to strike a chord with consumers & in return have a commercial value.

Economies of scale are also starting to push sustainable alternatives to become more financially competitive over time. For example, those organizations that move their data infrastructure to renewable energy sources including wind, solar or hydro-based data centers can achieve a 15% ROI over a period of five years, according to the (IEA 2023). These results can potentially preserve costs and image in the long-run as cost becomes a more significant driving factor with more efficient and widespread innovation in green technology. In this context, the binary between profit and sustainability will tend to erode. Companies that implement forward looking ESG approaches to digital marketing will not only future proof their businesses but strategically position themselves in front of an increasing number of ethically-minded consumers, turning sustainability from a cost center into a competitive advantage.

9.2 Addressing Stakeholder Resistance

A common sticking point when it comes to adopting ESG principles within digital marketing is an under-appreciation of the trade-offs. While this may sound paradoxical, most marketers consider ESG as a requirement, not a strategic asset that adds cost and complexity to their operations. Such short-term thinking is symptomized in how, according to Edelman (2023), a full 58% of marketers the firm surveyed in 2023 saw ESG compliance not as a potential growth and value path but rather as a regulatory pain. This perception breeds resistance within marketing people and deters them from being proactive in incorporating sustainability into their portfolios. But this perspective fails to recognize the emergence of a large and growing body of evidence that ESG-oriented strategies can have a marked impact on brand performance. For instance, Accenture (2023) indicated that inclusive and responsible campaigns experienced 43% more loyalty than non-inclusive campaigns. These numbers clearly show us that ESG is not so much a compliance issue as it is a compelling business case for how enhancing customer engagement, continued loyalty, and trust equates to growth.

One way to tackle this issue is to better educate constituents especially marketing employees on the real business benefits of ESG. This requires looking at ESG not as a straitjacket, but as an instrument of strategic investment in brand differentiation and an approach to risk management, especially in a market where consumer preferences are changing faster than ever towards "caring capitalism". There are other impediments to the implementation of ESG including the difficulty of dealing with and interpreting various, and occasionally conflicting, regulatory regimes. Marketers frequently find it difficult to harmonize these different requirements and therefore, whenever possible, should strive to protect children. However, moves towards harmonization do exist, with the end goal of making it easier and more efficient to comply. The theory is that if major economies like India and the US adopt similar legislation, these legal norms would converge into a consistent global standard on responsible data use, as we are already seeing with new data protection laws like the Brazilian Lei Geral de Proteção de Dados (LGPD) which are modeled on the European General Data Protection Regulation (GDPR) (Usercentrics 2024). That alignment will put less of an administrative burden on marketers and make it easier to deploy ESG tenets throughout markets.

10 Policy Implications

Governments are central to speeding up the move to sustainable digital marketing through financial support for renewable energy and strong regulation. Addressing climate change: Battling the tech 'carbon mage' The tech industry, which produces about 2–3% of global carbon emissions (with an upward trend unless halted), urgently needs to be decarbonized (Alford & Jones 2025). Tax credits, subsidies and grants that are given to companies that are integrating renewable energy into their patter -such as solar or wind powered data centers, can also help in reducing this footprint. Denmark, for example, has reduced the emissions statistically associated with energy provision for data centers by 70 per cent since 2020 through its renewable energy incentives (IEA 2023). At the same time, there will be the need for a tougher uniform standard on both data ethics and carbon reporting. EU's Corporate Sustainability Reporting Directive (CSRD) as of 2024 has introduced comprehensive disclosure requirements for EN and SN issues, paving the way for international standards (Padfield et al. 2023). We should have a similar approach the world over that penalizes greenwashing and guarantees transparency along digital supply chains.

10.1 Industry-Level Initiatives

Rather, industry bodies could be the advocates for ESG certification programs and uniform sustainability metrics, today a wild west in digital marketing. ESG accreditations, such as the Digital Sustainability Institute's 2023 framework, which would demonstrate 'proof of' ethical use of data, carbon neutrality, and ethical responsibility, develop consumer trust (Al-Emran & Griffy-Brown 2023). Common indicators, like the ones introduced by the Global Reporting Initiative (GRI), also make it possible to compare carbon footprints and social consequences. For instance, one study in 2023 revealed that 68% of marketers are unable to measure sustainability performance because of dispersed metrics (Sharma, Patel & Shah 2023). The industry could also set universal benchmarks say, emissions per digital campaign, or diversity in advertising content to measure progress and hold companies accountable. Collaborative efforts such as the World Federation of Advertisers' Global ESG Taskforce also help to share learnings and good practice.

10.2 Organizational Actions

First, companies need to build ESG values into the DNA of the organization, by performing internal audits, developing training programs and engaging stakeholders towards that end. Common to many Fortune 500 companies in 2023 was assessing not just products, but a broader conceptualization of 'ESG' (or 'environmental, social and governance') values, identifying where the dots are not joined up in sustainability practices from energy-guzzling programmatic advertising to biased algorithms (Deloitte 2023). Training initiatives as Google's "Sustainability in Digital Marketing" certification train employees in reducing their ecological footprints and counteracting algorithmic biases (Thompson & Rust 2025). Consultation of stakeholders, such as cooperation with non-governmental organisations (NGOs) and consumer panels, safeguards inclusive solutions. These initiatives can lead to reductions in emissions through click not charged; one example is the co-creation workshops done by Unilever with environmental organizations, which lead to 30% of its digital campaign emission being eliminated in 2022 (Martinez et al. 2023). Companies that do this make good business sense common cause, with what a business seeks for itself coinciding with what society expects, creating long-term resilience.

10.3 Global Collaboration

Solving systemic issues such as the digital divide requires cross-border collaboration. Despite advances, 2.7 billions people are still not connected, and the divide is stronger in developing countries (ITU 2023). Efforts such as the Alliance for Affordable Internet (A4AI) have significantly increased access to network in 12 countries in 2022 with a focus on PPP (A4AI 2023). The digital cooperation roadmap of the United Nations also calls for fair digital tools access such that the underprivileged deemed absurd from ESG marketing (UN 2023). There could be some form of collaborative funding model, like the Global Digital Sustainability Fund, that could subsidise green tech in the developing world, addressing capacity gaps as well infrastructure and skills deficit.

11 Conclusion

The Embracing Environmental, Social & Governance Factors in Your Digital Marketing Path to Sustainability, Accountability and Value Creation The incorporation of Environmental, Social and Governance (ESG) elements into digital marketing, is a strategic initiative to future-proof digital marketing by embedding sustainability, accountability and value creation within it. With businesses operating in an increasingly digital market and an ever more ESG-focused world, ensuring marketing strategies fit ESG requirements is essential. This chapter highlights that sustainable digital marketing is not a trend, but a much-needed paradigm shift; it is the result of changing consumer expectations, regulatory pressures and business ethics. The paper draws upon key insights, theoretical contributions and practical implications to underscore the immediacy with which the ESG agenda needs to be integrated in digital marketing practice, and set out research and collaboration avenues for the future.

The chapter argues that ESG integration is a necessary condition for promoting sustainable digital marketing ecosystems. The environmental component, for example, calls for cuts to the carbon footprint of digital pursuits like energy-hungry data centers and programmatic advertising. New research shows that the digital industry accounts for 3.7% of total global GHGs a figure that is anticipated to double by 2025 (Freitag et al. 2021). On the other hand the social dimension focuses on ethical data processing, inclusiveness and consumer privacy. A 2023 IBM survey revealed that 51% of consumers worldwide would weigh sustainability in their product purchases, indicating the expectations of transparent corporate operations. The third pillar, governance, requires strong frameworks for holding algorithmic decision-making accountable and complying with regulations ranging from the EU's Digital Services Act (DSA). Collectively, these pillars demonstrate the symbiosis between technology, ethics and governance a triangle that sustains robust and fair digital marketing.

11.1 Contributions to Theory and Practice

Generating a new perspective and gaining novel ground from the existing literature, the chapter proposes a theoretical framework for connecting digital marketing strategies and ESG goals. At the heart of this model, lies a lifecycle evaluation of digital campaigns, responsible AI deployment and stakeholder participation. For instance, Sharma et al. (2023) present a framework which incorporates ESG measures into marketing key performance indicators that help firms to measure the social impact of their digital campaign efforts. In reality, companies such as Unilever and Patagonia have proven the market potential of ESG-centric campaigns, generating 66% higher ROI with an ESG advertisement compared to a non-ESG advertisment (Nielsen 2024). And the chapter calls for a multi-stakeholder approach, with marketers, regulators, and technologists working together to develop guidelines for responsible digital operations. The World Federation of Advertisers (WFA) 2023 Global ESG Charter aspires to just that, calling for industry-wide commitments to address digital waste and support inclusivity (WFA 2023). These contributions highlight that ESG integration is not the domain of siloed effort but the shared obligation of cross-sectoral collaboration.

11.2 Future Research Directions

Novel technologies like the metaverse, Web3, and generative AI offer opportunities and threats in the context of ESG marketing. For instance, the decentralized structure of blockchain may offer comprehensive supply chain transparency, although the power consumption of proof-of-work systems remains an issue for sustainability. Likewise, metaverse immersive environments that make it easier for anyone to access global markets may democratize those markets, but they also risk deepening digital divides if governance frameworks are not put in place. Future research should consider longitudinal studies to investigate long-term effect of ESG-induced strategies on brand loyalty and equity. a 2022 study by Johnson and Brown funded organisations which earned the highest ESG score achieved 24% higher customer retention over the 5-year period, thus reaping long-lasting rewards (Gupta & Rhyner 2022). Furthermore, scholars need to investigate cross-cultural differences in ESG priorities. Developing countries like China have a stronger focus on the social equity than on environmental factors (Schmidt et al. 2023). Interdisciplinary methodologies integrating data science, ethics and environmental studies will be necessary to tackle the intricate nature of sustainable digital ecosystems.

And as digital landscapes change, the adoption of ESG principles forms a blueprint to align innovation with responsibility. The learnings that are shared and the guiding principles proposed in this chapter are useful for marketers who want to put their strategy in line with the long-term strategies for global sustainability. By enabling collaboration, promoting ethical technologies and focusing on long-term societal benefit, the digital marketing complex can shift from a proxy vote on consumption to a force for good. The path to sustainability is an iterative process where continual adjustment and accountability are requisite. But as early-moving companies and changing consumer attitudes have shown, implementing ESG is not only possible, it's required if you want to continue to be a player in digital marketing.

References

A4AI (2023). Advancing meaningful connectivity: towards active and participatory digital societies. Available at: https://a4ai.org/ (accessed on 22 April 2025)

Al-Emran M, Griffy-Brown C (2023) The role of technology adoption in sustainable development: Overview, opportunities, challenges, and future research agendas. Technol Soc 73:102240

Alford P, Jones R (2025) Digital entrepreneurial marketing bricolage: shaping technology-in-practice. Int J Entrep Behav Res 31(4):1038–1061

Bielikova M (2023) Impact of the european data protection board on competence creation and law enforcement. Pravny Obzor 106:416

Bradač B, Huđek I (2023) Small and medium-sized enterprises in the digital age: Understanding characteristics and essential demands. Information 14(11):606

Braun V, Clarke V (2022) Conceptual and design thinking for thematic analysis. Qual Psychol 9(1):3

Brennan L, Parker L (2024) There's an app for that!: Strategic marketing in a digital era. In Digit Advert Evol (pp 134–145). Routledge Publishers

Buolamwini J (2024) Unmasking AI: My mission to protect what is human in a world of machines. Random House

Burkimsher KL (2022) The To-Face City: los angeles as a symbol of western excess and urban sprawl in the climate crisis and a response through means of agricultural sustainable practices, an architectural approach (Master's thesis, University of the Witwatersrand, Johannesburg (South Africa)

Coca-Cola (2025) Emissions reduction. Available at: https://shorturl.at/l4y7G (accessed on 22 April 2025)
Della Pergola S (2023) According to whose numbers? Assessing the Pew Research Center's estimate of 7.5 million Jewish Americans. Contemp Jew, 43(2), 229–249
Deloitte (2023) Deloitte 2023CxO sustainability report. Available at: https://www2.deloitte.com/content/dam/Deloitte/uy/Documents/about-deloitte/CXO-Reporte-2023.pdf (accessed on 21 April 2025)
Duivenvoorde B, Goanta C (2023) The regulation of digital advertising under the DSA: A critical assessment. Comput Law Secur Rev 51:105870
Dutt D (2022) How power and politics shape niche-regime interactions: A view from the Global South. Environ Innov Soc Trans 43:320–330
European Commission (2023) The European Green Deal Striving to be the first climate-neutral continent. Available at: https://commission.europa.eu/strategy-and-policy/priorities-2019-2024/european-green-deal_en (accessed on 23 April 2025)
Eccles RG, Krzus MP (2010) One report: integrated reporting for a sustainable strategy. John Wiley and Sons, Inc, New York
Edelman H (2023) Triple Bottom Line framework as a competitive edge in international B2B environment
EDPB (2023) EDPB Annual Report 2023. Avalable At: https://www.edpb.europa.eu/our-work-tools/our-documents/annual-report/edpb-annual-report-2023_en (accessed On 22 April 2025)
Freitag C, Berners-Lee M, Widdicks K, Knowles B, Blair G, Friday A (2021) The climate impact of ICT: A review of estimates, trends and regulations. *arXiv preprint* arXiv:2102.02622
Global Sustainable Investment Alliance (2022) Global sustainable investment review. Available at: https://www.gsi-alliance.org/members-resources/gsir2022/ (accessed on 26 APRIL 2025)
Global E-waste Monitor (2024) Global e-Waste monitor 2024: electronic waste rising five times faster than documented e-waste recycling. Available at: https://shorturl.at/QwJeb (accessed on 24 April 2025)
Google Sustainability (2023) Environmental report 2023. Available at: https://www.gstatic.com/gumdrop/sustainability/google-2023-environmental-report.pdf (accessed on 23 April 2025)
Gupta S, Rhyner J (2022) Mindful application of digitalization for sustainable development: the digitainability assessment framework. Sustainability 14(5):3114
Haug S, Braveboy-Wagner J, Maihold G (2021) The 'Global South'in the study of world politics: examining a meta category. Third World Quarterly 42(9):1923–1944
Heaven WD (2023) ChatGPT is everywhere. Here's where it came from. MIT Technol Rev, 10
Hutson J, Coble K, Kshetri N, Smith A (2023) Exploring the intersection of digital marketing and retail: Challenges and opportunities in AI, privacy, and customer experience. Confronting Secur Priv ChallS Digit Mark, 50–72
IBM (2021) IBM watson advertising accelerator expands to bring AI to Fast-Growing OTT media and streaming landscape. Available at: https://newsroom.ibm.com/2021-04-29-IBM-Watson-Advertising-Accelerator-Expands-to-Bring-AI-to-Fast-Growing-OTT-Media-and-Streaming-Landscape (accessed on 28 March 2024)
IEA (2023) Data centres and data transmission networks. Available at: https://www.iea.org/energy-system/buildings/data-centres-and-data-transmission-networks (accessed on 24 April 2025)
Israel Ansah E (2025) Exploring sustainable AI efficient, ethics and future marketing practices. Int J Innov Sci Res Technol 10(3):1845–1860
ITU (2023) Measuring digital development—facts and figures 2023. Available at: https://shorturl.at/EEEcP (accessed on 23 April 2025)
Lopes A, Carvalho AR (2025) Determinants of voluntary adoption of sasb standards to disclose sustainability issues: an international perspective. Rev Produção E Desenvolv 11(1):e724–e724
Luccioni AS, Strubell E, Crawford K (2025) From efficiency gains to rebound effects: The problem of jevons' paradox in AI's polarized environmental debate. In Proceedings of the 2025 ACM Conference on Fairness, Accountability, and Transparency (pp 76–88). https://dl.acm.org/doi/10.1145/3715275.3732007

Majcher K, Beck THL, Bedre Defolie Ö, Bosoer L, Botta M, Brogi E, Streinz TFW (2024) Charting the digital and technological future of Europe: what priorities for the European Commission in 2024–2029?. European University Institute

Mäkitie T, Hanson J, Damman S, Wardeberg M (2023) Digital innovation's contribution to sustainability transitions. Technol Soc 73:102255

Martínez-Peláez R, Ochoa-Brust A, Rivera S, Félix VG, Ostos R, Brito H, Mena LJ (2023) Role of digital transformation for achieving sustainability: mediated role of stakeholders, key capabilities, and technology. Sustainability 15(14):11221

Mckinsey & Company (2023, June 26) Beyond belt-tightening: How marketing can drive resiliency during uncertain times. https://www.mckinsey.com/capabilities/growth-marketing-and-sales/our-insights/beyond-belt-tightening-how-marketing-can-drive-resiliency-during-uncertain-times

McKinsey & Company (2025) Unlocking the next frontier of personalized marketing. Available at: https://www.mckinsey.com/capabilities/growth-marketing-and-sales/our-insights/unlocking-the-next-frontier-of-personalized-marketing (accessed on 22 April 2025)

Michael Grothaus (2023) Digital privacy had some big wins in 2023 thanks to Apple, Google, and Meta. Available at: https://www.fastcompany.com/90997582/digital-privacy-had-some-big-wins-in-2023-thanks-to-apple-google-and-meta (accessed on 29 March 2024)

Microsoft (2024) Microsoft's fresh approach to accessibility powered by inclusive design. Available at: https://www.microsoft.com/insidetrack/blog/microsofts-fresh-approach-to-accessibility-powered-by-inclusive-design/ (accessed on 21 April 2025)

MIT Technology Review (2025) The download: stereotypes in AI models, and the new age of coding. Available at: https://www.technologyreview.com/2025/04/30/1115952/the-download-stereotypes-in-ai-models-and-the-new-age-of-coding/ (accessed on 27 April 2025).

Moher D, Liberati A, Tetzlaff J, Altman DG (2009) Preferred reporting items for systematic reviews and meta-analyses: the PRISMA statement Ann. Intern Med 151(4):264–269

Nannini L, Bonel E, Bassi D, Maggini MJ (2024) Beyond phase-in: assessing impacts on disinformation of the EU Digital Services Act. AI and Ethics, 1–29

Neuvonen R (2023) Social media councils as self-regulatory bodies. in perspectives on platform power, public values and the potential of social media councils: research report Europe (pp. 49–55)

Nielsen JA, Elmholdt KT, Noesgaard MS (2024) Leading digital transformation: A narrative perspective. Public Adm Rev 84(4):589–603

Padfield RW, Dales A, Mishra J, Smith T (2023) Digital disruption: Towards a research agenda for sustainability and business in a digital world. A Research Agenda for Sustainability and Business, 185–204

Page MJ, McKenzie JE, Bossuyt PM, Boutron I, Hoffmann TC, Mulrow CD, Moher D (2021) Updating guidance for reporting systematic reviews: development of the PRISMA 2020 statement. J Clin Epidemiol 134:103–112

PCAF (2023) PCAF report 2023 report on financed carbon emissions in aareal bank's commercial real estate financing portfolio. Available at: https://shorturl.at/Bi884 (accessed on 22 April 2025)

Peattie K, Belz FM (2025) Perspectives on marketing strategy and sustainability. In Routledge Companion Mark Sustain (pp. 32–44). Routledge

Pew Research Center (2023) A sampling of overarching views on digital change. Available at: https://www.pewresearch.org/internet/2023/06/21/a-sampling-of-overarching-views-on-digital-change/ (accessed on 24 April 2025)

P&G (2023) Climate transition Action plan. Available at: https://s1.q4cdn.com/695946674/files/doc_downloads/esg/PG_CTAP.pdf (accessed on 21March 2024)

Phan-Le NT, Brennan L, Parker L (2024) An Integrated model of the sustainable consumer. Sustainability 16(7):3023

Reed C, Wynn M, Bown R (2025) AI in digital marketing: towards an analytical framework for revealing and mitigating bias. Big Data Cogn Comput 9(2):40

Rendtorff JD (2023) Responsible Business and Integrated Stakeholder Reporting: towards a stakeholder model for integrated reporting of ESG and SDG. Measuring sustainability and CSR: from reporting to decision-making. Springer International Publishing, Cham, pp 49–59

Research, Markets (2023) Digital advertising and marketing global market to reach $1.5 Trillion by 2030: influencer marketing is powerful weapon for digital marketing teams. Available at: https://shorturl.at/gVy4V (accessed on 22 April 2025)

Sachs LE, Toledano P, Mardirossian N (2023) Finance for zero: redefining financial-sector action to achieve global climate goals

Sarkar M (2023) Environmental sustainability under E-commerce: a holistic perspective. Eur J Dev Stud 3(3):1–6

SASB (2022) SASB Standards connect business and investors on the financial effects of sustainability. Available at: https://shorturl.at/jBGAp. (accessed on 22 April 2025)

Schmidt MC, Veile JW, Müller JM, Voigt KI (2023) Industry 4.0 implementation in the supply chain: a review on the evolution of buyer-supplier relationships. International journal of production research, 61(17), 6063–6080

Sharma D, Patel P, Shah M (2023) A comprehensive study on Industry 4.0 in the pharmaceutical industry for sustainable development. Environ Sci Pollut Res, 30(39), 90088–90098

Shehata Nakhil S (2024) Modern urgencies creating a market advantage: Sustainable business models in the fashion industry-a Patagonia Case Study

Singhania M, Bhan I, Chadha G (2024) Sustainable investments: a scientometric review and research agenda. Manag Financ 50(1):266–294

The Shift Project (2024) what virtual worlds for a sustainable Real world? Available at: https://theshiftproject.org/wp-content/uploads/2024/03/The-Shift-Project-What-virtual-worlds-for-a-sustainable-real-world-Final-report-March-2024_ENG.pdf (accessed on 25 April 2025)

Thelisson E, Mika G, Schneiter Q, Padh K, Verma H (2023) Toward responsible AI use: Considerations for sustainability impact assessment. *arXiv preprint* arXiv:2312.11996

Thompson BS, Rust S (2025) Sustainable development of seafood supply chains via blockchain technology: innovation adoption and implementation by businesses and entrepreneurs. Sustain Dev

Tironi E, Sunkel G (2000) The modernization of communications: The media in the transition to democracy in Chile. Democr Media: Comp Perspect 22:165–194

UN (2023) Roadmap for digital cooperation. Available at: https://www.un.org/digital-emerging-technologies/content/roadmap-digital-cooperation (accessed on 28 April 2025)

Unilever (2023) How AI and digital help us innovate faster and smarter. Available at: https://www.unilever.com/news/news-search/2023/how-ai-and-digital-help-us-innovate-faster-and-smarter/ (accessed on 23 April 2025)

Unitar (2023) Global e-Waste Monitor 2024: Electronic Waste Rising Five Times Faster than Documented E-waste Recycling. Available at: https://shorturl.at/ED0Wx (accessed on 26 April 2025)

Usercentrics (2024) Brazil's General Data Protection Law/Lei Geral de Proteção de Dados (LGPD)—an overview. Available at: https://usercentrics.com/knowledge-hub/brazil-lgpd-general-data-protection-law-overview/ (accessed on 28 March 2024)

Wollan R, Davis P, De Angelis F, Quiring K (2017) Seeing beyond the loyalty illusion: It's time you invest more wisely. Dublin: Accent Strat

World Economic Forum (2024) Transforming urban logistics: sustainable and efficient last-mile delivery in cities. Available at: https://shorturl.at/gNxyE (accessed on 26 April 2025)

World Federation of Advertisers (2023) Global media charter 2023. https://www.wfanet.org/globalmediacharter2023

Financial Inclusion and Climate Resilience: An Empirical Analysis for Sustainable Development Planning

Sumera Mehmood and Firdous Ahmad Malik

Abstract This paper investigates the complex association between sustainable development and financial inclusion in Pakistan. Hence, the nature of the research undertaken in this study is one that integrates economic development, the economy, the environment, and renewable energy. The current state of financial inclusion in Pakistan links its status to the different aspects of the development issue. They provide some recommendations to address the financial inclusion gap in Pakistan particularly in the sectors of fossil fuel, public investment in renewable energy, CO2 emission, low carbon imports and exports. The strategies proposed above are not only capable of improving the level of financial inclusion but also capable of improving the economic, environmental, and social development.

Keywords Financial inclusion · Climate resilience · Sustainable development · Renewable energy · Pakistan

1 Introduction

The use of financial services by individuals and firms is referred to as financial inclusion. It is especially relevant to those who can be referred to as the 'unbanked' or 'underbanked' that is, those who are excluded from the financial system (World Economic Forum 2021). This is the case for a large and large population of the people in Pakistan. When we look at the reduced-access context of financial services, it is evident that efforts to compel people and companies to seek those services leads to poverty reduction (State Bank of Pakistan 2021a, b). And we notice that the services delivered are good channels through which people can be given the channels they require to achieve the "economic freedom." It is imperative to identify the factors

S. Mehmood (✉)
Iqra University, Karachi, Pakistan
e-mail: sumera.mashhood@iqra.edu.pk

F. A. Malik
Department of Management, University of People, Pasadena, USA

F. A. Malik et al. (eds.), *Aligning ESG Goals with the UNs Sustainable Development Goals Planning*, World Sustainability Series,
https://doi.org/10.1007/978-3-032-01715-4_9

that cause this financial exclusion so that policy measures can be created that afford proper solutions to such systems. Previous studies have established that the level of income, education, employment, and urban/rural status are some of the factors that affect financial inclusion in Pakistan. There are few researches already which try to see how these factors affects the financial inclusion of men and women in a country like Pakistan (Allen et al. 2016). This knowledge gap reduces the ability to understand the multifaceted underlying processes that define financial inclusion in Pakistan and prevent the design and implementation of female-sensitive interventions to address Pakistani women's challenges (Khattak & Aslam 2022).

The level of public investment in financial inclusion in Pakistan is still significant, which means that there is a need for specific measures and actions. It has implications for the country's economic development and the attainment of the SDGs, especially SDG 7 and 13 on gender. Renewable energy is playing a great role in changing the environment and the economy (Mader 2018). The level of financial inclusion in Pakistan is still very low. Currently, only about one-fifth having formal banking facilities and only 21 percent of the adults have bank accounts according to the State Bank of Pakistan (State Bank of Pakistan 2021a, b). Some of the factors include; poor infrastructure, low level of financial literacy, and socio-economic differences between male and female and between the rural and urban population (Al-Mulali & Ozturk 2016). This section will present a current state of financial inclusion in Pakistan in terms of statistics, current activities, and challenges (Fig. 1).

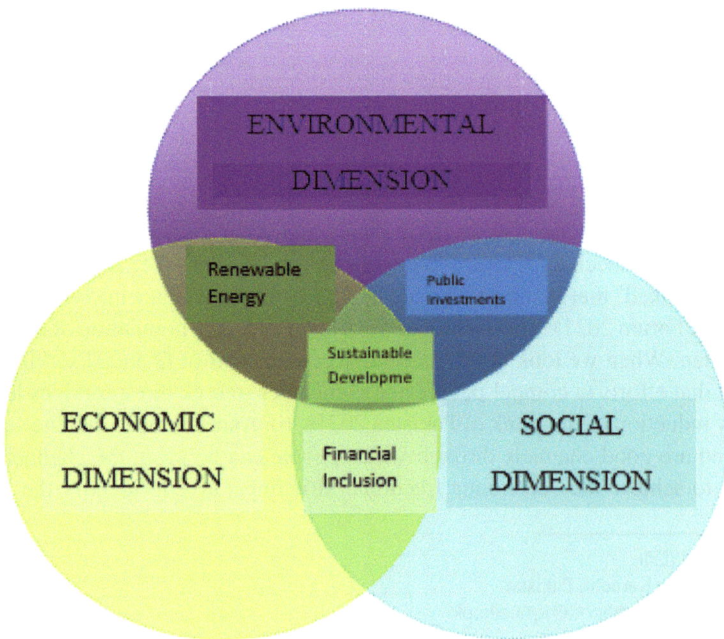

Fig. 1 *Source(s)* Author A framework connecting financial inclusion and sustainable development

Financial inclusion is the ability of the population to access the formal financial services desired by them through convenient and economical ways. It has a significant role in the development process especially in the developing countries such as Pakistan. The correlation between financial inclusion and sustainable development is evident when financial inclusion policies are aligned with the existing key economic and social frameworks for sustainable development (International Renewable Energy Agency 2023; Ali & Malik, 2023). These structures are important for the extension of financial services to the financially excluded population and for the delivery of services to the financially included population. This is too important and too obvious an interrelationship to keep within the exclusive precincts of specialists in either field. It is also important to note that the current discussions on the link between financial inclusion and sustainable development should involve a much wider audience (Financial Stability Board 2017; IRENA, 2023).

Financial inclusion empowers the marginalized groups to access the formal financial system hence promoting economic growth, poverty reduction and social inclusion (Beck et al. 2007). In Pakistan, despite the various measures that have been taken, there are still some barriers that hinder the achievement of the financial inclusion, which in turn affects the sustainable development goals.

The synergies between financial inclusion as a key to sustainable development will be seen in the sort of economic and social returns that heedless benefits users, business entities and governments in the achievement of sustainable development. This implies that financial inclusion will promote savings, credit, and insurance to avoid risks, education, health, and investments for self-employment (Alola 2019). Companies get to access credit facilities that help in the expansion of businesses, innovation, and creation of employment opportunities. In return, governments can use financial inclusion to improve tax revenues, minimize the cost of delivering welfare, and support inclusive economic policies, which will create a sustainable development environment (Bölük & Mert 2014).

Over the past few years, the concept of financial inclusion has emerged as one of the most important factors for the development of the economy in Pakistan. Research has shown that there is a positive relationship between the growth of the financial sector and the enhancement of the economic situation. However, the access to these services is still a challenge to a large population of the population especially women and those in the rural areas hence the issue of inequality in the distribution of the economic gains.

Another key issue is that there has been complain historically low rates of women's financial inclusion. According to the Global Findex 2017, it was established that women are less likely to own or use mobile money or other formal financial services compared to men (Bekhet & Othman 2018). Such differences do exist based on the socio-economic status including income level, employment, and education status, meaning that there is a rationale for specific health programs.

The Government of Pakistan has realized the importance of financial inclusion for sustainable development and therefore launched the National Financial Inclusion Strategy (NFIS) in 2015. The NFIS targeted at least 50% of the adult population to be financially included in the formal financial sector. Nevertheless, the achievement of these goals has been slow and still lags behind the set targets (Ali & Malik 2023).

Digital financial platforms, while promising, also face limitations in Pakistan's context. These are challenges such as; low internet penetration, low literacy levels on the use of electronic gadgets, and weak legal frameworks. As it was mentioned by Asad et al. (2020), FinTech is capable of improving such sectors as agriculture by providing credit access for smallholder farmers, yet it is important to consider these factors.

Access to financial services is considered as one of the key factors for sustainable development in Pakistan. However, structural barriers are still in the way of fairly providing the necessity to all people universally. Reducing the gender gap, enhancing the digital environment, and implementing favorable policies are crucial to turn financial inclusion into a tool for inclusive development (Bashir & Javed 2020). Therefore, more studies are required to establish the relationship between financial inclusion or sustainable environment and the establishment of barriers that hinder their integration.

2 Literature Review

Access to financial services has emerged as a critical factor in the development process, especially in the developing countries such as Pakistan. In this regard, financial inclusion means the provision of financial services to the excluded groups especially the low income earners (Muhammad & Rashid 2019). This paper aims to establish the relationship between financial inclusion and sustainable development in the economic, social, and environmental contexts, with reference to the studies conducted by Beck et al. (2007) in Pakistan.

Pakistan is among the least financially included countries in the world, with only 21% of the adult population having a bank account, which is the second lowest in the world. Such restriction diminishes the opportunity that had been granted to concepts of financial systems in developing the growth capabilities. According to Bigirimana and Hongyi (2018), the level of financial inclusion is directly proportional to the rate of economic development in any country. Similar to many other developing countries, Pakistan has initiated its National Financial Inclusion Strategy in May, 2015 (State Bank of Pakistan 2015).

There is no universally agreed definition of financial inclusion but one of the most commonly used definitions is the one provided by the Rangarajan Committee (2008) which defines financial inclusion as the provision of timely and affordable financial services and credit to the excluded and low income groups. Financial inclusion is generally defined as the provision of financial services such as banking, credit, savings and insurance products to the needy majority with the aim of enhancing on their quality of lives and economic status (World Bank 2014).

The literature thus highlights the role of financial enablers in the advancement of the economy The role of bank in economic development The importance of promotion of the economic development The role of financial service in improvement of the economy Financial inclusion enables credit, investment funds, savings instruments, and insurance to be offered to individuals and firms, thus promoting economic growth (Nathaniel & Khan 2020). In Pakistan, a large number of people work in the informal sector, and the exclusion from the formal financial sector hinders engagement in economic activities.

Some of the studies show that financial inclusion fosters economic growth in the following ways. According to Phiri (2021), it is associated with the overall financial sector development that directs funds to research, innovation, and high return projects. Theoretical roots of what now paralysis the global economy and the traditional finance theory have been seen as early as in Schumpeter's (1934).

According to Beck et al. (2007), inclusion facilitates the flow of savings and channeling of capital to productive activities that enhance trade and business. However, Sadiq et al. (2021) note that financial exclusion is highly detrimental to small and medium enterprises (SMEs), which are significant contributors to the Pakistan economy. They suggest mobile banking, agency banking, and micro finance based financial products in order to create employment and fight poverty.

Bashir & Javed (2020) substantiate that financial inclusion is pro-cyclical as it helps economic risk mitigation, consumption smoothing, and access to funds for emergencies. Hassan and Ali (2023) note that although digital finance has advanced, especially in the urban areas, the uptake in the rural areas is still low because of low financial literacy and lack of trust in the formal financial institutions. Even schemes like agent banking and mobile wallet solutions are still in their infancy due to the absence of such a structure.

Pakistan has its peculiarities: more than half of the population has no access to financial services (Zulfiqar et al. 2016), the rural population is significantly behind, and women are still excluded from the financial system. However, the government has introduced measures like the "Asaan Account" and the "Kamyab Jawan Program", but measures like the transaction taxes such as withdrawal of over Rs. 50,000 in 2015 might have demotivated the people to engage in formal financial sector (Nasir et al. 2022).

Environmental sustainability is one of the subcategories of sustainable development. A closer look at the relation between finance and environmental problems shows that the attempt to meet the need of using financial services can bear positive results, thus contributing to the improvement of the ecological situation by investing in the green technologies and projects. In particular, Pakistani citizens living in vulnerable populations face severe environmental issues including air pollution, water scarcity, and climate change yet they cannot access 'basic' finance products (Sadiq et al. 2021).

Sustainable development can only be achieved if all the stakeholders make a concerted effort towards the achievement of this goal. This section focuses on the role of financial inclusion in sustainable development (State Bank of Pakistan 2021a,

b) Financial inclusion is part of the solution as well as a pathway to sustainable development. When people and enterprises have access to adequate financial resources, they can invest in sustainable endeavors, for instance, the use of renewable energy (Saidi et al. 2018). Sustainable development has many aspects, but in this paper, I concentrate on the environmental sustainability aspect because financial inclusion can positively impact it in a direct and meaningful manner.

According to Karim and Noor (2021), financial inclusion can be used to advance environmental sustainability by providing green financial products to the business and consumers. These products can enhance the use of renewable energy sources, usage of efficient energy structures and other Climate Change mitigation technology. Their study also revealed that Pakistan's financial sector has started financing renewable energy projects especially in the rural areas where solar energy is used for home and small business uses.

According to Ali and Malik (2023), the incorporation of environmental sustainability in the financial services sector can assist Pakistan in achieving its climate objectives. The authors claim that financial institutions should diversify their investment portfolio by investing in climate bonds and sustainable infrastructure. Thus, the integration of financial inclusion with environmental goals in Pakistan can help to build a more sustainable economy for the country's development.

Hussain et al. (2022) also expand on the topic of green finance in Pakistan and the importance of policy measures that would encourage the use of green technologies. They also point out that when financed environment is linked to other forms of financing, notably financial inclusion it is possible to tackle immediate problems such as water scarcity, management of wastes and pollution hence promoting environmental sustainability.

Social sustainability as the aspect that is concerned with equity, equality, and general well-being of people is the other principle in sustainable development. Financial inclusion can go a long way in eradicating social imparity, especially in developing such as Pakistan including women, rural communities and the minorities who are financially excluded. Expanding how income and power are distributed favours the vulnerable stand to benefit in the process. Financial inclusion placed is in how it looks at innovative financial sectors, most especially composed of microfinance institutions, microcredit in lending, and diverse banking approaches applicable to modern society. This section will focus on some of the most important problems (State Bank of Pakistan 2021a, b). The first is about half the world's population. It is an economic problem, of course, but much of the exclusion is also due to social and cultural factors. It's a jigsaw with many parts; however, to start some kind of an action on the economic plane, first there is need for basic financial services for women. That is one of the case studies that this paper will discuss. Another piece in the push for financial inclusion in education (The Financial Times 2023).

Khattak and Aslam (2022) argue that financial inclusion is important for poverty alleviation and social inclusion in Pakistan. According to the authors, financial inclusion enables individuals to access financial services that help them to manage their finances, invest in education, health, and shelter, and hence improve their standard of living. They observe that microfinance institutions that offer small loans to the low-income households have played a significant role in helping many people to come out of the vice.

The women in Pakistan are still restricted from accessing financial services due to cultural and social barriers. As stated by Javed and Aftab (2023), the financial inclusion has provided women with the opportunity to access savings and credit products but there are still gaps. The authors state that mobile money and digital banking empower women to overcome physical and social constraints and gain more financial independence (United Nations Development Programme, 2019). Furthermore, Khan and Fatima (2021) examine the impact of mobile banking on the disadvantaged groups. They observe that mobile financial services offer an opportunity for the poor, especially women and those in rural areas, to access financial services. This is because through their research they discovered that mobile money has expanded access to basic banking in the regions of Pakistan where the usual amenities are not so available.

3 Objectives and Hypothesis

To examine the relationship between financial inclusion and sustainable development indicators (economic growth, CO_2 emissions, public investment in renewable energy) in Pakistan.

To assess the impact of mobile banking services on financial accessibility among disadvantaged groups, particularly women and rural populations in Pakistan.

H1: Financial inclusion has a significant positive impact on sustainable development in Pakistan, including economic, environmental, and social dimensions.

H2: Mobile banking services significantly increase financial accessibility for disadvantaged groups, especially women and rural residents, in Pakistan (Tables 1, 2).

4 Methodology

SDGs are the United Nations' development agenda for the next fifteen years to improve the living standards of people in the world which was passed in 2015. Most of these goals such as; SDG 7 Affordable and Clean Energy and Climate Change (SDG 13) are closely linked to financial inclusion. In the contemporary world striving to attain sustainable development and poverty eradication, the importance of financial inclusion has been acknowledged (World Bank 2020) (Table 3).

Following Pesaran and Shin (1998), the three models to study the long-run relationship between variables are presented below.

Table 1 Renewable energy and economic growth

Author(s)	Method	Period	Country	Results
Apergis and Payne (2010)	Panel Cointegration & Causality Tests	1985–2005	OECD Countries	Positive long-run relationship between renewable energy and economic growth
Sadorsky (2009)	Panel Data Analysis	1980–2005	G7 Countries	Renewable energy consumption positively affects economic growth
Shahbaz et al. (2020)	ARDL Bounds Testing	1990–2015	Renewable Energy Leading Countries	Renewable energy enhances economic growth
Bhattacharya et al. (2016)	Panel Data Analysis	1991–2012	38 Countries	Renewable energy positively impacts economic growth
Bölük and Mert (2014)	Panel Cointegration	1980–2009	G7 Countries	Renewable energy consumption has a strong effect on GDP
Pao and Fu (2013)	VAR & Causality Analysis	1980–2010	Brazil	Unidirectional causality from economic growth to renewable energy
Inglesi-Lotz (2016)	Panel Data Regression	1990–2010	34 OECD Countries	Positive relationship between renewable energy and GDP
Al-Mulali and Ozturk (2016)	Environmental Kuznets Curve (EKC) Model	1980–2012	MENA Region	Renewable energy helps reduce environmental degradation and supports GDP growth
Alola (2019)	Time Series Analysis	1970–2016	USA	Renewable energy contributes to long-term economic sustainability
Omri and Kahouli (2014)	Panel VAR Model	1990–2010	MENA Region	Positive bidirectional causality between renewable energy and GDP
Chen et al. (2020)	Panel Data Estimation	1995–2017	Emerging Economies	Renewable energy consumption boosts economic growth
Bekhet and Othman (2018)	Dynamic ARDL Model	1980–2015	Malaysia	Renewable energy helps achieve sustainable development
Destek and Sarkodie (2019)	Environmental Kuznets Curve	1990–2015	Global	Energy efficiency and renewable energy help GDP growth while reducing pollution
Saidi et al. (2018)	GMM Estimation	1990–2016	OECD Countries	Renewable energy positively affects GDP and reduces CO2 emissions
Nathaniel and Khan (2020)	Panel Cointegration	1990–2018	ASEAN Countries	Renewable energy helps economic growth and reduces environmental impact
Payne (2012)	Granger Causality	1949–2006	USA	No significant causality between renewable energy and GDP

(continued)

Table 1 (continued)

Author(s)	Method	Period	Country	Results
Stern and Kander (2012)	Structural VAR	1850–2009	UK	Renewable energy consumption positively influences economic growth
Menyah and Wolde-Rufael (2010)	Toda-Yamamoto Causality	1965–2007	South Africa	No causality between renewable energy and GDP
Marques and Fuinhas (2012)	Panel Data Analysis	1990–2007	EU Countries	Renewable energy boosts GDP growth but with some variations
Kahia et al. (2017)	VECM Model	1980–2012	MENA Region	Bidirectional causality between renewable energy and GDP
Bhattacharya et al. (2017)	Dynamic Panel Estimation	1990–2013	45 Countries	Renewable energy is a significant factor in GDP growth
Singh et al. (2019)	ARDL Bounds Testing	1980–2018	India	Renewable energy supports GDP and reduces carbon emissions
Doğan and Seker (2016)	Panel Data Approach	1990–2012	EU Countries	Renewable energy consumption fosters GDP growth
Rafindadi and Ozturk (2017)	VECM & Granger Causality	1980–2014	Japan	Long-term causality between renewable energy and economic growth
Wang et al. (2020)	Panel Quantile Regression	1990–2017	China	Renewable energy helps economic expansion but has nonlinear effects
Sebri and Ben-Salha (2014)	Panel Cointegration	1971–2010	BRICS Countries	Renewable energy consumption positively impacts economic growth
Fotourehchi (2017)	ARDL & Johansen Cointegration	1985–2015	Developing Countries	Renewable energy significantly boosts GDP
Eren et al. (2019)	FMOLS & DOLS	1990–2016	Emerging Markets	Strong linkage between renewable energy and GDP growth
Abid (2016)	Panel ECM	1995–2013	African Countries	Renewable energy supports sustainable economic growth
Yildirim et al. (2012)	Toda-Yamamoto & Granger Causality	1980–2009	Turkey	No significant relationship found between renewable energy and GDP

Model 1: Renewable Energy and Economic Growth

$$\Delta REt = \mu + \sum \beta i \Delta REt - i + \sum \beta j \Delta GDPt - j + \alpha 1REt - 1 + \alpha 2GDPt - 1 + \varepsilon 1, t$$

$$\Delta GDPt = \phi + \sum \beta i \Delta REt - \mathrm{i} + \sum \beta j \Delta GDPt - \mathrm{j} + \alpha 1\mathrm{GDPt} - 1 + \alpha 2\mathrm{REt} - 1 + \varepsilon 2, \mathrm{t}$$

Table 2 Literature review of renewable energy, CO_2 emissions, and economic growth

Author	Method	Period	Country	Results
Abid (2016)	Panel Data Analysis	1971–2012	Sub-Saharan Africa	Renewable energy consumption reduces CO2 emissions while supporting economic growth
Al-Mulali and Ozturk (2016)	Panel Cointegration & Causality	1980–2012	MENA Countries	Trade and renewable energy reduce CO2 emissions, supporting the environmental Kuznets curve
Alola (2019)	ARDL & Granger Causality	1980–2016	United States	Renewable energy consumption supports economic growth while reducing emissions
Apergis and Payne (2010)	Panel Cointegration & Causality	1985–2005	OECD Countries	Bidirectional causality between renewable energy and economic growth
Balsalobre-Lorente et al. (2018)	ARDL Bounds Test	1990–2015	EU Countries	Renewable electricity and economic growth reduce CO_2 emissions
Bekhet and Othman (2018)	VECM	1971–2015	Malaysia	Renewable energy consumption reduces carbon emissions in the long run
Bhattacharya et al. (2016)	Panel Data Analysis	1990–2012	Top 38 Countries	Strong evidence that renewable energy consumption drives economic growth
Bölük and Mert (2014)	ARDL	1960–2010	Turkey	Renewable energy significantly reduces CO_2 emissions

(continued)

Table 2 (continued)

Author	Method	Period	Country	Results
Chen et al. (2020)	Time-Series Analysis	1990–2017	China	Renewable energy negatively impacts CO_2 emissions while supporting economic growth
Dogan and Seker (2016)	FMOLS & DOLS	1990–2012	Top Renewable Energy Countries	Renewable energy mitigates CO_2 emissions in developed economies
Eren et al. (2019)	Panel Data Analysis	1990–2016	Emerging Markets	Financial development and renewable energy positively impact economic growth
Fotourehchi (2017)	Panel Regression	1990–2014	Developing Countries	Renewable energy consumption promotes economic growth
Inglesi-Lotz (2016)	Panel Cointegration	1990–2012	34 OECD Countries	Renewable energy positively affects economic growth
Kahia et al. (2017)	VECM	1980–2013	MENA Net Oil Importers	Renewable energy reduces CO_2 emissions and fosters economic growth
Koçak and Şarkgüneşi (2017)	Panel Data Analysis	1990–2016	Developing Countries	Renewable energy consumption has a significant impact on economic growth
Magazzino (2016)	Cointegration & Granger Causality	1970–2012	Italy	Bidirectional causality between renewable energy and economic growth
Marques and Fuinhas (2012)	GMM Estimation	1990–2007	European Countries	Public policies supporting renewable energy drive economic growth

(continued)

Table 2 (continued)

Author	Method	Period	Country	Results
Menyah and Wolde-Rufael (2010)	Toda-Yamamoto Causality	1965–2006	South Africa	Renewable energy reduces CO_2 emissions without hindering economic growth
Nathaniel & Khan (2020)	Panel Data Analysis	1990–2016	ASEAN Countries	Trade and urbanization influence renewable energy consumption and CO_2 emissions
Omri & Kahouli (2014)	Simultaneous Equations	1990–2011	MENA Countries	Strong link between energy consumption, economic growth, and CO_2 emissions
Pao & Fu (2013)	Cointegration & Granger Causality	1980–2010	Brazil	Renewable energy contributes to long-term economic growth
Rafindadi & Ozturk (2017)	ARDL & Granger Causality	1971–2012	Nigeria	Financial development and renewable energy enhance economic growth
Sadorsky (2009)	Panel Data Analysis	1990–2006	Emerging Economies	Renewable energy consumption is a key driver of economic growth
Saidi et al. (2018)	Panel Cointegration	1990–2014	19 Emerging Economies	Renewable energy reduces CO_2 emissions and fosters economic growth
Sebri & Ben-Salha (2014)	Panel Cointegration	1980–2010	BRICS Countries	Renewable energy consumption mitigates CO2 emissions in BRICS nations
Shahbaz et al. (2020)	ARDL & FMOLS	1990–2018	China	Technological innovation and renewable energy investment lower CO_2 emissions

(continued)

Table 2 (continued)

Author	Method	Period	Country	Results
Singh et al. (2019)	Panel Regression	1990–2015	Global Panel	Renewable energy investment is crucial for sustainable economic growth
Stern and Kander (2012)	Growth Rate Model	1971–2008	Global	Renewable energy reduces CO_2 emissions while ensuring long-term growth
Wang et al. (2020)	Panel Cointegration	1990–2017	Various Countries	Renewable energy is essential for carbon neutrality and economic growth

Table 3 Summary statistics

Variable	Obs	Mean	Std. Dev	Min	Max
Renewable energy(−1)	8	1.248476	1.864162	0.0031	3.998882
Ff_co2_emissions_per_cap	8	−0.351725	0.872186	−1.397564	0.515730
Ff_co2_emissions_per_cap(−1)	8	0.936607	0.948938	0.052500	2.232647
Low_carbon_export_$	8	−0.214570	0.632305	−1.133097	0.286500
Low_carbon_export_$(−1)	8	0.489322	0.691598	0.106854	1.525979
Low_carbon_import_$	8	−0.428924	0.903018	−1.775777	0.109500
Re_public_investment_$	8	0.260641	0.416164	0.000782	0.874600

Model 2: CO_2 Emissions and Economic Growth

$$\Delta GDPt = \mu + \sum \beta i \Delta GDPt - \mathrm{i} + \sum \beta j \Delta FF\_\mathrm{CO_2t} - \mathrm{j} + \alpha 1\mathrm{GDPt} - 1 + \alpha 2\mathrm{FF\_CO_2t} - 1 + \varepsilon 1, \mathrm{t}$$

$$\Delta FF\_\mathrm{CO_2t} = \mu + \sum \beta i \Delta GDPt - \mathrm{i} + \sum \beta j \Delta FF\_\mathrm{CO_2t} - \mathrm{j} + \alpha 1\mathrm{FF\_CO_2t} - 1 + \alpha 2\mathrm{GDPt} - 1 + \varepsilon 2, \mathrm{t}$$

Model 3: Renewable Energy and CO_2 Emissions

$\Delta REt = \mu + \sum \beta i \Delta REt - i + \sum \beta j \Delta FF\_CO_2t - j + \alpha 1REt - 1 + \alpha 2FF\_CO_2t - 1 + \varepsilon 1,t$

$$\Delta FF\_CO_2t = \mu + \sum \beta i \Delta REt - i + \sum \beta j \Delta FF\_CO_2t - j + \alpha 1FF\_CO_2t - 1 + \alpha 2REt - 1 + \varepsilon 2, t$$

Extending the access to the formal financial services to people and communities is viewed as a way to assist them and enhance their financial well-being, especially in the developing countries where a large portion of the population remains unbanked. Financial inclusion, which is the extent to which individuals and households can access financial services, is widely known to have a direct relationship with the economic growth particularly in the developing country like Pakistan. It extends to payment services, financial literacy, and other forms of insurance and credit facilities especially to the needy and the poor (World Bank Group, 2018).

These services are essential in the process of ensuring economic security and human development since the excluded groups can be assisted to integrate into the economy. This approach safeguards the economy and fosters its growth through integration of the excluded groups into the relevant sector, human and economy development (World Economic Forum, 2021). In addition, while financial inclusion is one of the most important and widely used goals of the SDG, in particular, to achieve SDG-7 (Affordable and Clean Energy) and SDG-13 (Climate action). The last goal is to foster the partnership for sustainable development at the global level. The 2030 Agenda is for all people of the world and what is encouraging is that the call to action is made to all countries including the developed and the developing world. It requires the involvement of the government, business organization and non-government organizations.

Sadiq et al. (2021) have identified that financial inclusion supports SDG 10 (Reduced Inequalities) because it provides equal chances for everyone to access financial services. Thus, Pakistan can play its role in enhancing the social and economic justice for the excluded groups, which is in the line with the national and international development agenda. The United Nations adopted the Sustainable Development Goals also referred to as the Global Goals in 2015 now as 17 goals for development by the year 2030. Most of these goals require financial inclusion; for instance; SDG 7 (Affordable and Clean Energy), and SDG 13 (Climate Action).

Wagner (2019) stated that financial inclusion helps in the attainment of several of the SDGs such as the SDG 1: No Poverty and SDG 10: Reduced Inequalities since it provides the disadvantaged groups with the right financial tools to improve their lives. By so doing, the vulnerable groups are in a better position to cope with shocks and undertake income-generating activities thus eradicating poverty and stabilizing the economy.

Shah and Ali (2023) opined that financial inclusion contributes to the achievement of SDG 8 or Decent work and Economic growth through enterprise among Female and rural persons. They have also established the extent to which the youth is able to access the digital finance services and the micro financing institutions in relation to capital, employment and economic development.

Javed and Khan (2024) have chosen financial inclusion as the variable that can help in achieving SDG 5 (Gender Equality) in Pakistan. They note that the mobile money and digital wallets have been a boon for women especially in the rural and conservative areas where physical cash services are limited. From their study, the measures that can be taken to address the gender gap in the utilization of financial services include increasing financial literacy and women's participation in the formal financial system.

Hussain et al. (2023) examine the link between financial inclusion and SDG 13 (Climate Action). They argue that Pakistan has to face growing environmental challenges and that green finance and environmental sustainability products such as climate loans and insurance are the solutions. Pakistan can contribute to the global climate goals and foster its economy by promoting green finance.

Sadiq et al. (2021) further argue that; mobile banking and other FinTech solutions offer opportunities for improving the BOP and other vulnerable groups' access to finance, through which the groups can participate in sustainable development. This inclusion directly relates to the Zero Hunger (SDG 2), Clean Water and Sanitation (SDG 6), and Responsible Consumption and Production (SDG 12) because proper finance improves farming, water and sanitation, and lastly, the use of sustainable resources.

In this regard, the role of financial inclusion in achieving the SDGs in Pakistan is an opportunity to address various development challenges comprehensively and sustainably. Pakistan can improve the financial access, reduce the inequality, and promote the global goals by improving the access to the financial services, particularly through the digital channels.

The present study seeks to investigate the relationship between FI and growth in the light of the following empirical model using annual time series data for the sample period of 2003–2023 The sample period has been chosen due to data availability. The dependent variable, renewable yearly energy, is obtained from IRENA. FI index (independent variable) is obtained from World Bank and constructed through various proxies of FI such as number of low carbon imports (LCI), low carbon export (LCE), Fossil Fuel CO_2 (FF CO_2), Public Investments in Renewable Energy (PIRE) in various projects on Renewable energy. The data for the year 2010–2011 are also obtained from the World Bank and IRENA reports. PIRE is computed on the basis of annual data of various major projects that have an impact on the economy.

The model evaluates the association between financial inclusion and sustainable development

$$\text{Financial Inclusion} = \text{f sustainable development}$$

$$\text{Sustainable Development} = \text{f financial inclusion}$$

5 Results and Discussion

Renewable Energy, Low Carbon Import $, FF Fossil CO_2 Tons, and FF CO_2 Emissions Per Capita all show **statistical significance** ($p < 0.01$), making them critical drivers in the analysis. Their relationships should be further examined and leveraged in decision-making or policy recommendations (Tables 4, 5, 6).

Method	Statistic	Prob.**	Cross-sections	Obs
Null: Unit root (assumes common unit root process)				
Levin, Lin & Chu t*	−2.17849	0.0147	5	103
Null: Unit root (assumes individual unit root process)				
Im, Pesaran and Shin W-stat	−2.97066	0.0015	5	103
ADF—Fisher Chi-square	30.4461	0.0007	5	103
PP—Fisher Chi-square	31.3981	0.0005	5	105
** Probabilities for Fisher tests are computed using an asymptotic Chi				
-square distribution. All other tests assume asymptotic normality				

Significant**Granger causality was found in:**
FF_CO_2_Emissions_Per_Cap → Renewable Energy (p = 0.0456)
RE_Public_Investment $ → Renewable Energy (p = 0.0337)
FF_CO_2_Emissions_Per_Cap → Low Carbon Import $ (p = 0.0213)
RE_Public_Investment $ → FF_CO_2_Emissions_Per_Cap (p = 0.0364)
Marginal Granger causality (weak evidence) was observed in:
Low Carbon Export $ → Renewable Energy ($p = 0.0786$)
Low Carbon Import $ → Renewable Energy ($p = 0.0734$)
Low Carbon Import $ → FF_CO_2_Emissions_Per_Cap ($p = 0.0660$)
Low Carbon Import $ ↔ Low Carbon Export $ (p = 0.0594 and 0.0920)
RE_Public_Investment $ → Low Carbon Import $ ($p = 0.0898$)

Fossil Fuel CO_2 Emissions Per Capita does not Granger Cause Renewable Energy: The p-value is 0.0456, which is significant ($p < 0.05$). This means that there is a relationship between FF_CO_2_Emissions_Per_Cap and Renewable Energy. Renewable

Table 4 Unit root test

	Level	First difference
Renewable energy	0.8424	0.0018
Low carbon import $	0.1494	0.0085
Low carbon export $	0.0772	
FF Fossil CO_2 Tons	0.5692	0.0045
FF CO_2 emissions per capita	0.3704	0.006
RE public investment $	0.0001	

Table 5 ARDL bounds test

Test statistic	Value	k
F-statistic	0.415322	4
Critical value bounds		
Significance	I0 Bound	I1 Bound
10%	2.45	3.52
5%	2.86	4.01
2.50%	3.25	4.49
1%	3.74	5.06
Test equation:		
Dependent variable: D(RENEWABLE_ENERGY)		
Method: least squares		
Sample: 2004 2020		
Included observations: 17		
Variable	Coefficient	Std. Error
D(FF_CO_2_EMISSIONS_PER_CAP)	−0.007001	0.023833
D(LOW_CARBON_EXPORT_$)	−0.100768	0.077156
C	0.095951	0.648427
FF_CO_2_EMISSIONS_PER_CAP	−0.070969	0.638066
LOW_CARBON_EXPORT_$(-1)	0.07926	0.086893
LOW_CARBON_IMPORT_$(−1)	−0.007861	0.070925
RE_PUBLIC_INVESTMENT_$(−1)	−0.006356	0.003908
RENEWABLE_ENERGY(−1)	−0.098093	0.172485
R-squared	0.096424	
Adjusted R-squared	0.104753	
S.E. of regression	0.030673	
Sum squared resid	0.008467	
Log likelihood	40.51856	
F-statistic	0.267453	
Prob(F-statistic)	0.61798	

Energy does not Granger Cause FF_CO_2_Emissions_Per_Cap: The p-value is 0.5007 which is greater than 0.05. In this case, there is no causality that can be established in this direction. RE renewable Public Investment $ does not Granger Cause FF_CO_2_Emissions_Per_Cap: The p-value is 0.0364, which is significant ($p < 0.05$). This means that RE_Public_Investment $ Granger causes FF_CO_2_Emissions_Per_Cap. The *p*-value of FFCO_2_Emissions_Per_Cap does not Granger Cause RE_Public_Investment $ is 0.7854, which is insignificant. Low Carbon Import $ does not Granger Cause Low Carbon Export $: The p-value is 0.0594, which is marginally significant ($p < 0.10$). This indicates that there is a very low level of causality evidence. Low

Table 6 Granger causality tests

Pairwise granger causality tests			
Sample: 2000 2023			
Null Hypothesis:	Obs	F-Statistic	Prob
FF_CO_2_EMISSIONS_PER_CAP does not Granger Cause RENEWABLE_ENERGY	21	0.53476	0.0456
RENEWABLE_ENERGY does not Granger Cause FF_CO_2_ EMISSIONS_PER_CAP		0.72254	0.5007
LOW_CARBON_EXPORT_$ does not Granger Cause RENEWABLE_ENERGY	19	0.95301	0.0786
RENEWABLE_ENERGY does not Granger Cause LOW_ CARBON_EXPORT_$		0.48568	0.6253
LOW_CARBON_IMPORT_$ does not Granger Cause RENEWABLE_ENERGY	19	0.26168	0.0734
RENEWABLE_ENERGY does not Granger Cause LOW_ CARBON_IMPORT_$		0.26789	0.7688
RE_PUBLIC_INVESTMENT_$ does not Granger Cause RENEWABLE_ENERGY	19	0.42341	0.0337
RENEWABLE_ENERGY does not Granger Cause RE_ PUBLIC_INVESTMENT_$		0.18323	0.3351
LOW_CARBON_EXPORT_$ does not Granger Cause FF_CO_2_ EMISSIONS_PER_CAP	18	0.02590	0.3858
FF_CO_2_EMISSIONS_PER_CAP does not Granger Cause LOW_CARBON_EXPORT_$		0.61743	0.5544
LOW_CARBON_IMPORT_$ does not Granger Cause FF_CO_2_ EMISSIONS_PER_CAP	18	0.14552	0.0660
FF_CO_2_EMISSIONS_PER_CAP does not Granger Cause LOW_CARBON_IMPORT_$		0.49197	0.0213
RE_PUBLIC_INVESTMENT_$ does not Granger Cause FF_ CO_2_EMISSIONS_PER_CAP	19	0.23718	0.0364
FF_CO_2_EMISSIONS_PER_CAP does not Granger Cause RE_ PUBLIC_INVESTMENT_$		0.24584	0.7854
LOW_CARBON_IMPORT_$ does not Granger Cause LOW_ CARBON_EXPORT_$	19	0.42912	0.0594
LOW_CARBON_EXPORT_$ does not Granger Cause LOW_ CARBON_IMPORT_$		0.84329	0.0920

(continued)

Table 6 (continued)

Pairwise granger causality tests			
Sample: 2000 2023			
RE_PUBLIC_INVESTMENT_$ does not Granger Cause LOW_CARBON_EXPORT_$	16	0.08156	0.9222
LOW_CARBON_EXPORT_$ does not Granger Cause RE_PUBLIC_INVESTMENT_$		0.25792	0.7772
RE_PUBLIC_INVESTMENT_$ does not Granger Cause LOW_CARBON_IMPORT_$	16	0.93992	0.0898
LOW_CARBON_IMPORT_$ does not Granger Cause RE_PUBLIC_INVESTMENT_$		0.62541	0.5530

Carbon Export $ does not Granger Cause Low Carbon Import $: The p-value is 0.0920, which is also marginally significant ($p < 0.10$). This suggests that there is a very low level of bidirectional causality (Table 7).

The p-value of 0.0364 is less than 0.05, therefore, we can conclude that re_public_investment_$ Granger causes ff_co$_2$_emissions_per_cap. The p-value of 0.092 is less than 0.1, which indicates that there is weak evidence that low_carbon_import_$ Granger causes low_carbon_export since it Granger causes ff_co$_2$_emissions_per_cap, further analysis should be done to establish causality and policy implications. Since there are not many significant relationships, try using different lag lengths or adding other variables to make the model more stable.

6 Policy Implications

This study also highlights the importance of financial inclusion in the process of sustainable development in Pakistan. The analysis demonstrates that the relationship between financial inclusion and four dimensions of economic development is quite intricate as environmental sustainability and social equity. The current status of financial inclusion in Pakistan reveals that there are still many barriers that are socio-economic, geographical and gender based. Male, lack of infrastructure and poor financial education are still areas of concern since they increase due to high illiteracy levels among individuals in the rural areas and especially women.

The proposed strategies include the deployment of investments in sustainable and renewable power systems, diversification out of fossil fuels and the encouragement of lesser carbon trade. These interventions are in line with the Sustainable Development Goals (SDGs) especially the economic growth (SDG 8), affordable and clean energy (SDG 7) and climate change (SDG 13). If these challenges are met, Pakistan can create a more open and effective financial environment that will improve the country's economic activity and promote sustainability of the environment and society.

Table 7 Pairwise granger causality tests

Null hypothesis	Obs	F-Statistic	Prob
ff_co2_emissions_per_cap does not Granger-cause renewable_energy	21	0.53476	0.2456
renewable energy does not granger cause ff_co2_emissions_per_cap		0.72254	0.5007
ff_fossil_co2_tons does not granger cause renewable_energy	21	0.27625	0.1349
renewable energy does not granger cause ff_fossil_co2_tons		0.38463	0.6868
low_carbon_export_$ does not granger cause renewable_energy	19	0.95301	0.1786
renewable_energy does not granger cause low_carbon_export_$		0.48568	0.6253
low_carbon_import_$ does not granger cause renewable_energy	19	0.26168	0.7734
renewable_energy does not granger cause low_carbon_import_$		0.26789	0.7688
re_public_investment_$ does not granger cause renewable_energy	19	0.42341	0.2737
renewable_energy does not granger cause re_public_investment_$		0.18323	0.3351
ff_fossil_co2_tons does not granger cause ff_co2_emissions_per_cap	21	1.40232	0.2747
ff_co2_emissions_per_cap does not granger cause ff_fossil_co2_tons		2.32567	0.1298
low_carbon_export_$ does not granger cause	18	1.0259	0.3858
ff_co2_emissions_per_cap does not granger cause		0.61743	0.5544
low_carbon_import_$ does not granger cause	18	0.14552	0.866
ff_co2_emissions_per_cap does not granger cause		0.49197	0.1213
re_public_investment_$ does not granger cause	19	0.23718	0.0364
ff_co2_emissions_per_cap does not granger cause re_public_investment_$		0.24584	0.7854
low_carbon_export_$ does not granger cause ff_fossil_co2_tons	18	0.59873	0.564
ff_fossil_co2_tons does not granger cause low_carbon_export_$		0.21837	0.8067
low_carbon_import_$ does not granger cause ff_fossil_co2_tons	18	0.333	0.7227

(continued)

Table 7 (continued)

Null hypothesis	Obs	F-Statistic	Prob
ff_fossil_co2_tons does not granger cause low_carbon_import_$		1.50457	0.2584
re_public_investment_$ does not granger cause ff_fossil_co2_tons	19	3.21393	0.071
ff_fossil_co2_tons does not granger cause re_public_investment_$		0.45473	0.6437
low_carbon_import_$ does not granger cause low_carbon_export_$	19	0.42912	0.6594
low_carbon_export_$ does not granger cause low_carbon_import_$		0.84329	0.092
re_public_investment_$ does not granger cause low_carbon_export_$	16	0.08156	0.9222
low_carbon_export_$ does not granger cause re_public_investment_$		0.25792	0.7772
re_public_investment_$ does not granger cause low_carbon_import_$	16	0.93992	0.1898
low_carbon_import_$ does not granger cause re_public_investment_$		0.62541	0.553

7 Conclusion

The study suggests a coordinated strategy, which has to include public spending increase, sound institutions, and measures sensitive to gender. The role that has been highlighted above will be monumental and therefore it will require the support of the government, the financial institutions and international organizations in order to see a marked improvement in the level of financial inclusion. In this way, Pakistan can open up opportunities for economic development, address environmental issues, and work towards a better future.

References

Abid M (2016) Impact of economic, financial, and institutional factors on CO_2 emissions: Evidence from Sub-Saharan Africa economies. Util Policy 41:85–94. https://doi.org/10.1016/j.jup.2016.06.009

Ali A, Malik MA (2023) Financial inclusion and environmental sustainability: A case study of Pakistan. J Sustain Financ 14(1):15–29

Allen F, Demirgüç-Kunt A, Klapper L, Martinez Peria MS (2016) The foundations of financial inclusion: Understanding ownership and use of formal accounts. J Financ Intermediation 27:1–30

Al-Mulali U, Ozturk I (2016) The investigation of environmental Kuznets curve hypothesis in the MENA region: The role of renewable energy and non-renewable energy. Renew Sustain Energy Rev 66:1192–1205. https://doi.org/10.1016/j.rser.2016.08.031

Alola AA (2019) The trilemma of trade, monetary and immigration policies in the United States: Accounting for environmental sustainability. Sci Total Environ 658:260–267. https://doi.org/10.1016/j.scitotenv.2018.12.212

Apergis N, Payne JE (2010) Renewable energy consumption and economic growth: Evidence from a panel of OECD countries. Energy Policy 38(1):656–660. https://doi.org/10.1016/j.enpol.2009.09.002

Asad M, Sharif MN, Aleemi AR (2020) Financial inclusion and economic development in Pakistan: A review. Pakistan Business Review 22(2):245–264

Balsalobre-Lorente D, Shahbaz M, Roubaud D, Farhani S (2018) How economic growth, renewable electricity and natural resources contribute to CO_2 emissions? Energy Policy 113:356–367. https://doi.org/10.1016/j.enpol.2017.10.050

Bashir T, Javed H (2020) The role of financial inclusion in promoting economic growth in Pakistan. J Econ Stud 11(3):112–128

Beck T, Demirgüç-Kunt A, Levine R (2007) Finance, inequality, and the poor. J Econ Growth 12(1):27–49

Bekhet HA, Othman NS (2018) The role of renewable energy in validating dynamic interaction between CO_2 emissions and GDP towards sustainable development in Malaysia. Energy Econ 72:47–61. https://doi.org/10.1016/j.eneco.2018.03.001

Bhattacharya M, Paramati SR, Ozturk I, Bhattacharya S (2016) The effect of renewable energy consumption on economic growth: Evidence from top 38 countries. Appl Energy 162:733–741. https://doi.org/10.1016/j.apenergy.2015.10.104

Bigirimana M, Hongyi X (2018) Research on relationship between financial inclusion and economic growth of Rwanda: Evidence from commercial banks with ARDL approach. Int J Innov Econ Dev 4(1):7–18. https://doi.org/10.18775/ijied.1849-7551-7020.2015.41.2001

Bölük G, Mert M (2014) Fossil & renewable energy consumption and economic growth: A panel data analysis for G7 countries (1980–2009). Renewable Energy 65:6–15. https://doi.org/10.1016/j.renene.2013.08.006

BRICS Countries and renewable energy development: empirical evidence from 1988–2017. (n.d.). Front Environ Sci. https://doi.org/10.3389/fenvs.2022.878394

Carney M (2015) Breaking the tragedy of the horizon—Climate change and financial stability. Bank of England Speech

Chaudhry IS, Malik S (2018) Financial inclusion and human development in Pakistan: An empirical analysis. Pak J Commer Soc Sci 12(1):23–44

Chen Y, Zhao J, Lai Z, Lu Y, Wang Z (2020) Does energy efficiency and renewable energy consumption contribute to economic growth? Empirical Evidence from Emerging Economies. Energies 13(7):1522. https://doi.org/10.3390/en13071522

Demirgüç-Kunt A, Klapper L, Singer D, Ansar S (2020) The Global Findex Database 2017: Measuring financial inclusion and the fintech revolution. World Bank Group

Destek MA, Sarkodie SA (2019) Investigation of environmental Kuznets curve for ecological footprint: The role of energy and financial development. Sci Total Environ 650:2483–2489. https://doi.org/10.1016/j.scitotenv.2018.10.017

Doğan E, Şeker F (2016) The influence of real output, renewable and non-renewable energy, trade and financial development on carbon emissions in the top renewable energy countries. Renew Sustain Energy Rev 60:1074–1085. https://doi.org/10.1016/j.rser.2016.02.006

Dupas P, Robinson J (2013) Why don't the poor save more? Evidence from health savings experiments. American Economic Review 103(4):1138–1171

Eren BM, Taspinar N, Gokmenoglu KK (2019) The impact of financial development and economic growth on renewable energy consumption: Empirical analysis of India. Sci Total Environ 663:189–197

Financial Stability Board (2017) Financial stability implications of financial inclusion: A policy overview

Fotourehchi Z (2017) Renewable energy consumption and economic growth: A panel data approach for developing countries. Renew Energy 95:678–686. https://doi.org/10.1016/j.renene.2016.04.078

Gonzalez-Castellanos A, Chaves-Ávila JP, Azevedo IM (2018) Economic impact of wind generation penetration in the Colombian electricity market. Energy Policy 117:350–365. https://doi.org/10.1016/j.enpol.2018.03.024

Gyimah D, Chen W, Antwi S (2023) Exploring the impact of renewable energy on economic growth and carbon emissions: A panel data approach. PLoS ONE 18(1):e0279234. https://doi.org/10.1371/journal.pone.0279234

Hara K (2022) Economics of 100% renewable power systems. *arXiv* preprint arXiv:2205.15451. https://arxiv.org/abs/2205.15451

Hassan S, Ali K (2023) Financial inclusion in Pakistan: Barriers and opportunities. Int J Financ Serv 8(4):35–49

Honohan P (2008) Cross-country variation in household access to financial services. J Bank Finance 32(11):2493–2500

Hussain S, Farooq U, Yousaf R (2022) Green finance and its impact on environmental sustainability in Pakistan. Int J Environ Stud 31(4):303–317

Hussain W, Ashiq M, Qazi MU, Ali M (2023) The impact of cognitive biases on the investment decision of individual investors: The role of risk propensity. J (FinTech) Sustain 2(1):47–72. https://doi.org/10.52461/jftis.v2i1.2068

Ikeda Y (2019) Power grid with 100% renewable energy for small island developing states—Nexus of energy, environment, and economic growth. *arXiv* preprint arXiv:1904.04597. https://arxiv.org/abs/1904.04597

Inglesi-Lotz R (2016) The impact of renewable energy consumption to economic growth: A panel data application. Energy Econ 53:58–63. https://doi.org/10.1016/j.eneco.2015.01.003

International Energy Agency (2023) Clean energy is boosting economic growth. IEA Comment. https://www.iea.org/commentaries/clean-energy-is-boosting-economic-growth

IRENA (2016) Renewable energy benefits: Measuring the economics. International Renewable Energy Agency. https://www.irena.org/-media/Files/IRENA/Agency/Publication/2016/IRENA_Measuring-the-Economics_2016.pdf

IRENA (2023) Record renewables growth fuels cost competitiveness—IRENA report. Reuters. https://www.reuters.com/business/energy/record-renewables-growth-fuels-cost-competitiveness-irena-report-shows-2024-09-24/

Javed R, Aftab M (2023) Gender gaps in financial inclusion: The role of digital banking in Pakistan. Financial Inclusion Review 9(2):24–38

Javed A, Khan F (2024) An overview of microfinance institutions and related developments in Pakistan. J Dev Policy Res Pract 7(1):1–30. https://doi.org/10.59926/jodprp.vol07/0

Kamran M, Qadir A (2021) Renewable energy and financial inclusion nexus in Pakistan. Energy Rep 7:5611–5622

Karim A, Noor N (2021) Sustainable development through financial inclusion: The Kamyab Jawan program in Pakistan. J Econ Policy 6(3):112–128

Khan HU, Ali MH (2020) Financial literacy and inclusion for economic development: Evidence from Pakistan. Int J Econ Financ Res 6(2):54–60

Khan S, Fatima S (2021) Mobile banking and financial inclusion: A study of rural Pakistan. Digit Financ Serv J 12(2):88–102

Khattak S, Aslam M (2022) Financial inclusion and social sustainability: The case of marginalized communities in Pakistan. Social Development Review 8(1):47–63

Kahia M, Aïssa MSB, Charfeddine L (2017) Renewable and non-renewable energy use–economic growth nexus: The case of MENA net oil importing countries. Renew Sustain Energy Rev 71:127–140. https://doi.org/10.1016/j.rser.2016.12.055

Klapper L, Lusardi A, Panos GA (2012) Financial literacy and the financial crisis. J Financ Econ 10(4):673–707
Koçak E, Şarkgüneşi A (2017) The renewable energy and economic growth nexus in Black Sea and Balkan countries. Energy Policy 100(Supplement C):51–57. https://doi.org/10.1016/j.enpol.2016.10.007
Mader P (2018) Contesting financial inclusion: Debates about meaning and measurement in development research. Dev Chang 49(2):461–484
Magazzino C (2016) The relationship between CO_2 emissions, energy consumption and economic growth in Italy. Int J Sustain Energy 35(9):844–857. https://doi.org/10.1080/14786451.2014.953160
Malik S, Mustafa K (2021) Gender and financial inclusion in Pakistan: Challenges and opportunities. Pak Econ Rev 40(3):73–90
Mckenzie D, Woodruff C (2014) What are we learning from business training and entrepreneurship evaluations around the developing world? World Bank Res Obs 29(1):48–82
Mdpi (2023) Renewable energy integration for sustainable economic growth: A bibliometric analysis. Sustainability, 15(20), 15030. https://doi.org/10.3390/su152015030
Menyah K, Wolde-Rufael Y (2010) CO_2 emissions, nuclear energy, renewable energy and economic growth in the US. Energy Policy 38(6):2911–2915
Muhammad N, Rashid S (2019) The role of fintech in financial inclusion in Pakistan. Asian Dev Policy Rev 7(1):16–27
Nathaniel SP, Khan SAR (2020) The nexus between urbanization, renewable energy, trade, and ecological footprint in ASEAN countries. J Clean Prod 272:122709. https://doi.org/10.1016/j.jclepro.2020.122709
Omri A, Kahouli B (2014) The nexus among foreign investment, domestic capital, and economic growth: Empirical evidence from the MENA region. Res Econ 68(3):257–263. https://doi.org/10.1016/j.rie.2014.04.002
Pao HT, Fu HC (2013) Renewable energy, non-renewable energy and economic growth in Brazil. Renew Sustain Energy Rev 25:381–392. https://doi.org/10.1016/j.rser.2013.05.004
Payne JE (2012) The causal dynamics between US renewable energy consumption, output, emissions, and oil prices. Energy Sources, Part B: Econo Plan Policy 7(4):323–330. https://doi.org/10.1080/15567249.2011.595248
Phiri A (2021) Renewable energy and economic growth hypothesis: Evidence from developing economies. Econ Res-Ekon Istraživanja 34(1):1241–1260. https://doi.org/10.1080/1331677X.2022.2121741
Rafindadi AA, Öztürk I (2017) Dynamic effects of financial development, trade openness and economic growth on energy consumption: Evidence from South Africa. Int J Energy Econ Policy 7(3):74–85
Rana Z, Shah A (2020) Evaluating the impact of financial inclusion on socio-economic development: Evidence from Pakistan. Econ J Emerg Mark 12(1):1–12
Ray S (2015) Financial inclusion and sustainable development: A study on the linkage. J Econ Sustain Dev 6(8):32–39
Renewable energy and economic growth: evidence from panel long-run causality analysis. (n.d.). ResearchGate. https://www.researchgate.net/publication/279569928
Reuters. (2024). The battle for net-zero emissions will be won or lost in emerging markets. Reuters News. https://www.reuters.com/sustainability/climate-energy/comment-battle-net-zero-emissions-will-be-won-or-lost-emerging-markets-2024-09-12/
Sachs JD (2015) The age of sustainable development. Columbia University Press
Sadiq R, Khan Z, Ali H (2021) The role of fintech in promoting economic sustainability in Pakistan. J Financ Technol 15(1):25–39
Sadorsky P (2009) Renewable energy consumption, CO_2 emissions, and oil prices in the G7 economies. Energy Economics 31(3):456–462. https://doi.org/10.1016/j.eneco.2008.12.010

Saidi K, Hammami S, Ozturk I (2018) The role of financial development and renewable energy in reducing environmental degradation in OECD countries. Environ Sci Pollut Res 25:14226–14241. https://doi.org/10.1007/s11356-018-1603-1

Schumpeter JA (1934) The theory of economic development: An inquiry into profits, capital, credit, interest and the business cycle (R. Opie, Trans.). Harvard University Press. (Original work published 1912)

Sebri M, Ben-Salha O (2014) On the causal dynamics between economic growth, renewable energy consumption, CO_2 emissions and trade openness: Fresh evidence from BRICS countries. Renew Sustain Energy Rev 39:14–23. https://doi.org/10.1016/j.rser.2014.07.033

Shahbaz M, Raghutla C, Chittedi KR, Jiao Z, Vo XV (2020) The effect of renewable energy consumption on economic growth: Evidence from the renewable energy country attractive index. Energy 207:118162. https://doi.org/10.1016/j.energy.2020.118162

Singh N, Nyuur R, Richmond B (2019) Renewable energy development as a driver of economic growth: Evidence from multivariate panel data analysis. Sustainability 11(8):2418. https://doi.org/10.3390/su11082418

State Bank of Pakistan. (2021a) A review of the current landscape of financial inclusion in Pakistan

State Bank of Pakistan. (2021b) Financial inclusion strategy 2021–2025: Enhancing access to financial services

Stern DI, Kander A (2012) The role of energy in the Industrial Revolution and modern economic growth. Energy J 33(3):125–152. https://doi.org/10.5547/01956574.33.3.5

The Financial Times. (2023) "It's boom time": Renewable growth is faster in the global south than in rich countries. Financ Times. https://www.ft.com/content/5af9367e-821c-4907-9f27-1371a928371e

The Guardian. (2022, November 19). Housework to green energy: The new markers of growth in rival metric to GDP. The Guardian News. https://www.theguardian.com/business/2024/nov/19/inclusive-income-rival-gdp-measuring-economic-growth

Town Carbajal R, Yekini D, Iroegbu S (2024) Social and economic impact analysis of solar mini-grids in rural Africa: A cohort study from Kenya and Nigeria. *arXiv* preprint arXiv:2401.02445. https://arxiv.org/abs/2401.02445

United Nations Development Programme. (2019) The role of financial inclusion in achieving the Sustainable Development Goals.

United Nations (2020) Sustainable development goals report.

Wagner M (2019) The role of financial inclusion in achieving sustainable development goals. Sustain Dev J 22(4):113–121

Wang L, Chang H-L, Rizvi SKA, Sari A (2020) Are eco-innovation and export diversification mutually exclusive to control carbon emissions in G-7 countries? J Environ Manag 270:110829. https://doi.org/10.1016/j.jenvman.2020.110829

World Bank Group (2018) Financial inclusion and development: Recent impact evidence Policy Research Working Paper

World Bank (2020) A global perspective on financial inclusion and sustainable development

World Economic Forum (2021) The future of financial systems: Insights on innovation and inclusion. World Economic Forum

Yildirim E, Saraç Ş, Aslan A (2012) Energy consumption and economic growth in the USA: Evidence from renewable energy. Renew Sustain Energy Rev 16(9):6770–6774. https://doi.org/10.1016/j.rser.2012.09.004

Zulfiqar K, Chaudhary MA, Aslam A (2016) Financial inclusion and its implications for inclusive growth in Pakistan. Pakistan Econ Soc Revi 54(2):297–325. https://www.jstor.org/stable/26616711

Financial Inclusion and Sustainability: Challenges, Curative Measures and Prospects

Ghousia Khatoon, Karwan H. Sherwani, Hussaini Bala, Silky Vigg Kushwah, Umair Zahid, Maleeha Ashraf, Firdous Ahmad Malik, and Armaya'u Alhaji Sani

Abstract Financial Inclusion is recognized as a great mechanism to accomplish economic development. It refers to the accessibility of formal financial support from banks and financial institutions to the adult members. It may include having savings account in a bank, availing loan facility from a bank or financial institutions or having an insurance to safeguard against risk. According to Global Findex database (2021), 1.7 billion adults across the globe are financially excluded and out of this 56% of them are from Asia. The percentage of women having a bank account in developing economies is only 9%. The financial exclusion has led to poverty, income

G. Khatoon (✉) · H. Bala · A. A. Sani
Accounting Department, Tishk International University, Erbil, Iraq
e-mail: ghousia.khatoon@tiu.edu.iq

H. Bala
e-mail: hussaini.bala@tiu.edu.iq

A. A. Sani
e-mail: armayau.sani@tiu.edu.iq

K. H. Sherwani
Business and Management Department, Tishk International University, Erbil, Iraq
e-mail: karwan.sherwani@tiu.edu.iq

S. V. Kushwah
Department of Finance, New Delhi Institute of Management, New Delhi, India
e-mail: dr.silkyviggkushwah@gmail.com

U. Zahid
Department of Business Management, Global Banking School, University of Suffolk, Ipswich, UK
e-mail: umair_zahid07@yahoo.com

M. Ashraf
Business School, The University of Law, Bristol, UK
e-mail: maleeha.ashraf.14@ucl.ac.uk

F. A. Malik
Department of Management, University of People, Pasadena, USA
e-mail: malikfirdouscrc@gmail.com

F. A. Malik et al. (eds.), *Aligning ESG Goals with the UNs Sustainable Development Goals Planning*, World Sustainability Series,
https://doi.org/10.1007/978-3-032-01715-4_10

inequality which has cascading effect on the economic development of a nation. One of the goals of United Nations under SDGs (Sustainability Development Goals) is to ensure inclusive growth so that no one is left behind. The existing literature underscores the importance of financial inclusion. Nevertheless there exists the knowledge gap relating to digital financial solutions, financial literacy as a tool to enhance financial inclusion, updating curriculum and sustainability. This piece of research aims to contribute towards the accomplishment of SDGs more specifically Sustainable Development Goals number 1, 4, 5 and 9 by addressing the crucial knowledge lacuna by uncovering the obstacles and to offer recommendations to policy makers to take the financial inclusion to the next level so that more and more people are financially included.

Keywords SDGs · Financial inclusion · Financial literacy · Poverty reduction · Quality education · Curriculum · Gender equality · Digital infrastructure

1 Introduction

Financial inclusion aims to offer financial services to diverse groups. It has evolved over time and is closely linked to economic and social progress. Currently, it is gaining significant momentum. Financial inclusion pertains to providing affordable access to basic formal financial services, including credit, savings, insurance, payments, and remittance facilities, to all the individuals and the absence of such services often leads people to resort to costly informal financial sources, which significantly impacts low-income groups in a disproportionate manner (Inoue 2019; Sarma & Pais 2011). The United Nations introduced financial inclusion in 2005 during the International Year of Microcredit, and it has since then thrived (Chen and Yuan 2021). The significance of financial inclusion as a fundamental channel in reducing poverty, and financial inequality, enhancing equitable prosperity is underscored by the World Bank (2022). Likewise, the United Nations Capital Development Fund (UNCDF) has also acknowledged the significant role of financial inclusion in advancing the objectives outlined in the 2030 Sustainable Development Goals (SDGs) (Desello & Agner 2023). The United Nations, through its SDGs adopted in 2015, recognized the importance of financial inclusion in achieving sustainable development. SDG Target 8.10 specifically aims to "strengthen the capacity of domestic financial institutions to encourage and expand access to banking, insurance, and financial services for all" (UNCTAD 2016). In the late 20th and early twenty-first centuries, several initiatives aimed at facilitating financial inclusion and it gained momentum due to the rise of Microfinance, the digital revolution, and evolution in policy and regulation. Statistics reveals that financial inclusion has improved from 62% in 2011 to 69% in 2017 globally (Sakanko et al. 2023).

The concept of financial Inclusion is considered as one of the important pillars of finance development, well-being and largely benefits citizens (Hererrias & Alvarez 2023) and therefore an index has been developed by Sarma (2008) named Index of

Financial Inclusion (IFI) taking into account three dimensions to measure IFI such as Accessibility, Availability and Usage of service of the banking system measured by scale of 0 to 1 and the results of the research shows that Norway, Belgium and Austria emerge as top three in the list of most financially inclusive countries (Sarma 2008). Moreover, many countries have made significant progress in financial inclusion with supportive policies and digital technology; there are various initiatives to support financial inclusion such as Grameen Bank in Bangladesh by Mohammad Younis, Agricultural initiatives by Japan, Community banks in USA (Chen & Yuan 2021).

Due to its contribution to the development of economy, there is great deal of necessity to explore the various areas of financial inclusion. The greatest motivation to take up this topic is to discover why people are financially deprived, What role does financial inclusion play to well-being of people?, what are the hurdles or challenges in its implementation?, what measures could be undertaken to enhance the inclusion of people financially? and what are its prospects?. This paper is based on the earlier literatures. This research is bifurcated into sections as follows: The first part covers the introduction to the topic. Part two throws light on the literature covering the concept of financial inclusion and its crucial importance. Part three depicts the objectives of the Study. Section 4 presents the discussions on challenges faced in the implementation of financial inclusion and remedial action under various SDGs. The focal point of Part 5 is to cover the Policy Implications and the last section concludes the paper.

2 Literature Review

The concept of financial Inclusion is clear from the following definitions. Financial inclusion refers to the extent to which the individuals and companies make use of financial services (World Bank 2014). According to another author financial inclusion is the state of the economy wherein everyone has access to financial services and no one is deprived of it (Amidžić et al. 2014). Another interesting and inclusive definition of financial inclusion according to Sahay et al. (2015) is that financial inclusion is the state of the economy wherein people have easy access to financial services and they are reasonably priced. According to United Nations the fundamental objective of financial inclusion should be ensuring that individuals have an access to wide range of financial services such as provision of savings account, provision of credit to the needy, making available to them the insurance policies, providing remittances facilities at an economical cost. According to the another researcher financial inclusion refers to provision of banking services at an economical cost to the large section of people belonging to poor section of the community or underprivileged groups (Leeladhar 2006). Financial inclusion is a means wherein the adults and businesses have an access to financial services such as meeting payments, possessing credit cards, saving for the future and having insurance to secure future and these services are provided to the end users in the most sustainable manner (Singh & Singh Kondan 2011). According to Reserve Bank of India, Financial Inclusion is the ways and means through which vulnerable stratum namely people with low income groups gain access to formal financial services such as credit, bank accounts, insurance etc., at a cost which could be afforded by these sections of people (https://rbi.org.in/scripts/BS_SpeechesView.aspx?Id=862.). Financial inclusion is a means to eradicate the obstructions which are a cause of financial exclusion. These obstructions include non availability of affordable financial services to the underprivileged groups such as access to credit, savings and deposits, payments, insurance etc. (Carbo et al. 2005).

Research has proved that the financial inclusion plays a pivotal role in the economic development of a nation. The following literature throws light on its significance. Financial inclusion is gaining more attention of countries for the benefits it offers such as financial inclusion leads to alleviation of poverty. A financial system that includes everyone helps to reduce exploitative informal credit sources like private money lenders and improves efficiency and welfare by offering safe saving options and various useful financial services (Sarma & Pais 2011). When families can use financial services, it helps them manage money and invest in education and health and businesses can also grow and create jobs when they have credit and to make the society more equal and provides chances for better economic situation (Furusawa 2016). According to empirical research conducted by Honohan (2007) and Kai & Hamori (2009), access to financial services inclusively reduced inequality in income. Moreover, recent research by Eid et al. (2023) shows that financial inclusion has an important role in the financial stability of banks. Moreover, financial inclusion leads to expansion in branches of banks, minimizing barriers to access financial services and overall economic growth (Bakar & Sulong 2018). In an another Study, the authors discovered that if the people are in a position to access financial services, it boosts their savings capacities. According to previous studies in the US it was discovered that if people have no access to banks accounts, it resulted in a number of detrimental effects (Lusardi 2010). For example, if a person does not have bank account, it is difficult to make payments through checks. Payments made through checks are more safer than the payments made in the form of cash as there are great possibilities that cash can be stolen (Gross et al. 2012). In an another Study the authors found that if an individual has a bank account, it results in creation of wealth and building the assets and this will serve as shock absorber or cushion during the rainy days or during the evening of his/her life i.e. it helps to meet expenses after retirement (Rhine et al. 2006). Financial inclusion mitigates the side effects of poverty. According to Mader (2018) there is a positive correlation between inclusive financial markets and reduction in poverty and inequalities. It means if the adults are financially included, it allows them to manage their payments, enable them to avail bank loans and also obtain insurance. Besides, financial inclusion not only helps the existing companies in their expansion but also becomes the leading cause of creation of new companies. This leads to creation of job and has a cascading effect on the savings (Ajide 2020). The study conducted by Ozili (2018) explores that financial inclusion gained momentum as it is contributing towards the achievement of SDGs. The very first Sustainable Development Goals of United Nations is to mitigate the poverty level. It lays focus on that if people are financially included, it will help in the establishment of new business and empowers people financially. SDG 2 aims to achieve zero hunger. It is discovered that if the peasants are provided financial assistance, they can use good quality seeds to achieve high yielding crops (Asongu and De Moor 2015). Likewise, similar research shows financial inclusion could be key for the inclusion of peasants in the financial services and development in the case of Zimbabwe (Ndhlovu & Mhlanga 2023).

3 Objectives

The fundamental objective of this research paper is to discover why people are financially deprived, What role does financial inclusion play to well-being of people, what are the hurdles or challenges in its implementation, what measures could be undertaken to enhance the inclusion of people financially and what are its prospects. This piece of research aims to contribute towards the accomplishment of SDGs more specifically Sustainable Development Goals number 1, 4, 5 and 9 by addressing the crucial knowledge lacuna by uncovering the obstacles and to offer recommendations to policy makers and also sheds light on the future prospects to take the financial inclusion to the next level so that more and more people are financially included.

4 Discussion

This section discusses the impediments in the way of financial inclusion and the curative measures to overcome these challenges. The impediments have been discussed under SDG 1(Poverty reduction), SDG 4 (Quality Education), SDG 5 (Gender Equality) and SDG 9 (Industry, Innovation and infrastructure), The discussion is as follows:

4.1 Reluctance to Serve the Underprivileged (SDG1 Poverty Reduction)

United Nation's SDG 1 aims at bringing down the poverty level. One of the challenges in the way of financial inclusion is the unwillingness of banks to serve the vulnerable groups as these groups have irregular incomes and cannot provide collateral. Banks do not consider financial obligation as a business opportunity, they see it as a burden. Banks perceive that due to their irregularity in income they cannot repay the loan and there are more possibilities of default. Moreover, though there were schemes to include adults financially, it did not take the shape as the delivery mechanism was poor (Varghese, & Viswanathan 2018).

When these vulnerable groups are excluded from the formal financial services, it leads to poverty. People in rural areas have no access to basic financial services such as bank account, credit cards, insurance and electronic platforms to make the payments. This puts a ceiling on the capability of people to devote time and money on financial education and health (Demirguc-Kunt et al. 2018).

A Study was conducted to gauge the extent to which adult population owns bank accounts. The respondents were asked if they have an account in a bank, post office or microfinance institution. Around 45% of them reported that they have an account. It was found that there was a positive correlation between economic growth and account ownership. It means Adults belonging to middle-income countries are more eager to have a bank account than the adults belonging to low-income countries. The respondents were asked the reason for not having the bank accounts and the answer was that they do not have adequate money to be deposited so that they can use it later. Lack of money was the most quoted reasons by the respondents across the globe. Other reasons include—(i) The distance is the barrier. Banks are located far away from the prospective account holders. About 47% of the respondents in Tanzania who were not having a bank account quoted the problem of distance (ii) Trust in the banks is another barrier (iii) the individual do not have required documents to open the bank accounts—The eligibility criterion imposed by banks restricts the workers working in rural and informal sectors to get included financially because these people may not have the proof of income or may lack the proof of address, proof of their nationality, (iv) according to some of the respondents opening and maintaining bank accounts is a costly affair (Allen et al. 2016).

Financial exclusion is one of the major challenges faced by the developing countries across the Globe. Ghana the Sub-Saharan region Africa is not the exception. 60% of the population of Ghana is still financially excluded. It means three in every five adults are financially excluded and unbanked. The factors responsible for financial exclusion comprise poverty, lack of financial literacy, people lack trust in the formal financial sector, tedious documentation and lack of social networking. The Study recommends that the policy makers need to play a crucial role in mitigating the impact of these factors which are impacting the financial inclusion negatively and boost those factors which have a positive effect on it. There has to be public–private partnership and derive the benefits of this synergy. This may bring the unbanked adults into the frame of financial inclusion (Akudugu 2013).

4.2 What Could Be Done?

4.2.1 Easy Accessibility

The research found that there is positive correlation between financial inclusion and better financial services. It means more and more adults will be financially included if the adults have easy accessibility to financial services. Accessibility includes lower cost of maintaining accounts, nearness to bank branches, stability of government policies and motivation to women and young adults to own bank accounts (Allen et al. 2016).

In order to boost access to formal financial services, the African Development Bank introduced ADFI (Africa Digital Financial Inclusion) in the year 2019. The aim of this facility is to provide easy access to formal financial services to millions of Africans. Correspondingly, 'Aadhaar' which is a biometric Identity system was launched by Government of India to enable citizens to gain easy access to formal financial services (Banerjee 2016).

Another study supporting the cause of financial inclusion discovered that providing access to financial services to the vulnerable groups results in driving the economic diversification in Nigeria. If people are included financially, it leads to eradication of poverty and ensure prosperity (Adeola and Evans 2017). In another study the authors discovered that financial inclusion has a positive influence on the economic growth of OIC (Organization of Islamic Cooperation) countries (Kim et al. 2018).

4.2.2 Overall Quality at Affordable Cost

In an another Study conducted to explore the relationship between financial inclusion and economic growth taking into consideration the data for the periods from 2004 to 2017 in the region of Sub - Saharan African nations proved that there is a positive correlation between financial inclusion and the economic growth. Since financial inclusion has positive influence on the economic development, the study stresses on the active role of policy makers in framing the policies favoring financial inclusion, specially the policies favoring the provision of quality financial services and the products at reasonable prices and this will have a positive impact on the inclusive growth (Adedokun & Ağa 2023). Another Study was conducted in the country of Nigeria to discover the relationship between the financial inclusion and inclusive growth. The study was conducted taking into account the data for the years ranging from 1981 to 2017. The Study discovered that in the both short run and long run the variables such as liquidity, loan to the rural people and increase in bank branches have a positive influence on the inclusive growth whereas the rate of interest hindered the growth. One of the measures suggested by this study is that there needs to be improvement with respect to the provision of financial services to inhabitants of the rural areas to enable them to become a cause of national productivity and to accomplish the reduction in inequalities (Afolabi 2020).

According to the Research carried out by Koomson et al. (2020) the results reveal that financial inclusion not only mitigates the poverty level but also resulted in reducing the inequalities in income in Africa. The research advocates that the policies pertaining to financial inclusion needs to be strengthened so that the poor people are financially included. The authors discovered that an increase in financial inclusion resulted in the mitigation of poverty level of the households in Ghana especially the households which were taken care by women and restricted them from getting exposed to poverty.

4.2.3 Lack of Financial Literacy (SDG 4 –Quality Education)

As per Target 1.4 of SDG 4, there is need to a stress on enhancing the relevant skills to help the youth to be employable or to empower them to be entrepreneurs which comprises financial literacy too. It is evident from the earlier literature that financial literacy is the edifice upon which the strong structure of financial inclusion is built. Higher level of financial literacy leads to high level of financial inclusion. Both financial literacy and financial inclusion are positively correlated. It means financial literacy has an effect on financial inclusion (Grohmann, Klühs & Menkhoff, 2018b, a). But nearly 1.7 billion adults lack access to financial services across the globe and one of the major hindering factors is lack of financial literacy (Demirguc-Kunt et al. 2018). People cannot make strong financial decisions as they lack financial literacy. Due to lack of financial education, people are not in a position to make informed decisions. Financial literacy empowers people to have access to various financial products and this enables them to choose the best financial products from the available ones.

Another Study reports that the adult population of Zimbabew is not aware of the financial services which are available to them and this restricts them from using the right financial product. Lack of sufficient information pertaining to various financial products and low level of financial literacy are the major challenges. In an another Study conducted in Iraq it is found that the major hurdle faced by the entrepreneurs to avail the financial services is the lack of financial literacy. They do not possess the basic financial education and because of this reason they do not know what the financial services are. Global financial literacy survey was carried out in the year 2014 by S & P (Standard & Poor's) to gauge the financial literacy among the Iraqis. It is discovered that the Adult's financial literacy level stood at 27% which is quite low (https://www.ilo.org/wcmsp5/groups/public/-arabstates/-robeirut/documents/publication/wcms_775892.pdf). The level of financial literacy is too low in Iraq. It has occupied 105th place out of 144 countries with regard to financial literacy rate (Jaber & Yasir 2021).

4.3 *What Could Be Done?*

4.3.1 Accelerate Financial Literacy

The remedial measure is to provide financial education to the users as this knowledge helps to know the pros and cons of various financial products and services and select the best as per need (Barugahara 2021).

Financial education with respect to insurance can help the households to manage well the financial risk to meet their emergencies. It can help peasants insure their crops to safe guard against natural calamities. It can help households to ensure their house property. Knowledge of financial education can help people to select better savings and credit options. It also guides them to invest wisely according to the level of their risk bearing capacity.

The financial institutions such as banks can support the cause of conducting financial literacy programmes. Having the financial literacy programmes in place is not sufficient, there is a great deal of necessity to monitor the implementation and measure the results. This exercise will help us to see how far we have succeeded in ensuring inclusive financial education. To accelerate the process of financial education there is a strong need to have public and private collaboration.

In another research the authors identified the factors which can help to incorporate the households into the frame work formal financial services sector. The respondents surveyed were from the four villages of West Bengal which is one of the States of India. The study aimed at exploring the effect of financial literacy and accessibility to mass media on inclusive growth. It is revealed from the results that there is a cascading effect in the sense the extent to which adults are financially included depends upon the extent of financial literacy. On the other hand the level of financial literacy depends upon the extent to which adults use the mass media. The study also revealed that adults who possess BPL card (Below Poverty Level) had more access to financial services compared to the casual laborers who work in agricultural and non-agricultural sectors (Chatterjee & Raychaudhuri 2021).

In an another study carried out by the researchers to gauge the impact of subscribers to internet on achieving economic growth though financial inclusion and SDGs. The results portray that there is no doubt lack of financial inclusion is the leading cause of economic growth. Attracting subscribers towards internet alone is not sufficient, subscribers have to be imparted financial literacy to enable them how to utilize digital platform to avail financial services. It is not enough government is spending lot of money on providing internet faculty, it has to increase financial literacy of the subscribers though promotional campaign and various other schemes (Saumya et al. 2022).

4.3.2 Update the Curriculum in Schools and Colleges (SDG 4-Quality Education)

The fundamental reason for unequal distribution of income and wealth, rich people becoming richer, poor people remaining poor is that there is lack of financial literacy. According to the surveys conducted at the national level portrays that the level of financial education is very low among the population of young adults. They are not in a position to select the financial products that suits their budget. The subject of financial literacy is not taught at schools and colleges. Financial education should start at the earliest at the school level. PISA (Program for International Student Assessment) is part of OECD program. It tests the abilities of youth in various areas such a mathematics, reading and science across sixty five countries. Introduction of similar financial literacy tests will act as a benchmark against which financial literacy of the young adults can be gauged. Based on the suggestions provided by OECD, there is an increase in the number of countries who are introducing financial education in the curriculum. But there is a need for strong political will, availability of resources needed to accomplish this goal. The countries that adopted the suggestions made by OECD and implemented financial education at the school level have shown remarkable progress (https://www.oecd.org/daf/fin/financial-education/FinEdSchool_web.pdf). To create an awareness amongst the individuals about the benefits of using bank accounts, savings, various financial products, the schools and colleges need to incorporate a course on financial literacy, digital payments, mobile banking, internet banking etc. in their curriculum (Barik & Sharma 2019).

4.3.3 Discrimination (SDG 5—Gender Equality)

Poor people residing in rural areas and women lack access to financial services. According to the Finscope survey (2014) people from urban area have more access to financial services compared to the people from rural areas. According to another study, gender inequalities is one of the causes of financial exclusion as women have less access to financial services such as owning a bank account than the men. Furthermore, it is difficult for women to obtain credit facility from the banks as they cannot offer collateral and also their credit history is not so good. From this it is clear that the policy makers ought to focus on women and people from rural population (Asuming et al. 2019). Women usually encounters obstruction with respect to owning assets, society related customs and their mobility is constrained.

Another Study quotes that though efforts were made to include people financially, it still remains a big challenge especially in Asian and African regions. It is noticed that the there is a gender gap when it comes to provision of these services. It means the benefits of financial inclusion were not provided equally between men and women, between rich and poor and between households of rural and urban areas (Demirgüç-Kunt et al. 2018). As a result many countries remain outside the purview of financial inclusion. For example in Nigeria even after the implementation of strategies pertaining to financial inclusion in 2012, 68% of them still are not financially included. In developing nations of Asia, more than one million population cannot access the formal financial services such as opening and maintaining bank accounts (Le, Chuc et al. 2019).

4.3.4 What Could Be Done?

Saying no to discrimination is the need of the hour. Since the financial exclusion rate is high in developing nations, they are taking initiatives to frame policies concerning discrimination to reduce gender inequalities and also enhance financial literacy. The authors recommend that the stringent policies against discrimination will go a long way in helping the groups which are prone to these discriminations and this will lead to economic development (Atkinson & Messy 2013).

Owning bank accounts empowers women and makes them financially independent. This is evident from Indian scenario wherein the benefits transferred directly to women's account through government schemes (not to their male partners) had a great effect. It enhanced their financial control, empowered them to work independently and increased the job opportunities (Field et al. 2021). In an another interesting research conducted in Nepal showed that the households which were financially controlled by women were able to spend 15% more on healthy food (Prina 2015).

Another significant contribution of financial inclusion is that it empowers women. In an empirical study in Nigeria by Kazeem and Ademola (2021) results show that there is a significant relationship between women's involvement in entrepreneurship and financial inclusion, which means that when financial inclusion increases, more women will be involved in starting new ventures. But the percentage of women having access to financial services is lesser than that of men. Research says that 46% of the men have an account in banks and financial institutions in comparison with women wherein the percentage stands at 37%.

4.3.5 Lack of Technology (SDG 9—Industry, Innovation and Infrastructure)

Technology forms the basic infrastructure for the improvement of financial inclusion. Lack of technology is a great hindrance in the way of financial inclusion. When the individuals have no easy accessibility to technology, it limits their access to banking transactions, availing credit facility, availing insurance facility and also they cannot availing the facility of mobile payments. When the individuals have no access to digital platforms such as internet connection and mobile phones as this is the case in rural areas or the areas where people have literally low income, it is a great barricade in the way of opening bank account by the individuals and carrying out transactions digitally (Demirgüç-Kunt et al. 2018).

4.3.6 What Could Be Done?

Provision of financial services through traditional platforms involves enormous cost as there is a need for physical infrastructure and it may not be possible to afford such infrastructure in the less developed economies. The solution to overcome this limitation is the provision of digital platform (Suri and Jack, 2016). There is a crucial need to boost financial markets through digital transformation in order to provide efficient and effective services. Digital transformation can ensure that the services are more accessible and convenient for the customers. But the implementation of digital transformation requires cyber security and strict actions against violations. There is a need to fill the digital skill gap and this could be achieved through fintech.

Use of technological breakthroughs is crucial measure to ensure financial inclusion. Technological breakthroughs boost the cause of financial inclusion. Mobile technology introduced by Kenya such as M-Pesa has penetrated so quickly that on an average the number of registrations to this app went beyond 5000 for a single day in 2007 and by the year 2007 around 7.9 million people registered to the facility

M-Pesa is a mobile Banking app which enabled individuals to transfer money and also make purchases without carrying hard cash.

There is a strong requirement for the government, private banks to come together and create awareness amongst the adults about the savings, investment and their importance. The central bank and government should make it obligatory for the banks to provide banking services electronically, In Iraq one of the major challenges in the way of financial inclusion is digital transformation (Musawy & Kashmoush 2022).

5 Policy Implications

To achieve remarkable economic progress, we need to harness the power of financial inclusion. There has to be great endeavor from the government, Banks both private and Public and other stakeholders to recognize the challenges hindering the cause of financial inclusion and address those challenges. The policy makers should stress on framing policies which should be intended to include the unbanked usually poor, financially illiterate, people from rural areas, women, low income groups etc. into the framework financial inclusion. It is imperative to cater to the needs of these groups by offering financial services at affordable cost.

6 Conclusion

This research was intended to study the concept of financial inclusion, its significance, challenges and prospects using the existing literature. The study made use of the literature pertaining to the developing economies with special reference to developing countries. Developing countries were selected as the financial inclusion in these economies is very low. From the above discussion it is clear that the factors responsible for financial exclusion are low level of financial literacy, lack of bank branches, and lack of nearness to banks, tedious documentations for opening bank accounts and availing credit, lack of accessibility to internet. Gender Discriminations, low level of income, lack of information about the financial products etc.

The major barricade in the way of financial inclusion is the lack of financial literacy. It is seen that around 1.7 Billion adults have no access to financial services across the Globe and the major hindering factor is lack of financial literacy. Due to lack of financial literacy people are not aware of various financial products available to them and thereby not able to take informed decisions. In a Study conducted by Allen et al (2016) only 45% of the adults have a bank account. The reasons quoted for not owning the bank accounts are that the adults have not money to deposit, banks are located far away, some of them do not trust bank and the documentation is tedious. The reason for the distrust in banks is that the people feel their money is not safe as it is evident from Zimbabwe wherein some of the commercial banks went into liquidation. Banks are willing to provide the loan facility only to those adults who can provide collateral as they are risk averse and do not want to take the risk of default. Discrimination is another major barrier in the way of financial inclusion. Women have less access to formal financial services than men, people from rural area have less access to financial services than the urban population. People from high income countries have more access than the adults from low income countries. In some of the countries the banking system is controlled by the state owned banks. For example in Iraq around 90% of the deposits and assets are under the control of commercial banks and private banks have no role to play. The percentage of SMEs who availed credit from the formal sector is minimal which less than 5% is and it is quite alarming. There has to be public–private partnership to resolve this issue.

Financial inclusion plays a crucial role in the mitigation of poverty, results in wealth creation and a major cause of enhancing the livelihoods of large number of people especially from the low income groups. There is an increase in the number of adults who own a bank account. It has increased from 51% in 2011 to 76% in the year 2021. Account ownership is vital to bring the unbanked adults within the framework of financial inclusion (Global Findex Database of 2021).

To put it nut shell, there is a strong need to focus on the strategies to enhance financial inclusion at the global level. Sensitizing the people about the importance of financial literacy, making them understand various financial products, improving the quality of financial services, boosting the required infrastructure, motivating the banks and financial institutions to focus on the gals of sustainable development, removing discrimination, strictly monitoring the implementation of policies or programs designed to empower unbanked will go a long way in boosting the financial inclusion.

Finally, the study suggests that enhancing financial inclusion particularly via digital channels, policy innovation, and inclusive financial instruments has the potential to be a strategic enabler of a range of Sustainable Development Goals (SDGs) such as poverty reduction, economic empowerment, gender equality, and climate action. But it must be achieved by incorporating financial inclusion into environmental and social governance (ESG) principles to prevent unintended side effects like over-indebtedness or unsustainable environment-damaging investments. The paper emphasizes that there should be concerted action to craft financial systems that are not only affordable and accessible but also responsible, resilient, and compatible with long-term development and sustainability objectives.

References

Adedokun MW & Ağa M (2023) Financial inclusion: A pathway to economic growth in Sub-Saharan African economies. Int J Finance & Economics, 28(3): 2712–2728

Adeola O, Evans O (2017) Financial inclusion, fnancial development, and economic diversifcation in Nigeria. J Dev Areas 51(3):1–15

Afolabi JO (2020) Impact of financial inclusion on inclusive growth: an empirical study of Nigeria. Asian J Econ Empir Res, 7(1), 8–14. https://doi.org/10.20448/journal.501.2020.71.8.14

Ahamed MM, Mallick SK (2019) Is financial inclusion good for bank stability? International evidence. J Econ Behav Organ 157:403–427

Ajide FM (2020) Financial inclusion in Africa, does it promote entrepreneurship? J Financ Econ Policy 12(4):687–706

Aker JC, Boumnijel R, McClelland A, Tierney N (2016) Payment mechanisms and antipoverty programs: Evidence from a mobile money cash transfer experiment in Niger. Econ Dev Cult Change 65(1):1–37

Akudugu MA (2013) The determinants of financial inclusion in Western Africa: Insights from Ghana. Res J Financ Account 4(8):1–7

Allen F, Demirguc-Kunt A, Klapper L, Peria MSM (2016) The foundations of financial inclusion: Understanding ownership and use of formal accounts. J Financ Intermediation 27:1–30

Amidžić G, Massara A & Mialou A (2014) Assessing countries' financial inclusion standing—A New Composite Index(IMF Working Paper No. 14/36. International Monetary Fund

Ashraf N, Karlan D, Yin W (2010) Female empowerment: Impact of a commitment savings product in the Philippines. World Dev 38(3):333–344

Asongu Simplice A, Lieven De Moor (2015) Recent advances in finance for inclusive development: a survey. Available online: https://econpapers.repec.org/paper/abhwpaper/15_2f005.htm

Asuming PO, Osei-agyei LG, Ibrahim J (2019) Financial inclusion in Sub-Saharan Africa: Recent trends and determinants. J Afr Bus 20(1):112–134. https://doi.org/10.1080/15228916.2018.1484209

Atkinson A, Messy FA (2013) Promoting fnancial inclusion through fnancial education, OECD/INFE evidence, policies and practice. Cross Ctry Evid, World Dev, 32, 1–54

Bakar HO, Sulong Z (2018) The role of financial inclusion on economic growth: theoretical and empirical literature review analysis. J Bus & Financ Aff, 7(4), 2167–0234. https://doi.org.https://doi.org/10.4172/2167-0234.1000356

Banerjee-Chatterjee T, Roy M, Raychaudhuri A (2021a) Financial inclusiveness and mobilization of household savings in rural area: are they related? J Quant Econ 19:223–249

Banerjee S (2016) Aadhaar: Digital inclusion and public services in India. World Dev Rep, 81–92

Banerjee-Chatterjee T, Roy M, Raychaudhuri A (2021b) Financial inclusiveness and mobilization of household savings in rural areas: are they related? J Quant Econ 19(2):223–249

Barik R, Sharma P (2019) Analyzing the progress and prospects of financial inclusion in India. J Public Aff 19(4):1–6. https://doi.org/10.1002/pa.1948

Barugahara (2021) Financial inclusion in Zimbabwe: Determinants, challenges, and opportunities. Int J Financial Research. 12(3): 261–270

Blumenstock J, Callen M, Ghani T (2018) Why do defaults affect behavior? Experimental evidence from Afghanistan. Am Econ Rev 108(10):2868–2901

Breza E, Kanz M, Klapper LF (2020) Learning to navigate a new financial technology: Evidence from payroll accounts (No. w28249). Natl Bur Econ Res

Capuano A, Ramsay I (2011) What causes suboptimal financial behaviour? An exploration of financial literacy, social influences and behavioural economics. U Melb Leg Stud Res Pap, (540)

Chen W, Yuan X (2021) Financial inclusion in China: an overview. Front Bus Res China 15(1):1–21. https://doi.org/10.1186/s11782-021-00098-6

CFI. (2023). Financial inclusion. Corp Financ Inst. https://corporatefinanceinstitute.com/resources/economics/financial-inclusion

Carbo S, Gardener EPM, Molyneux P (2005) Financial Exclusion. Palgrave MacMillan, England

Demirguc-Kunt A, Klapper L, Singer D, Ansar S (2018) The Global Findex Database 2017: Measuring financial inclusion and the fintech revolution. World Bank Publications

Demirguc-Kunt A, Klapper L, Randall D (2014) Islamic finance and financial inclusion: Measuring use of and demand for formal financial services among Muslim adults. Rev Middle East Econ Financ 10(2):1–42
Desello J, Agner MG (2023) Financial inclusion and the role of financial literacy in the Philippines. Int J Econ Financ 15(6): 27–43 https://doi.org/10.5539/ijef.v15n6p27
Eid QMA, Al Houl MAA, Alqudah MTS, Almomani MAA (2023) The role of financial inclusion in the stability of islamic banks. Int J Prof Bus Rev, 8(4): 1–19 https://doi.org/10.26668/businessreview/2023.v8i4.1214
Field E, Pande R, Rigol N, Schaner S, Troyer Moore C (2021) On her own account: How strengthening women's financial control impacts labor supply and gender norms. Am Econ Rev 111(7):2342–2375
Florence B (2021) Financial inclusion in zimbabwe: determinants, challenges, and opportunities. Int J Financ Res. 12(3):261–270
Furusawa M (2016) Financial inclusion: Bridging economic opportunities and outcomes. IMF. https://www.imf.org/en/News/Articles/2016/09/20/sp092016-Financial-Inclusion-Bridging-Economic-Opportunities-and-Outcomes
Grohmann A, Klühs T, Menkhoff L (2018b) Does financial literacy improve financial inclusion? Cross country evidence. World Dev 111:84–96
Grohmann A, Klühs T, Menkhoff L (2018a) Does financial literacy improve financial inclusion? Cross Ctry Evid, World Dev 111:84–96
Gross MJ, Hogarth M, Schmeiser MD (2012) Use of financial services by the unbanked and underbanked and the potential for mobile financial services adoption. Fed Reserv Bull 98, 4
Gulati A, Jain S, Satija N (2014) Rising farm wages in India—The 'pull'and 'push'factors. J Land Rural Stud 2(2):261–286
Kashmoush MB (2022) Digital financial inclusion in Iraq: reality, challenges, and investigation requirements. J Posit Sch Psychol 6(2):3455–3470
Herrerias R, Alvarez C (2023). Financ Behav Degrees Financ Incl. https://doi.org/10.2139/ssrn.3717100
Honahan, P (2007) Cross-country variation in household access to financial services. World Bank Paper Conference on "Access to Finance," Washington, DC.
Inoue T (2019) Financial inclusion and poverty reduction in India. J Financ Econ Policy 11(1):21–33. https://doi.org/10.1108/JFEP-01-2018-0012
Islam KJ, Simpson W (2018) Payday lending and microcredit: two faces of the same problem? J Int Dev 30(4):584–614
Jaber FH, Yasir MH (2021) Determinate main challenges that facing financial inclusion in Iraq - an evaluation study. Acad Account Financ Stud J 25(5):1–16
Jack W, Suri T (2014) Risk sharing and transactions costs: evidence from Kenya's mobile money revolution. Am Econ Rev 104(1):183–223
Kai H, Hamori S (2009) Microfinance and inequality. Munich Personal Repec Archive (MPRA) Paper, 1757: 1–57
Kazeem B, Ademola A (2021) Women entrepreneurs and financial inclusion. Int J Soc Manag Sci 8(1):90–102
Khan F, Siddiqui MA, Imtiaz S (2022) Role of financial literacy in achieving financial inclusion: A review, synthesis and research agenda. Cogent Bus Manag 9(1):2034236. https://doi.org/10.1080/23311975.2022.2034236
Kim D-W, Yu J-S, Hassan MK (2018) Financial inclusion and economic growth in OIC countries. Res Int Bus Financ 43:1–14
Koomson I, Villano RA, Hadley D (2020) Effect of financial inclusion on poverty and vulnerability to poverty: Evidence using a multidimensional measure of financial inclusion. Soc Indic Res 149(2):613–639. https://doi.org/10.1007/s11205-019-02263-0
Leeladhar V (2006) Taking banking services to the common man: financial inclusion', RBI Bulletin, 73–7
Lusardi A (2010) Americans' financial capability. Rep Prep U S GovMent Financ Cris Inq Comm (FCIC)

Le TH, Chuc AT, Taghizadeh-Hesary F (2019) Financial inclusion and its impact on fnancial efciency and sustainability, empirical evidence from Asia. Borsa Istanbul Rev 19(4):310–322
Mader P (2018) Contesting financial inclusion. Dev Chang 49(2):461–483
Muri Wole Adedokun & Mehmet Ağa (2023) Financial inclusion: A pathway to economic growth in Sub-Saharan African economies, Int J Financ & Econ, 28(3): 2712–2728
Mialou A, Amidzic G, Massara A (2017) Assessing countries' financial inclusion standing–a new composite index. J Bank Financ Econ 2(8):105–126
Ndhlovu E, Mhlanga D (2023) Peasant financial inclusion for inclusive development in Zimbabwe. In: Mhlanga D, Ndhlovu E (eds) Econ Incl Post-Indep Africa Adv Afr Econ, Soc Polit Dev. Springer, Cham. https://doi.org/10.1007/978-3-031-31431-5_13
Prina S (2015) Banking the poor via savings accounts: evidence from a field experiment. J Develop Econ 115(C): 16–31
Ozili PK (2018) Impact of digital finance on financial inclusion and stability. Borsa Istanbul Rev 18:329–340
Pomeranz D, Kast F (2022) Savings accounts to borrow less: Experimental evidence from Chile." J Hum Resour. Ahead of print. https://doi.org/10.3368/jhr.0619-10264R3
Poonam, Chaudhry, A (2016) Financial inclusion in India: A state level study. SSRG Int J Econ Manag Stud, 3(3): 01–04
Rhine S, Greene W, Toussaint-Comeau M (2006) The importance of check-cashing businesses to the unbanked: Racial/ethnic differences. Rev Econ Stat 88(1):146–157
Sarma M (2008) Index of financial inclusion (2008). ICRIER Working Paper 215
Sarma M, Pais J (2011) Financial inclusion and development. J Int Dev 23(5):613–628. https://doi.org/10.1002/jid.1698
Sahay MR, Cihak M, N'Diaye MP, Barajas MA, Mitra MS, Kyobe MA, Yousefi MR (2015) Financial inclusion: can it meet multiple macroeconomic goals?. International Monetary Fund
Sakanko MA, Yahaya SU, Abdullahi S (2023) Financial literacy and financial inclusion. Zakariya J Soc Sci 2(1):1–10
Sarva M (2014) Financial literacy: important for growth and inclusion. International. Journal Innov Res Dev 3(4):214–221
Sharma D (2016) Nexus between financial inclusion and economic growth. J Financ Econ Policy 8(1):13–36
Shetty S, Hans V (2018) Women empowerment in India and financial inclusion barriers. Int J Manag Sociol Humlties 9(3):344
Siddik MNA (2017) Does financial inclusion promote women empowerment? Evidence from Bangladesh. Appl Econ Financ 4(4):169–177
Silvia P (2015) Banking the poor via savings accounts: Evidence from a field experiment. J Dev Econ 115(1):16–31
Singh D, Singh H, Sandhu M (2017) New market entry strategies: public and private sector banks in India. J Indian Manag 15(2):79–86
Singh K, Singh Kondan A (2011) Fnancial inclusion, development and its determinants, an empirical evidence of Indian States. Asian Econ Rev 53(1):115
Singh S, Gautam RS, Agarwal B, Pushp A, Barge P, Rastogi S (2022) Impacts of financial inclusion on sustainable development in India. J Asian Financ, Econ Bus 9(10):235–242
Simkhada NR (2013) Problems and prospects of the cooperative sector in Nepal for promoting financial inclusion. Enterp Dev Microfinance 24(2):146–159
Srijanani D (2012) Financial inclusion: Taking banking services to the common man. Int J Manag & Bus Stud 2(3):54–58
Suri T & Jack W (2016) The long-run poverty and gender impacts of mobile money. Science, 354(6317): 1288–1292
Swamy V (2011) Financial inclusion in India: an evaluation of the coverage, progress and trends. IUP J Financ Econ 9(2):7–26

Tanmoyee Banerjee-Chatterjee & Malabika Roy & Ajitava Raychaudhuri (2021). Financial inclusiveness and mobilization of household savings in rural Area: Are they related?, J Quantitative Economics, 19(2): 223–249

Varghese G, Viswanathan L (2018) Financial inclusion: opportunities, issues and challenges. Theor Econ Lett 8:1935–1942. https://doi.org/10.4236/tel.2018.811126

Unctad (2016) UNCTAD DGFF2016 SDG goal 8.10—banking, finance and insurance. UNCTAD DGFF 20.https://stats.unctad.org/Dgff2016/prosperity/goal8/target_8_10.html

World Bank (2014) Global financial development report (2014): Financial Inclusion; World Bank: Washinton. DC, USA

Yangdol R, Sarma M (2019) Demand-side factors for financial inclusion: A cross-country empirical analysis. Int Stud 56(2–3):163–185

Zhang Q, Posso A (2017) Thinking inside the box: A closer look at financial inclusion and household income. J Dev Stud, 1–16. https://doi.org/10.1080/00220388.2017

Volatility Analysis and Forecasting of SENSEX and USD/INR Exchange Rate: An Empirical Study Using GARCH (1,1) Models

Palak Chitlangiya, Parul Bhatia, Sudhi Sharma, Mayank Goyal, and Shab Hundal

Abstract This study explores the volatility dynamics and forecasting capabilities of the BSE SENSEX and the USD/INR exchange rate using the Generalized Autoregressive Conditional Heteroskedasticity (GARCH) (1,1) model. These two financial indicators serve as vital benchmarks for assessing economic stability and investor confidence in India. Utilizing daily data from April 2020 to March 2025a period marked by both post-pandemic recovery and global economic uncertaintythe research begins with descriptive statistics and stationarity testing via the Augmented Dickey-Fuller (ADF) method. ARIMA models are employed to estimate mean behaviour, while GARCH (1,1) captures time-varying volatility and persistence. The results reveal significant volatility clustering and long memory in both markets, with exchange rate fluctuations displaying slightly greater persistence. Diagnostic checks, including the Ljung-Box test, ARCH LM test, and Sign Bias test, confirm model adequacy. The findings suggest that GARCH-based models are effective tools for volatility forecasting and can aid investors, policymakers, and financial analysts in strategic risk assessment and decision-making.

P. Chitlangiya · S. Sharma · M. Goyal
Fortune Institute of International Business, New Delhi, India
e-mail: 26-palak.chitlangiya@fiib.edu.in

S. Sharma
e-mail: sudhi.sharma@fiib.edu.in

M. Goyal
e-mail: 26-mayank.goyal@fiib.edu.in

P. Bhatia (✉)
Shri Vishwakarma Skill University, Haryana, India
e-mail: parul.bhatia@svsu.ac.in

S. Hundal
JAMK University of Applied Sciences, Jyväskylä, Finland
e-mail: shabnamjit.hundal@jamk.fi

F. A. Malik et al. (eds.), *Aligning ESG Goals with the UNs Sustainable Development Goals Planning*, World Sustainability Series,
https://doi.org/10.1007/978-3-032-01715-4_11

Keywords GARCH (1,1) · Volatility forecasting · Time series analysis · BSE SENSEX · ARIMA model

1 Introduction

In today's highly interconnected and unpredictable global financial environment, understanding market fluctuations has become crucial for investors, policymakers, and researchers. Volatility describes how much the price of a financial asset fluctuates over time and is often used to gauge market uncertainty, risk levels, or major changes in the economy. In the case of a developing economy like India, two important indicators that reflect these financial movements are the BSE SENSEX and the USD/INR exchange rate.

The BSE SENSEX includes a selection of well-established companies listed on the Bombay Stock Exchange and acts as a key indicator of the overall performance of India's stock market. As a leading benchmark index, it provides valuable information about the financial stability and investor confidence in the country's corporate sector.

At the same time, the USD/INR exchange rate directly influences trade, inflation, and capital flows, as India heavily depends on imports and foreign investment. Changes in these indicators often reflect broader shifts in market sentiment, global events, and domestic economic policies. As such, analysing their volatility not only helps in understanding market behaviour but also supports better financial planning and policy formulation.

While the traditional time series models such as ARIMA are well-suited for identifying trends and seasonal patterns, they are not equipped to effectively model volatility clusteringa frequent occurrence in the financial time series where period of high volatilities are followed by similarly volatile intervals (Mandelbrot 1963). To address this shortcoming, Engle (1982) introduced the ARCH model, which was later extended to the more flexible GARCH model by Bollerslev (1986). These models are specifically designed to capture and forecast time-varying volatility, making them especially useful in financial time series analysis, where the presence of volatility clusters is almost ubiquitous.

The GARCH (1,1) model, in particular, has gained popularity due to its simplicity and reliability in capturing volatility persistence in financial returns (Brooks 2019). It allow the conditional variance to be a function of previous periods' squared residuals and lagged conditional variances, making it capable of modelling persistent market shocks. This dynamic modelling of variance helps identify how the long-lasting impact of market shocks might be, thereby offering valuable information for risk assessment and hedging strategies.

The motivation for this research stems from the unprecedented volatility witnessed in Indian financial markets, especially during and after the COVID-19 pandemic. Beginning in early 2020, the global financial system was subjected to immense pressure due to lockdowns, global supply chain disruptions, geopolitical conflicts, and fluctuating investor sentiments. India, being a large emerging market, experienced notable turbulence in both its equity and currency markets. The BSE SENSEX and USD/INR exchange rate exhibited sharp reactions to global and domestic news, fiscal stimuli, monetary interventions through the Reserve Bank of India, and uncertainty over vaccination rollouts and economic recovery timelines. This period, spanning from April 2020 to March 2025, serves as an ideal window for empirical analysis of financial volatility.

This study adopts a structured methodology, beginning with descriptive statistics to understand the fundamental characteristics of the return series. Subsequently, the Augmented Dickey-Fuller (ADF) test is used to check for stationarity, as non-stationary time series can lead to spurious regression results (Dickey and Fuller 1979). Once stationarity is confirmed through appropriate transformations such as differencing or logarithmic adjustments, the ARIMA model is employed to detect deterministic patterns and trends. However, to explicitly model the stochastic nature of volatility, the GARCH (1,1) model is applied to the residuals of the ARIMA model.

Several diagnostic tests are utilized to assess the goodness of fit and robustness of the model. These include the Ljung-Box test for autocorrelation in residuals (Ljung and Box 1978), the ARCH LM test to detect remaining heteroskedasticity (Engle 1982), and the Sign Bias test proposed by Engle and Ng (1993), which evaluates the asymmetric impact of positive and negative shocks. These tests not only ensure that the model accurately captures the volatility dynamics of the financial series but also verify whether it holds practical value for forecasting and strategic decision-making.

In addition to its analytical depth, what enhances the relevance of this study is its dual focus on both the equity market (SENSEX) and the currency market (USD/INR). Each of these markets reacts differently to economic eventsequities are often driven by corporate earnings, investor sentiment, and policy measures, while exchange rates are influenced more directly by global trade, interest rate differentials, and foreign capital flows. Comparing their volatility patterns can help investors in portfolio diversification, policymakers in designing targeted interventions, and multinational businesses in planning currency risk hedges (Mishra and Mohanty 2020). Moreover, this study provides a comparative perspective on how shocks propagate across financial sectors and which market shows greater resilience or sensitivity to economic disturbances.

Furthermore, the findings of this study can inform algorithmic trading strategies and financial engineering practices. In recent years, quantitative trading firms and institutional investors have increasingly relied on volatility modelling to optimize execution timing and risk-adjusted returns. The predictive capabilities of GARCH models allow for adaptive responses to market regimes, enabling proactive risk control. Additionally, currency traders, corporate treasurers, and hedge fund managers can leverage such volatility insights to structure options and futures contracts more effectively. This study highlights the important link between academic research and real-world financial practice, making a meaningful contribution to both the scholarly community and market participants.

Overall, the research aims to bridge financial theory and practical use by applying advanced statistical methods to analyse volatility trends in two key Indian financial indicators. Using the GARCH (1,1) model, the study strives to deliver accurate forecasts while offering deeper insights into market behaviour amid a changing economic environment.

2 Literature Review

Volatility stands as one of the most important and extensively studied phenomena in financial markets. It refers to the degree of variation of a financial instrument's returns over time and serves as a proxy for financial risk. In emerging economies such as India, where markets are more sensitive to global shocks, policy changes, and political instability, understanding volatility patterns is essential for investors, policymakers, and analysts. Traditional time-series models assume constant variance over time, which is often violated in financial time series. To address this, Engle (1982) introduced the Autoregressive Conditional Heteroskedasticity (ARCH) model, which accounts for time-varying variance in financial data. Subsequently, Bollerslev (1986) generalized this model into the GARCH (Generalized ARCH) model, which considers both past squared returns and past variances to predict current volatility. The GARCH family of models has since become a standard tool in volatility modelling across global and emerging markets.

Since the liberalization of the Indian economy in 1991, the country's financial markets have experienced significant growth and have become increasingly connected with global markets. This transformation has created a demand for reliable models capable of explaining and forecasting market behavior. Numerous researchers have utilized GARCH models and their variations to analyze volatility within India's stock markets. For example, Karmakar (2005) examined the Bombay Stock Exchange and found notable volatility clustering, particularly during key policy announcements. Likewise, Nath and Samanta (2003) applied the GARCH (1,1) model to daily return data from the National Stock Exchange, observing persistent volatility and gradual shock dissipationpatterns typical of emerging markets.

In a comparative analysis of Indian and international stock markets, Mukherjee and Mishra (2005) noted that India's markets tend to show higher volatility and reduced predictability relative to developed markets. They attributed these differences to factors such as lower liquidity, frequent changes in policy, and shifting investor sentiment. Their study also highlighted that GARCH models generally produced better forecasts for Indian stock returns than alternative methods like ARIMA and Exponential Smoothing. Additionally, they emphasized the importance of accounting for structural breakssuch as those caused by the 2008 global financial crisiswhen modelling volatility, as these events can significantly alter market dynamics.

Beyond equities, research has also focused on volatility within India's foreign exchange market. Tripathy (2011) used GARCH models to investigate the relationship between volatility in the forex market and stock indices, uncovering evidence of two-way volatility spillovers between the USD/INR exchange rate and major indices such as the Nifty 50. This suggests a level of integration that makes joint modelling necessary. Furthermore, Aftab et al. (2016) explored exchange rate volatility of the Indian rupee through both symmetric and asymmetric GARCH frameworks, concluding that the rupee's value against the dollar is particularly sensitive to global shocks and external market conditions, more so than domestic stock indices.

Several empirical studies have highlighted the responsiveness of Indian financial markets to external shocks. Choudhry (1996), for instance, analysed the impact of the Asian Financial Crisis and observed a significant rise in market volatility in India during that period. His findings demonstrated that conventional GARCH models were capable of effectively capturing such heightened fluctuations. More recently, Bhowmik and Wang (2021) explored the implications of global disturbances, specifically the COVID-19 pandemic, on India's financial landscape. Using high-frequency data spanning two decades (2000–2020), they documented extraordinary volatility surges in both the equity and foreign exchange markets. Their analysis confirmed the efficacy of GARCH (1,1) and EGARCH models in modelling the extreme variability observed during crisis periods.

To account for the differential impact of positive and negative news on market volatilitycommonly referred to as the leverage effectresearchers have increasingly turned to asymmetric volatility models. The EGARCH model proposed by Nelson (1991) and the TGARCH model developed by Zakoian (1994) are particularly suited for capturing this phenomenon. These models prove especially valuable in emerging markets such as India, where investors often respond more sharply to unfavourable news or developments.

In a detailed study focusing on different sectors, Pandey (2005) employed both EGARCH and TGARCH models on Indian industry indices and found clear evidence of asymmetric volatility. His findings revealed that industries such as information technology and banking were especially vulnerable to negative market developments, supporting the notion that unfavourable news tends to cause a greater increase in volatility.

Apart from stock indices and exchange rates, volatility modelling has been extended to commodities, interest rates, and sectoral indices in India. Mohanty and Nandha (2006) used GARCH models to analyse oil price shocks and their impact on the Indian stock market. They discovered that volatility in crude oil prices had a strong spillover effect on market indices, particularly in oil-dependent sectors like transportation and energy. Gupta and Basu (2011) studied interest rate volatility using GARCH models and found that unexpected rate changes through the Reserve Bank of India significantly altered return variances, reinforcing the importance of policy uncertainty in modelling Indian financial dynamics.

Many studies have integrated macroeconomic variables into GARCH frameworks to better explain volatility behaviour. Tripathi and Seth (2014) conducted a detailed analysis on the relationship between macroeconomic indicators such as inflation, industrial production, and the money supply and their impact on Indian stock markets. Using GARCH-M and EGARCH models, they concluded that these indicators had statistically significant effects on both returns and volatility. Similarly, Chittedi (2012) examined the long-run relationship between exchange rates and stock prices in India using cointegration and GARCH models and found strong linkages, suggesting that macroeconomic stability is crucial for maintaining low market volatility.

More recent research has focused on volatility spillover across asset classes and international markets. Behera and Rath (2008) found that volatility in global indices such as the S&P 500 and FTSE had a measurable impact on Indian stock indices. They used multivariate GARCH models to capture volatility transmission and noted that the correlation between global and Indian markets increased significantly after 2005, post full capital account convertibility for Foreign Institutional Investors (FIIs). Similar findings were presented by Das and Mishra (2018), who examined cross-market volatility transmission between BRICS countries and India. They found that volatility shocks originating in China and Brazil were often transmitted to Indian markets within two to three trading days, reinforcing the need for multivariate modelling frameworks such as BEKK-GARCH and DCC-GARCH.

In terms of model performance, GARCH (1,1) remains one of the most commonly used and robust models for capturing volatility in Indian markets due to its parsimony and ability to capture volatility clustering. However, models such as EGARCH and GJR-GARCH are increasingly preferred for their ability to account for asymmetry in response to positive and negative shocks (Sharma and Vipul 2020). Model comparisons in Indian literature have shown that while GARCH (1,1) performs well during normal periods, EGARCH models are better suited for turbulent phases, such as during demonetization in 2016 or the onset of the pandemic in 2020.

Moreover, the model performance has been tested across different time horizons-daily, weekly, and monthly dataand across sectors. Rao and Naik (2017) conducted a sectoral volatility analysis for the Indian market and concluded that volatility persistence varies considerably across sectors, with financial and energy sectors showing higher persistence than FMCG and healthcare. They highlighted that GARCH-type models provide a meaningful framework to understand volatility behaviour, especially when tailored to the specific characteristics of different asset classes and macroeconomic contexts.

3 Objectives and Hypothesis

3.1 Objectives of the Study

The study is designed to achieve the following focused objectives:

1. **To analyse the volatility behaviour** of the BSE SENSEX and USD/INR exchange rate using GARCH (1,1) models.
2. **To forecast short-term volatility** for both financial series based on the estimated GARCH models.
3. **To test the stationarity** of the financial return series using the Augmented Dickey-Fuller (ADF) test.

4. **To perform diagnostic testing** to validate the model fit and forecasting accuracy.

These objectives aim to offer clear insights into market risk behaviour and volatility dynamics in the post-pandemic economic environment.

3.2 *Hypothesis of the Study*

Based on the objectives, the key hypotheses are framed as follows:

Null Hypotheses (H_0):

- H_{01}: The return series of BSE SENSEX and USD/INR are non-stationary.
- H_{02}: There is no significant volatility clustering in the BSE SENSEX and USD/INR returns.
- H_{03}: The GARCH (1,1) model does not significantly capture volatility behaviour in the BSE SENSEX and USD/INR returns.

Alternative Hypotheses (H_1):

- H_{11}: The return series of BSE SENSEX and USD/INR become stationary after transformation.
- H_{12}: There is significant volatility clustering in the BSE SENSEX and USD/INR returns.
- H_{13}: The GARCH (1,1) model significantly captures volatility behaviour in the BSE SENSEX and USD/INR returns.

4 Methodology

This study follows a systematic quantitative research methodology based on time series econometrics to analyse and forecast volatility patterns of the stock market, specifically focusing on the BSE SENSEX and USD/INR exchange rate. The methodology is carefully designed to ensure that each stagestarting from data collection to model diagnosticsis logical, robust, and justified in light of academic best practices (Brooks 2019). The approach is divided into structured stages: data sourcing, data transformation, stationarity testing, modelling the mean equation, modelling conditional volatility, forecasting, diagnostics, and comparative analysis.

4.1 *Data Source and Description*

The empirical analysis is based on daily closing price data for two financial series: BSE SENSEX (benchmark stock index in India) and USD/INR exchange rate (value of Indian Rupee against US Dollar). The dataset spans from April 1, 2020, to March 31, 2025. This five-year period was selected intentionally to capture the post-COVID economic recovery, central bank monetary policy shifts (such as RBI's tightening and rate hikes), rising global inflation episodes, volatility due to geopolitical tensions (e.g., the Russia-Ukraine war), and global financial market corrections. Thus, the chosen timeframe provides a rich environment of both normal and stressed market conditions, allowing a comprehensive volatility behaviour study (Tripathy 2011).

The data was extracted from reliable financial data sources such as Investing.com and RBI databases and compiled in Excel. Prior to modelling, basic data checks were conducted to ensure accuracy and completeness.

4.2 Software and Tools Used

To perform accurate and efficient analysis, the study employs the R programming language, known for its advanced capabilities in time series econometrics. Several R packages were utilized for specific functions: The t-series package was used for conducting the Augmented Dickey-Fuller (ADF) test for stationarity; the forecast package facilitated ARIMA model estimation; rugarch and fGarch were employed for GARCH (1,1) volatility modeling; the FinTS package helped perform ARCH LM tests and residual diagnostics; and ggplot2 was used for visualizing returns, volatility forecasts, and residual behaviour. The selection of R ensures reproducibility and access to a broad range of sophisticated modelling options with transparency.

4.3 Data Transformation and Cleaning

Financial time series data often encounter issues such as missing values, non-normality, and heteroskedasticity. Therefore, several preprocessing steps were undertaken before modelling (Tsay 2010). Days with missing price data, typically due to holidays or technical gaps, were removed to maintain a continuous time series without introducing artificial smoothness. Subsequently, logarithmic returns were computed instead of modelling raw prices. The formula used for calculating log returns is: where is the return on day, and are the closing prices on days and respectively, and denotes the natural logarithm. This transformation stabilizes the variance and transforms exponential growth patterns into linear forms, thereby making the data more suitable for statistical modelling.

4.4 Stationarity Testing: Augmented Dickey-Fuller (ADF) Test

Before using any forecasting models, it is essential to verify that the data series is stationary. Stationarity means that key statistical characteristics like the mean and variance do not change over time. To test this, the Augmented Dickey-Fuller (ADF) test is commonly used, where the null hypothesis assumes the presence of a unit root (indicating non-stationarity), while the alternative hypothesis suggests that the series is stationary.

The test results indicated that both the original SENSEX and USD/INR price series were non-stationary (p-values > 0.05). However, after differencing and converting to log returns, the series achieved stationarity, with p-values falling below the 0.05 threshold. Consequently, all subsequent modelling was conducted on stationary return series (Enders 2014).

4.5 Mean Equation Modelling Using ARIMA

Although volatility was the principal focus, it was necessary to model and remove any autocorrelation in returns prior to volatility modelling. ARIMA models were employed for this purpose (Box et al. 2008). For the BSE SENSEX returns, an ARIMA(1,1,0) model with drift was found optimal. Meanwhile, for the USD/INR returns, a more complex ARIMA(3,0,2) model was selected based on criteria such as AIC, BIC, and residual diagnostics. By removing serial correlation through ARIMA, it was ensured that subsequent GARCH modelling would purely capture volatility effects and not mean structure dynamics.

4.6 Volatility Modelling: GARCH (1,1)

After removing the mean effects, the next stage focused on modelling the time-varying variance, known as conditional heteroskedasticity, using the GARCH (1,1) model introduced by Bollerslev (1986). The model can be described as follows: the conditional variance depends on a long-term average variance, the influence of shocks from the previous day (the ARCH component), and the impact of the prior day's variance (the GARCH component). A higher value of the ARCH term indicates that recent news has a strong effect on volatility, while a larger GARCH term suggests that volatility shocks tend to last longer. When the sum of these parameters nears one, it signals a high level of volatility persistence. The model parameters were estimated through maximum likelihood estimation (MLE), under the assumption that the error terms follow a normal distribution.

4.7 Forecasting

Once the GARCH models were appropriately fitted, they were used to forecast volatility for the next 15 trading days. The forecasts included point predictions of expected volatility as well as 80% and 95% confidence intervals to account for forecast uncertainty. These forecasts provide valuable insights for expected market risk, aiding in asset allocation, hedging decisions, and Value-at-Risk estimation (Hull 2015).

4.8 Diagnostic Checking of GARCH Models

To ensure the reliability and adequacy of the GARCH models, several diagnostic tests were conducted. The Ljung-Box test was applied to standardized residuals to verify the absence of autocorrelation, while the ARCH LM test checked for any remaining ARCH effects in residuals. Additionally, the Nyblom Stability test evaluated the constancy of GARCH parameters over time, and the Sign Bias test assessed whether positive and negative shocks impacted volatility differently (Engle and Ng 1993). Across all tests, p-values exceeded the 0.05 significance level, indicating well-specified models with no significant misspecifications.

4.9 Comparative Analysis: SENSEX Versus USD/INR Volatility

Finally, a comparative analysis between the two series was conducted by examining the GARCH parameter estimates, volatility forecast behaviour, and diagnostic results. It was observed that USD/INR returns exhibited slightly more persistent volatility than SENSEX returns, with a higher sum. Moreover, stock market volatility appeared more reactive to domestic economic and political events, whereas currency market volatility responded more significantly to international and macroeconomic developments (Karmakar 2005).

These findings underscore the necessity for investors to consider asset-specific volatility characteristics when designing portfolio management and hedging strategies.

5 Results and Interpretation

This chapter provides a detailed exploration of the results obtained through the econometric modelling of BSE SENSEX and USD/INR exchange rate volatility. The analysis follows a systematic approach, beginning with descriptive statistics, stationarity testing, mean modeling using ARIMA, volatility modelling using GARCH (1,1), diagnostic checking, forecasting, comparative analysis, and ending with key interpretations and implications for stakeholders. The findings are interpreted in the context of financial theory and market behaviour, ensuring that both statistical and practical relevance are emphasized.

5.1 Descriptive Statistics

A preliminary statistical analysis helps to understand the fundamental behaviour of financial returns before advancing into complex econometric modelling (Table 1).

The result shows that the mean returns of both series are close to zero, which aligns with the stylized facts of financial time series that daily returns are usually centred around zero (Tsay 2010). SENSEX returns are characterized by a higher standard deviation, indicating greater variability and risk compared to USD/INR returns. The negative skewness observed in both series suggests a higher probability of extreme negative returns. In practical terms, this implies that market participants may face more frequent sharp declines rather than sharp increases. Furthermore, both return series exhibit negative kurtosis values, indicating platykurtic behaviour. This finding is somewhat unusual, as financial return series typically show leptokurtosis (fat tails). However, the presence of platy kurtosis here could result from periods of stability and relatively fewer extreme shocks during the sample period.

Understanding these statistical properties sets the stage for correctly specifying models that account for volatility clustering, skewness, and other return dynamics (Brooks 2019).

Table 1 Presents the key descriptive statistics of the daily return series for both BSE SENSEX and USD/INR, spanning from April 2020 to March 2025

Statistic	BSE SENSEX returns	USD/INR returns
Minimum	–0.0545	–0.0165
Maximum	0.0526	0.0128
Mean	0.000478	–0.000038
Median	0.000503	–0.000117
Standard deviation	0.0108	0.0029
Skewness	–0.2255	–0.0845
Kurtosis	–0.4753	–1.5513

5.2 Stationarity Testing: ADF Test

To ensure the time series data met the requirement of stationarity, the Augmented Dickey-Fuller (ADF) test was conducted on both the price and return series (Table 2).

In the ADF test results for the price series was not accurate which indicates that the series was non-stationary in its original form. This non-stationarity in price data is common in financial markets (Enders 2014). However, after converting the data into log returns, the series showed stationarity, demonstrated by significantly negative ADF test statistics and p-values below 0.01. Ensuring stationarity means that key statistical properties like the mean and variance remain stable over time, which is essential before applying models such as ARIMA and GARCH.

5.3 Mean Modelling: ARIMA

Once stationarity was confirmed, the conditional mean of the return series was modelled to eliminate any linear patterns that might affect the accuracy of volatility estimates.

For the BSE SENSEX return series, the ARIMA(1,1,0) model with drift was selected. The AR(1) coefficient was estimated at –0.5272, suggesting a mild negative autocorrelation where positive returns are likely to be followed by slight negative returns. This mean-reverting behaviour is not uncommon in financial time series, especially in equity markets where short-term corrections are frequent (Box et al. 2008).

The USD/INR return series was best modelled by ARIMA(3,0,2). The presence of multiple autoregressive and moving average terms reflects a more complex return-generating process. Currency returns often exhibit more intricate dependence structures due to external shocks, central bank interventions, and macroeconomic announcements (Kumar 2020).

Model residuals from both ARIMA models exhibited white noise properties, indicating that the conditional mean structure was appropriately captured and that any remaining patterns in the data would likely be captured by volatility modelling.

Table 2 Summary of the results

Series	Test statistic	p-value	Stationarity
BSE SENSEX prices	–1.9311	0.3246	Non-stationary
USD/INR prices	–2.8128	0.0671	Non-stationary
BSE SENSEX returns	–22.486	< 0.01	Stationary
USD/INR returns	–11.781	< 0.01	Stationary

Table 3 Presents the estimated parameters

Parameter	BSE SENSEX GARCH (1,1)	USD/INR GARCH (1,1)
ω (Constant)	0.000001381	0.0000000163
α (ARCH term)	0.1169	0.0497
β (GARCH term)	0.8480	0.9490
Persistence (α + β)	0.9649	0.9987

5.4 Volatility Modelling: GARCH (1,1)

The GARCH (1,1) model was applied to the residuals of the mean-adjusted returns to model the conditional variance (Table 3).

The findings reveal that both series display high volatility persistence, as the sum of α and β is close to 1. However, the persistence in the USD/INR series (0.9987) is much higher than in the BSE SENSEX series (0.9649). High persistence indicates that volatility shocks decay slowly over time, meaning that a large movement in returns today could influence volatility several periods into the future (McAleer 2014).

The ARCH term (α) is larger for SENSEX, suggesting that new information impacts equity market volatility more immediately compared to the currency market. Conversely, the larger GARCH term (β) for the USD/INR exchange rate, this suggests that previous volatility significantly affects current volatility levels, which aligns with observations seen in currency markets worldwide (Sadorsky 2012).

5.5 Diagnostic Checking

Several diagnostic tests were conducted to validate the adequacy of the GARCH models. The Ljung-Box Q-test applied to the standardized residuals revealed no significant autocorrelation, confirming the residuals were effectively whitened. The ARCH LM test also failed to detect additional ARCH effects, supporting the notion that the GARCH (1,1) model successfully captured the time-varying volatility (Engle 2001).

Furthermore, the Nyblom Stability test showed no structural instability in the GARCH parameters, implying that the model coefficients remained stable over the sample period. While a minor sign bias was observed in the USD/INR series, it was statistically insignificant, meaning that the model responded similarly to both upward and downward shocks in volatility (Engle and Ng 1993).

5.6 Volatility Forecasting

The GARCH (1,1) models were used to generate 15-day ahead forecasts of conditional volatility for both series. The BSE SENSEX forecast indicated that volatility would remain relatively stable if no major market disruptions occurred. In contrast, the USD/INR forecast suggested a slight upward trend in conditional variance, reflecting potential growing uncertainty in currency markets due to macroeconomic or geopolitical factors.

An important observation from the forecasts was that the confidence intervals widened as the forecast horizon extended. This widening is typical because uncertainty naturally increases when forecasting further into the future (Hull 2015).

Forecasted volatility provides practical insights for portfolio managers and risk analysts. For instance, stable volatility in equities suggests a favourable environment for deploying leverage, whereas increasing volatility in currency markets may require hedging strategies to mitigate exchange rate risks.

5.7 Comparative Analysis

The analysis highlights that while both markets demonstrate significant volatility clustering, the currency market exhibits nearly perfect volatility persistence. Stock markets, being influenced by firm-specific earnings, investor sentiment, and government policies, tend to react more immediately to shocks but revert to normalcy faster compared to currency markets. On the other hand, the foreign exchange market is influenced by slower-moving macroeconomic forces, resulting in highly persistent volatility (Tripathy 2011) (Table 4).

This differentiation has strong implications for market participants. Investors in equities may find short-term opportunities during periods of increased volatility, while those dealing with currencies must be prepared for prolonged periods of elevated risk.

Table 4 Summarizes the key comparative metrics between BSE SENSEX and USD/INR

Metric	BSE SENSEX	USD/INR
Volatility persistence ($\alpha + \beta$)	0.9649	0.9987
Immediate shock reaction (α)	0.1169	0.0497
Volatility clustering strength	High	Very high
Risk forecast trend	Stable	Increasing

5.8 Key Interpretations and Implications

The empirical findings offer several important insights for financial markets. First, the higher volatility persistence in the USD/INR series suggests that external shocks such as oil price fluctuations, global interest rate changes, or geopolitical tensions have lasting impacts on the currency market. Consequently, policymakers must remain proactive, using monetary interventions and communication strategies to stabilize exchange rates (Kumar 2020).

Second, for investors, understanding volatility clustering can enhance portfolio optimization. High and persistent volatility in currencies implies that currency hedging strategies are not only necessary but must be maintained over longer periods. Equity investors, on the other hand, can adopt tactical asset allocation strategies, adjusting their exposures based on observed short-term volatility patterns.

Finally, while the GARCH (1,1) model adequately captures the core volatility structure, minor asymmetries observed, particularly in the USD/INR series, suggest that future research should explore asymmetric volatility models like EGARCH or GJR-GARCH (Nelson, 1991). Such models could better capture the different impacts of positive and negative news on volatility.

Overall, the study reinforces the necessity of tailored risk management approaches for different asset classes, based on their inherent volatility behaviour and market sensitivities.

6 Policy Implications

The findings from this study provide meaningful insights for policymakers aiming to ensure financial stability. In a developing country like India, where the stock market and currency exchange rates are deeply influenced by both international capital flows and local economic factors, understanding how volatility behaves is key to crafting well-informed and effective policies.

One key finding is the strong persistence of volatility in the USD/INR exchange rate, indicating that currency shocks tend to have prolonged effects. This highlights the need for the Reserve Bank of India (RBI) to implement timely and well-planned interventions. Such actions should aim not only to reduce short-term volatility but also to shape market expectations over the longer term. Research has demonstrated that central banks can successfully control exchange rate fluctuations through a mix of direct market operations and transparent communication strategies (Dominguez 2006). Hence, transparency about policy intentions can help stabilize expectations and reduce the persistence of volatility shocks.

Secondly, the relatively quicker mean-reversion of stock market volatility compared to currency volatility implies that equity market shocks are less persistent. However, during periods of extreme eventssuch as global financial crises or pandemicsequity markets may still exhibit unusually high volatility clustering. Policymakers, particularly those involved in financial regulation like SEBI, must ensure that circuit breakers and volatility controls are dynamically responsive. Static thresholds might not suffice during periods of heightened market stress, and a more flexible, data-driven approach may be warranted to protect retail investors without distorting market functioning (Claessens and Kodres 2014).

Thirdly, the findings emphasize the critical role of macroeconomic stability in mitigating persistent volatility. In particular, stable inflation, manageable fiscal deficits, and a credible monetary policy framework contribute significantly to anchoring both equity and currency markets. India's inflation-targeting framework, for example, has been associated with lower financial market volatility since its adoption (Patra 2017). Policymakers should therefore prioritize macroeconomic fundamentals as a proactive strategy for financial stability, rather than relying solely on reactive interventions.

Additionally, the increasing integration of Indian financial markets with global systems implies that domestic policymakers must closely monitor external risk factors. Volatility spillovers from advanced economies, commodity price shocks, and global liquidity shifts have the potential to amplify domestic volatility. A macroprudential surveillance system that tracks these external risks in real-time could greatly enhance the RBI's and SEBI's ability to pre-emptively respond to emerging threats (Basu 2017).

Finally, financial literacy initiatives must be strengthened. Persistent volatility, particularly in currency markets, suggests that investors and businesses exposed to foreign exchange risks need better awareness of hedging instruments and strategies. Government-led programs, in collaboration with financial institutions, could help promote the adoption of risk management practices across sectors.

In summary, policy frameworks must be multidimensionalcombining intervention, macroeconomic discipline, regulatory flexibility, international risk monitoring, and financial educationto effectively manage volatility and safeguard financial market resilience.

7 Conclusion

This study has comprehensively analysed the volatility dynamics of the BSE SENSEX and the USD/INR exchange rate over the period from April 2020 to March 2025 using advanced time series econometric techniques. Employing ARIMA models for mean adjustment and GARCH (1,1) models for volatility estimation, the analysis offers several important conclusions relevant for both academic researchers and financial practitioners.

The descriptive statistics revealed that while both return series exhibited small mean returns, equity markets were more volatile than currency markets. Negative skewness and platykurtic behaviour were observed, highlighting the asymmetric nature of financial returns and the tendency for occasional extreme movements.

Stationarity testing confirmed that price series were non-stationary but became stationary upon transformation into returns, thereby validating the appropriateness of ARIMA and GARCH modeling frameworks. The ARIMA results showed that equity returns followed a simpler autoregressive structure compared to the more complex patterns in currency returns, consistent with theoretical expectations regarding different asset classes.

Volatility modeling using GARCH (1,1) revealed significant volatility clustering and high persistence in both markets. However, the persistence was much stronger in the USD/INR exchange rate, suggesting that currency volatility is stickier and more challenging to control. This finding aligns with the broader empirical literature on emerging markets, where currency volatility often reflects deeper macroeconomic vulnerabilities and external shocks (McAleer 2014; Sadorsky 2012).

Diagnostic tests confirmed that the GARCH models were well-specified, with no remaining autocorrelation or ARCH effects in residuals. The forecasts for volatility indicated a stable outlook for the SENSEX under normal conditions but an upward trend in volatility for USD/INR, raising potential concerns for risk managers and policymakers.

The comparative analysis underscored the need for asset-specific strategies. Equity investors can adopt more tactical adjustments based on shorter-term volatility dynamics, while currency-exposed entities must prepare for prolonged periods of elevated risk.

The policy implications derived from the study suggest that regulatory frameworks must be flexible, macroeconomic stability must be preserved, external risks must be actively monitored, and financial literacy must be promoted. Especially in a globally interconnected environment, proactive and coordinated policy responses are critical to maintaining financial stability.

Finally, while the GARCH (1,1) model proved effective, minor asymmetries observed in the data point to potential areas for future research. More sophisticated models such as EGARCH, TGARCH, or multivariate GARCH could capture asymmetric and cross-market volatility spillovers more precisely.

In conclusion, this research reinforces the vital role of volatility modeling in understanding market behaviour, guiding investment strategies, informing regulatory frameworks, and ensuring macro-financial stability in an increasingly complex global environment.

Appendix

Figure 1 shows the time series plot of the USD/INR exchange rate's daily returns during the study period. The return series, labelled as "rusd," varies around a mean of zero, which is common for financial return data and a sign of a stationary process after proper transformation. One of the major visual aspects of the plot is volatility clustering—episodes of high volatility are succeeded by matching episodes of elevated fluctuations, and quieter periods are marked by chronically low volatility. This activity is best observed in the initial half of the series, where higher values of returns occur more frequently and with greater magnitude, especially within time indices 0–600. However, the latter half of the series (roughly from index 600 to 1200) is marked by a precipitous reduction in return variance, which indicates a period of relative stabilization of the market. These kinds of patterns are typical of financial time series and confirm the suitability of using GARCH-type models, which are particularly tailored to account for time-varying volatility. Further, the change in volatility intensity could reflect underlying structural changes in the foreign exchange market, which could be due to macroeconomic interventions, policy shifts, or recovery in the global economy after COVID-19. In general, the story highlights the dynamic volatility of exchange rate fluctuations in the Indian case and offers a robust empirical justification for capturing this behaviour through the GARCH (1,1) model.

Figure 2 displays the daily return time series plot of the BSE SENSEX index, denoted as "rsesnsex," for the given study duration. The return series has oscillations around a zero mean, a characteristic feature of transformed financial return data representing a stationary process. One of the striking features of the plot is the occurrence of volatility clustering—where volatility of high returns is more likely to

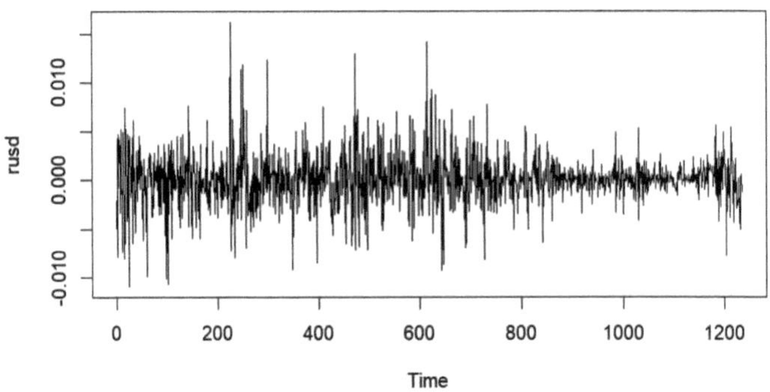

Fig. 1 Daily return fluctuations of USD/INR highlighting volatility clusters in the Indian forex market

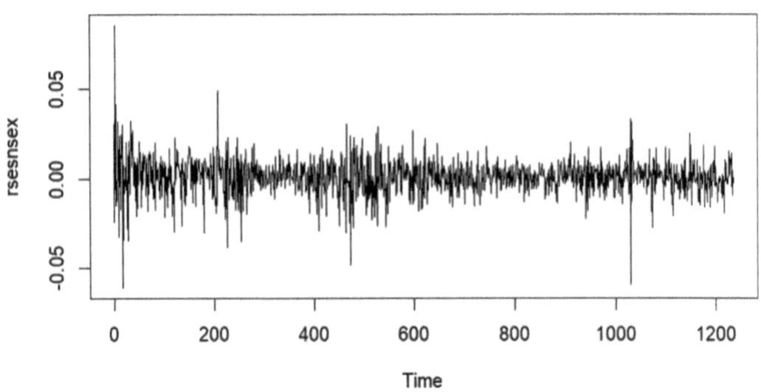

Fig. 2 BSE SENSEX daily returns

be followed by additional high volatility, and low fluctuation periods last in a similar manner. Clustering is very evident in the early part of the series, particularly in the first 200-time units, which represent the initial pandemic phase. During this time, the market saw intense responses to global economic shocks as evidenced by extreme spikes in returns both above and below the zero line. As the timeline goes on beyond this stage, the return volatility tends to decline slowly, indicating stabilization of market sentiment, most probably due to economic recovery efforts, growing investor confidence, and policy interventions. But transient spikes above time index 1000 show that the market still remains sensitive to sudden news and shocks. The pattern corroborates the insufficiency of constant-variance models like ARIMA in capturing the richness of financial volatility. Rather, clustering and persistence of variance make the GARCH (1,1) model plausible for modelling time-varying conditional variance of equity market returns. The following graph illustrates the pattern of conditional variance. In general, this return plot highlights the dynamic nature of the Indian equity market and emphasizes the significance of strong volatility modelling in financial analysis.

Figure 3 shows the autocorrelation function (ACF) plot for the BSE SENSEX return series. The plot shows that the return series autocorrelations decay very quickly and fall within the 95% confidence limits after lag 1, which means that there is no linear autocorrelation in the return series at any lag greater than the first lag. The presence of a pronounced spike at lag 1 followed by a rapid fall to near-zero values is a hallmark of a white noise process, which indicates that the return series mean is not predictable using prior values. This attests to the weak-form efficiency of the market, such that previous returns do not contain meaningful information for predicting future returns. But while the ACF of the returns is not highly autocorrelated, this does not exclude the existence of autocorrelation in the squared returns, which is frequently indicative of volatility clustering—a phenomenon that is better described

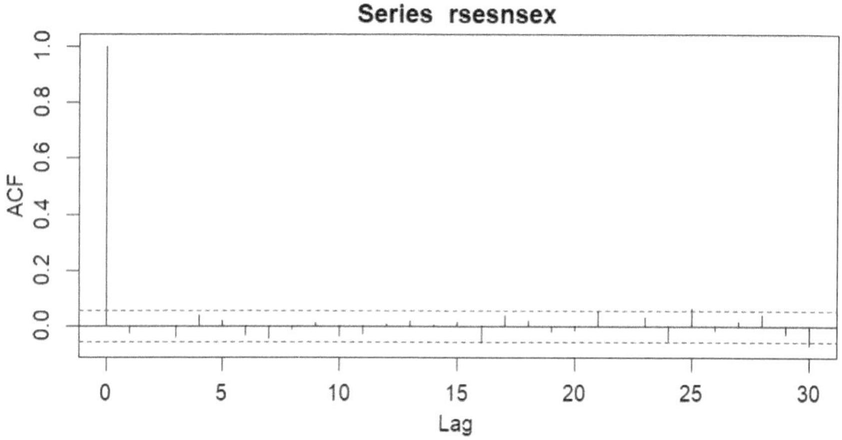

Fig. 3 Autocorrelation Function (ACF) for BSE SENSEX returns

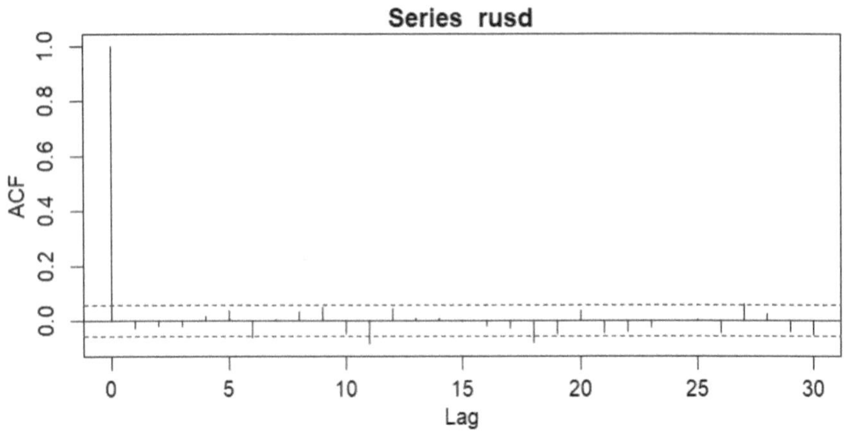

Fig. 4 Autocorrelation pattern of USD/INR returns suggesting random walk characteristics with short-term lag effects

by ARCH or GARCH models. Thus, while the return series per se may look serially uncorrelated, more investigation into the conditional heteroskedasticity (say, using the ACF of squared returns or the ARCH LM test) must be conducted in order to know if a GARCH (1,1) model can adequately model the volatility pattern.

Figure 4 is the autocorrelation function (ACF) plot of the USD/INR return series. The plot reveals that the return series has little serial correlation since all the autocorrelation values after lag 1 fall within the 95% confidence interval. This quick decline in autocorrelation implies that the return series is similar to a white noise process, where returns are mostly uncorrelated with their previous values. The sharp peak at lag 1 can be indicative of short-run market response or microstructure, but its quick decay suggests no high or long-lasting autocorrelation in the series. This result is in line with the weak form of efficient market hypothesis, which suggests that historical exchange rate returns do not allow for predictive ability of future returns. But, similar to the case of equity returns, the absence of autocorrelation in the series of returns does not exclude the presence of volatility clustering. Thus, one needs to continue testing the squared returns or run heteroskedasticity tests to determine whether a GARCH model can be used to capture the conditional variance behaviour in the USD/INR exchange rate.

References

Aftab M, Ismail M, Noor M (2016) Exchange rate volatility and macroeconomic performance in Pakistan: An empirical investigation. Int J Econ Financ Issues 6(2):385–389

Basu P (2017) Financial stability in India: Frameworks and practices. Reserv Bank India Occas Pap 38(2):1–20

Behera HK, Rath BN (2008) Volatility spillover between stock and foreign exchange market: Indian evidence. IUP J Appl Econ 7(3):36–47
Bhowmik R, Wang S (2021) Stock market volatility during COVID-19: Evidence from the GARCH models. Financ Res Lett 38:101872
Bollerslev T (1986) Generalized autoregressive conditional heteroskedasticity. J Econ 31(3):307–327. https://www.sciencedirect.com/science/article/abs/pii/0304407686900631?via%3Dihub
Bombay Stock Exchange (2020–2025) Market data archives. https://www.bseindia.com/
Box GEP, Jenkins GM, Reinsel GC (2008) Time series analysis: Forecasting and control, 4th ed. Wiley
Brooks C (2019) Introductory econometrics for finance, 4th ed. Cambridge University Press
Chittedi KR (2012) Do macroeconomic variables influence stock prices in India? Int J Mark Technol 2(3):212–219
Choudhry T (1996) Stock market volatility and the Asian financial crisis: Evidence from emerging equity markets. J Int Financ Mark, Inst Money 6(4):331–347
Claessens S, Kodres L (2014) The regulatory responses to the global financial crisis: Some uncomfortable questions. IMF Working Papers, vol 2014, no 46, pp 1–30. https://doi.org/10.5089/9781475582218.001
Das RC, Mishra S (2018) Volatility spillover between BRICS stock markets. Glob Bus Rev 19(2):480–495
Dickey DA, Fuller WA (1979) Distribution of the estimators for autoregressive time series with a unit root. J Am Stat Assoc 74 (366):427–431. https://www.jstor.org/stable/2286348?origin=crossref
Dominguez KM (2006) When do central bank interventions influence intra-daily and longer-term exchange rate movements? J Int Money Financ 25(7):1051–1071
Enders W (2014) Applied econometric time series, 4th ed. Wiley
Engle RF (1982) Autoregressive conditional heteroscedasticity with estimates of the variance of United Kingdom inflation. Econometrica 50(4):987–1007. https://www.jstor.org/stable/1912773?origin=crossref
Engle RF, Ng VK (1993) Measuring and testing the impact of news on volatility. J Financ 48(5):1749–1778. https://onlinelibrary.wiley.com/doi/10.1111/j.1540-6261.1993.tb05127.x
Engle RF (2001) GARCH 101: The use of ARCH/GARCH models in applied econometrics. J Econ Perspect 15(4):157–168. https://www.aeaweb.org/articles?id=10.1257/jep.15.4.157
Gupta R, Basu P (2011) Interest rate volatility and the transmission of monetary policy: Evidence from India. Appl Financ Econ 21(15):1117–1132
Hull JC (2015) Risk management and financial institutions, 4th ed. Wiley
IMF (2021–2024) World economic outlook reports
Investing.com (2020–2025) USD/INR historical exchange rates
Karmakar M (2005) Modeling conditional volatility of the Indian stock markets. Vikalpa 30(3):21–37
Kumar S, Mukhopadhyay C (2002) Volatility of Indian stock market: A study of select scripts. IIM Ahmedabad Working Papers, Indian Institute of Management
Kumar S (2020) Dynamics of stock market volatility: Evidence from India. Asian Econ Financ Rev 10(6):591–603
Ljung GM, Box GEP (1978) On a measure of lack of fit in time series models. Biometrika 65(2):297–303
Mandelbrot B (1963) The variation of certain speculative prices. J Bus 36(4):394–419. https://www.jstor.org/stable/2350970
McAleer M (2014) Asymmetry and leverage in conditional volatility models. Econometrics 2(3):145–150. https://www.mdpi.com/2225-1146/2/3/145
Mishra AK, Mohanty RK (2020) Volatility spillover between foreign exchange and stock markets in India: An empirical analysis. Glob Bus Rev 21(1):176–190
Mohanty SK, Nandha M (2006) Oil risk exposure: A comparison of the BRIC countries. Resour Policy 31(2):114–121

Mukherjee S, Mishra RK (2005) Stock market efficiency in the post-liberalization era: Evidence from India. Int J Bus Manag 10(2):124–133
Nath GC, Samanta GP (2003) Dynamic relation between exchange rate and stock price—a case for India. Prajnan 32(1):1–11
Nelson DB (1991) Conditional heteroskedasticity in asset returns: A new approach. Econometrica 59(2):347–370. https://www.jstor.org/stable/2938260?origin=crossref
Pandey IM (2005) Volatility models and their performance in Indian stock markets. J Risk Financ 6(2):28–41
Patra MD (2017) A monetary policy for all seasons. RBI Bulletin, Reserve Bank of India
Rao S, Naik B (2017) Sector-wise volatility analysis of Indian stock market using GARCH models. Indian J Econ Dev 13(1):199–204
Reserve Bank of India (2021–2025) Monetary policy reports. https://www.rbi.org.in/
Sadorsky P (2012) Correlations and volatility spillovers between oil prices and the stock market. Energy Econ 34(1):212–218. https://www.sciencedirect.com/science/article/abs/pii/S0140988311002337?via%3Dihub
Sharma R, Vipul (2020) Forecasting stock market volatility: An application of asymmetric GARCH models. J Quant Econ 18(2):273–295
Tripathi V, Seth R (2014) Stock market volatility and macroeconomic fundamentals: An empirical analysis of the Indian economy. Int J Econ Financ 6(6):235–242
Tripathy N (2011) Stock market volatility and macroeconomic variables: Evidence from India. Bus Econ Res 1(2):1–12. https://ideas.repec.org/a/sae/globus/v15y2014i2p291-316.html
Tsay RS (2010) Analysis of financial time series, 3rd ed. Wiley
World Bank (2021) Global economic prospects: COVID-19 impact and recovery. https://www.worldbank.org/en/publication/global-economic-prospects
Zakoian J-M (1994) Threshold heteroskedastic models. J Econ Dyn Control 18(5):931–955

Empowering Women's Financial Security Through ESG- SDG Alignment

Tarika Singh Sikarwar, Nidhi Jain, Firdous Ahmad Malik, and Shafali Mangal

Abstract This study investigated the effects of financial literacy, saving behavior, and financial management, on retirement confidence of young- and middle-adult working women in India, specifically women aged 21 to 51 from various educational and occupational backgrounds. Although the results show that financial literacy is of lesser importance than saving behavior and financial management, a combination of all three is necessary to build retirement confidence. Together, they provide greater financial readiness for women. This has ramifications for financial planners, policy-makers and educators, who can develop targeted initiatives in line with SDG 8 (Decent Work and Economic Growth) to assist women's retirement planning. The paper also connects this topic to the wider agenda of ESG (Environmental, Social and Governance) goals and the UN SDGs, particularly SDG 5 (Gender Equality) and SDG 8. It illustrates how corporates can focus on systemic barriers faced by women by providing inclusive retirement plans, financial literacy, equal pay and workplace policies. As such, they can help empower women while maximizing ESG performance and building inclusive, sustainable development.

Keywords SDG5 · SDG 8 · Gender equality · Economic growth · Financial literacy · Saving behavior · Financial management · Retirement confidence · Working women · Psychology · Asia

JEL Codes D10 · D14 · D19

T. S. Sikarwar (✉) · N. Jain (✉) · S. Mangal
Prestige Institute of Management and Research, Gwalior (M.P.), India
e-mail: tarika.singh@prestigegwl.org

N. Jain
e-mail: nidhi.jain@prestigegwl.org

F. A. Malik
Department of Management, University of People, Pasadena, CA, USA

F. A. Malik et al. (eds.), *Aligning ESG Goals with the UNs Sustainable Development Goals Planning*, World Sustainability Series,
https://doi.org/10.1007/978-3-032-01715-4_12

1 Introduction

Retirement is the stage when an individual permanently leaves the workplace. After retirement, expenses do not stop, but the income of an individual stops permanently after leaving the workplace. However, their expenses will continue, and individuals will need to continue their lives depending solely on their accumulated savings. The main source of retirement income is savings; therefore, proper planning during the pre-retirement stage is essential for a secure retirement.

Retirement can be a tough period for both women and men. The life expectancy of an individual (women = 77.2 years, men = 72.3 years; as per statista.com) and the minimum retirement age in India is 60 years. Thus, the rest of the years left after retirement will be a traumatic phase for retirees, and accumulating funds for retirement life is a necessity for every individual. Financial preparedness for retirement has a wide scope, and knowledge is important to each individual. Since we consider financial planning as a protective factor for retirees, apart from that, there is also a need for understanding the benefits of savings in the life of an individual. To avoid a financial crisis after retirement, learning financial literacy, having savings behavior, and good financial management play a crucial role even for those who are literate and working in their respective departments, but are not aware of pre-retirement planning.

Demographic-based studies of men and women have suggested that women display poor retirement confidence as compared to men of the same age (Malroutu and Xiao 1995; Quick and Moen 1998; Joo and Pauwels 2002). Contrary to this work done by Hershey et al. (2010; Mutran et al. 1997), Schellenberg and Ostrovsky (2008) indicated more retirement confidence in women as compared to non-married women. Education also affects confidence level for retirement as educated women are found to be more confident than lower educated ones (Helman et al. 2012; Joo and Pauwels 2002). There is a gender gap in employment, which results in inequalities in retirement in the US. Similar work in South Africa also projected an association between financial preparations for retirement and the attitudes of individuals towards retirement. The Retirement Confidence Survey indicates a gap exists between men's and women's retirement confidence. A study of retirement confidence done amongst working women of government agencies found low confidence in Malaysian women. Results of a worker's survey done in the United States displayed their poor retirement confidence since 2009 (Helman et al. 2012).

Retirement issues have customarily been considered to relate to the male population, yet progressively, the significance to the female population has been perceived (Weaver 1994). A few variables may add to retirement confidence, and they may vary for people. As cited recently in one of the articles in Economic Times, "There is an unhappy bond between women and Money in India" (ET, 2nd March 2020). There exists a gender pay gap, financial illiteracy, and family responsibilities for working women, leading to a lack of freedom to earn and invest. They are very concerned about the retirement money they will have post-retirement (ET, 7th March 2015).

There are several factors other factors which affect retirement confidence including Financial literacy (Lusardi 2019), Saving Behaviour (Moorthy et al. 2012), Financial Management (Sabri et al. 2015), Demographic and Socioeconomic Factors (Ekerdt et al. 2000), marital status (Sundén and Surette 1998; Szinovacz and DeViney 2000), no. of dependents (Weaver 1994), Ethnicity (Caputo 2000), Racial inequality, Income (Malroutu and Xiao, 1995), Attitudinal and Behavioral Factors (Joo and Grable 2000; Pauwels, 2001), Investment strategies (Bajtelsmit and Bernasek 1997; Jianakoplos and Bernasek 1998; Hinz et al. 1997), retirement investments (Pauwels 2001), Financial knowledge (McCarthy and Turner 2000), workplace education (Garman et al. 1999; Joo and Grable 2000).

ESG Compliant organizations are increasingly expected to measure and enhance the Financial Literacy, retirement readiness, and economic empowerment of women employees, aligning with SDG. In many Asian Countries, women face informal work structures, pay gaps, and limited pension coverage diminishing retirement confidence.

Hence, the planned work intends to contribute to the work done concerning retirement confidence of the women's population in Asian Countries. With changing times, women are more educated, financially independent, and working, but do they have retirement confidence? what contributes to this retirement confidence is yet to be explored, especially for working women in India. The motivation behind this investigation is to look at the components influencing retirement confidence in the working population of women, as no study is available exploring this topic. Particularly, the work is done to know the role of financial literacy in saving behavior and financial management on the confidence of working women in India.

The retirement confidence of young and middle-adult working women in Asia is closely tied to the higher aim of meeting ESG (Environmental, Social and Governance) targets through the United Nations Sustainable Development Goals (SDGs). For example, SDG 5 (Gender Equality) and Goal 8 (Decent Work and Economic Growth) create a level playing field for women to access and benefit from financial resources, provide employment opportunities, and pursue sustained economic growth. The Social component (S) in ESG supports these Goals by requiring organizations to adopt workplace practices that are in the long-run interests of women employees financial wellbeing. In many Asian countries, working women are still faced with informal work arrangements, wage gaps and low pension coverage contributing to diminished retirement confidence. Corporates are in a unique position to implement their ESG strategies through their SDG targets. For instance, through offering an inclusive retirement plan, financial education to expand women's financial literacy, and equitable pay and opportunities of advancement to mitigate the gender pay gap as well as inclusive practices in maternity and returns to work policies. These actions will not only help women economically but will also improve ESG performance for the company and attract social responsible investors that participate in sustainable and inclusive growth. Improving retirement confidence among working women is not only a social need, but also an environmental step toward achieving global sustainability goals.

The investigation begins with a clarification of hypothetical ideas utilized in the work and comprehension of the momentum research work accessible. In light of this, investigation theories are planned and checked utilizing the proper approach. The outcomes and conversation part portray the factual tests applied and derivations drawn. Further introduced are the end, restrictions, and ramifications of the investigation.

2 Review of Factors Influencing Retirement Confidence

A survey cited by https://www.insidehighered.co in the year 2015, that women professor lacks confidence in retirement planning and are eager to learn more about it.

3 Financial Literacy

Retirement planning should be done early to avoid any adverse situation in later phases of life (Zhou 2003; Bender 2012; Masud et al. 2012). One of the most cited factors in determining retirement confidence is financial literacy, which has proved to have a positive association with it (Lusardi and Mitchell 2008). Financial knowledge, heavy efforts are needed for planning for retirement. As it is a time-consuming process, it is better to start planning in the early 20 s and 30 s (Financial Planning Association 2011).

"Financial literacy" incorporates knowledge, education, ability, competence, and responsibility at the same time. The focus is either on knowledge or on the ability to use the knowledge and even people's self-*confidence* towards their financial actions (Lusardi and Mitchell 2014).

Different languages and regions deal with different expressions of financial literacy (Zait and Bertea 2014). For the USA and Australia, the dominant expression is "financial literacy"; for Great Britain and Canada, it is "financial capability" (Orton 2007; Remund 2010). The OECD uses "financial literacy" without clearly defining the concept (OECD 2005). All these make comparisons different, especially when studies have been conducted in different countries and different languages. The most used explicit definitions found for the concept of financial literacy include: "ability to read, analyze, manage and communicate about the personal financial conditions that affect material well- being" (Chen and Volpe 1998; Vitt et al. 2000; Cude et al. 2006; Huston 2010; Cude et al. 2006); "basic knowledge necessary for people to survive in the modern society"; "capability to understand key financial concepts necessary to function in the normal American society" (Bowen 2002); "ability to manage the situation of cash and payments, knowledge about opening a saving account and obtaining a credit, basic understanding of health and life insurance, ability to compare offers and plan for future financial needs" (Emmons 2005); "ability to use knowledge and manage financial resources for a good financial well-being throughout the whole life" (JumpStart Coalition, 2007, cited by Huston 2010); "ability to make informed judgments and effectively take decisions concerning money" (ANZ 2005); "measure of the degree to which a person understands key financial concepts and has the necessary ability and confidence to manage own finances through short term decisions and long term planning, taking into consideration the economic events and changing conditions" (Remund 2010).

The absence of financial education influences women's overall training levels and makes women especially susceptible to poor practices in money-related management and retirement planning. Low financial literacy profoundly corresponds with most proportions of money-related misery and financial helplessness. Even though monetary proficiency is exceptionally low, working women look for financial incentives sparingly. Such discoveries are irksome given the special conditions that women face, for example, expanded periods out of the workforce and a longer future.

Professional advice combined with workplace financial education was found to be significant in predicting retirement confidence. Retirement planning should be personalized depending on an individual's situation (Kim et al. 2005a, b; Navickas et al. 2014). Financial literacy is a major concern towards achieving retirement success and retirement security as an important variable towards planning (Habila 2015; Agunga 2018a, b). Financial knowledge has been found to influence financial well-being, and this implies that employees' financial well-being could be enhanced by increasing their financial knowledge (Kamakia 2017).

According to Huston (2010), financial literacy is a skill that can help people make financial decisions effectively. Financially literate individuals are expected to possess a basic understanding of financial concepts such as interest rate, inflation rate, compound interest, and risk (Huston 2010). As retirement planning is a complex process that requires a certain level of financial knowledge, there is no doubt that financial literacy and retirement confidence are highly correlated (Lusardi 2009; Lusardi and Mitchell 2008; Mullock and Turcotte 2012; Van Rooij et al. 2012). Financially literate people tend to be more confident in their retirement preparation as compared to those who are financially illiterate (Mullock and Turcotte 2012). This is as those who are financially literate understand the importance of retirement planning and will hence accumulate wealth as preparation for retirement. In contrast, financially illiterate people tend to overestimate their savings and pension income (Alessie et al. 2011). Although their savings and pension income is low, they still believe that they can afford their expenses during retirement (Alessie et al. 2011).

Across different countries (Lucardi 2019), there is a difference in financial literacy amongst men and women, which varies with different levels of income and age too. When asked finance-related questions, women know fewer answers than their male counterparts. This is evident in the US, Europe, Switzerland, Canada, and Asia (Kalmi and Ruuskanen 2017).

Despite the importance of financial literacy for retirement preparedness, much past research has indicated that most people are financially illiterate (Lusardi et al. 2010; Samy et al. 2008). Furthermore, women tend to report lower financial literacy than men (Lusardi and Mitchell 2008). This will make women more vulnerable to the negative consequences compared to men. In other words, poor financial literacy among women might put them at a higher risk of suffering from poor retirement confidence than men.

Working women experience issues putting something aside for the present moment and getting ready for their retirement. These difficulties are related to and likely exacerbated by the way that working women have low monetary proficiency and utilize proficient financial counsel sparingly.

The following proposition can be set from the above research:

4 Financial Management

Financial Management means planning, organizing, directing, and controlling the financial activities such as procurement and utilization of funds of individuals and enterprises. It can be understood as a strategy to manage finances by individuals. Higher retirement confidence comes from good financial management, wherein one of the components is retirement planning. Daily financial management leads to more positive retirement confidence, as those who save money also contribute a portion to retirement planning.

For women, contributing is a deep-rooted procedure, and following the investment is just as significant as making the correct investment. The right financial management is the key to confidence. Higher retirement confidence can be earned from better financial management and practicing proper cash & credit management (Kim et al., 2005a, b; Hilgert et al. 2003; Parrotta and Johnson 1998). There are studies available indicating a positive relationship between good financial management and greater retirement confidence (Kim et al. 2005a, b; Russell and Stramoski 2011; Taylor-Carter et al. 1997). Sabri et al. (2015) found a lesser direct impact of financial literacy on retirement confidence by controlling for financial management practices and was significant indirect relation was found between financial literacy and retirement confidence.

Awareness of retirement confidence is low in many people, especially in women. Women in Malaysia, for example, tend to invest less of their contributions to qualified retirement plans in equity assets compared to men (VanDerhei and Olsen, 2000). Education also leads to retirement confidence in women (American Savings Education Council (ASEC), n.d./2001). It is more when education is employee-sponsored (Alcon 2000). Catrambone (1998) has indicated that lack of knowledge acts as a barrier to retirement planning, especially to older women (Alcon 1999).

Different investment patterns adopted by women, as compared to men, lead to inadequate fund availability upon retirement to women; (Bajtelsmit et al. 1999). Further, the sources of income may prove to be inadequate as well at retirement.

Quickly changing financial markets and expanding individual obligations, specifically for retirement income, require on decision making in informed decision-making. Money-related information is a fundamental instrument for making great financial decisions. In difficult times, working women can borrow from the account set aside for retirement. This excessive debt leaves little room for savings, has serious financial consequences, and leaves women in a helpless financial position. Working women also deal with critical financial situations post-divorce or the death of a spouse. Difficulties presented by an absence of money-related information are intensified when people think they have high monetary information. This wonder applies somewhat to the number of inhabitants in working women. A lion's share of working women believe themselves to be profoundly financially literate and fit for making great everyday choices. Reliance on unbiased financial advice may help in lowering problems associated with low financial literacy.

The following proposition can be set from the above research:

5 Saving Behaviour

'Saving', this word was clarified by Keynes as the excess of income over what is spent on consumption. In detail, saving can also be defined as a portion of disposable income which excess after the consumption of consumer goods.

$$Saving = income - consumption$$

Saving Behaviour includes an understanding of the saving habits of people in the country (Nayak 2013). Saving behavior differs from country to country, the level of development of a nation, and financial policies (Loayza et al. 2000). The world's average saving rate has been declining since the first oil shock and through the early 1990s (Loayza et al. 2000).

The influence from family, peers, and financial literacy plays an important role in shaping savings behavior (Jamal et al. 2015; Patel and Kumar 2017). These researchers observed that the financial well-being of an individual depends upon their financial behavior, which in turn depends upon attitude towards personal financial planning and the financial literacy of an individual. Saving is influenced by current real income, demographic effects, and the real rate of return. Households with higher income uncertainty tend to save more. Further, households with higher levels of lifetime income have a higher lifetime saving rate. Under the life-cycle hypothesis, household saving varies with age. With the further assumption of overlapping generations, demographic changes such as the aging of a baby boom can have important implications for saving rates (Braun 2008). McKinnon and Shaw (1973) showed that financial liberalization led to higher interest rates which equating the demand and supply of savings. The higher interest rates lead to increased savings and financial intermediation in improving the efficiency of savings and investment. The higher real interest rates increase the extent of financial intermediation, which in turn raises the rate of economic growth in developing countries (Balassa 1981).

Issahaku found how individuals respond differently to savings when the interest on deposits changes. Lusardi (2008) wrote that the policies that aim to stimulate saving and financial security after retirement should consider a variety of incentives, including how to decrease informational barriers and simplify decision-making. Kim et al. (2005a, b), Russell and Stramoski (2011), and later Helman et al. (2012) in their work had written that retired people mainly depend upon their savings post-retirement, and thus there exists an association between retirement confidence and savings. The irony is that most people do not realize the importance of saving for retirement (Russell and Stramoski 2011) and do not plan for the savings required post-retirement (Levin 1998). Especially, women need to understand the importance of savings for their retirement (Hayes and Parker 1993).

Retirement confidence is based on the extent to which an individual has a positive attitude and readiness for future retirement (Kim et al. 2005a, b), which is gained through retirement funds who have. A study done in Malaysia shows that confident people are the ones who believe to have accumulated savings for a financially secure retirement life (Moorthy et al. 2012).

Based on the above review, the following hypothesis can be postulated:

Hence, the analysis of the existing literature discussed above may lead to the conclusion that financial literacy, advice, good financial & saving behavior, planning, and financial management may help to reduce financial vulnerability and increase confidence towards retirement.

The following hypothesis can be put forward:

6 Objectives and Hypotheses

Objectives

- To design and develop instruments for measuring the Influence of Financial Literacy, Saving Behaviour, Financial Management, and Retirement Confidence
- To study the Influence of Financial Literacy on Retirement Confidence.
- To study the effect of Saving Behaviour on Retirement Confidence.
- To study the effect of Financial Management on Retirement Confidence.
- To study the influence of Financial Literacy, Saving Behaviour, and Financial Management on Retirement Confidence.

Hypotheses

- *Ha1: There is an influence of financial literacy on Retirement Confidence.*
- *Ha 2: There is an Influence of Financial Management on Retirement Confidence.*
- *Ha3: There is an influence of saving behaviour on Retirement Confidence.*
- *Ha4: There is an influence of financial literacy, saving behavior, financial management, and Retirement Confidence.*

7 Research Methodology

The study uses the data from the survey done on working women of the Gwalior region from April 2019 to July 2019. Working women considered were employed in the service sector at the executive level or above executive positions. They belonged to different educational levels and age groups (21–51 years). These women are in young and middle-aged adulthood (PsycNET). These women are either in the early stages of their careers or are at an advanced stage of their careers. They show a different level of involvement in their retirement (Ekerdt et al. 2000). Initially, 650 questionnaires were circulated, out of which 609 were returned. Out of those 609 questionnaires, 28 questionnaires were rejected again since they were incomplete. Lastly, the study proceeded with a 581 respondent sample.

8 Results and Discussion

8.1 Descriptive Findings

Summary Statistics of Respondents.

Table 1 shows the number of cases analyzed and the summary statistics of women respondents.

Table 2 shows that the highest number of respondents are of age between 21 and 31 years, and the least number of age group in the 42–51 years.

Table 3 shows that the highest number of respondents have an undergraduate degree, and the least number of respondents have a Ph.D. as their highest degree.

Graphical representation of the Sample.

Figures 1 and 2, respectively, display the age-wise and education-wise categorization of respondents.

Table 1 Summary statistics of respondents

Statistics	Age	Education
Cases analyzed	581	581

Table 2 Frequency summary of age of respondents

Cases	Frequency	Percent
21–31 years	333	57.3
32–41 years	222	38.2
42–51 years	26	4.5
Total	581	100.0

Table 3 Frequency summary of education of respondents

Cases	Frequency	Percent
UG	296	50.9
PG	258	44.4
Ph. D	27	4.6
Total	581	100.0

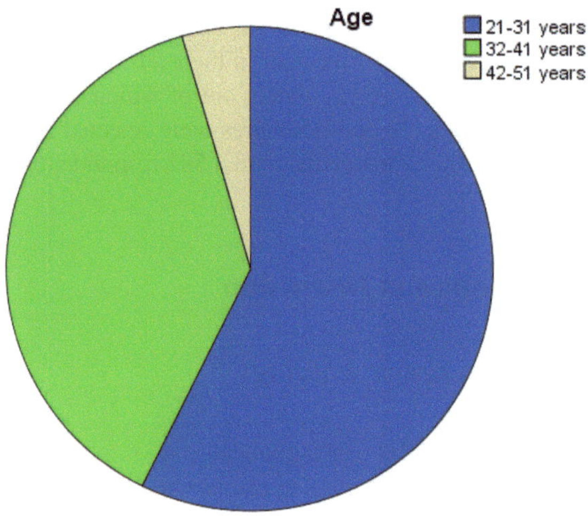

Fig. 1 Distribution of respondents by age group

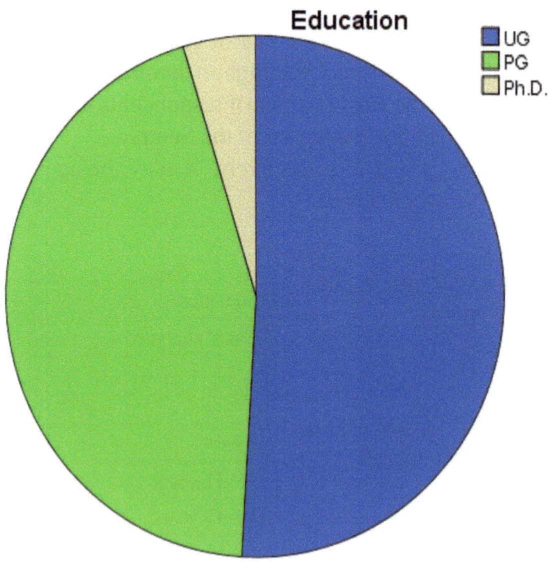

Fig. 2 Distribution of respondents by educational qualification

Table 4 Reliability figures

Name of variables	Cronbach's alpha'	No. of items
Financial literacy	0.791	18
Saving behaviour	0.670	16
Financial management	0.763	24
Retirement confidence	0.749	19

9 Technique

For data collection, questionnaires for the Influence of Financial Literacy (self-designed), Saving Behaviour (Based on Brown and Taylor 2016), Financial Management (Based on Kien Phan et al., 2021), and Retirement Confidence (based on Sabri and Teo 2014) were designed and used. The responses were taken on a Likert-type scale of *1 to 5, where 1 stands for minimum agreement and 5 stands for maximum agreement*. The reliability of the instruments was checked through Cronbach's Alpha (refer to Table 4).

10 Inferential Findings

To determine the influence of financial literacy, saving behavior, and financial management on the Retirement Confidence of working women, linear regression analysis was applied. The influence of each variable is checked by itself, and then the combined effect of all the variables is seen. This is done to understand the change in the influence of variables in the presence of other variables. The results summary is in Table 5.

Refer to Table 5, the *F*-value is insignificant (3.623 @5.8%) when the causation of Financial Literacy and Retirement Confidence is seen. 1.8% variance in Retirement Confidence is seen because of Financial Literacy, indicating the role of other factors too in explaining Retirement confidence. The *Beta* value of -0.135 shows that financial literacy explains poorly the retirement confidence in working women. The negative value shows that it has a reverse relationship with retirement confidence. The beta value as tested through the *t*-value (-1.903@ 5.8%) is insignificant (refer to Table 5).

The *F*-value is insignificant (1.788@18.3%) when the causation of Saving Behaviour and Retirement Confidence is seen. 0.9% variance in Retirement Confidence is seen because of Saving Behaviour, indicating the role of other factors too in explaining Retirement confidence. The *Beta* value of 0.095 shows that Saving Behaviour explains poorly the retirement confidence in working women. The beta value as tested through the *t*-value (1.337 @18.3%), is insignificant (refer to Table 5).

The *F*-value is significant (87.128@0%) when the causation of Financial Management and Retirement Confidence is seen. 30.7% variance in Retirement Confidence

Table 5 Statistics for regression analysis

Dependent variable	Independent variable	R square value	F value and sig	Unstandardized coefficient Beta	Standardized coefficients	T value and significance	Hypothesis rejected/ not-rejected
Retirement confidence	Financial literacy	0.018	3.623 @5.8%	–0.266	–0.135	–1.903@ 5.8%	Ha1: Rejected
Retirement confidence	Saving behaviour	0.009	1.788@18.3%	0.089	0.095	1.337 @18.3%	Ha2: Rejected
Retirement confidence	Financial management	0.307	87.128@0%	0.378	0.554	9.334 @0%	Ha3: Not Rejected
Retirement confidence	Influence of financial literacy, saving behaviour, financial management	Adjusted_R_ Square = 0.535	223.561@0%	Financial_Literacy = 0.070 Saving_Behaviour = 0.166 Fianacial_ Management = 0.463	Financial_Literacy = 0.099 Saving_Behaviour = 0.147 Financial_ Management = 0.594	Financial_ Literacy = 2.938@0% Saving_ Behaviour = 4.207@0% Fianacial_ Management = 17.553@0%	Ha4: Not rejected

is seen because of Financial Management, indicating the role of other factors too in explaining Retirement confidence. The *Beta* value of 0.554 shows that Financial Management explains the retirement confidence in working women in a better manner. The beta value as tested through the *t*-value (9.334 @0%) is significant (refer to Table 5).

To determine the significant predictors (Financial Literacy, Saving Behaviour, Financial Management) of Retirement Confidence of working women, multiple regression was applied. The *F*-value is significant (223.561@0%) when the causation of all variables and Retirement Confidence is seen. 53.5% variance in Retirement Confidence is seen because of Financial Literacy, Saving Behaviour, and Financial Management, indicating significant explanatory power of these variables together on Retirement confidence. The *Beta* value of Financial Literacy is 0.099, and Saving Behavior is.147, Financial Management is of 0.594 shows that Financial Management explains the retirement confidence in working women in a better manner than Financial literacy and saving behavior. The beta value as tested through *t*-value (Financial Literacy = 2.938@0%, Saving Behaviour = 4.207@0%, Financial Management = 17.553@0%) is significant for all variables (refer to Table 4).

11 Discussion

When the results of this study are viewed through an ESG-integrated lens and through the lens of the United Nations Sustainable Development Goals (in particular SDG 5 and SDG 8) the discussion offers three important observations about financial behaviors which impact the retirement confidence of working women in Asia. The study results indicate that when financial literacy and saving behavior are examined separately, they have little impact on retirement confidence and in the case of financial literacy, we sometimes found a reverse relationship. This means that although women may be financially literate, or they may have a saving habit, it does not mean they have retirement confidence. Financial management also emerged as the most influential factor. Women that actively and intentionally managed their financial lives reported significantly higher retirement confidence, which emphasizes the need for women to have applied financial skills rather than financial knowledge alone.

The combined effect of financial literacy, ability to save and financial management is important when examined together because it requires a more holistic view: none of these three solutions can help to promote major retirement confidence alone, however, the three together can create a robust foundation. Results like these reiterate the importance of ESG aligned strategies that advance the issues of women's wealth that are financially empowering rather than a strictly narrow educative or saving program. This is even more important in some regions where women are at an economic disadvantage caused by wage inequality, coverage exclusion in pensions and other societal factors.

Numerous global studies suggest mixed results. For instance, studies from Kenya and Malaysia suggest that while financial literacy may directly not lead to retirement confidence, financial management practices may act as a mediator. In India, women are, in the main, aware of the need to save, but the "system" continues to undermine women's long term financial security (e.g., the gender wage potential, investment exposure), etc.

Evidence also suggests that women tend to be reluctant to seek financial advice, as well as making personal investment decisions, due to sociocultural reasons.

These results call for ESG-based interventions at the corporate level and at the policy level, which would involve financial management training, ensuring retirement options were inclusive of retirement plans, and developing training for women entrepreneurs (and men) regarding empowerment. These strategies would tap into the particular areas of concern facing women regarding preparedness for retirement, but would also, satisfy a company's overall ESG obligations—as well as more broadly—United Nations sustainability development goals (SDG's). Overall, the need to assist women's retirement-confidence is not only a local, individual, and social need, but a strategic priority for inclusive sustainable economic development as part of women's confidence to engage and support their involvement in their confirmation of their own retirement plan.

12 Conclusion

To summarize, improving retirement readiness for working women requires a comprehensive approach that combines financial literacy, saving behavior and the personal finance management change environment. In the context of shifting social cultures in India, such as more women being single, divorced or widowed and all earning potential household income, these combined factors are particularly critical. For example, the 2011 Census found that there were over 74 million single women in India, and this demographic grew by a whopping 39% from 2001 to 2011 (Government of India 2011). Accordingly, women must take action concerning retirement regardless of age, as studies have shown age is not a differentiating factor in retirement preparedness. In addition, determining how individual actions can align with people's broader ESG (Environmental, Social and Governance) objectives and across the UN SDG's (Specifically achievements on SDG 5, and SDG 8), is a significant collaboration. We can help address this knowledge gap for women by allowing financial literacy learning to take place at schools, workplaces and other local community groups. Knowledge, and planning ahead may stop women from being very low financially knowledgeable and acting ahead! Schools are immeasurably valuable in: (i) exposing young girls to basic financial concepts earlier, (ii) taking steps toward underrepresented demographics and (iii) magnifying the cost/efforts associated with acquiring financial knowledge at a later stage of life. Evidence indicates that women, once aware of the financial knowledge gap, are further advanced in enhancing their financial literacy in this space.

On a policy and corporate level, financial advisors and corporations need to appreciate the contributing factors that effect women's retirement confidence to proffer effective care for women's retirement planning with a gendered lens. Given financial management was found to be the most important factor predicting retirement confidence, it needs to be included in the planning efforts. Therefore, governments should take action, with partners such as financial institutions like the Reserve Bank of India, to create inclusive, economical, and accessible retirement saving schemes for working women. It is also important to note while this study found that three variables together could explain a great deal of retirement confidence (about 70%), factors such as income, financial dependents, urban or rural, and type of employment may also have an effect on retirement confidence. These dimensions will need to be researched in follow-up studies with women workers in sectors such as manufacturing, to build a more comprehensive knowledge. Ultimately, creating policy and programs in the areas assessed in the current study will not only improve financial security for individual women, it will address global ESG and SDG improvements to support inclusive, equitable, and sustainable economic development for women.

13 Policy Implications

Consistent with the overarching themes around ESG and United Nations Sustainable Development Goals, particularly SDG 5 and SDG 8, there are several key policies to improve the retirement confidence of working women. The government should develop comprehensive retirement policies, inclusive of women, that builds a positive and proactive attitude among women on financial planning for life after work. Enhanced retirement education and information programs should be introduced to raise awareness to support working women to understand options available and execute better. These programs must include substantial financial literacy education and new investment opportunities, especially for women who are educators and professional women, with an eye to empower women financially for life. Not only could these assistance programs ensure secure financial futures for individuals but created a more inclusive and sustainable economy following ESG principles and global development goals.

References

Agunga M (2018a) Effect of financial literacy on financial preparedness for retirement among permanent and pensionable employees in state-owned corporations in Nairobi, Kenya. Doctoral dissertation, Kenyatta University

Agunga M (2018b) Effect of financial literacy on financial preparedness for retirement among permanent and pensionable employees in state-owned corporations in Nairobi, Kenya. Doctoral dissertation, Kenyatta University

Alcon A (1999) Financial planning and mature women. J Financ Plan 12(2):82–88

Alcon A (2000) The efficacy of employer-sponsored financial education programs and activities: What impact on women's financial situation in old age? (Unpublished doctoral dissertation). Brandeis University, Waltham, Massachusetts
Alessie R, van Rooij M, Lusardi A (2011) Financial literacy, retirement preparation and pension expectations in the Netherlands, DNB Working Paper No. 289. https://www.dnb.nl/media/2p4dbqft/working-paper-289.pdf
Aluodi E, Njuguna A, Omboi B (2017) Effect of financial literacy on retirement preparedness among employees in the insurance sector in Kenya. Int J Bus Manag 12(10):242–253. https://doi.org/10.5539/ijbm.v12n10p242
Anokye A, Frimpong S, Boadu M (2017) Financial literacy and financial planning: Implication for financial well-being of retirees. Bus Econ Horiz 13(2):224–236. https://doi.org/10.15208/beh.2017.17
ANZ (2005) ANZ survey of adult financial literacy in Australia. ANZ Bank and AC Nielsen. http://www.thesmithfamily.com.au/webdata/resources/files/ANZreportMay03.pdf
Arora A (2016) Assessment of financial literacy among working Indian women. Business Analyst 36(2):219–237
Bahl S (2012) Investment behaviour of working women of Punjab. Arth Prabhand: J Econ Manag 1(6). ISSN 2278-0629
Bajtelsmit V, Bernasek A (1997) Why do women invest differently than men? Financ Couns Plan 7:1–10. https://doi.org/10.2139/ssrn.2238
Balassa B (1981) The newly-industrializing developing countries after the oil crisis. Weltwirtschaftliches Archiv 117(1):142–194
Bender KA (2012) An analysis of well-being in retirement: The role of pensions, health, and 'voluntariness' of retirement. J Socio-Econ 41(4):424–433. https://doi.org/10.1016/j.socec.2011.05.010
Bomikazi Z (2017) Exploring the retirement attitudes of pre-retired individuals in the Eastern Cape, South Africa. Int J Manag Appl Sci 3(9):19–24. ISSN: 2394–7926
Bowen CF (2002) Financial knowledge of teens and their parents. J Financ Couns Plan 13(2):93–102
Brown M (2016) Early influences on saving behaviour: Analysis of British panel data. J Bank Finance 62:1–14
Bucher-Koenen T, Lusardi A, van Rooij M (2017) Women, confidence, and financial literacy, GFLEC Working Paper Series, WP 2016–1. Global Financial Literacy Excellence Center
Caputo RK (2000) The availability of traditional and family-friendly employee benefits among a cohort of young women. Fam Soc 81(4):422–436
Catrambone K (1998) Women face barriers in retirement. Pension & Investments 26(11):34
Chen H, Volpe RP (1998) An analysis of personal financial literacy among college students. Financ Serv Rev 7(2):107–128
Cude BJ, Lawrence FC, Lyons AC, Metzger K, LeJeune E, Marks L, Machtmes K (2006) College students and financial literacy: What they know and what we need to learn. In: Eastern family economics and resource management association annual conference proceedings, pp 102–109
Das SC, Kumar V (2016) Discretionary savings and investment behaviour of middle class households in India: Empirical evidence study. Int J Appl Mark Manag 1(1):26–34
Ekerdt DJ, Kosloski K, DeViney S (2000) The normative anticipation of retirement by older workers. Res Aging 22(1):3–22. https://doi.org/10.1177/0164027500221001
Emmons WR (2005) Consumer-finance myths and other obstacles to financial literacy. Federal Reserve Bank of St. Louis, Supervisory Policy Analysis Working Papers, 24
Financial Planning Association (2011) Planning for the stages of retirement. http://www.plannersearch.org/life-events/documents/6-7-11%20retirement%20brochure.pdf
Garman TE, Kim J, Kratzer C, Brunson B, Joo S-H (1999) Workplace financial education improves personal financial wellness. Financ Couns Plan 10:79–88
Grace D, Weaven S, Ross M (2010) Consumer retirement planning: An exploratory study of gender differences. J Cetacean Res Manag 13(2):174–188. https://doi.org/10.1108/13522751011032601

Habila M (2015) Influence of financial education on retirement security: Evidence from the state of Illinois. MPRA Paper. https://mpra.ub.uni-muenchen.de/73988/
Haque A, Zulfiqar M (2015) Women's economic empowerment through financial literacy, financial attitude and financial wellbeing. Int J Bus Soc Sci 7(3):78–88
Haruna I (2011) Determinants of saving and investment in deprived district capitals in Ghana: A case study of Nadowli in the Upper West Region. Curr J Soc Sci 4:1–12
Hayes CL, Parker M (1993) Overview of the literature on pre-retirement planning for women. J Women Aging 4(4):1–18. https://doi.org/10.1300/J074v04n04_01
Helman R, Copeland C, VanDerhei J (2012) The 2012 retirement confidence survey: Job insecurity, debt weigh on retirement confidence, savings. EBRI Issue Brief/Empl Benefit Res Inst 369(1):5–32
Hershey DA, Henkens K, van Dalen HP (2010) What drives retirement income worries in Europe? A multilevel analysis. Eur J Ageing 7(4):301–311. https://doi.org/10.1007/s10433-010-0167-z
Hilgert MA, Hogarth JM, Beverly SG (2003) Household financial management: The connection between knowledge and behavior. Fed Reserv Bull 89:309–322
Hinz RP, McCarthy DD, Turner JA (1997) Are women conservative investors? Gender differences in participant-directed pension investments. In Gordorn MS, Mitchell OS, Twinney M (eds) Racial inequality in men's employment and retirement earnings, pp 431–438. The Sociological Quarterly, 38
Hogan R, Perrucci CC (2007) Black women: Truly disadvantaged in the transition from employment to retirement income
Huston SJ (2010) Measuring financial literacy. J Consum Aff 44(2):296–316
Jamal AAA, Ramlan WK, Mohidin R, Karim M, Osman Z (2015) The effects of social influence and financial literacy on savings behavior: A study on students of higher learning institutions in Kota Kinabalu, Sabah. Int J Bus Soc Sci 6(11):110–119
Jianakoplos NA, Bernasek A (1998) Are women more risk averse? Econ Inq 36(4):620–630
Joo S, Grable JE (2000) A retirement investment and savings decision model: Influencing factors and outcomes. Consum Interes Annu 46:43–48
Joo S, Pauwels V (2002) Factors affecting workers' retirement confidence: A gender perspective. J Financ Couns Plan 13(2):1–10
Kalmi P, Ruuskanen O-P (2017) Financial literacy and retirement planning in Finland. J Pension Econ Financ 17(1):1–28. https://doi.org/10.1017/S1474747217000270
Kamakia M (2017) Financial literacy and financial wellbeing of public sector employees: A critical literature review. Eur Sci J 13(16):233–249. https://doi.org/10.19044/esj.2017.v13n16p233
Kaur B, Hassan Z (2018) Impact of age, gender, income, education and financial literacy towards retirement planning among Generation 'Y' in Malaysia. Int J Educ, Learn Train 3(1):30–53. https://doi.org/10.24924/ijelt/2018.11/v3.iss1/30.53
Kien Phan T, Hong Thuy Nguyen T, Ha Dang T, Thuan Tran V, Ngoc Le K (2021) Non-financial factors affecting the operational performance of hospitality companies: Evidence from Vietnam. Prob. Perspect. Manag. 19(4):48–62. https://doi.org/10.21511/ppm.19(4).2021.05
Kim J, Garman ET, Quach A (2005a) Workplace financial education participation and retirement savings by employees and their spouses. J Pers Financ 4(3):92–108
Kim J, Kwon J, Anderson E (2005b) Factors related to retirement confidence: Retirement preparation and workplace financial education. J Financ Couns Plan 12(2):77–89
Klatt M (2009) An assessment of women's financial literacy, Master's thesis. University of Wisconsin-Stout
Levin LM (1998) Are assets fungible? Testing the behavioral theory of life-cycle savings. J Econ Behav Organ 36:59–83
Loayza N, Schmidt-Hebbel K, Servén L (2000) What drives private saving across the world? Rev Econ Stat 82(2):165–181
Lusardi A (2009) The importance of financial literacy. NBER Report 2:1–7
Lusardi A (2019) Financial literacy and the need for financial education: Evidence and implications. Swiss J Econ Stat 155(1). https://doi.org/10.1186/s41937-019-0027-5

Lusardi A, Mitchell OS (2005) Financial literacy and planning: Implications for retirement well-being. In: Mitchell OS, Lusardi A (eds) Financial literacy: Implications for retirement security and the financial marketplace. Michigan Retirement Research Center Research Paper No. WP 2005–108. Oxford University Press

Lusardi A, Mitchell OS, Curto V (2010) Financial literacy among the young. J Consum Aff 44(2):358–380

Malroutu YL, Xiao JJ (1995) Perceived adequacy of retirement income. Financ Couns Plan 6:17–24

Masud J, Haron S, Gikonyo L (2012) Exploring economic status of the elderly in Peninsular Malaysia using net flow and net worth. Int J Humanit Soc Sci 2(17):154–160

McCarthy DD, Turner JA (2000) Pension education: Does it work? Does it matter? Benefits Q 16(1):64–72

McKinnon R (1973) Money and capital in economic development. Brookings Institution

Mitchell O, Lusardi A (2008) Planning and financial literacy: How do women fare? Am Econ Rev 98(2):413–417. https://doi.org/10.2139/ssrn.1094097

Moorthy MK, Chelliah TD, Sien CS, Leong LC, Kai NZ, Rhu WC, Teng WY (2012) A study on the retirement planning behavior of working individuals in Malaysia. Int J Acad Res Econ Manag Sci 1(2):54–72

Mullock K, Turcotte J (2012) Financial literacy and retirement saving, Working Paper 2012–01. Department of Finance Canada. http://www.fin.gc.ca/wp/2012-01-eng.asp

Mutran EJ, Reitzes DC, Fernandez ME (1997) Factors that influence attitudes toward retirement. Res Aging 19(3):251–273. https://doi.org/10.1177/0164027597193001

Navickas M, Gudaitis T, Krajnakova E (2014) Influence of financial literacy on management of personal finances in a young household. Verslas: Teorija ir Praktika, 15:32–40. https://doi.org/10.3846/btp.2014.04

Nayak S, Sethi N, Bhujabal P, Mallick A (2016) Determinants and pattern of saving behavior in rural households of Western Odisha: A micro-level analysis. J Econ Policy Res 12:50–61

Nguyen TNM (2015) Factors affecting personal financial management behaviors: Evidence from Vietnam. In: Proceedings of the Second Asia-Pacific conference on global business, economics, finance and social sciences (AP15Vietnam Conference) (Paper ID: VL532). https://globalbizresearch.org/vietnam_proceedings/pdf/VL532.pdf

Organisation for Economic Co-operation and Development (OECD) (2005) Improving financial literacy: Analysis of issues and policies. https://doi.org/10.1787/9789264012578-en

Orton L (2007) Financial literacy: Lessons from international experience. Canadian Policy Research Networks. https://www.deslibris.ca/ID/213658

Parrotta JL, Johnson PJ (1998) The impact of financial attitudes and knowledge on financial management and satisfaction of recently married individuals. Financ Couns Plan 9(2):59–74

Patel AT, Kumar S (2017) A study of awareness, attitude and factors influencing personal financial planning for residents of Gujarat. Gujarat Technological University

Pauwels VW (2001) Factors related to women's confidence regarding their preparedness for retirement. In Hogarth JM (ed) Proceedings of the association for financial counseling and planning education. Orlando, FL, pp 232–239

Quick HE, Moen P (1998) Gender, employment and retirement quality: A life course approach to the differential experiences of men and women. J Occup Health Psychol 3(1):44–64

Remund D (2010) Financial literacy explicated: The case for a clearer definition in an increasingly complex economy. J Consum Aff 44(2):276–295. https://doi.org/10.1111/j.1745-6606.2010.01169.x

Russell K, Stramoski S (2011) Financial management practices and attitudes of dental hygienists: A descriptive study. J Dent Hyg 85:340–347

Sabri MF, Teo J (2014) The influence of financial literacy, saving behaviour, and financial management on retirement confidence among women working in the Malaysian public sector. Asian Soc Sci 10(14):40. https://doi.org/10.5539/ass.v10n14p40

Sabri MF, Teo J, Othman MA, Rahim HA (2015) Financial literacy, financial management practices, and retirement confidence among women working in government agencies: A mediation model. J Dev Areas 49(6):405–412. https://doi.org/10.1353/jda.2015.0092
Samy M, Tawfik H, Huang R, Nagar A (2008) Financial literacy of youth: A sensitivity analysis of the determinants. Int J Econ Sci Appl Res 1:55–70
Schellenberg G, Ostrovsky Y (2008) The retirement plans and expectations of older workers. Can Soc Trends 86:11–34
Sundén AE, Surette BJ (1998) Gender differences in the allocation of assets in retirement savings plans. Am Econ Rev 88(2):207–211
Szinovacz ME, Deviney S (2000) Marital characteristics and retirement decisions. Res Aging 22(5):470–498. https://doi.org/10.1177/0164027500225002
Taff L, Junk V, Fox L, Einerson M (1997) Preretirees' perception of retirement income. Financ Couns Plan 8(2):41–56
Taylor M, Cook K, Weinberg C (1997) Planning and expectations of the retirement experience. Educ Gerontol 23(3):273–288. https://doi.org/10.1080/0360127970230306
Transamerica Center for Retirement Studies (2014) The changing face of retirement: Women balancing family, career & financial security. https://www.transamericacenter.org/docs/default-source/global-survey-2014/tcrs2014_sr_global_women_report.pdf
Van Rooij MCJ, Lusardi A, Alessie RJM (2012) Financial literacy, retirement planning and household wealth. Econ J 122(560):449–478
VanDerhei J, Olsen K (2000) Social security investment accounts: Lessons from participant-directed 401(k) data. Financ Serv Rev 9(1):65–78. https://doi.org/10.1016/S1057-0810(00)00056-1
Vitt LA, Anderson C, Kent J, Lyter DM, Siegenthaler JK, Ward J (2000) Personal finance and the rush to competence: Financial literacy education in the U.S. Institute for Socio-Financial Studies, Middleburg, VA
Weaver DA (1994) The work and retirement decisions of older women: A literature review. Soc Secur Bull 57(4):3–24
Zait A, Bertea PE (2014) Financial literacy—Conceptual definition and proposed approach for a measurement instrument. J Account Manag 4(3):37–42
Zhou J (2003) When the presence of creative coworkers is related to creativity: Role of supervisor close monitoring, developmental feedback, and creative personality. J Appl Psychol 88(3):413–422. https://doi.org/10.1037/0021-9010.88.3.413

Reimagining Development: Environmental Sustainability and the Global Green Transition

Ridhima Sharma, Rohit Bansal, Afzal Sayed Munna, and Jihene Mrabet

Abstract The dynamics of the global green shift and the necessity of rethinking development with an eye toward environmental sustainability are the foci of this article. The study employs a qualitative mixed-methods strategy that incorporates literature review, comparative case studies (including the EU, China, and Costa Rica), and expert interviews against a background of increasing environmental challenges and systemic inequities. Various governance structures' institutionalization of sustainability is examined, as are the ways in which financial and technological systems align with environmental goals, and the impact of socio-cultural dynamics and justice concerns on policy outcomes. It is clear from the results that sustainable development is inherently context-dependent, depending on factors such as political institutions, cultural values, and economic systems. Costa Rica exemplifies community-based environmental management, whereas China demonstrates the efficacy of state-led change and the European Union exemplifies innovative regulations and international cooperation. Successful transitions are facilitated in all cases by inclusive government, legal requirements, and public engagement. The study stresses that systems thinking, adaptive governance, and justice-centered frameworks are essential for achieving sustainability in the long run. Implications for policy include the need for holistic leadership, inclusive funding and expertise, regionalized international frameworks, and proactive resilience preparation. Rather than being seen as merely an environmental necessity, the green transition presents a remarkable developmental opportunity to reshape society in a way that prioritizes equity, ecological

R. Sharma (✉)
Vivekananda Institute of Professional Studies-TC, New Delhi, India
e-mail: sharma.ridhi1311@gmail.com

R. Bansal
Rockford College, Sydney, Australia

A. S. Munna
University of Hull, Hull, London, England
e-mail: a.munna@hull.ac.uk

J. Mrabet
Amity University, Dubai, United Arab Emirates
e-mail: jmrabet@amityuniversity.ae

F. A. Malik et al. (eds.), *Aligning ESG Goals with the UNs Sustainable Development Goals Planning*, World Sustainability Series,
https://doi.org/10.1007/978-3-032-01715-4_13

integrity, and intergenerational responsibility. This book contributes to the academic and policy discourse by offering a nuanced, evidence-based view of many future directions that might lead to equity and sustainability.

Keywords Sustainability · Environment · Green · Transition · Institutional resilience · Governance

1 Introduction

The environmental disasters, increasing social inequalities, and rapidly shifting economic landscape of the twenty-first century have brought the globe to a crossroads. Resource loss, greenhouse gas emissions, and ecological imbalance have all been exacerbated by the conventional development paradigm, which is mostly centered on GDP. This approach has neglected the complex relationship between human development and environmental limits on occasion, leading to unsustainable practices and systemic flaws. Currently, development is undergoing significant changes. Constant progress must conform to the planet's natural constraints, which is becoming increasingly apparent to academics, lawmakers, and members of civil society. Climate change is becoming more and more apparent in the form of severe weather events, higher sea levels, and disruptions to food and water security (Al Amin et al. 2025). A worsening of existing inequalities and a threat to global stability, these events disproportionately affect already-vulnerable populations. Sustainable development emerged as a framework to address this issue, with the goal of achieving a balance between social inclusion, environmental responsibility, and economic progress. The Brundtland Commission has described economic, environmental, and social integration as "development that meets the needs of the present without compromising the ability of future generations to meet their own needs," and it is an important component of this definition (Allam and Cheshmehzangi 2024). The challenge remains, though, in turning this concept into workable policies and processes. The global green transition is a game-changing initiative that seeks to redirect development pathways toward sustainability. This calls for an extensive shift away from fossil fuel-based systems and toward inclusive, low-carbon economies that are efficient with their resources. Besides being a financial or technical revolution, this transformation also necessitates substantial changes in societal values, institutional frameworks, and forms of governance (Almusaed et al. 2025). Environmental sustainability is not an afterthought, but an essential requirement for human flourishing in the long run, and this is the driving force behind this objective. A plethora of international frameworks and agreements stress the magnitude and critical nature of the green transition. United Nations Sustainable Development Goals (SDGs) on climate change, renewable energy, sustainable cities, and responsible consumption, adopted in 2015, provide a comprehensive roadmap for global action (Balcı et al. 2025). The Paris Agreement aims to limit global warming to far below 2 °C and has an aspirational target of 1.5 °C; hence, it necessitates widespread decarbonization of all industries (Benner 2025). Sustainability is one of these global commitments, and countries are picking different ways to incorporate it into their development goals. National green growth efforts, such as China's Ecological Civilization project and the European Union's Green Deal, aim to align economic policies with environmental aims. All approaches recognize the need for institutional transformation, even though their aims and scope vary (Chand 2025). Technological advancements have made the green shift possible. Improvements in renewable energy, energy storage, and smart grid technologies are transforming energy systems, while digital tools like blockchain, AI, and the Internet of Things (IoT) enhance resource efficiency and environmental monitoring. At the same time, more and more people are looking for alternatives to the traditional "take-make-dispose" model of production and consumption, and the circular economy is gaining traction as a result. Reimagined development also requires a shift in the economy (Cole and Hamilton 2024). Massive investments in environmentally friendly infrastructure, environmentally conscious jobs, and robust supply chains are necessary for a green economic transition. Green bonds, ESG investments, and carbon pricing are financial tools that greatly assist in channeling capital toward sustainable enterprises (Dukelow and Murphy 2022). However, a meaningful transition that protects communities and workers affected by structural changes is essential for preserving social cohesiveness and fairness. The social and political dimensions of sustainability are equally important. Information dissemination, environmental education, and cultural narratives all have a role in shaping individual behavior, policy decisions, and collective action (Eversole and Gambetta 2021). Incorporating indigenous knowledge systems, youth movements, and civil society projects into sustainability efforts helps to make it more inclusive and responsive to context. Transitioning to sustainable development is not without its challenges. Institutional laggards, political opponents, and corporate interests might all slow down the process (Ghosh and Pearson 2025). Low- and middle-income nations often face challenges when trying to pursue environmentally friendly policies due to disparities in financial resources and technological availability. Additionally, the possibility of greenwashing—false or cosmetic assertions of environmental responsibility—undermines trust and accountability. The global green movement does, however, present opportunities like never before. It lays forth a plan to increase innovation, decrease climate threats, and strengthen energy security (Grant 2024). A more robust, equitable, and prosperous future is possible for nations when growth is viewed through the lens of environmental sustainability. The world green transition is being bolstered by policy innovations, technological breakthroughs, economic restructuring, and socio-cultural transformations; this research seeks to examine these key components of this changing agenda (Gurieff et al. 2020).

2 Literature Review

The "environmental sustainability" refers to actions taken in harmony with nature that reduce waste and protect our planet's natural resources for future generations. The systematic shift away from low-carbon, resource-efficient, socially inclusive systems and toward economies based on fossil fuels is known as the global green transition (Hashemi 2025). The Sustainable Development Goals (SDGs) and the Paris Agreement provide a framework for making these concepts a reality. A review of the existing literature reveals the multifaceted nature of the global green transition and its effects on development theory and practice (Hasselbalch & Lund). The literature delves into a wide range of disciplines, including political ecology, sustainability science, environmental economics, and development studies, all of which provide light on the challenges and opportunities associated with rethinking development. The limitations of traditional theories of development are being called into question by one main line of research. Unrestrained economic growth is inherently unsustainable, according to some scholars, on a finite planet (Joseph et al. 2025). Their research laid the groundwork for steady-state economics, which has since evolved to highlight ecological limits within which people may safely operate, and for the concept of planetary boundaries. Discussion of sustainable development's ascent to mainstream paradigm status is abundant in academic and policy-oriented journals. The Brundtland Report is a popular resource for anyone looking for a definition of sustainable development that places an emphasis on equity between generations (1987). Multidimensional methods that integrate environmental, social, and economic goals have been supported by subsequent research that has expanded this concept to encompass participatory governance and social inclusion (Komninos and Panori 2025). The transition to green and low-carbon economies has been the subject of considerable study. Studies conducted by the International Energy Agency (IEA), the Intergovernmental Panel on Climate Change (IPCC), and the United Nations Environment Program (UNEP) provide a wealth of empirical data and models about decarbonization, energy efficiency, and the deployment of renewable energy sources (Machin 2025). Carbon pricing, clean energy subsidies, and regulatory regulations are some of the policy instruments highlighted in these articles as ways to hasten the transition. The majority of the literature on sustainable transitions focuses on the dissemination and innovation of technology (Marquardt et al. 2025). To analyze the relationships of technological niches, socio-technical regimes, and wider surroundings in order to promote systematic transformation, the multi-level perspective (MLP) offers a framework. This approach demonstrates the merit of tailored experimentation and institutional harmony when applied to sectors such as energy, transportation, and agriculture. The concept of a circular economy has gained considerable traction recently, with backing from academic works in industrial ecology and the Ellen MacArthur Foundation (McKenzie et al. 2015). To improve resource efficiency and cut down on waste, this literature stresses the need to rethink consumption and industrial processes. Several countries, such as Finland, China, and the Netherlands, have incorporated circular economy principles into their national agendas, but with varying degrees of success and adaptation. Investments and green money are also discussed extensively in the literature (Miglani et al. 2025). Academic journals including Ecological Economics and Sustainable Finance, as well as the World Bank and the International Monetary Fund, have conducted studies that look at the potential of financial markets to support environmental objectives. The capacity of green bonds, environmental, social, and governance (ESG) standards, and impact investment to motivate private funding for environmental change is examined. However, concerns around greenwashing and the lack of uniform standards remain paramount. Research on the green shift's social, cultural, and justice dimensions is growing (Mohamed and Akande 2025). The disproportionate impact of climate change and environmental degradation on already-vulnerable populations is a primary focus of environmental justice research. Research on the concept of a "just transition" has been conducted more recently. This approach seeks to ensure that the shift to a green economy is equitable and inclusive, particularly for communities and workers that rely on firms that produce a lot of carbon (Mullally et al. 2023). Institutions and governments have been the subject of significant study regarding potential ways to facilitate sustainable transitions. The literature on environmental governance stresses the importance of participatory and polycentric forms of government (Nag et al. 2025). The impact of institutional frameworks, policy consistency, and stakeholder engagement on environmental project outcomes is demonstrated via case studies spanning the globe. There are still many gaps and challenges, despite the amount of research (O'Donovan 2024). There has to be more research that bridges the gap between theory and practice and is interdisciplinary and integrative. In addition, much of the content originates in the North, which raises concerns about how accurately it represents other regions and whether or not the models used are applicable to a wide range of socioeconomic and environmental contexts (Owojori and Erasmus 2025). Moving forward, research must prioritize inclusivity, contextual awareness, and the co-production of knowledge with affected communities. The literature as a whole stresses how challenging it is to envision growth in a way that is environmentally sustainable. It stresses the importance of integrated technical, financial, institutional, and cultural solutions. This analysis lays the groundwork for future research on the shifts in policy, technology, economics, and society and culture that characterize the global green transition.

Governments throughout the world are implementing environmentally friendly policies, including carbon pricing, subsidies for renewable energy, and investments in green infrastructure (Plangger et al. 2025). Regulatory frameworks like the European Green Deal and China's Ecological Civilization demonstrate different approaches to integrating sustainability into national development goals. Solar, wind, and battery storage are renewable energy technologies that are crucial in reducing dependence on fossil fuels. Innovative concepts in the circular economy, such as resource recovery, reuse, and recycling, further encourage sustainable industrial growth (Prabhakar 2025). Digital technologies, such as the Internet of Things (IoT) and artificial intelligence (AI), enable efficient monitoring of the environment and management of resources. In a green economy, environmental sustainability and economic stability are of the utmost significance (Pujun 2025). Industry shifts necessitate low-carbon approaches in sectors such as energy, transportation, and agriculture. While green bonds and ESG investments direct capital toward environmentally friendly endeavors, fair transition regulations shield underprivileged workers and communities from financial harm (Ramsarup and Mohamed 2022). Sustainable social ideals are defined in part by public information, educational opportunities, and cultural narratives. In addition to shifting governmental discourse, environmental justice is strongly supported by youth activism, indigenous knowledge systems, and grassroots movements (Reid et al. 2021). Despite gathering steam, the green movement faces obstacles such as political resistance, unequal access to green technologies, and the risk of greenwashing (Samuel and Richie 2023). Nevertheless, it also presents opportunities for innovation, job creation, and the establishment of a more equitable international system.

3 Research Methodology

Using a mixed-methods qualitative research strategy, this article delves into the mechanics of reevaluating development in light of environmental sustainability and the worldwide green transition. The method is grounded on interpretivist epistemology and aims to understand complex socio-environmental events from various perspectives and contexts. Expert interviews, case studies, and a literature review are all part of the study's multi-stage approach. A thorough and comprehensive examination of the data is ensured by this triangulated approach. A thorough literature review was conducted using academic databases such as Scopus, Web of Science, and Google Scholar. Among the keywords are "sustainable development," "green transition," "climate policy," "circular economy," together with "environmental justice." From 1990 to 2025, a total of scholarly articles, policy reports, and institutional publications were analyzed, with a focus on empirical findings, theoretical models, and policy frameworks. To illustrate different regional perspectives on sustainability, three case studies were selected:

- European Union Fit for 55 and Green Deal
- The Ecological Civilization Model in China
- Costa Rica's policy for decarbonization and ecological preservation.

The policy originality, adaptability to different types of governance, and availability of recorded outcomes were the determining factors in the selection of these examples. In each case, we looked at the institutional framework, policy instruments, implementation challenges, and socio-environmental outcomes. Examining how concepts of environmental sustainability are shifting global development paradigms through the green transition is the primary objective of this study.

The specific objectives are

1. To examine theoretical frameworks that are used to back up discussions on sustainable development and green transition.

2. To evaluate new approaches to low-carbon, inclusive development that have been put into place at the national and regional levels.
3. Investigate the potential role of institutional frameworks, financial resources, and technological advancements in facilitating systemic shifts toward sustainability.
4. To look at the global green transition from a sociocultural and justice perspective, paying specific attention to disadvantaged communities.
5. To identify viable strategies for envisioning growth and potential roadblocks by integrating cross-contextual ideas from literature, case studies, and stakeholder perspectives.

4 Analysis

Using a mixed-methods qualitative research strategy, this article delves into the mechanics of reevaluating development in light of environmental sustainability and the worldwide green transition. The method is grounded on interpretivist epistemology and aims to understand complex socio-environmental events from various perspectives and contexts. A multi-stage design incorporating case studies and literature review forms the basis of the research. In terms of data collection and processing, this strategy ensures both breadth and depth. The study's findings are organized into four primary subject areas that emerged from the approach: policy efficacy, technological integration, socio-cultural influence, and governance dynamics. This research draws on literary criticism as well as insights from the three case studies (European Union, China, and Costa Rica).

5 Efficient and Innovative Policymaking

It became clear that the caliber of policymaking has a significant impact on the effectiveness of green transition strategies. Over time, the Green Deal has become the de facto model for comprehensive climate policy in the EU. To achieve carbon neutrality by 2050, it integrates funding mechanisms, regulatory requirements, and cross-sectoral strategies. The European Climate Law and the Emissions Trading System (ETS) were praised by stakeholders for their strong measures. Members' varying levels of administrative readiness and financial resources make it difficult to implement policies consistently. Environmental objectives in China's Ecological Civilization model are centrally planned and monitored by measurable indicators, revealing a highly integrated and centralized policy framework. Significant accomplishments include the control of pollution, the rapid expansion of renewable energy sources, and the extensive planting of trees. There are limitations to transparency and participatory government that interviewees did point up, which can restrict society responsibility and feedback loops. A novel strategy based on citizen participation and environmental protection is on display in Costa Rica. The Decarbonization Plan focuses on a low-emission economy and is based on community engagement, legislative requirements, and intersectoral collaboration. Costa Rica serves as an example of how to balance development with conservation, despite its limited financial and technological resources. This is due to the country's consistent policies and high level of public trust.

6 Combining Technology and Finances

Economic and technological strategies have a significant role in making the green transition a reality. In the European Union, synergies have been created across industries via the application of green industrial strategy, digital technologies, and the deployment of renewable energy. Innovations such as green mobility, smart grids, and hydrogen energy are expanding rapidly, despite the fact that national financial differences hinder equitable implementation. The massive investments made by China in renewable infrastructure and technology have allowed the country to become the global leader in electric vehicles, wind power, and solar power. It has been demonstrated that state-led industrial planning is effective in lowering prices and improving availability. However, issues like overcapacity, the effects of mineral extraction on the environment, and the working conditions in green supply chains continue to cause disagreements. Costa Rica's economic strategy is based on green energy, ecotourism, and sustainable practices. It is now possible to balance environmental objectives with financial resources by using hydropower and carbon sequestration in forests. The importance of economic diversity is underscored by its susceptibility to external shocks and reliance on a few industry.

7 Issues of Justice and Social-Cultural Dynamics

Societal values, cultural narratives, and legal systems influence how green transitions are seen and implemented. With a focus on assisting workers and undeveloped communities, the European Union has institutionalized justice concerns through the Just Transition Mechanism. Energy affordability and engagement remain gaps in low-income areas. Because of the government's monopoly on environmental policymaking, the Chinese people have little say in the matter. Relocation due to infrastructure projects is still a common concern, and policies frequently prioritize national goals above local needs. Environmental litigates and grassroots NGOs are becoming more involved, which is a slow but steady trend. Costa Rica exemplifies one of the most comprehensive strategies for achieving environmental fairness. One program that promotes equity is Payments for Environmental Benefits (PES), which pays landowners, including indigenous people, for environmental benefits. Social legitimacy and cohesion are enhanced by the incorporation of public input and environmental education into policy creation.

8 Institutions and Governance: Trends

The outcomes of environmental policy are substantially influenced by the structures of government. Although the EU's polycentric governance model is complicated and has coordination issues, it allows for freedom of thought and innovation. Governments' divergent understandings of EU regulations impact uniformity. Thanks to its centralized strategy and robust budgeting tools, China is able to respond quickly and deploy extensively. On the other hand, adaptive learning and context-specific innovation might be hindered by limited local autonomy. Integrative governance in Costa Rica exemplifies the efficacy of horizontal collaboration; it encompasses ministries, local governments, and civil society. Legal frameworks make government more approachable and participatory by establishing environmental rights and long-term planning. Analyzing the three models highlights the diversity of thought and the importance of tailoring strategies to different political and social contexts. China exemplifies the potential of state-led transformations, the European Union (EU) highlights the necessity of global collaboration and innovative ecosystems, and Costa Rica highlights the triumph of inclusive, trust-based development. The role of social narratives and education in promoting behavior change is one of the overarching themes, as is the necessity of aligning long-term goals with short-term policy instruments. In summary, the data demonstrate that successful green transitions are environment-dependent and multi-faceted. They stress the need of well-designed policies, cultural legitimacy, technological capability, and resilient institutions. Countries may gain knowledge from each other by working together to create inclusive, future-oriented sustainability policies and by tailoring best practices to specific regional contexts.

9 Discussion

This study's findings highlight the contextual sensitivity and complexity of the global green transition. Various paths shaped by institutional frameworks, governing logics, cultural narratives, and technological capacities are illustrated in the case studies of the European Union, China, and Costa Rica. Environmental justice, transition theory, and sustainable development are broader theoretical frameworks that these findings are situated within. It is clear from the comparative analysis that there is more than one way to achieve sustainability. The benefits and downsides of every paradigm are directly related to its underlying sociopolitical structure. The European Union is a model of integrationism and regulation, propelled by legal goals, market instruments, and inventive ecosystems. Nevertheless, it emphasizes the challenges of achieving policy consistency in multi-tiered governmental systems, particularly in systems characterized by economic inequality (Scheja and Kim 2024). Chinese Ecological Civilization exemplifies the potential of state-centric systems for coordinating massive environmental change. The advantages of centralized control are demonstrated by its rapid adoption of eco-friendly technologies and the execution of ecological objectives. Nevertheless, the democratic legitimacy and sustainability of such policies are called into doubt by the lack of participatory procedures and the underemphasis on distributive justice (Tang 2025). An example of how national identity and governance may mirror environmental care is Costa Rica's bottom-up, rights-based approach. Its emphasis on education and conservation, along with its participatory processes and constitutional protections, provides a remarkable model for inclusive sustainability. Scale, financial dependence, and structural vulnerabilities are also highlighted in the tale as limitations. An significant cross-cutting conclusion is the role of governance design. Institutional frameworks that promote openness, stakeholder engagement, and policy integration typically lead to more robust and adaptable changes. Costa Rica's intersectoral collaboration, China's top-down mobilization, and the European Union's polycentric administration are all instructive examples in their own right. However, public trust, performance monitoring, and established legal frameworks all tend to work together to improve outcomes (Wamane 2025). When it comes to technology, the green shift isn't only about deploying gear; it also requires institutional support, financial mechanisms, and social acceptance. Costa Rica's example highlights the equally crucial imperative of connecting technology with ecological ideals and social requirements, even while China and the EU lead in technical scale and innovation. This lines up with what is said in the literature on socio-technical transitions, which highlights the way in which technology, culture, and policy are all evolving together. The findings also have implications for the literature on environmental justice. The Just Transition Mechanism in the European Union and the PES program in Costa Rica are two examples of efforts to include equity into sustainability planning (Wells 2025). On the other hand, China's approach highlights the tension between rapid modernization and social inclusion. All of these variations stress the need for a justice-based view of sustainability that accounts for procedural, distributive, and recognitional factors. Education, culture, and narrative framing emerged as underappreciated yet crucial components of sustainable development. The emphasis on environmental education in Costa Rica and the shifting environmental discourse in China are two examples of how one's worldview may motivate or constrain actions. According to the expanding literature on sustainability cultures, this is in line with the idea that environmental action's symbolic and emotional components should be closely interacted with. The interplay between regional flexibility and international norms is a common thread. Localization determines the influence of international accords like the SDGs and the Paris Agreement, which affect national goals. The internal heterogeneity of the EU, the developmental state of China, and the normative alignment of Costa Rica with world aspirations all demonstrate different ways in which global–local dynamics impact implementation (Yu et al. 2024). Supporting post-development and decolonial critiques, this study highlights alternative development trajectories that challenge growth-centric conceptions. Costa Rica's environmental regulations and emphasis on well-being are indicative of buen vivir and degrowth. At the same time, in increasingly technologically advanced contexts, the likelihood of greenwashing and elite capture casts doubt on the veracity of sustainability promises. Finally, it's brought out that in a culture where uncertainty is on the rise, resilience thinking is essential.

Climate change, the COVID-19 pandemic, and geopolitical unpredictability have all brought attention to the need for governance that is flexible, inclusive, and proactive. Because of this, systems thinking and integrated planning are invaluable tools for navigating complex, non-linear changes. The global green movement must be viewed as an inclusive and ever-evolving phenomenon (Nag et al. 2025). Players' abilities to learn and adapt, as well as their potential for multidimensional sectoral collaboration, will determine how far they can develop; there is no silver bullet. Environmentally conscious, socially just, and institutionally robust routes must be co-created by knowledge institutions, civic society, and policymakers (Mohamed and Akande 2025). The global green transition has the potential to ignite a movement toward a more caring and sustainable world if it places an emphasis on diversity, equity, and systemic interconnectedness.

10 Policy Implications

For development institutions, civil society, international organizations, and national governments attempting to operationalize sustainable development through the global green transition, the study's findings have several significant policy implications. The case studies show that moving towards environmental sustainability is not an easy task; it depends on the specifics of the situation and calls for policies that are inclusive, proactive, and methodical. That sustainability must be institutionalized throughout all domains of influence ranks high among the most important insights. Comprehensive legislative frameworks and autonomous monitoring are necessary to support legally binding biodiversity and climate change targets set by national governments. Subnational and local governments also need the flexibility and resources to tailor these strategies to their own circumstances. Since decentralization has fostered innovation and responsiveness, the local implementation of conservation measures in Costa Rica exemplifies its efficacy. Because of this, a multi-tiered government that uses both horizontal and vertical coordination is essential. Ensuring a fair and inclusive transformation turned out to be another essential policy necessity. Economic dislocations caused by decarbonization operations necessitate targeted measures to protect vulnerable communities. These must be democratic decision-making processes, initiatives to diversify the economy, and programs to reskill individuals in order to guarantee procedural fairness. The EU's Just Transition Mechanism establishes benchmarks, whereas Costa Rica's Payments for Environmental Services (PES) program exemplifies how equity may be included into environmental incentive schemes. Local socioeconomic realities should inform how policymakers organize these models. Financial and economic systems that are in sync with sustainability objectives are of equal importance. To encourage public and private investment, governments should implement sustainability disclosure regulations, include climate risk into macroeconomic planning, and make use of green bonds other sustainability-linked securities. International financial institutions are vital in helping developing nations by providing technical assistance and concessional loans to fill capacity and access shortages. These monetary shifts are crucial for bolstering long-term resilience and driving structural transformation. Technology and innovation play a significant role in enabling the green transition. However, due to disparities in access and aptitude, fair standards for technology transfer are necessary. Clean technology research and development partnerships between the public and commercial sectors, open innovation platforms, and local manufacturing incentives in impoverished regions are crucial. China's state-led innovation model demonstrates the potential for rapid technological adoption; nevertheless, measures must be taken to address unintended consequences, such as over-reliance on resources and harm to the environment. Another significant influence is responsibility, monitoring, and transparency. Reliable data systems and transparent government policies are essential for monitoring progress and building public trust. The establishment of participatory auditing systems, standardized indicators, and real-time environmental monitoring requires collaboration between international organizations, civil society, and governments. These efforts can help prevent greenwashing and boost the trustworthiness of environmental initiatives. Culture and schooling also play significant roles in molding the public's view of sustainability. Beyond technical solutions, policymaking must take into account societal norms and ethical standards. Interaction with indigenous and faith-based communities, as well as traditional storytelling and environmental education, may help foster a shared sense of environmental responsibility. What Costa Rica has accomplished in incorporating sustainability into its national character via storytelling and education is particularly noteworthy in this regard. A second area of strategic emphasis is the effective regional implementation of global frameworks like the Paris Agreement and the SDGs. While these models do provide global objectives, how they are interpreted and implemented at the national and subnational levels determines how effective they are. Transnational learning platforms, localization toolkits, and the integration of nationally determined contributions (NDCs) with local plans are crucial for making sure that global ambitions become real actions. In this era of complex, interconnected dangers, policymakers should be led by resilience and systems thinking. While revealing the shortcomings of existing institutions, the COVID-19 pandemic emphasized the need for unified, future-oriented administration. A combination of climate-resilient infrastructure, cross-sectoral scenario planning, and institutional coordination mechanisms can lessen systemic fragility and increase adaptability. Now more than ever, stronger international cooperation is required to address the world's most pressing environmental issues. Resources insecurity, biodiversity loss, and climate change are examples of transboundary concerns that need global solutions. Unified environmental standards, data sharing, joint research, and a fair climate finance architecture must be advanced by policy. To ensure fairness and inclusiveness in global governance, international discussions should revolve on the concept of shared but separate responsibilities. Taken together, this research's policy implications highlight the need for visionary leadership and structural reform to achieve sustainable development through the green transition. Negotiating complexity while welcoming collaboration, equality, and innovation is a challenge for policymakers. As the green transition signifies not just a shift in technology and finance but also a profound shift in mindsets and institutions, it presents a unique opportunity to reconsider development in the sake of building a more equitable and resilient society. The policy implications of this study suggest that proactive government action and systemic reform are necessary conditions for long-term economic expansion. In order to craft ambitious yet adaptable programs, policymakers need to engage with complexity, equity, and creativity. There is a once-in-a-generation opportunity to reimagine growth through the global green transition so that human prosperity does not compromise environmental integrity.

11 Conclusion

A new understanding of progress via environmental sustainability is essential to Earth's survival. One solution to the problem of how to accommodate both human development and environmental limits is the global green transition. For global growth to be based on sustainability and for structural transformation to occur, there must be collaboration between policymakers, businesses, and civil society. An urgent environmental need and a once-in-a-generation opportunity for global green transformation have come together. A multifaceted and context-specific approach incorporating government, technology, finance, and culture is required, according to this study's findings, to reinterpret development through the lens of environmental sustainability. Case studies of the EU, China, and Costa Rica show that sustainable transitions are not uniform but rather result from unique institutional configurations, socioeconomic objectives, and cultural narratives in each region. Changes in economic policy or advances in technology will not be sufficient to achieve sustainability, as this study makes clear. The alternative is a demand for structural change, for ecological principles to permeate all levels of society's value system, economic foundation, and governmental structure. When supported by a consistent strategy and inclusive engagement, different forms of government may yield long-term benefits; examples of this include the legal and regulatory leadership of the European Union, the rapid infrastructural development of China, and the grassroots environmental stewardship of Costa Rica. As it turned out, equality and justice were the bedrock of this new development paradigm. Changes that don't take social inclusion into account can worsen existing inequalities or give rise to new forms of exclusion. Fair transition, pursuing environmental goals with social protection and democratic governance, must thus be at the core of national and global policy objectives. Inclusive planning frameworks, ecosystem payments, and transition funds are crucial tools for incorporating justice into the transition process. Additionally crucial is the manner in which global frameworks like the Paris Agreement and the Sustainable Development Goals are interwoven with local and national realities. Although these models provide a framework that any community may follow, their success will hinge on how effectively they are tailored to meet the needs of individual areas. Thus, localization is best understood as a process of co-creation, not only as an implementation, that values local knowledge, adapts global tools, and encourages shared ownership.

Ultimately, the green transition is more than just a means to reduce greenhouse gas emissions; it's a paradigm shift toward a more equitable and sustainable future for all generations. It makes us reevaluate progress not in terms of material gain but as the welfare of communities within natural limits. This study highlights the moral and practical necessity for strong leadership, widespread cooperation, and unwavering dedication to sustainability in order to realize such a goal.

References

Al Amin M, Chakraborty A, Baldacci R (2025) Industry 5.0 and green supply chain management synergy for sustainable development in Bangladeshi RMG industries. Clean Logist Supply Chain 14:100208

Allam Z, Cheshmehzangi A (2024) Sustainable futures and green new deals. Palgrave Macmillan, pp 1–137

Almusaed A, Almssad A, Alasadi A, Al-Asadi F (2025) Sustainable reimagining of urban habitats: Navigating Iraq's housing crisis through socio-technological and environmental approaches. Ain Shams Engineering Journal 16(2):103280

Balci O, Almohammad D, Kiliç C, Singh R (2025) Harnessing emerging technologies for green sustainability: Innovations, challenges, and future pathways. Clim Neutrality Smart Eco-Innov Environ Sustain, 209–225

Benner M (2025) Imaginaries: The ideational foundation of sustainability transitions. Tijdschr Voor Econ En Soc Geogr

Chand R (2025) Reimagining agriculture for growth, poverty, nutrition, and sustainability. Agric Econ:e70015

Cole LB, Hamilton EM (2024) Transition interior design: Reimagining practice and education for a post-carbon society. J Sustain Res 6(2)

Dukelow F, Murphy MP (2022) Building the future from the present: Imagining post-growth, post-productivist ecosocial policy. J Soc Policy 51(3):504–518

Eversole R, Gambetta N (2021) Reimagining development institutions for the SDG era: Pathways to impact. Partnerships for the Goals. Springer International Publishing, Cham, pp 1001–1012

Ghosh E, Pearson LJ (2025) Rethinking economic foundations for sustainable development: A comprehensive assessment of six economic paradigms against the SDGs. Sustainability 17(10):4567

Grant K (2024) How green is your practice? A process for auditing and reimagining career guidance practice for a greener world. J Natl Inst Career Educ Couns 53(1):89–104

Gurieff N, Green D, Koskinen I, Lipson M, Baldry M, Maddocks A, Doroodchi E (2020) Healthy power: Reimagining hospitals as sustainable energy hubs. Sustainability 12(20):8554

Hashemi S (2025) Advancing healthy and sustainable environmental stewardship: Reimagining strategies for air, water, food, and waste management. Sustainability 17(4):1345

Hasselbalch J, Lund-Larsen M Reimagining growth futures: Overcoming the false binary between green growth and degrowth. Available at SSRN 5096015

Joseph OO, Kevin-Alerechi E, Olumide OJ, Sunday D, Fagbohun O, Olanrewaju S (2025) Reimagining green financing with AI: A technological approach to sustainability. J Artif Intell Mach Learn & Data Sci 3(1):2346–2352

Komninos N, Panori A (2025) Smart carbon–neutral development: Embracing complexity with multi-level governance and convolution. World Dev 192:107026

Machin A (2025) The European Green Deal and the democratic imagination. J Eur Integr 47(2):155–171

Marquardt J, Pfeiffer F, Blum M, Daw TM, Dugasseh FA, Heitzig J, Tønder L et al (2025) Reconciling democracy and sustainability: three political challenges and the role of democratic innovations. Sustain: Sci, Pract Policy 21(1):2504239
McKenzie M, Bieler A, McNeil R (2015) Education policy mobility: Reimagining sustainability in neoliberal times. Environ Educ Res 21(3):319–337
Miglani R, Rawat G, Joshi S, Singh V, Parveen N, Siddiqui U, Bisht SS et al (2025) The green revolution reimagined: Agricultural biodiversity and sustainable food security. In: Biotechnological innovations for sustainable biodiversity and development. CRC Press, pp 331–342
Mohamed A, Akande AE (2025) Waqf-led buildings and green infrastructure role in environmental sustainability: understanding critical gaps in current research landscape. Manag & Sustain: Arab Rev
Mullally G, O'Neill M, de Bhailís D, Tuohy B, Breen M, Duggan A, Ní Loinsigh E (2023) Walking, talking, [Re-] imagining socio-ecological sustainability: Research on the move/moving research. Ir J Sociol 31(1), 37–62
Nag A, Ashish A, Rathore AS (2025) Urban waterfronts and tourism: reimagining cityscapes through ecological design. In: Integrating architecture and design into sustainable tourism development. IGI Global Scientific Publishing, pp 433–470
O'Donovan N (2024) Turning less into more: Measuring real GDP growth in the green transition. Ecol Econ 224:108293
Owojori OM, Erasmus LJ (2025) Public–private partnerships as catalysts for green infrastructure: a three-pronged analysis of economic, environmental, and institutional factors. Front Sustain Cities 7:1591278
Plangger K, Montecchi M, de Ruyter K, Keeling DI, Scott ML, Dahl DW (2025) "We could be heroes": Reflections on reimagining marketing strategies for a better world. J Acad Mark Sci:1–9
Prabhakar A (2025) A Sustainable and inclusive economic development: A global imperative. J Recycl Econ & Sustain Policy 4(1):1–16
Pujun C (2025) Green skills in logistics vocational education: A comparative study of curriculum integration in China and Germany. Int J Train Development
Ramsarup P, Mohamed N (2022) Reimagining skills planning for a low carbon and sustainable future. African Books Collective, Just transit low carbon future South Africa. Oxford, pp 357–381
Reid A, Dillon J, Ardoin N, Ferreira JA (2021) Scientists' warnings and the need to reimagine, recreate, and restore environmental education. Environ Educ Res 27(6):783–795
Samuel G, Richie C (2023) Reimagining research ethics to include environmental sustainability: a principled approach, including a case study of data-driven health research. J Med Ethics 49(6):428–433
Scheja E, Kim KB (2024) Looking back, redefining the future: an overview of the state of play of green structural transformation and the factors critical to its success. Rethink Econ Transform Sustain Incl Dev:1–24
Tang J (2025) Challenges and breakthroughs in the green economic transformation of double landlocked countries: A study on Uzbekistan's sustainable development path. J Interdiscip Insights 5(1):51–59
Wamane GV (2025) A "new deal" for a sustainable future: enhancing circular economy by employing ESG principles and biomimicry for efficiency. Manag Environ Qual: Int J 36(4):930–947
Wells PE (2025) New world order: reimagining mobility beyond 2035. In: How the global automotive industry stole our green mobility future. Edward Elgar Publishing, pp 135–160
Yu H, Zahidi I, Chow MF, Liang D, Madsen DØ (2024) Reimagining resources policy: Synergizing mining waste utilization for sustainable construction practices. J Clean Prod 464:142795

Green Transition and Human Resource Management–Sustainable Development Through Effective Management of Green Collar Workers

Dimitrina Stefanova, Catalin Popescu, and Valentin Vasilev

Abstract The overall economic and social development of the world in the coming years is deeply linked to the development of policies and practices in the field of green economy and green change. This is a fact as a result of the expansive increase in practices such as recycling, waste management, water supply management; social responsibility and many other new aspects in management. Of course, the results in these processes depend on the effective work with human resources. The role of Green Collar Workers and the development of key competencies and skills related to green economy, green logistics, green finance, green production, green energy, waste management, sustainable social health in these employees is becoming increasingly important. The chapter attempts to present key ideas and practices related to improving the effectiveness of Green Collar Workers management and presents practices in this area that have proven their effectiveness.

Keywords Green transition · Climate change · Sustainable development · Green collar workers · Effective management · Green practices

---

The original version of the chapter has been revised. A correction to this chapter can be found at https://doi.org/10.1007/978-3-032-01715-4_23

D. Stefanova (✉)
South-West University "Neofit Rilski", Blagoevgrad, Bulgaria
e-mail: dimi@swu.bg

C. Popescu
Department of Business Administration, Petroleum-Gas University of Ploiesti, Ploiesti, Romania
e-mail: cpopescu@upg-ploiesti.ro

V. Vasilev
Higher School of Security and Economics, Plovdiv, Bulgaria
e-mail: valentin.vasilev@vusi.bg

F. A. Malik et al. (eds.), *Aligning ESG Goals with the UNs Sustainable Development Goals Planning*, World Sustainability Series,
https://doi.org/10.1007/978-3-032-01715-4_14

1 Introduction

Every aspect of life has gone through continuous transformation during the past ten years. Numerous crises of various types have prompted organizational leadership to look for adaptable solutions that would help them find a way to achieve sustainability and growth. This is mostly true for the security field and the associated organizations.

Change is frequently viewed as a bad thing, as something that puts security at risk. However, change is meant to inspire growth, improvement, and the overcoming of shortcomings. We must thus learn to successfully manage it and see it as a necessity.

Since this is closely related to contemporary ideas of the information society and the growth of the online space, including the "fine" adjustment to changes in society at all levels, it is clear that all of this focuses attention and efforts on the development of new theoretical frameworks that will assist the management of organizations in managing these processes.

Effectively implementing change management principles and models is crucial for any firm in this situation. Furthermore, the emergence of new paradigms—some of which are directly tied to an emphasis on green management—is linked to crises and the crisis context of every management decision.

2 Literature Review

An initial step in exploring the proposed field consists of consulting relevant researches and studies, the value of which is highlighted by the significant number of citations. Global shifts towards sustainability and ecological responsibility have led to a reconfiguration of organizational strategies, bringing the green transition to the forefront of political and business agendas. More and more specialized articles emphasize the essential role of human resource management (HRM) in supporting this transition, by aligning employee skills with sustainable development goals (Jabbour and Santos 2008; Renwick et al. 2013).

Green HRM refers to human resource practices that promote environmentally sustainable behaviours among employees and integrate environmental considerations and approaches into the human resources function (Renwick et al. 2018). Researchers have explored how green recruitment, training, performance management, and reward systems contribute to a culture of sustainability (Jackson et al. 2011). Organizations that practice green HRM are more likely to see improvements in environmental performance and employee engagement in pro-environmental behaviours (Daily et al. 2007).

The content of the chapter refers to a new concept of workers in relation to the new developments of society. Sustainable development and the green transition have forced human society to create the so-called green collar workers. The concept of green collar workers emerged in response to the need for a workforce equipped to support the transition to a low-carbon economy. These workers operate in roles that directly contribute to environmental sustainability, such as renewable energy technicians, sustainability analysts and green construction professionals (Worldwatch Institute Report 2008). Research highlights the need for HRM to adapt to the unique training, development and career planning requirements of this workforce (Strietska-Ilina et al. 2011).

Several studies have examined the skill sets required for green jobs, often focusing on the integration of technical, environmental and soft skills. The importance of lifelong learning, adaptability and interdisciplinary knowledge is frequently highlighted (International Labour Organization 2019). The role of HRM in promoting these skills through specific learning and development strategies is well documented, although more empirical evidence is needed, especially in non-technical sectors. Technological, legal, social and environmental developments oblige competent entities to find immediate, concrete and appropriate solutions to the current and future needs of society.

On the other hand, HRM alignment with sustainability goals has been correlated with improved organizational outcomes, including innovation, employer branding, and long-term profitability (Zibarras and Coan 2015). The literature also shows that employees are more likely to engage in achieving environmental goals when sustainability is integrated into organizational culture and human resource policies (Boiral et al. 2015). Based on a broad and fairly comprehensive search, it can be stated that few studies provide longitudinal data on the impact of green HRM initiatives. Moreover, the specific challenges of managing green collar workers—such as balancing technical competence with environmental ethics—are still underexplored (Rosca and Stancu 2024).

Overall, the literature indicates a solid theoretical basis for integrating HRM and sustainability. Empirical studies, especially those targeting employees in the green economy, are still in their early stages of development. This chapter aims to make a relevant contribution to this evolving conceptual framework by investigating the ways in which human resource management practices can be implemented to support the green transition through strategic green workforce management.

3 Objectives

This chapter aims to generate enough info and useful findings for the selected topic. In fact, in the chapter will be found answers to several essential questions:

1. What is the role of Human Resource Management (HRM) in facilitating the green transition within organizations?
2. What are the key competencies and training needs of green collar workers in various fields?
3. How can HR practices improve the performance of organizations during the green transition?

4. Can a relationship be identified between sustainable HRM strategies and employee engagement in green initiatives?
5. What could be the challenges and/or opportunities in recruiting, retaining and developing green collar talent?

The analyses, syntheses and descriptions proposed in the Discussion section are intended to find pertinent answers to these questions. In addition, in the Conclusion section will be highlighted main findings related to the chapter objectives and core analysis.

4 Research Methodology

A series of empirical and theoretical analyses, a deductive technique, and a descriptive-correlative approach are all combined in the structure of scientific research. The research approach chosen by the authors combines several scientific methods. The study of the legislative base, strategic documents and analyses is of key importance for summarizing the main challenges related to the topic being developed. Another method is the identification of good practices and ideas (benchmarking) showing key results in Bulgaria and Romania and which can serve as a point from which readers can search for and identify practices relevant to their organizations. Thirdly, the comparative approach is chosen, which is applied in the identification of effective models for human resource management, with a focus on "green collars" and management through values. The main focus of the methodology is the identification of good practices and ideas that will serve researchers and practitioners to find and apply similar innovative approaches in human resources management. Of course, in terms of research, a detailed literature study on the topic will be made, and key characteristics are derived from current scientific publications. In this way, a scientific focus is achieved in summarizing some of the recommendations and conclusions in the scientific development.

5 Discussion

(a) Changes and their management in the context of green challenges for organizations

Ecology is associated with complex challenges in an interconnected world according to the conclusions made in the main report of the European Environment Agency—"The Environment in Europe—State and Outlook 2020 (SOER 2020)", which addresses a number of factors related to the environment—biodiversity loss, the impact of climate change and the overconsumption of natural resources. The report also aims to provide solutions to the problems that have arisen, including guidelines to be followed. The report presents a snapshot of the state of the EU countries in terms of environmental protection; this report is the basis for the creation of the European Green Deal. The fight against to the climate change requires the implementation of multifaceted policies.

Globally, biodiversity has been the subject of COP-type climate summits (Conferences of the Parties to the UN Convention on Climate Change) and international treaties for many years that establish conservation and protection norms for the elements of nature and for life on Earth. In 2022, almost 200 countries signed the Montreal Agreement, committing to protect at least 30% of the planet's land and waters by 2030, in a treaty that should do for biodiversity what the Paris Agreement did for emissions.

In the European Union, only 15% of habitats are in good condition, and fish, bee and butterfly populations are in sharp decline. In response, the EU is proposing the Nature Restoration Regulation, a key part of the European Green Deal, with ambitious targets for 2030: expanding protected areas to 30% of land and sea surface, restoring 20% of degraded ecosystems, planting 3 billion trees and reducing pesticide use by 50%. Currently, protected areas cover 26.4% of the EU's land and 12.1% of its seas. Climate change is not just about environmental sustainability; it also causes negative effects on social welfare and economic continuity. For example, Climate change has been a huge retarder for economic development.

According to Deloitte's estimates, the estimated total value of economic losses in 2070 is thought to be 178 trillion US dollars. Therefore, employment, living standards and welfare levels will drop significantly. Every employment area and sector bear its share of the effects of climate change. However, some sectors are directly exposed to climate change impacts. According to Deloitte's Job Vulnerability Index, "Agriculture; Energy; Heavy industry and manufacturing; Transport; Construction" sectors are the sectors that should be at the forefront in the fight against climate change (Deloitte 2022).

On the other hand, the transition to a zero-emissions economic model can protect the labour market from the risks associated with climate change and create up to 300 million new jobs globally by 2050.

In addition to being ecologically conscious, green leaders strive for stability and the establishment of ambitious yet attainable objectives. At the manager-subordinate level, green leaders insist on empathic communication and creating a peaceful communication process (Popescu et al. 2024a, b, c).

By offering a strategic and normative framework that regulates fundamental rights and responsibilities while allowing for sufficient flexibility in implementing innovations, it is possible to shift leadership toward sustainable development and the "greening" of management.

A sustainable future also depends on addressing today's environmental issues. Six problems and directions can be highlighted in this way. The expansion of the population is straining ecosystems and natural resources. The world's biodiversity is at risk due to overfishing and mining. A sustainable future depends on the circular economy and renewable energy sources. Addressing climate change requires social awareness and group efforts. The so-called "green professions" or "green collars" are in short supply and continue to be unpopular; At every educational level, there is still a lack of adequate preparation and attention to environmental issues.

In an effort to address global issues, the European Union implemented a variety of environmental protection regulations.

The European Green Deal serves as an illustration of this, demonstrating its importance in the economic recovery from the COVID-19 pandemic. Reducing net greenhouse gas emissions by at least 55% is its primary objective. Energy, agriculture, finance, transportation, the environment and seas, industry, research, and innovation are the primary sectors that are affected.

Because of the implementation of the European Green Deal, the use of energy from renewable energy sources has increased. As a result, greenhouse gas emissions in the EU have decreased and the economy has grown (Popescu et al. 2024a, b, c).

The European Union's member states should make investments in sustainable development while keeping to the European Green Deal's priorities if they hope to meet the goal of making Europe the first continent to be climate neutral by 2050.

For this reason, the National Environmental Strategy 2021–2030 is one of the key environmental protection strategic documents that Bulgaria has ratified. Limiting climate change and preparing for it are two of the document's top concerns. Limiting greenhouse gas emissions from all aspects of socioeconomic life will enable its implementation.

Another strategy that the Green Deal is using to make the European Union's economy more sustainable is "From Farm to Fork". Its main goal is to protect biodiversity, increase organic farming, which will lead to achieving a circular economy.

In order to accelerate the processes introducing energy efficiency and achieving climate neutrality, the European Parliament adopts a revised Energy Efficiency Directive (EU) 2023/1791.

Member States sought to hold to the fundamental idea of "energy efficiency first" with the measures outlined therein, which primarily affects buildings and transportation. The "energy efficiency first" tenet must be upheld at the municipal, state, and federal levels in order to be successful. In this sense, all towns with a population of more than 45,000 are required under the previously mentioned law to create their own heating and cooling programs in order to lower carbon emissions. Plans for heating and cooling municipalities in Bulgaria should be in accordance with the Republic of Bulgaria's 2021–2030 Integrated Energy and Climate Plan. To achieve sustainable energy, the strategic document's primary goals are to promote low-carbon development by promoting the use of energy from renewable sources and the use of innovative technologies.

The New European Bauhaus program is being developed by the European Commission to create climate-neutral "smart cities" and sustainable places. The policy is being implemented in order to achieve the European Green Deal's goal of making Europe the first continent to achieve carbon neutrality. The initiative's primary tenets center on designing inclusive, beautiful, and sustainable environments. All project proposals must align with the principles of the New European Bauhaus in order to establish them.

The Earth Overshoot Day is a significant event that pertains to the estimation of the amount of resources that the Earth can replenish after depleting them in a single year. The primary causes of the quick depletion of renewable resources are deforestation, production, carbon-emitting activities, and, last but not least, population feeding.

Global Footprint Network provides information for each country regarding the occurrence of the ecological debt day. Anyone who wants to determine their ecological footprint and personal ecological debt day can do so on the organization's website. Once data has been entered, the ecological debt day of a specific city can be determined.

Topics covered at the 50th anniversary meeting in Davos included the circular economy, climate change, and the new difficulties brought on by the depletion of natural resources. The meeting's established goals and pledges include that 60% of household garbage in the EU must be recycled by 2030; Bulgaria and a number of other nine nations have been allowed a five-year extension to fulfil this target. By 2035, 65% of household garbage should be recycled, with the remaining 10% going to landfills.

Assistance to institutions and people is provided by various non-governmental and independent organizations at the global and European level. One such organization is WWF—an independent conservation organization that has offices in 100 countries. The organization works together with individuals, governments and businesses to achieve more sustainable and timely goals. The organization has three main pillars of work. Preserving biodiversity worldwide. Making certain that the sustainable use of natural resources that are renewable. Encouraging the elimination of excessive and wasteful consumption as well as pollution.

In Bulgaria and Romania, independent non-governmental environmental organizations have been founded with the goal of promoting an egalitarian and ecologically sustainable world free from human and natural exploitation. The following are the organizations' primary areas of focus. Food and Agriculture—GMO. Zero waste—getting rid of plastic products, creating a deposit system, Air; and Energy and climate.

An exhibition of ecological insulations was held along with one of the organization's most recent projects, "Renovation, but ecologically," which focuses on insulating residential buildings. For instance, a number of environmental protection-related projects are listed under the "Missions for Nature" section of the volunteer website "TimeHeroes". Through the platform, they have released a volunteer program that enables you to report illegal logging in the nation by submitting an application.

A number of organizations are involved in the volunteer platform by seeking volunteers to implement nature conservation actions, such a volunteer action is the "Digital Map of Sofia's Trees" by the "One Tree" organization. This is possible through an especially platform designed with the help of volunteers, the organization aims to register trees, which will help to preserve them in the future.

"Urban Gardening-Sofia"—"Garden for Friendship" is a physical space and solidarity garden of "Food, Not War—Sofia", where citizens and guests of the capital can grow their own food, develop, create and help people in need with the produce grown in the garden.

Over the past 25 years, Romania has undergone significant economic development toward a Western European model, resulting in growing pressure on natural ecosystems. The restitution of nationalized forests has further intensified this pressure, leading to widespread illegal logging across tens of thousands of hectares. Additional threats, including micro-hydropower plants, unregulated tourism infrastructure, poaching, and overhunting, have negatively affected biodiversity, particularly in the Carpathian Mountains.

The context of the Făgăraş Mountains, a Natura 2000 site, illustrates the systemic challenges faced by many protected areas in Romania, affected by chronic underfunding and obstacles generated by economic development.

To foster a nature-based economy around Carpathia National Park, the Conservation Carpathia Foundation, with support from EEA Grants, conducted a study on ecosystem value and business opportunities in the Făgăraş Mountains. The project aimed to assess the socio-economic and ecological context, evaluated current commercial activities, and proposed pathways for building an economy that restores ecosystem services and safeguards biodiversity. Additionally, it was analyzed the existing political framework and offer policy recommendations to encourage conservation-based businesses both in the Făgăraş Mountains and across Romania's protected areas.

In Romania, protected areas cover 23.4% of the land area, slightly below the European Union average (26.1%), while 21.4% of the maritime area is protected, significantly above the European average (12.3%). Romania is home to over 533 species protected under European legislation and 87 habitat types, reflecting remarkable biodiversity. In total, Romania has over 1,550 protected areas, of which around 950 are regulated by national legislation, and around 600 are included in the European Natura 2000 network.

The Danube Delta and the Razim-Sinoe Complex stand out as the most valuable Natura 2000 sites in Romania, recording the highest number of protected species and habitats. These are followed by other important areas, such as the "Porţile de Fier" Natural Park, the Domogled-Valea Cernei Natural Park and the Apuseni Natural Park.

However, although Romania has a rich biological diversity, access to updated and centralized data remains limited. Most of the information comes from the European Biodiversity Information System (BISE), which aggregates data from multiple sources, in the absence of an integrated national platform. Currently, the data available on the websites of the Ministry of Environment or the National Agency for Protected Natural Areas (ANANP) are often outdated, with many documents dating back to 2013–2014.

The main information on biodiversity comes from the National Strategy for Biodiversity Conservation 2014–2020, which raises questions about their timeliness and relevance. Other statistics, such as those relating to the over 6.6 million hectares of forests covering almost 29% of the national territory, are available through the National Institute of Statistics, but are not integrated into a single system.

Initiatives such as the RO-ADAPT platform, intended to centralize this data, have not been finalized. This lack of coordination has been officially recognized, including in the National Climate Change Adaptation Strategy 2022–2030, which emphasizes the need to improve knowledge of biodiversity and develop interconnected databases.

On the other hand, as is known, Romania excels in the oil and gas industry, both through its tradition of over 150 years and through its important oil and gas resources. The oil and gas sector plays an important role in the global economy, both through its contributions to the income of states and through the significant number of jobs supported—both direct and indirect, in private companies or in the budgetary system.

The energy transition will mean, over time, a decrease in the consumption of conventional fuels, but not a decrease in the importance that this field has in the economy. This is because, in Romania, and at an international level, it is necessary to adopt new technologies and investments in the production and infrastructure for the use of renewable energy (Popescu and Gabor 2021).

Moreover, these transformations are possible using the financial capacity, business experience and, last but not least, the qualifications and professionalism of people working in oil and gas companies. These companies support the energy transition, both in the case of Romania and globally. Support is based on size and financial strength, long-term relationships with customers and stakeholders in the energy sector, the capacity to develop new technologies (hydrogen, then carbon dioxide capture, transport, storage and use, etc.). An important asset of companies in the oil and gas sector is the qualified human resource, which will be involved in new projects aimed at decarbonization and energy production from renewable sources (Popescu et al. 2024a, b, c).

These solutions were summarized in a 2023 McKinsey analysis, by identifying key issues and adequate addressing means (Kienzler et al. 2023):

(1) *Decarbonization solutions*. Oil and gas companies have initiated projects—at various stages of development—aimed at decarbonization, using the accumulated experience and technical solutions already implemented. These can be used both to reduce the carbon footprint of the sector and, as well, of industry or transport. Actions and measures related to the evaluation and efficiency of integrated national energy and climate change plans, the design of strategies for the integration of the energy sector into other sectors, the reworking of regulations regarding the design and operation of trans-European energy networks, the generation of initiatives for the energy renovation of buildings etc. are targeted.

(2) *Carbon capture, transport, storage and use*. This technology involves capturing carbon dioxide resulting from various industrial processes, transporting and storing it, safely and permanently, in onshore or offshore geological formations. The experience accumulated by companies, as well as the qualifications and capabilities of employees can be capitalized by implementing this technology. In the case of Romania, launching this type of project would also represent a form of technological leadership in the region.

(3) *Geothermal energy*: Geothermal energy is a form of renewable energy, derived from the heat emanating from the interior of the earth through rocks and underground fluids. It is used to produce electricity and heat. Its use is still limited in Romania, the installed power being used in thermal applications. An example

of good practice is the Municipality of Oradea, where open loop technology is used with the help of thermal water, where the contribution of geothermal energy to the district heating network increased from 5% to 15–20%, thanks to European-funded projects. The beneficiaries are 14,000 Oradea residents, in over 6,000 apartments.

(4) *Hydrogen production and transportation.* The capacities and knowledge accumulated by companies, both in gas transportation and storage, are also relevant for the production and transport of hydrogen. This is due to similar requirements, determined by the chemical specifics: both natural gas and hydrogen are flammable gases, which are maintained under pressure and require careful management.

(5) *Development of offshore renewable energy projects.* Oil and gas companies have experience in offshore natural gas production projects and have the capacity to develop integrated projects, including renewable energy generation, such as wind energy, which in turn can be used to produce green hydrogen.

(6) *EV charging.* This component targets downstream companies. Given the EU requirements to achieve a certain density of charging stations for electric and hydrogen vehicles, the current power grids will be diversified by installing dedicated charging facilities for electricity or alternative fuels.

Encouraging public awareness and engagement, "examine householders' attitudes, motivations and behaviours towards recycling, potential food waste segregation and prevention activities" To achieve this, various events can be organized with citizens and public opinion surveys can be conducted (Bostanci et al. 2024).

The activities of all of us are significant, and the little things we do in our daily lives have an effect on our shared home, planet Earth, even though the efforts of big organizations, enterprises, and institutions are channelled in the field of ecology.

In the context of a management environment that is continuously evolving, these difficulties are a part of the new face of change management. These issues pertain to the efficient administration of human resources in every organization, with a renewed focus on social responsibility, causal management, and the organization's mission and vision.

(b) Effective Human Resource Management—from Innovation to Sustainable Results in an Environmental Context

The new role of human resources, with the growing discussion of the so-called "green collars" and their growing significance for successful policies and outcomes in businesses, is a significant aspect of the changes in the management environment.

Accordingly, "green human resource management" refers to the categorical relationship between the management of people and the difficulties of the changing environment. Its significance has expanded in an impressive way. Green Human Resource Management (GHRM) is an emerging field that emphasizes the integration of environmental management practices within Human Resource Management (HRM) functions to enhance organizational sustainability. This review synthesizes existing literature to explore the breadth of GHRM practices across various HRM functions (Rosy et al. 2025).

It can be said that the term green collar workers is still developing in the literature. The concept of green collar workers is related to the green economy and 2030 Sustainable Development Goals. There is a clear difference between green collar workers and others (white, blue, etc.). In this point, the definition of green jobs and green collar workers is directly based on sustainability. The literature can give some definitions or explanations for green jobs and green collar workers (Yıldırım and Vasilev 2024). For example, the Bureau of Labour Statistics (BLS) provides two definitions for green jobs as below:

- "Jobs in businesses that produce goods or services that focus on environmental benefit and protect natural resources";
- "Jobs related to less use of natural resources, more environmentally friendly production processes and other business processes", can be considered as green work.

The significance of "green collars" in the context of general human resource management is highlighted in this chapter. In the upcoming years, the specificity and emphasis on policies and procedures pertaining to them will be especially crucial.

A holistic approach to the "green idea" in the organization is a concept that recognizes the interconnectedness of all aspects of the life and activity of the organization. This approach acknowledges that all of these factors are interrelated and have an impact on employees, customers, counterparties, and the organization's well-being (Vasilev et al. 2024). The challenges associated with green human resource management may be the challenge of the next decade in the field of management. In this context, sustainable development plays an increasingly important role in the development of organisations and specifically in the management of human. Increasingly, young job seekers are particularly concerned about the organisation is positioning in terms of green and other ethical standards. Therefore, HR managers should pay attention to green management (Ilieva and Vasilev 2024).

In order to address to all these issues, HR departments must better define what skills these green collar workers must possess. It is worth mentioning at least some of them:

- to possess the necessary skills to work in a sector that directly or indirectly contributes to environmental protection;
- to be aware of the impact of their activity on the environment and to act responsibly in this regard;
- to collaborate effectively in interdisciplinary teams and in sustainability-oriented contexts;
- to be adaptable and flexible as green technologies evolve rapidly and require constant updating of knowledge.

Thus, the job descriptions for this job must include several requirements that the holder of that position must have:

- ecological awareness, in the sense of a permanent concern for environmental protection;

- social responsibility, i.e. the permanent orientation of the employee towards the common good and sustainability;
- collaboration skills, i.e. the ability to work in various multidisciplinary teams;
- orientation towards continuous learning, in the idea that green technologies require permanent updating of knowledge;
- critical thinking and problem-solving skills, as creativity and innovation are needed in supporting sustainable processes.
- **Motivation through values and green-collar management**

In business life, in any organizational environment, processes occur in accordance with the values that dominate the community. Through them, not only the character and specificity of organizational culture are revealed, but also the priorities that organizational management follows. These characteristics of workplaces are especially significant in the context of general motivation, employee engagement on the relationship between workplace spirituality and employee job performance (Gurugamage et al. 2022).

It is challenging to provide a categorical definition of value in general or in an organizational environment in particular because people's priorities must also vary as their lives change. Nevertheless, values can be described as follows, regardless of differing opinions:

- All business declarations, policies, procedures, and instruments that guarantee the effective application of moral and accountable management are included in values management;
- Beyond only adhering to legal requirements, values management seeks to establish a genuine culture of moral principles and raise employees' personal accountability.

It raises the questions of which values dominate company culture and how to identify the most crucial management value traits. These, in our opinion, are especially significant when it comes to the motivation of the so-called "green collars" and are a good place to start when creating organizational policies and initiatives that aim to achieve this. On the other hand, The relationship between HRM and the digital circular economy is critical for the business to achieve sustainability goals and improve the environmental and social performance of the business. This chapter should be useful for practitioners and policymakers to drive the transition towards a digital circular economy and integrate digital circular economy principles across the different sectors (Singh et al. 2024).

Management characterizes modern management, which started to emerge as a key instrument for the success, development, and operation of companies in the middle of the twentieth century, by values. In this sense, it is intriguing to consider how "green collars" and sustainable entrepreneurship are associated through policies that support the growth of important initiatives pertaining to the wider implementation of shared values-based practices that safeguard the environment. Leaders need to challenge these stereotypes by fostering a culture of equality and inclusion. The absence of adequate resources—whether financial, educational, or organizational—restricts women's growth (Serafimova et al. 2025).

For example, since achieving sustainability necessitates intricate strategies and innovative solutions, sustainable entrepreneurship encounters challenges and difficulties while fulfilling social, economic, and ecological goals simultaneously. To generate systemic impact and long-term sustainable value, it is vital to strengthen partnerships between sectors (private sector, public institutions, non-profit organizations and government authorities). Integrating digital technologies into these partnerships offers entrepreneurs the opportunity to amplify public engagement and direct support for achieving strategic sustainability goals. Increasing awareness through various communication channels will reinforce the ecosystem for sustainable entrepreneurship (Gigauri et al. 2025). Its characteristic features are:

- management by values, not by goals;
- efficiency flourishes through honesty and integrity;
- people strive for excellence, are predisposed to work and expect recognition;
- the relationship between the boss and subordinates is democratic and the latter demand love and respect for him;
- a new role for the organization;
- cooperation in formal organizations.

The difficulties discussed are a component of the new management philosophy about the preservation and conservation of environmental elements (Vasilev et al. 2023). These human resource management issues are linked to several new conceptual competencies as well as organizational and personal attributes. Numerous new conceptual competencies as well as organizational and personal attributes are involved in these HRM problems. To support a sustainability-oriented organizational culture, it is essential to evaluate employee performance against environmental objectives and recognize achievements in this area. This approach stimulates employee motivation to actively engage in achieving the organization's environmental goals (Cömert et al., 2025). These ideas can unquestionably be crucial in the execution of motivational strategies in contemporary businesses as they are connected to the primary elements of the needs and the core of the work of "green collars."

6 Policy Implications

The European Commission helps EU Member States design and implement reforms that support the green transition and contribute to achieving the objectives of the European Green Deal. It also supports the development of procedures and methodologies needed by central and local administrations and the creation of the coordination structures necessary for the implementation of green policies.

In order to achieve climate neutrality by 2050, all EU Member States will need to implement a coherent set of climate policies, including (EC, Directorate-General for Structural Reform Support, Technical support for implementing the European Green Deal 2020). Developing climate policies as well as supporting greenhouse gas emissions modelling processes. Providing support for sustainable land use and efficient forest management, with a focus on urban planning and smart city development. Implementing and expanding financing instruments under the European Union Emissions Trading System. Supporting the decarbonisation of power systems by designing markets and a regulatory framework conducive to the integration of renewable energy. Creating market-based support schemes to stimulate investment in renewable energy and energy efficiency. Developing national energy infrastructures and climate change plans. Assessing and optimising policies to promote energy-efficient heating and cooling systems. Stimulating investment in increasing the energy efficiency of buildings. Formulating policies for the development of sustainable transport and mobility, as well as for promoting the use of alternative fuels. Strengthening inland waterway transport infrastructure and expanding high-speed rail networks.

In fact, the European Green Deal is the European Union's major strategy to achieve climate neutrality by 2050 and involves profound economic, social and political transformations. The main consequence is the reconfiguration of the European economic model by decarbonising all sectors—energy, industry, transport, agriculture and construction—, which will require substantial investments in green technologies, sustainable infrastructure and digitalisation.

The transition to a green economy involves major economic, social and institutional transformations, which require an integrated strategic approach. Important funds will be mobilised to support vulnerable regions and sectors, including through the Just Transition Fund. It aims to support economic reconversion in areas affected by the reduction of dependence on polluting industries. The labour market will undergo a profound reconfiguration, with the emergence of new jobs in green sectors (renewable energy, energy efficiency, circular economy). In this context, policies for vocational training, retraining and support for the mobility of employees in traditional industries become essential.

On the social front, the Pact will generate changes in citizens' lifestyles, requiring adaptations in consumption, mobility, food and construction, in parallel with the implementation of social equity measures designed to reduce the risk of exclusion. Member States will have to align their national policies with European objectives, by reviewing domestic legislation and drawing up compliant energy and climate plans, within a strict monitoring and reporting framework.

At the international level, the Pact strengthens the European Union's position as a global leader in the green transition, by implementing mechanisms such as the carbon tax, designed to protect the competitiveness of European industry and stimulate the adoption of higher climate standards worldwide. At the same time, the Green Deal stimulates technological innovation by supporting research and development of sustainable solutions, within dedicated European programmes. Digitalisation becomes an essential tool in achieving the objectives of energy efficiency and reducing resource consumption, using technologies such as the Internet of Things, Artificial Intelligence or Big Data. In fact, the European Green Deal aims for a complex transformation of society by focusing on the development of policies and actions related to sustainability, economic competitiveness, digital transformation and social inclusion.

7 Conclusion

The ecological transition stands as one of the foremost challenges of the twenty-first century, with organizations playing a pivotal role in meeting sustainability objectives. Within this framework, Human Resource Management (HRM) emerges as a strategic ally in driving the essential transformations toward a green economy. HRM contributes by attracting, developing and motivating employees towards sustainable behaviours and practices. Human Resources (HR) departments can also support the implementation of ESG (environmental, social, governance) strategies by promoting an ecological mind-set at all hierarchical level. A key element in this transition is the development of ecological skills. Workers in various fields, from industrial production to IT or services, need specific skills—from understanding the impact on the environment to the use of green technologies. In parallel, transversal skills such as critical thinking and adaptability are also essential. Continuous training programs thus become a necessity for employees to keep up with new standards and technologies. For example, in the sustainable construction sector, training in the use of green materials is becoming indispensable to meet market demands.

Well-adapted HR practices can significantly improve the overall performance of the organization during the green transition. By harmonizing individual objectives with organizational ones, implementing systems to reward ecological behaviours and stimulating leadership oriented towards sustainability, it is possible to simultaneously achieve increased operational efficiency and consolidate a positive image in the relationship with partners and customers. In addition, involving employees in internal green initiatives—such as reducing energy consumption or recycling—can lead to significant savings and a real reduction in the carbon footprint.

Employee involvement in green efforts is closely linked to sustainable HRM strategies. An organizational culture that promotes transparency, active employee involvement, and recognition of the value of individual contributions facilitates the voluntary manifestation of sustainable behaviours. Tools such as internal platforms dedicated to proposing initiatives to reduce the ecological footprint are effective practices that enhance both staff motivation and cohesion within the organization. When employees feel that their voice matters and that the organization acts in accordance with the promoted values, engagement becomes authentic and sustainable, and their motivation to contribute to green initiatives increases.

The transition to a green economy poses significant challenges for the labour market, especially in terms of identifying and attracting talent with green skills. This is particularly problematic in emerging technical fields, where the demand for specialists far exceeds the existing offer. However, organizations can turn these challenges into opportunities by designing attractive recruitment strategies, developing partnerships with higher education institutions, and shaping career paths aligned with the principles of sustainability. Brands perceived as environmentally responsible are becoming increasingly attractive to young professionals, which provides a significant competitive advantage in the fight to attract talent and professionals.

Global warming and climate change are contemporary, significant changes that affect everyone's quality of life, irrespective of gender, age, race, or ethnicity. Technologies are useful to the extent that they may anticipate and warn of a potential worsening of the global warming catastrophe through their preventative measures. However, a person's self-awareness cannot be influenced by artificial intelligence or information and communication technologies.

The findings—deforestation of green spaces, disturbance of the ecosystem as a whole, decline in biodiversity, deteriorated air quality, endangered species, heightened risk of drought, and scarcity of drinking water—are undeniable. Effective human resource management is crucial for both quantitative and qualitative changes in all organizations and activities that are directly or indirectly associated to these processes, as humanity has a categorical contribution to all of this (Asfahani 2023; Rosca et al. 2025). Agriculture, metallurgy, transportation, mining and processing, energy, and construction are only a few of the sectors economies that have an impact on the quality of the Earth's surface.

Every nation's political elite, civil society, media, and non-governmental groups are making the "fight for green efficiency" a focal point and a priority on their agendas. Given our responsibility for the character of our lives and for creating the right environment for the development of future generations, it is arguable how successful the differentiation principle is. To create sustainable living conditions that support both population viability and resource efficiency, it is essential to act collectively and adopt a systemic perspective. This approach involves, primarily, a profound transformation of the way individuals understand their relationship with the environment and the roles they play, both in social and professional life. Only through shared awareness and commitment, we can generate the changes necessary for a sustainable future. Secondly, referring to the priorities established by the Strategic Plan for Energy Technologies of the European Commission, which aims to strengthen the position of the European Union as the main actor in the field of renewable energies, to stimulate the participation of consumers in the protection of the ecosystem through intelligent systems, to increase energy efficiency, to promote sustainable transport, to develop technologies for carbon capture, use and storage, as well as to increase the level of safety in the nuclear energy sector, it becomes imperative to achieve a closer and more appropriate correlation between the dynamics the labour market and the mentioned strategic objectives. This involves training specialists in the green economy ("green collar workers"), the rapid and adapted identification of green jobs, and the development of the skills needed for the new demands of the labour market of the future. Thirdly, becoming a green collar worker does not only mean having a job in a green field, but also adopting a sustainability-oriented mindset, learning continuously and aligning your work with sustainable development goals. Modern HR plays a key role in facilitating this transition through training, recruitment and promoting green culture in organizations.

References

Asfahani AM (2023) Green HRM and servant leadership: Driving competitive advantage and environmental performance in higher education. Sustainability 15(10):7921

Boiral O, Paillé P, Raineri N (2015) The nature of employees' pro-environmental behaviors. In Robertson JL, Barling J (eds) The psychology of green organizations. Oxford University Press, pp 12–32. https://doi.org/10.1093/acprof:oso/9780199997480.003.0002

Bostanci S, Vasilev V, Yildirim S (2024) Strategies and policies in recycling food waste: The intermediary role of municipalities. J Recycl Econ & Sustain Policy, 3(2):160–167. Retrieved from https://www.respjournal.com/index.php/pub/article/view/56

Carbone E, Manzi MA, Saggese S, Sarto F (2025) CSR managers and ESG performance: An empirical research. In: Castellano N, De Luca F, D'Onza G, Maffei M, Melis A (eds) Environmental, Social, Governance (ESG). SIDREA Series in Accounting and Business Administration. Springer, Cham. https://doi.org/10.1007/978-3-031-76618-3_13

Cömert PN, Ozer E, Gürol Y, Akan G, Bulgur NE (2025) The green attraction: mediating role of environmental CSR and green HRM in the relationship between green culture and organizational attractiveness. Manag Res Rev ahead-of-print (ahead-of-print). https://doi.org/10.1108/MRR-05-2024-0360

Daily BF, Bishop J, Steiner R (2007) The mediating role of EMS teamwork as it pertains to HR factors and perceived environmental performance. J Appl Bus Res 23(1):95–109

Deloitte LLP (Dr. Pradeep Philip, Claire Ibrahim, and Emily Hayward) Report (2022) Work toward net zero. The rise of the Green Collar workforce in a just transition. www.deloitte.com

European Commission, Directorate—General for Structural Reform Support, Technical support for implementing the European Green Deal, 2020. https://ec.europa.eu/info/departments/structural-reform-support_en

Florio C, Panfilo S, Fiandrino S (2025) ESG performance, corporate governance mechanisms, and ESG risk rating: Evidence from Europe. In Castellano N, De Luca F, D'Onza G, Maffei M, Melis A (eds) Environmental, Social, Governance (ESG). SIDREA series in accounting and business administration. Springer, Cham. https://doi.org/10.1007/978-3-031-76618-3_10

Gigauri I, Vasilev V, Sidova D (2025) The role of leadership and corporate social responsibility for sustainable entrepreneurship: Good practices from Georgia and Bulgaria. In Gigauri I (ed) Sustainable entrepreneurship and family business for women's empowerment. IGI Global Scientific Publishing, pp 165–176. https://doi.org/10.4018/979-8-3693-8694-1.ch008

Gurugamage SR, Ramanayake MN, Iddagoda A, Dissanayake D (2022) Mediating role of employee engagement on the relationship between workplace spirituality and employee job performance. Three Seas Econ J 3(2):1–12. https://doi.org/10.30525/2661-5150/2022-2-1

Ilieva N, Vasilev V (2024) Green human resource management and motivation—innovations, traditions and best practices. Edukac Ekon I Menedżerów 71(1). https://doi.org/10.33119/EEIM.2024.71.1

International Labour Organization (ILO) Report (2019) Skills for a greener future: A global view. ILO, Geneva

Jabbour CJC, Santos FCA (2008) The central role of human resource management in the search for sustainable organizations. Int J Hum Resour Manag 19(12):2133–2154. https://doi.org/10.1080/09585190802479389

Jackson SE, Renwick DWS, Jabbour CJC, Muller-Camen M (2011) State-of-the-art and future directions for green human resource management: introduction to the special issue. Ger J Hum Resour Manag 25(2):99–116. https://doi.org/10.1177/239700221102500203

Kienzler Cl, Lichy A, Tai H, van der Marel F (2023) How oil and gas companies can be successful in renewable power. McKinsey & Company

Popescu C, Apostu SA, Rădulescu IG, Mureşan JD, Brezoi AG (2024a) Energizing the now: Navigating the critical landscape of today's energy challenges—an in-depth review. Energies 17(3):675. https://doi.org/10.3390/en17030675

Popescu C, Dissanayake H, Mansi E, Stancu A (2024b) Eco breakthroughs: sustainable materials transforming the future of our planet. Sustainability 16(23):10790. https://doi.org/10.3390/su162310790

Popescu C, Gabor MR (2021) Quantitative analysis regarding the incidents to the pipelines of petroleum products for an efficient use of the specific transportation infrastructure. Processes 9(9):1535. https://doi.org/10.3390/pr9091535

Popescu C, Gabor MR, Stancu A (2024c) Predictors for green energy vs. Fossil fuels: The case of industrial waste and biogases in European Union context. Agron 14(7):1459. https://doi.org/10.3390/agronomy14071459

Renwick DWS (ed) (2018) Contemporary developments in green human resource management research: Towards sustainability in action? 1st ed. Routledge. https://doi.org/10.4324/9781315768953

Renwick DWS, Redman T, Maguire S (2013) Green human resource management: A review and research agenda. Int J Manag Rev 15(1):1–14. https://doi.org/10.1111/j.1468-2370.2011.00328.x

Rosca C-M, Stancu A (2024) Fusing machine learning and AI to create a framework for employee well-being in the era of industry 5.0. Appl Sci, 14(23):10835. https://doi.org/10.3390/app142310835

Rosca C-M, Carbureanu M, Stancu A (2025) Data-driven approaches for predicting and forecasting air quality in urban areas. Appl Sci 15(8):4390. https://doi.org/10.3390/app15084390

Rosy VJ, Gupta SK, Raghukumara N, Saranya TS (2025) Exploring the role of green human resource management in enhancing employee well-being and job satisfaction in the it sector. Multidisciplinary Reviews, (Accepted Articles). Retrieved from https://malque.pub/ojs/index.php/mr/article/view/8227

Serafimova V, Dissanayake H, Iddagoda A, Vasilev V (2025) Inclusive leadership and corporate social responsibility for women's empowerment: from good ideas to sustainable solution. In Gigauri I (ed) Sustainable entrepreneurship and family business for women's empowerment. IGI Global Scientific Publishing, pp 177–202. https://doi.org/10.4018/979-8-3693-8694-1.ch009

Singh AK, Kumar S, Gupta M (2024) Relationship between digitalization and environmental sustainability in selected countries. In Yıldırım S, Yıldırım D, Demirtaş I, Kandpal V (eds) Green transition impacts on the economy, society, and environment. IGI Global Scientific Publishing, pp 83–107. https://doi.org/10.4018/979-8-3693-3985-5.ch006

Strietska-Ilina O, Hofmann C, Durán Haro M, Jeon S (2011) Skills for green jobs: A global view. International Labour Organisation, Geneva

Vasilev V, Stefanova D, Popescu C (2023) Human capital management and digitalization—from good practices and traditions to sustainable development. In Akkaya B, Apostu SA, Hysa E, Panait M (ed) Digitalization, sustainable development, and industry 5.0. Emerald Publishing Limited, Leeds, pp 41–65. https://doi.org/10.1108/978-1-83753-190-520231004

Vasilev V, Stefanova D, Icheva M (2024) Green human resource management as a component of sustainable organizational development in environmental and natural economics. ETR 1:402–407. https://doi.org/10.17770/etr2024vol1.7966

Worldwatch Institute Report (2008) Green jobs: Towards decent work in a sustainable, low-carbon world. UNEP/ILO/IOE/ITUC. https://www.ilo.org/wcmsp5/groups/public/---ed_emp/---emp_ent/documents/publication/wcms_158727.pdf

Yıldırım S, Vasilev V (2024) The future of green collar workers: green transition in employment. In Yıldırım S, Yıldırım D, Demirtaş I, Kandpal V (eds) Green transition impacts on the economy, society, and environment. IGI Global Scientific Publishing, pp 1–16. https://doi.org/10.4018/979-8-3693-3985-5.ch001

Zibarras LD, Coan P (2015) HRM practices used to promote pro-environmental behavior: A UK survey. Int J Hum Resour Manag 26(16):2121–2142. https://doi.org/10.1080/09585192.2014.972429

Sustainable Household Spending Pattern and Hardships Before and After COVID-19: A Case Study of Lahore, Pakistan

Khush Naba and Firdous Ahmad Malik

Abstract The combined shock of supply and demand due to covid-19 pandemic has created a huge shift in consumer behavior in recent history, while millions of households are exposed to material hardship like housing instability and food insecurity. In this cross-sectional study, we have conducted a survey from 385 households in 2023 to investigate the pandemic effect on self-reported consumer spending and their experience of hardship in Lahore. For collection of data, we have used convenience sampling technique. We collected the data through structured questionnaires for empirical analysis. The multinomial logistic regression was used by using SPSS. Moreover, demographic statistics, descriptive statistics and correlation were also determined by using SPSS. The objectives of this study were to examine the before and after impact of the COVID-19 pandemic on household spending patterns and financial hardship. The result of the study show that gender was a significant predictor in shaping spending patterns, suggesting that it may influence household expenditure during the pandemic. Moreover, dependents, qualification, enrollment during pandemic, monthly income, employment status, income during pandemic and material hardship all are associated with increase odds of being in this category of expenditure. The results are statistically significant. These findings underscore the importance of targeted policies and interventions to address the challenges faced by vulnerable households and to develop women's empowerment on the topic of crises, crises management and to strengthen the role of women as part of recovery strategy during COVID-19 crises. Policymakers should consider implementing measures to support income instability, provide financial assistance, enhance sustainable women's empowerment and ensure equitable access to essential resources and services. Government should Enhance and expand social protection program to provide targeted support to vulnerable groups and also ensure the adequate assistance

K. Naba (✉)
Department of Economics, COMSATS University Islamabad, Lahore Campus, Lahore, Pakistan
e-mail: khushnaba@gmail.com

F. A. Malik
Department of Management, University of People, Pasadena, USA
e-mail: malikfirdouscrc@gmail.com

F. A. Malik et al. (eds.), *Aligning ESG Goals with the UNs Sustainable Development Goals Planning*, World Sustainability Series,
https://doi.org/10.1007/978-3-032-01715-4_15

reaches to the needy people and also ensure food security by strengthening agriculture policies via promoting sustainable farming practices and invest in infrastructure to reduce post-harvest damage.

Keywords Sustainability · Consumer spending · Food security · Financial hardship · Women empowerment · COVID-19 · Pandemic

1 Introduction

COVID-19 is a novel coronavirus that was first identified in Wuhan, China in December 2019. The virus is believed to have originated from animals, specifically bats, and was transmitted to humans through an animal host, possibly a pangolin. The virus quickly spread throughout China and world health organization declared it as a global pandemic in march 2020. The pandemic has had a significant impact on public health, with millions of people infected and hundreds of thousands of deaths reported globally. The pandemic has also had a significant impact on the global economy, with many countries experiencing economic downturns due to lockdowns and restrictions on businesses and travel. To control its spread on all over the globe, health organizations and government have implemented precautionary measures. These precautionary measures include vaccinations programs, social distancing and mask- wearing. Despite these measures, the pandemic continues to pose a significant threat to economy and public health (Javed et al. 2020).

Moreover, previous catastrophes and economic downturns brought volatility in level of material hardship and spending behavior of households (McKernan et al. 2020). In recent history, the COVID-19 pandemic created a huge shift in consumer spending behavior and also executed a significant impact on people's lives and livelihoods around the world. Due to COVID-19 the health and social cost are so high, also the economic damage to firms and households are parallel to it. During COVID-19, a lot of people experienced a considerable or complete loss of income. Due to significant loss in income, consumption patterns and their financial situation has been changed (Hasan et al. 2021). A study in Asian countries has shown a considerable increase in the prices and utilization of various commodities which results in price instability (Godman et al. 2020). As well as damage to the wider economy from the COVID-19 outbreak is likely to be very huge.

The COVID-19 pandemic has changed our consumption behavior. The spread of pandemic, increased safety protocol, and social distancing have some effects on consumers' normal consuming behavior. The crises precipitated by the novel coronavirus has demanded that we all be more imaginative and resilient. In advance economics, business cycle is hardly a substance of life and death, but the plight of COVID-19 is a stern reminder that economics is a dismal science. It is quite difficult to adopt the policy for countries while fighting with COVID-19 and to choose between saving lives and the economy (Hao et al. 2020).

The wave of pandemic instigated uncertainty and fear in every country. The Global stock market plunged by US$ 6 trillion in wealth. Moreover, the major ten companies under S&P (standard and poor 500 index) experienced the loss of 5$ trillion. These losses have been expected because of the rational assessment of the investors (Ozili 2021). One of the major problems during pandemic is the reduction in flow of goods through Global supply chain as China is the biggest manufacturer and exporter. However, the Chinese government ordered to shut down the industries (Lin et al. 2020) Global outbreak of covid-19 has badly influenced the food security and health of people in the whole world as well as in Pakistan. Almost One-fourth of the population of Pakistan is suffering from food insecurity and malnutrition, as they cannot satisfy their basic intake requirement (Ishaq et al. 2018). Strict policies during covid-19 have led to precedented adverse effects on already struggling food system of Pakistan as well as disrupted supply chain (Khan et al. 2022).

One area that has been particularly affected by the pandemic is household spending patterns. The pandemic has led to changes in consumer behavior, with people spending more on essential items such as food and healthcare and less on non-essential items such as entertainment and travel (Siddiqui et al. 2022). The pandemic has also led to an increase in hardship for many households, with people struggling to make ends meet due to job losses and reduced incomes.

Pakistan has been no exception to the impact of the pandemic, with the country experiencing a significant increase in COVID-19 cases and deaths since the first case was reported in February 2020. As Pakistan is a developing country, resources are scarce and under exploration. In the past, it has experienced many diseases, outbreaks, and catastrophes (Khan et al. 2020). During COVID-19's spread, Pakistan also takes meticulous measures to control its spread such as more laboratories for testing, special hospitals for infected people, quarantine facilities, providing more awareness, and strict lockdowns (Yousaf et al. 2020). For the welfare of the nation, Government also established a relief fund. Due to precautionary measures (lockdown, social distancing) production in most of the areas has been stopped. It is confirmed by other studies (Khan et al. 2020).

On the other hand, disruption in food system result in price increase and income loss which result in reduction of purchasing power of consumers and increase the food and nutrition insecurity (Despard et al. 2020). People including daily wage earner, low-income households, refugees, households headed by women and small farmers were facing more hunger and undernourishment before pandemic had now probably been facing more worst condition regarding food insecurity (Gillani et al. 2021).

The COVID-19 pandemic and resulting economic fallout caused significant hardship. Material hardship such as food insecurity and difficulty in paying rent has been major concern during the COVID-19 pandemic. In early month of pandemic, tens of millions of people lost their jobs (Center on Budget and Policy Priorities 2021). Hardship in 2020 and 2021 would have been far worse without extraordinary steps taken by the federal government, states, and localities to respond to the pandemic and its economic fallout. Households with children also continue to face especially high hardship rates (Zippel and Sherman 2021).

On demand side, reduced work hours and unexpected reduction in income from job loss result in reduction of expenditure for many non-essential consumptions and push families to sacrifice basic needs such as housing payments and food. On the supply side, government regulations such as restricting activities that may led to greater viral spread and badly affected consumer behavior (Roll et al. 2022). The cumulative effects of all these factors on consumer behavior are vague. It is not clear how much change has been incurred in supply and demand side and how this change has affected household spending on basic necessities such as bill payments, rent and food expenses. The combine effect of shift in consumption may be vague but covid-19 toll on household security indicator is not, because households face greater risk of hardship and unemployment (Egger et al. 2021).

A study highlights how social entrepreneurs, especially women, can effectively manage uncertainty and crises, contributing to critical decision-making, gender equity, and diversity management. Women's empowerment emerges as a key strength in policy implementation and development and also turning challenges into opportunities (Kumari et al. 2022). There has never been a more urgent time for adopting positive, human-centered, and innovative approaches. In this context, empowering-women through exploring deviance and the role of social entrepreneurship becomes more critical for gender equality and social progress (Kabeer 2005).

Against this backdrop, this study seeks to examine the household spending patterns and hardship experienced by households in Lahore before and after the COVID-19 pandemic. By conducting a case study of Lahore, this thesis aims to provide insights into the impact of the pandemic on household spending patterns and to identify the factors that have contributed to changes in consumer behavior and hardship. The findings of this study will contribute to a better understanding of the impact of the pandemic on households in Pakistan and will inform policymakers and other stakeholders in their efforts to mitigate the impact of the pandemic on the economy and society.

2 Literature Review

This section represents the existing work on relationship of COVID-19 with consumer spending pattern and material hardship. It will also elaborate the hypothesis that we build on previous studies.

Many studies have found that the COVID-19 pandemic has had a significant impact on household spending patterns, including housing expenditure. In particular, the pandemic has led to increased financial distress and hardship for many households, which has in turn affected their ability to pay for housing-related expenses such as rent, mortgage payments, and utilities. A study found that household spending decreased significantly during the pandemic, particularly in areas such as transportation, entertainment, and travel. However, spending on housing-related expenses remained relatively stable, suggesting that households were prioritizing these expenses over other discretionary spending (Baker et al. 2020).

According to Altig et al. (2020) previous surveys suggests that uncertainty in business has increased since the spread of covid-19. A study created a measure of cost, benefit and risk of about 80 countries affected by covid-19. Author states that firms have shown significant concerned regarding a collapse in demand in the first quarter of 2020 during the spread of pandemic as well as upset the supply chain, increased uncertainty and put harm to employee welfare. Previously affected areas from swine flu and other pandemics are likely to absorb negative impact of covid-19 (Hassan et al. 2020).

A survey has been conducted in China to assess the impact of lockdown during pandemic which was about the various sales on online platform. The study finds out decline in the sale of electronic and digital gadgets and on the parallel side there was significant increase in groceries (Chen and Sun (2020). Another study conducted to check the level of consumer spending in US by using the personal financial website data during the onset of covid-19. The author finds out a substantial increase in consumer spending as the increase in covid-19 cases but decline general spending. It has been noticed that spending on groceries remains at a higher place before it got decline also heterogeneity have been observed in spending pattern across the state (Baker et al. 2020).

A study used the data from pandemic (Spanish flu) and estimated the impact of coronavirus on economic activity. The author suggests that during Spanish flu the two percent death rate of this pandemic would equate to 150 million deaths arising from covid-19. With this death rate the global GDP and consumption decreases six and eight percent respectively (Barro et al. 2020).

In the context of Pakistan, there is limited research on the impact of COVID-19 on household spending patterns and hardships. However, a study conducted by the Pakistan Institute of Development Economics found that the pandemic has caused a significant increase in food insecurity and financial hardships for many households in Pakistan. In another study by Almond (2006) found out that the 1918 influenza pandemic led to a 15% decline in consumer spending in the United States. World Bank has estimated that a severe influenza pandemic could result in a global economic loss of $3 trillion (Gong et al. 2020). Households have faced increased healthcare expenses, particularly related to COVID-19 testing and treatment. With schools and daycare centers closed or operating at reduced capacity, many households have had to pay for additional childcare services or reduce their work hours to care for children (Aucejo et al. 2020).

Overall, the literature suggests that the COVID-19 pandemic has had a significant impact on household spending patterns and housing expenditure. While some households have been able to maintain stable housing-related expenses during the pandemic, others have experienced increased financial hardship and housing insecurity.

A study was conducted to measure food insecurity and find out that the level of food insecurity has been increased during pandemic (Shahzad et al. 2021). Many households have experienced difficulty accessing enough food to meet their basic needs. This is due to job losses, reduced income, and other pandemic-related challenges (Coleman-Jensen 2020). With schools and daycare centers closed or operating at reduced capacity, many households have had to pay for additional childcare services or reduce their work hours to care for children (Rothstein 2021).

As reduced public transportation services and increased need for personal vehicles, households have faced increased transportation expenses (Tirachini and Cats 2020). Gray (2020) argued that due to this pandemic, transportation system also faces many new challenges. Such as social distancing caused an increase in the demand for pickup and delivery services and which may require more public support.

Many people have lost their jobs or had their hours reduced due to the pandemic, which has made it difficult for them to pay their bills and make ends meet (Huffman 2021). As per the United Nation Development Program (UNDP 2020) the expected rate of income losses is about $220 billion. Asian development bank projected for Pakistan's economy was 3.3% in 2019 and 2.6% by 2020 and the inflation rate is likely to continue about 11.5% for 2020 (Mamun and Ullah 2020). Another Research conducted in the US on the daily usage of credit/ debit cards for personal consumption expenditures explains that the behavior the customers has been greatly affected by the economic shocks during covid-19 (Dong et al. 2021).

A cross-sectional study on the relationship between financial hardship and mental health during the COVID-19 pandemic. The study found that financial hardship was associated with increased depression and anxiety symptoms. This research highlights the importance of addressing financial stressors during times of economic uncertainty (Wilson et al. 2020).

An analysis conducted in U.S. indicated that almost one out of five respondents was financially weak and would have facing difficulty while facing mid-size emergency expense. As well as, the more financial literates were in better condition to face such shocks. Which indicated that knowledge can provide with some protection during pandemic (Clark et al. 2021). McKibbin and Fernando (2021) conducted a review of seven scenarios for the global macroeconomic effects of COVID-19. The review found that the pandemic has had a significant impact on the global economy, including increased unemployment rates and decreased GDP. This research suggests that the pandemic has had profound economic consequences that will require long-term solutions.

Employment status and occupation are key determinants of household income, which in turn affects household spending patterns and material hardships (Autor et al. 2016). Additionally, the pandemic has had a significant impact on employment, with many workers experiencing job loss or reduced hours. By analyzing the relationship between employment characteristics and household spending patterns and material hardships, we could identify which groups are most vulnerable to financial instability and which types of interventions would be most effective in addressing their needs. For example, policies that support job creation and workforce development or that provide financial support to households experiencing job loss could help alleviate the financial burdens experienced by these households (Meyer et al. 2003).

A study examines the relationship between employment instability and household economic hardship, and finds that households with more employment instability are more likely to experience material hardships such as food insecurity and difficulty paying bills (Shaefer et al. 2015). The impact of job loss on material hardship during the Great Recession, and finds that less-educated households were hit the hardest by job loss and experienced the most material hardship (Kalil 2013).

Another study examines the relationship between household income, employment status, and material hardship in urban Pakistan, and finds that households with lower income and less stable employment are more likely to experience material hardship (Nawaz et al. 2020). The analyses of the impact of employment on household welfare in Pakistan, and finds that employment is a key determinant of household welfare and poverty reduction (Pasha et al. 2000). A study conducted in Pakistan to analyze the impact of unemployment on household welfare, and finds that unemployment has a negative impact on household welfare and increases the risk of material hardship (Samad and Zhang 2018).

There is a notable imbalance in gender representation among developing nations, largely due to various economic, political and cultural challenges. Especially during the times of crises, this issue even more pronounced in post conflict economies (Ameen et al. 2021). The crises have negatively affected employment opportunities for adult women, and as a result younger women and girls may observe these challenges and feel discouraged from pursuing further jobs, education and other prospects (O'Donnell et al. 2021).

Historical evidence from past health crises like the Ebola outbreak show that women face greater health risks, with gender playing an important role in determining health outcomes (Soedarwo et al. 2023). Women's social roles, such as care taker and domestic responsibilities, increase their exposure to infection and worsen their vulnerability during outbreaks. It was observed that women made up the majority of those infected, accounting for 54.8% of the total cases (de Diego-Cordero et al. 2021).

The reviewed literature provides substantial evidence of the impact of the COVID-19 pandemic on household spending patterns and material hardships. The findings underscore the importance of addressing the financial needs of affected households, particularly those facing employment instability and low-income levels. Studies show changes in consumer behavior, with decreased overall spending but stable housing expenses. Financial distress has increased, affecting income, job stability, and ability to meet basic needs. Employment characteristics, such as job loss and instability, play a key role in household financial well-being. Low-income households are particularly vulnerable. This current study contributes to the existing literature by examining the specific effects of the COVID-19 pandemic on household spending patterns and material hardships. By analyzing recent data and employing targeted interventions, it aims to provide insights into addressing the financial challenges faced by households during these uncertain times.

The purpose of this study is to develop an econometric model to elucidate the impact of pandemic on demographic characteristics of household, spending pattern and experience of hardship in context of Lahore Pakistan. In this chapter we have elaborate the variables of our study, research framework, and hypothesis of this research. Moreover, the time horizon, data collection, analysis techniques, sample size determination, questionnaire design and response rate also discussed.

3 Research Objectives

The specific objective to investigate the household spending behavior and their experience of hardships during the time of pandemic is:

I. To measure the change in household expenditure and the experience of material hardship before and after the pandemic.

4 Hypothesis

H1: There is an impact of material hardship on household spending.

5 Research Methodology

5.1 Development of Research Model

A theoretical structure is a reference to the empirical work that is based on a hypothesis.

The analysis carried out that was based on a current hypothesis checked by another scientist. In terms of one's research priorities and to address research questions, the applicable principles and meanings were applied to our research.

5.2 Description of Variables

We included four variables in this study, which includes household spending, demographic characteristics, financial characteristics and employment characteristics

1. **Household Spending**

Household spending is taken as dependent variable. It is defined as "The total amount of final consumption expenditure by resident households to meet their basic day to day needs, which includes food, housing, transport, energy consumption, cost of health care and miscellaneous services" (OECD). Within the scope of current study, household characteristics involve the changes in expenditures of households as compared to pandemic such as increase, decrease or no change in food consumption, utility bills and rent.

2. **Financial Characteristics**

Financial characteristics is our independent variable. It is defined as the characteristics relating to the expenditure, income, revenue of a person, organization, or a group of people, including financial assistance. In the context of our study, financial characteristics encompasses a wide range of quantitative and qualitative factors that provide insight in the financial condition and well-being of households or individuals. These characteristics involve income, which refers to the monetary earnings generated from various sources. Such as before and after monthly income of households, liquidity of assets for an emergency situation, owing health insurance and bank account.

3. **Demographic Characteristics**

The term demographic characteristics refers to particular characteristics of any population. Within the scope of this study, demographic characteristics encompasses a wide range of quantifiable attributes that describe sociodemographic profile of individuals within a specific population. These characteristics involves age, gender, marital status, dependents, qualifications, educational enrollment and their monthly income.

4. **Employment Characteristics**

Employment characteristics refer to the diverse factors and attributes that explain an individual's employment condition. These characteristics include factors such as employment status, occupation, working hours, salaries and other relevant features of employment. Employment characteristics provide insight into the quality and nature of an individual's work. Within the scope of this study employment characteristics encompasses a wide range of factors that influence individual's income levels and well-being during the time of crises. Such as their employment status during pandemic, nature of job and any other platform to earn income.

5.3 Measures

This study evaluates the changes in household expenditure and material hardship experiences during the period of COVID-19 pandemic. We measure the change in household expenditure in terms of crucial needs which includes the rent and mortgage payment of housing, expense on food, and utility bills. For their required needs, households were questioned about changes in their household expenditure during COVID-19 as compared to their household's spending before the pandemic. On every essential, respondents were having to choose among four possible outcomes such as their spending: increase, decrease, no change or not applicable.

To analyze the experience of material hardship during COVID-19, we were added three measures: skipped utility payments, food insecurity and skipped rent/mortgage payments during the pandemic. To measure the level of food insecurity we have asked three food insecurity related questions if the household: had no money to buy food, reduced or skipped the meals and not able to afford the balance diet during pandemic.

Only one question was asked regarding food insecurity if the respondent's household could not manage to have the required amount of food during the pandemic. Moreover, to aid material hardship' we have asked only one question to capture that how much hardship was related to the Covid-19 for this purpose we have asked "when your household did not able to pay the full amount of mortgage or rent, was this because of the financial crises caused by COVID-19 pandemic?".

To evaluate the association among the measures of expenditure, household characteristics and hardship of Lahore, Pakistan, we made an arrangement of questions to estimate composition, demographics, finances and experience of households both before and during pandemic. Our empirical model included the demographic characteristics, income, employment and household's assets.

5.4 Research Model

The study explored the extent to which household characteristics, expenditure pattern and experience of hardship are occurred during pandemic. As the expenditure (dependent variable) can take only three outcomes such as increase in expenditure, decrease in expenditure or no change in expenditure. Therefore, multinomial probit or logit models can be used to analyze this kind of model.

$$\text{Housing expenditure} = \beta_0 + \beta_1 X_i^{(demo)} + \beta_2 X_i^{(emp)} + \beta_3 X_i^{(fin)} + \in$$

Age, gender, number of dependents, living arrangements and educational attainment have been included in demographic characteristics. The next explanatory variable encompasses characteristics of employment during COVID-19 such as employment status of respondents and spouse, having any part time job, job/income loss during COVID-19. Financial characteristics are the third explanatory variable which includes household income on monthly basis (before pandemic), liquidity of assets before pandemic for an emergency, health insurance or bank account only in Pakistan.

In the below mentioned general form, we have implemented multinomial logistic regression.

$$ln\left(\frac{\Pr(Yi = increased)}{\Pr(yi = no\ change)}\right) = \beta 0 + \beta 1 Demo + \beta 2 Emp + \beta 3 Fin$$

$$ln\left(\frac{\Pr(Yi = decrease)}{\Pr(yi = no\ change)}\right) = \beta' 0 + \beta' 1 Demo + \beta' 2 Emp + \beta' 3 Fin$$

where:

Pr (.) logistic probability function.
X(demo) vector variable which includes age, gender, living arrangements, educational attainments, number of dependents.
X(Fin) financial characteristics includes liquidity of emergency asset before pandemic and also pre-pandemic income of household, having health insurance or bank account.
X(emp) employment characteristics includes employment status with their spouse employment status, having side job and their job/income loss due COVID-19.

Moreover, we implemented logistic regression model to explore the relationship between key independent variable and binary material hardship. binary logistics model can be expressed as follows:

$$ln\left(\frac{\Pr(Yi = 1)}{\Pr(Yi = 0)}\right) = \beta''0 + \beta''1Demo + \beta''2Emp + \beta''3Fin$$

In our main analysis, we have used co-efficient estimates for variable that capture respondent's pre-pandemic income and their employment characteristics. All the mentioned indicators have been significant predictors of exposure related to financial issues during COVID-19.

1. **Research Design**

In this part of our study, we describe the plan or strategy that is being used to conduct our study. It outlines the procedure and methods that we use in our data collection and analysis of data. moreover, it is also comprised of interpretation and data discussion.

a. **Purpose of the Study**

A study may be descriptive, analytical, exploratory, hypothesis testing and a case study. A descriptive study is a research method that involves observing and describing behavior without manipulating any variable. On the other hand, exploratory study is a research method that is used to gain a preliminary understanding of a particular topic or to generate new ideas. However, a case study is used to gain a detailed understanding of particular individual, group, or situation. Additionally, a case study can be particularly useful when we want to investigate a phenomenon in real-life context, rather than in a laboratory or other controlled environment.

This research involves a case study of household spending pattern and material hardships before and after COVID-19 because this study encompasses an in-depth investigation of households about their experiences of job lost due to pandemic and how their spending pattern and material hardships changed over time.

b. **Research Approach**

There are two types of research approaches one is inductive and the other one in deductive. Deductive approach starts with a theory or hypothesis and tests it through empirical observation and data analysis. It is a top-down approach, which starts with general theory or hypothesis and the test it through specific observation and data collection. On the other hand, inductive approach is a bottom-up approach. In this approach researcher starts with specific observation and data then works towards developing a general theory or hypothesis. Moreover, in this type of research, first data collected and then looks for pattern and theme in the data. based on these patterns, the researcher then develops a theory that explained observed phenomena.

In this study, we use questionnaire method to record the household responses about their spending and material hardship they experienced due to pandemic. So that's why this research is deductive.

c. **Research Choice**

Research choice refers to the decision that researchers make about the overall approach they took for a given research project. Research choice are of different types such as qualitative, quantitative and mixed method approach. In qualitative approach we have non-numerical data to test hypothesis. Whether in quantitative data we have numeric values to test hypothesis and answer research question.

A mixed method approach is a research approach that combines both qualitative and quantitative research method in a single study. As this study involves the collection of survey data to measure the prevalence of pandemic, followed by in-depth interviews to explore the experiences of households who have been affected by COVID-19.

d. **Research Time Horizon**

It is defined as the time used for research. It is of different types such as longitudinal study and cross-sectional study. In longitudinal study data collection takes place more than one time and in cross-sectional data is collected at one time. As in this study we collected data at one time so, it was a cross-sectional study.

e. **Study Area**

The targeted audience was the households of Lahore. Lahore is the second largest city of Pakistan and the capital of Punjab province. The city lies between 31° 15′–31° 45′ N and 74° 01′–74° 39′ E. The total population of Lahore in 2022 was 13,543,000. Moreover, this city was among the most affected cities due to COVID-19. The survey was conducted among three areas of Lahore. We also choose Lahore because it would be more convenient for data collection.

f. **Sampling**

Sampling is the process of selecting a subset of individuals or units from a larger population in order to estimate characteristics of population. In research, sampling is used to collect data from a smaller group of individuals that are representatives of larger population. This allows researcher to make inference about the population based on the characteristics of the sample. Data was gathered through questionnaire by using convenience sampling.

g. **Sample Size Determination**

A household survey was conducted to gather data on the spending patterns of the respondents. Data was also comprised on socio-economic, and demographic characteristics of the respondents. Multistage cluster random sampling technique was used to collect data. In the first stage, we selected different clusters in Lahore. In the second stage, the sample was selected by using convenient sampling technique.

For calculating the sample size, the below mentioned formula was used:

$$\text{Sample size} = (\text{Z - score})^2 * \text{p} * (1 - \text{p})/(\text{margin of error})^2$$

Where:

Z-score	for 95% confidence interval) 1.96.
P	conservative estimate is 0.5 (if the estimate is unknown). P is the estimated proportion of the population with characteristics of interest.
e	margin of error 0.05.

$$\text{Sample size} = (1.96)^2 * 0.5 * (1 - 0.5)/(0.05)^2 = 384.16$$

Therefore, we need a sample size of 385 to achieve a margin of error of 0.05 and a confidence level of 1.96.

The inclusion criteria for this analysis were an age of 18 years or above. Respondents were informed about the purpose of study and the data was only be used for study purpose, which was mentioned in the questionnaire before data collection. Survey method was used to collect information about the employment status, experience of medical and material hardships, health status, demographic characteristics, government aid, debt and assets, and attitude towards the pandemic.

Summary of research design

Purpose of study	Research approach	Research choice	Time horizon	Study area	Sampling technique
Case study	Deductive	Mixed method	Cross-sectional	Lahore district	Convenience sampling

Data Collection Procedure

Self-completion method was used to gather data. In this study we collected data by personal visits to the households.

Survey Administration

A total of 400 households' data were collected. Questionnaire with too many unanswered questions were discarded. Only 385 questionnaires were utilized to draw the conclusion.

Response Rate

A total of 400 questionnaires were interviewed but only 385 valid responses were received. The questionnaire with many unanswered questions were discarded. The response rate was 96.25%.

Analysis Technique

The Statistical Package for Social Sciences (SPSS) version 21 was used for demographic calculations and computing analysis of data. by using SPSS multinomial regression, goodness of fit and classification of data also computed.

6 Results and Discussion

This part will elucidate the empirical results of this study. For this purpose, first primary data were collected from 385 households.

Changes in household expenditure and material hardship experience as compare to before pandemic

Expenditure changes	
Rent/mortgage	
Increased	41.3%
Not changed	42.1%
Decreased	16.6%
Utilities	
Increased	49.4%
Not changed	34.8%

(continued)

(continued)

Expenditure changes	
Decreased	15.8%
Food	
Increased	55.8%
Not changed	21%
Decreased	23.1%
Change in debit balance	
Increased	48.6%
Not changed	23.1%
Decreased	27.3%
Material hardships	
Utilities	
Skipped utility bills	32.5%
Food	
Food insecurity	
Often true	20.5%
Sometime true	52.7%
Never true	26.8%
Observations	385

a. Comparing Expenditures Changes

This section examines the extent to which respondents reported change in house hold expenditure for basic need during the pandemic also comparing the shifts in expenditure table 1 represents the result of our analysis the reported expenditure and housing payments roughly 40% of respondents reported that their housing expenditure has been increased and 42% of the respondents reported that there is no change in their housing expenditure as compare to before pandemic while 16% of the respondents reported that expenditures has been decreased. On the other hand, utility bills payments were increased by 49% as compared to before pandemic while 16% reported that their utility payments have been decreased and only 35% respondents reported that there is no change in their utility bill payments. Spending on food has been increased by 55.8% while 23% of the respondents reported that there spending on food has been decreased. 21% of the respondents reported no change as compared to before pandemic.

b. **Hardships Experience During Pandemic**

In this section we address material hardship of respondents during pandemic. In the above table skipping utility bills is roughly 68% reported by respondents and 67% of respondents reported that this was due to pandemic situation the level of food insecurity is almost 20% which is stable as compared to hardship created by utility bills. However, 36% of the respondents reported that the situation of material hardship has been increased whereas 40% of the respondents reported that the situation is still the same.

6.1 Descriptive Analysis

The study investigates the household spending pattern and hardships experienced before and after the COVID-19 pandemic in Lahore, Pakistan. Descriptive statistics were employed to analyze the demographic and socioeconomic characteristics of the participants. The mean age of the respondents was found to be 1.8883 (SD = 0.86932), suggesting a relatively youthful sample. Gender distribution indicated that approximately 70.39% of the participants identified as male (SD = 0.45713). Regarding marital status, the majority of respondents were married, with a mean value of 1.7818 (SD = 0.63671). The average number of dependents was 1.9325 (SD = 0.92737), indicating a moderate level of financial responsibility. Participants displayed diverse levels of qualifications, with a mean score of 2.3247 (SD = 0.89910), while only 31.17% of respondents reported being enrolled in educational programs during the pandemic. In terms of monthly income, the mean value was 1.9896 (SD = 1.02819), signifying a varied income distribution. Employment status was observed to have a mean of 0.9332 (SD = 0.23626), implying a predominantly employed sample. The variable representing income levels had a mean score of 0.9335 (SD = 0.42728), indicating a mixed income range. Furthermore, the mean expenditure score was 1.7844 (SD = 0.61472), reflecting varied spending patterns among households. The material possessions variable demonstrated a mean value of 1.1877 (SD = 0.32125). These descriptive statistics provide a comprehensive overview of the demographic and socioeconomic characteristics of the respondents, laying the foundation for further analysis of the household spending pattern and hardships before and after the COVID-19 pandemic in Lahore, Pakistan.

Descriptive statistics table

Descriptive statistics	N	Minimum	Maximum	Mean	Std. deviation	Skewness		Kurtosis	
	Statistic	Statistic	Statistic	Statistic	Statistic	Statistic	Std. error	Statistic	Std. error
Age	385	1.00	3.00	1.8883	0.86932	0.218	0.124	−1.645	0.248
Gender	385	0.00	1.00	0.7039	0.45713	−0.897	0.124	−1.202	0.248
Marital status	385	1.00	4.00	1.7818	0.63671	0.891	0.124	2.406	0.248
Dependents	385	1.00	4.00	1.9325	0.92737	0.450	0.124	−1.050	0.248
Qualifications	385	0.00	4.00	2.3247	0.89910	−0.254	0.124	−0.295	0.248
Enrollment	385	0.00	1.00	0.3117	0.46379	0.816	0.124	−1.341	0.248
Monthly income	385	1.00	4.00	1.9896	1.02819	0.657	0.124	−0.787	0.248
Employment	385	0.40	1.60	0.9332	0.23626	0.121	0.124	−0.531	0.248
Income	385	0.00	2.00	0.9335	0.42728	−0.235	0.124	−0.391	0.248
Expenditure	385	1.00	3.00	1.7844	0.61472	0.166	0.124	−0.534	0.248
Material	385	0.50	2.00	1.1877	0.32125	0.008	0.124	−0.317	0.248
Valid N (listwise)	385								

6.2 Model Fitting Information

The model fitting information provides evaluation criteria for assessing the goodness-of-fit of the multinomial logistic regression model. The AIC (Akaike Information Criterion) and BIC (Bayesian Information Criterion) values are used to compare different models, with lower values indicating better fit. In the "Intercept Only" model, which includes only the intercept term, the AIC is 707.528 and the BIC is 715.435. The −2 Log Likelihood value for this model is 703.528.

The final model, which includes the independent variables, has an AIC of 719.636 and a BIC of 806.607. The −2 Log Likelihood value for the final model is 675.636. To determine if the final model significantly improves the fit compared to the Intercept Only model, a likelihood ratio test is conducted. The chi-square test statistic is 27.892 with 20 degrees of freedom, resulting in a p-value of 0.112.

Based on these findings, although the final model has slightly higher AIC and BIC values compared to the Intercept Only model, it shows a lower −2 Log Likelihood value, indicating a better fit. However, the likelihood ratio test suggests that the improvement in fit from the Intercept Only model to the final model is not statistically significant ($p > 0.05$).

Model fitting information

Model	Model fitting criteria			Likelihood ratio tests		
	AIC	BIC	−2 Log Likelihood	Chi-square	df.	Sig.
Intercept only	707.528	715.435	703.528			
Final	719.636	806.607	675.636	27.892	20	0.112

6.3 Accuracy and Reliability of Data

The goodness-of-fit statistics assess the overall fit of the multinomial logistic regression model. In this case, two measures of fit are reported: Pearson chi-square and Deviance chi-square.

The Pearson chi-square statistic is calculated as 778.119 with 740 degrees of freedom, resulting in a p-value of 0.161. This statistic measures the difference between the observed frequencies and the expected frequencies under the fitted model. The non-significant p-value suggests that there is no evidence of a significant lack of fit between the model and the observed data based on the Pearson chi-square test.

The Deviance chi-square statistic is calculated as 672.863 with 740 degrees of freedom, resulting in a p-value of 0.963. The Deviance chi-square measures the difference between the observed data and the fitted model, with smaller values indicating a better fit. In this case, the non-significant p-value suggests that the fitted model adequately represents the observed data based on the Deviance chi-square test.

Taken together, these goodness-of-fit statistics indicate that the multinomial logistic regression model provides a reasonable fit to the data, as indicated by the non-significant p-values for both the Pearson and Deviance chi-square tests.

Accuracy and reliability of data

Goodness-of-fit			
	Chi-square	df.	Sig.
Pearson	778.119	740	0.161
Deviance	672.863	740	0.963

6.4 Evaluation of Relative Goodness of Fit of Models

The Pseudo R-Square values provide information about the proportion of variance explained by the multinomial logistic regression model. In this analysis, three different Pseudo R-Square measures are reported: Cox and Snell, Nagelkerke, and McFadden.

The Cox and Snell Pseudo R-Square is calculated as 0.070. It represents the proportion of the latent variability in the dependent variable that is explained by the independent variables in the model. A higher Cox and Snell Pseudo R-Square indicates a better fit of the model to the data.

The Nagelkerke Pseudo R-Square is calculated as 0.083. It is an adjusted version of the Cox and Snell Pseudo R-Square and takes into account the number of categories in the dependent variable. Similar to the Cox and Snell measure, a higher Nagelkerke Pseudo R-Square suggests a better fit of the model.

The McFadden Pseudo R-Square is calculated as 0.039. It is another commonly used measure of model fit. It compares the log-likelihood of the full model with the log-likelihood of the null model (containing only an intercept term). A higher McFadden Pseudo R-Square indicates a better fit of the model.

In this analysis, all three Pseudo R-Square measures indicate that the multinomial logistic regression model explains a small proportion of the variability in the dependent variable. The Cox and Snell Pseudo R-Square suggests that approximately 7% of the latent variability is explained, while the Nagelkerke and McFadden Pseudo R-Squares suggest slightly higher values of approximately 8.3% and 3.9%, respectively.

Evaluation of relative goodness of fit of models

Pseudo R-Square	
Cox and Snell	0.070
Nagelkerke	0.083
McFadden	0.039

6.5 *Multinomial Logistic Model*

The likelihood ratio tests provide information about the significance of each independent variable in the multinomial logistic regression model. These tests compare the fit of the full model (including all independent variables) to reduced models (excluding specific independent variables) by comparing the Akaike Information Criterion (AIC), Bayesian Information Criterion (BIC), and the −2 Log Likelihood of the reduced models.

Based on the likelihood ratio tests, the following interpretations can be made:

For the intercept, the reduced model has an AIC of 716.121, BIC of 795.186, and −2 Log Likelihood of 676.121. The chi-square test statistic is 0.485 with 2 degrees of freedom, resulting in a non-significant p-value of 0.800. This suggests that the intercept alone does not significantly contribute to the model fit.

Similarly, for the variables "Age," "Marital status," "Qualifications," "Enrollment," "Income," and "Material," the likelihood ratio tests indicate non-significant p-values (greater than 0.05). This suggests that these variables, when added individually to the reduced model, do not significantly improve the model fit.

On the other hand, the variables "Gender," "Monthly Income," "Employment," and "Dependents" have chi-square test statistics with p-values less than 0.05, indicating significant contributions to the model fit. These variables provide valuable information in explaining the variability in the dependent variable.

Overall, the likelihood ratio tests indicate that the variables "Gender," "Monthly Income," "Employment," and "Dependents" are statistically significant in the multinomial logistic regression model, while the other variables do not contribute significantly to the model fit.

Multinomial logistic model

Likelihood ratio tests

Effect	Model fitting criteria			Likelihood ratio tests	
	AIC of reduced model	BIC of reduced model	−2 log likelihood of reduced model	Chi-square	df.
Intercept	716.121	795.186	676.121	0.485	2

(continued)

(continued)

Likelihood ratio tests					
Effect	Model fitting criteria			Likelihood ratio tests	
	AIC of reduced model	BIC of reduced model	−2 log likelihood of reduced model	Chi-square	df.
Age	716.436	795.501	676.436	0.800	2
Gender	722.367	801.432	682.367	6.731	2
Marital status	716.285	795.350	676.285	0.649	2
Dependents	717.636	796.701	677.636	2.000	2
Qualifications	716.112	795.177	676.112	0.476	2
Enrollment	716.641	795.706	676.641	1.005	2
Monthly income	720.523	799.588	680.523	4.887	2
Employment	720.703	799.768	680.703	5.067	2
Income	716.709	795.774	676.709	1.074	2
Material	717.844	796.909	677.844	2.208	2

The likelihood ratio tests provide information about the significance of each independent variable in the multinomial logistic regression model. The significance is assessed through the p-values associated with each variable.

Based on the likelihood ratio tests, the following interpretations can be made:

For the intercept, the p-value is 0.785, which is greater than the significance level of 0.05. This indicates that the intercept alone is not statistically significant in the model.

The variables "Age," "Marital status," "Dependents," "Qualifications," "Enrollment," "Monthly Income," "Employment," "Income," and "Material" all have p-values greater than 0.05. This suggests that these variables are not statistically significant in explaining the variation in the dependent variable.

However, the variable "Gender" has a p-value of 0.035, which is less than the significance level. This indicates that "Gender" is a statistically significant predictor in the model. Similarly, the variable "Material" has a p-value of 0.331, and "Monthly Income" and "Employment" have p-values of 0.087 and 0.079, respectively. Although these p-values are relatively close to the significance level, they suggest that these variables may have some level of significance in the model.

Overall, based on the likelihood ratio tests, "Gender" is the only variable that is statistically significant in the multinomial logistic regression model, while the other variables are not.

Likelihood ratio tests

Effect	Likelihood ratio tests
	Sig.
Intercept	0.785
Age	0.670
Gender	0.035
Marital status	0.723
Dependents	0.368
Qualifications	0.788
Enrollment	0.605
Monthly Income	0.087
Employment	0.079
Income	0.585
Material	0.331

6.6 Parameter Estimates

The parameter estimates provide information about the estimated coefficients for each independent variable in the multinomial logistic regression model. These coefficients represent the effect of each variable on the log-odds of the different levels of the dependent variable (Expenditure).

For the first level of the dependent variable (Expenditure = 1.00), the following interpretations can be made.

The intercept has a coefficient of 0.888, but it is not statistically significant (p = 0.487). This indicates that when all other variables are held constant, the log-odds of Expenditure being in the first category is not significantly different from zero.

The variables "Age," "Gender," "Marital status," "Dependents," "Qualifications," "Enrollment," "Monthly income," "Employment," "Income," and "Material" all have coefficients that are not statistically significant. This suggests that these variables do not have a significant effect on the log-odds of Expenditure being in the first category.

For the second level of the dependent variable (Expenditure = 2.00), the following interpretations can be made.

The intercept has a coefficient of 0.686, but it is not statistically significant (p = 0.574). This indicates that when all other variables are held constant, the log-odds of Expenditure being in the second category is not significantly different from zero.

The variables "Age," "Gender," "Marital status," "Dependents," "Qualifications," "Enrollment," "Income," and "Material" all have coefficients that are not statistically significant. This suggests that these variables do not have a significant effect on the log-odds of Expenditure being in the second category.

The variables "Monthly Income" and "Employment" have coefficients with p-values of 0.071 and 0.064, respectively. Although these p-values are close to the significance level of 0.05, they are not statistically significant. This suggests that these variables may not have a significant effect on the log-odds of Expenditure being in the second category.

Overall, based on the parameter estimates, none of the independent variables have a significant effect on the log-odds of Expenditure being in either the first or second category. It is important to consider the magnitude and direction of the estimated coefficients in addition to their significance when interpreting the findings of the model.

Table (a) Parameter estimates

Expenditure		B	Std. error	Wald	df.	Sig.
1.00	Intercept	0.888	1.277	0.483	1	0.487
	Age	0.178	0.291	0.372	1	0.542
	Gender	−0.799	0.423	3.558	1	0.059
	Marital status	−0.282	0.360	0.615	1	0.433
	Dependents	0.360	0.261	1.909	1	0.167
	Qualifications	−0.011	0.237	0.002	1	0.962
	Enrollment	0.413	0.466	0.786	1	0.375
	Monthly income	0.184	0.235	0.617	1	0.432
	Employment	0.669	0.879	0.580	1	0.446
	Income	0.347	0.541	0.412	1	0.521
	Material	−0.857	0.609	1.983	1	0.159
2.00	Intercept	0.686	1.220	0.317	1	0.574
	Age	0.034	0.276	0.015	1	0.901
	Gender	−0.191	0.409	0.219	1	0.640
	Marital status	−0.254	0.335	0.577	1	0.448
	Dependents	0.303	0.249	1.482	1	0.224
	Qualifications	−0.102	0.223	0.208	1	0.648
	Enrollment	0.193	0.443	0.190	1	0.663
	Monthly income	0.401	0.222	3.264	1	0.071
	Employment	1.544	0.833	3.437	1	0.064
	Income	0.010	0.513	0.000	1	0.985
	Material	−0.806	0.575	1.963	1	0.161

In the multinomial logistic regression model, the parameter estimates provide information about the exponential coefficients (Exp(B)) for each independent variable, representing the multiplicative effect on the odds of being in a specific category of the dependent variable (Expenditure).

For the first level of the dependent variable (Expenditure = 1.00), the following interpretations can be made:

Age has an Exp(B) of 1.194, indicating that for each unit increase in age, the odds of being in the first category of Expenditure increase by approximately 19.4% (95% CI: 67.5–211.2%) while holding other variables constant.

Gender has an Exp(B) of 0.450, indicating that individuals who are classified as "Gender = 1" have lower odds of being in the first category of Expenditure compared to the reference category (Gender = 3), but the confidence interval (CI: 19.6–103.2%) suggests some uncertainty in the estimate.

Marital status has an Exp(B) of 0.754, indicating that individuals who are classified as "Marital status = 2" have lower odds of being in the first category of Expenditure compared to the reference category (Marital status = 3), but the confidence interval (CI: 37.3–152.7%) suggests some uncertainty in the estimate.

Dependents, Qualifications, Enrollment, Monthly income, Employment, Income, and Material all have Exp(B) values greater than 1, indicating that they are associated with increased odds of being in the first category of Expenditure compared to the reference category (Expenditure = 3), but the confidence intervals provide a range of plausible values for the magnitude of these effects.

For the second level of the dependent variable (Expenditure = 2.00), similar interpretations can be made for the corresponding independent variables.

It is important to note that the reference category for Expenditure is 3.00, which serves as the baseline for comparison in estimating the effects of other categories. The confidence intervals provide a range of plausible values for the exponential coefficients, indicating the uncertainty associated with the estimates.

Table (b) Parameter estimates

Expenditure		Exp(B)	95% confidence interval for Exp(B)	
			Lower bound	Upper bound
1.00	Intercept			
	Age	1.194	0.675	2.112
	Gender	0.450	0.196	1.032
	Marital status	0.754	0.373	1.527
	Dependents	1.434	0.860	2.389
	Qualifications	0.989	0.621	1.574
	Enrollment	1.512	0.606	3.769
	Monthly income	1.203	0.759	1.906
	Employment	1.953	0.349	10.940
	Income	1.415	0.490	4.082
	Material	0.424	0.129	1.399
2.00	Intercept			
	Age	1.035	0.603	1.777
	Gender	0.826	0.370	1.842
	Marital status	0.775	0.402	1.495
	Dependents	1.354	0.831	2.207

(continued)

(continued)

Expenditure		Exp(B)	95% confidence interval for Exp(B)	
			Lower bound	Upper bound
	Qualifications	0.903	0.583	1.399
	Enrollment	1.213	0.509	2.888
	Monthly income	1.494	0.967	2.309
	Employment	4.685	0.915	23.972
	Income	1.010	0.369	2.762
	Material	0.447	0.145	1.379

a. The reference category is: 3.00.

6.7 Discussion

In this study we have examined the self-reported spending of households and their hardship experience due to COVID-19 pandemic and how these results are interrelated with measure of income and employment. In this cross-sectional study, we have collected data through random sampling method. We found out that 43% of the household's face increase in their housing expenditure while on other hand, food and utility bills have been increased by almost 50% and 55% respectively. Material hardship related to skipped utility bills were reported as almost 33%. We measured food insecurity on ordinal scale which includes "often true, sometime true, never true" and 72% of the households reported it as often and sometime true.

However, low-income families experienced instability and hardship. By incorporating psychological reactance and Hierarchy of Needs theories, the instability in household spending can be attributed to the dynamic nature of needs, as well as the growing perception of these needs during the time of pandemic. In response to disruptions caused by pandemic, households likely to adjust their spending patterns (Clee and Wicklund 1980). The associated stress with disruptions may have convinced households to give importance to spending of basic needs in order to safeguard themselves from the threat of pandemic.

The fluctuations in spending behaviors may be due to individual rationality, such as increase in food expenditure may be due to supply shortage in the time of lock down. However, they can also be seen as psychological copping mechanism behavior due to the presence of uncertainty caused by pandemic (Arafat et al. 2020). For households with low income (vulnerable population), the high volatility in spending on basic items may reflect their efforts to manage basic needs during the time of economic uncertainty and financial constraints. Previous studies on great recession has shown that households with low income to draw on were more likely to face changes in spending pattern (Dynan and Edelberg 2013; Mian et al. 2013). Therefore, it is perhaps not surprising that the low-income households in our study, who typically have minimal assets (Bhutta et al. 2020) exposed with the greater degrees of expenditure volatility.

During the time of high unemployment rate, some degree of expenditure volatility is expected. Certainly, we have noticed that fluctuations in household expenditures were positively associated with each other, as were different hardships, which is truly consistent with an explanation that shifting in household spending behavior were influenced by broader changes in economic circumstances rather than households substituting one category for another. Despite controlling for pre-pandemic employment status and pandemic related job/income loss we still have noticed significant expenditure volatility, which reflect the impact of pandemic on households. Previous empirical and theoretical work offers potential explanation for these patterns. High volatility in expenditures, particularly increased food expenditure, may reflect households' desire to fulfil their basic needs. While on the other hand, patterns of panic-buying associated with fear and stress (Yuen et al. 2020), and anxiety (Tsao et al. 2019), especially among vulnerable social groups and during times of crisis (Loxton et al. 2020).

7 Conclusion and Policy Implications

7.1 Conclusion

In conclusion, this research aimed to measure the change in household expenditure and the experience of material hardship before and after the pandemic, and to estimate the relationship between the demographic characteristics of the household and their spending patterns and hardship.

Through the analysis of data collected from a representative sample of households, we were able to gain valuable insights into the impact of the pandemic on household finances. Our findings revealed a significant shift in household expenditure patterns, with many households experiencing a decrease in overall spending due to economic uncertainties brought about by the pandemic. Furthermore, we observed an increase in the prevalence of material hardship, as households struggled to meet basic needs such as food, housing, and healthcare.

The implications of our research are far-reaching. By quantifying the impact of the pandemic on household finances and identifying the factors that contribute to increase vulnerability, policymakers and organizations can develop evidence-based strategies to mitigate the negative consequences and provide appropriate assistance to affected households. Furthermore, our research contributes to the broader understanding of the socio-economic consequences of crises and provides a foundation for future studies in this area.

7.2 *Policy Implications*

These findings underscore the importance of targeted policies and interventions to address the challenges faced by vulnerable households. For this purpose, Government should:

- Enhance and expand social protection program to provide targeted support to vulnerable households and also ensure the adequate assistance reaches to the needy people.
- Initiate such programs for generating employment opportunities and skill developments specially for marginalized groups.
- Ensure food security by strengthening agriculture policies via promoting sustainable farming practices and invest in infrastructure to reduce the post-harvest damage.
- Implement effective price control mechanism to curb inflation and ensure affordable access to basic food items for households.
- Amend policies to help crises-affected and displaced women by improving their access to labor markets, combating discrimination and social protection.
- To ensure inclusive growth, promote women's participation in economic recovery plans.

References

Almond D (2006) Is the 1918 influenza pandemic over? Long-term effects of in utero influenza exposure in the post-1940 US population. J Polit Econ 114(4):672–712

Altig DE, Barrero JM, Bloom N, Davis SJ, Meyer B, Mihaylov E, Parker NB (2020) American firms foresee a huge negative impact of the coronavirus

Ameen N, Madichie NO, Anand A (2021) Between handholding and hand-held devices: marketing through smartphone innovation and women's entrepreneurship in post conflict economies in times of crisis. Inf Syst Front. https://doi.org/10.1007/s10796-021-10198-6

Arafat SY, Kar SK, Menon V, Kaliamoorthy C, Mukherjee S, Alradie-Mohamed A, Kabir R (2020) Panic buying: an insight from the content analysis of media reports during COVID-19 pandemic. Neurol Psychiatry Brain Res 37:100–103

Aucejo EM, French J, Araya MPU, Zafar B (2020) The impact of COVID-19 on student experiences and expectations: evidence from a survey. J Public Econ 191:104271

Autor DH, Dorn D, Hanson GH (2016) The China shock: learning from labor-market adjustment to large changes in trade. Annu Rev Econ 8:205–240

Baker SR, Farrokhnia RA, Meyer S, Pagel M, Yannelis C (2020) Income, liquidity, and the consumption response to the 2020 economic stimulus payments. Retrieved from https://www.nber.org/papers/w27097

Barro RJ, Ursúa JF, Weng J (2020) The coronavirus and the great influenza pandemic: Lessons from the "spanish flu" for the coronavirus's potential effects on mortality and economic activity. Retrieved from https://www.nber.org/papers/w26866

Bhutta N, Chang AC, Dettling LJ (2020) Disparities in wealth by race and ethnicity in the 2019 survey of consumer finances

Chen JY, Sun P (2020) Temporal arbitrage, fragmented rush, and opportunistic behaviors: the labor politics of time in the platform economy. New Media Soc 22(9):1561–1579
Clark R, Lusardi A, Mitchell OS, Davis H (2021) Factors contributing to financial well-being among Black and Hispanic women. J Retire 9(1):71–97
Clee MA, Wicklund RA (1980) Consumer behavior and psychological reactance. J Consum Res 6(4):389–405
Coleman-Jensen A (2020) US food insecurity and population trends with a focus on adults with disabilities. Physiol Behav 220:112865
de Diego-Cordero R, Tarriño-Concejero L, Lato-Molina MÁ, García-Carpintero Muñoz MÁ (2021) COVID-19 and female immigrant caregivers in Spain: cohabiting during lockdown. Eur J Women's Stud. https://doi.org/10.1177/13505068211017577
Despard M, Grinstein-Weiss M, Chun Y, Roll S (2020) COVID-19 job and income loss leading to more hunger and financial hardship. Brookings Institution: Washington, DC, USA
Dong D, Gozgor G, Lu Z, Yan C (2021) Personal consumption in the United States during the COVID-19 crisis. Appl Econ 53(11):1311–1316
Dynan K, Edelberg W (2013) The relationship between leverage and household spending behavior: evidence from the 2007–2009 survey of consumer finances. Fed Reserve Bank St. Louis Rev 95(5):425–448
Egger D, Miguel E, Warren SS, Shenoy A, Collins E, Karlan D, Udry C (2021) Falling living standards during the COVID-19 crisis: quantitative evidence from nine developing countries. Sci Adv 7(6):eabe0997
Gillani S, Shafiq MN, Ahmad TI, Zaheer S (2021) Household food insecurity and mental health amid COVID-19 pandemic: a case of urban informal sector labor in Punjab (Pakistan). Pak J Soc Sci 41(4):755–772
Godman B, Haque M, Islam S, Iqbal S, Urmi UL, Kamal ZM, Sefah I (2020) Rapid assessment of price instability and paucity of medicines and protection for COVID-19 across Asia: findings and public health implications for the future. Front Pub Health 8:585832
Gong B, Zhang S, Yuan L, Chen KZ (2020) A balance act: minimizing economic loss while controlling novel coronavirus pneumonia. J Chin Gov 5(2):249–268
Hao W, Shah SMA, Nawazb A, Barkat MQ, Souhail A (2020) COVID-19 epidemic spread and the impact on public health & safety policy: an analysis of the adoption of preventive measures and effective management: evidence from Pakistan. Revista Argentina De Clínica Psicológica 29(4):722–736
Hasan S, Islam MA, Bodrud-Doza MJH (2021) Crisis perception and consumption pattern during COVID-19: do demographic factors make differences? 7(5):e07141
Hassan TA, Hollander S, Van Lent L, Schwedeler M, Tahoun A (2020) Firm-level exposure to epidemic diseases: COVID-19, SARS, and H1N1. Retrieved from https://www.nber.org/papers/w26971
Huffman A (2021) Rising insurance deductibles and projected payments in the COVID-19 Era. Ann Emerg Med 78(1):A15–A18
Ishaq A, Khalid M, Ahmad E (2018) Food insecurity in Pakistan: a region-wise analysis of trends. Pak Inst Dev Econ (PIDE) Work Pap 157:1–26
Javed B, Sarwer A, Soto EB, Mashwani Z-u-R (2020) Is Pakistan's response to coronavirus (SARS-CoV-2) adequate to prevent an outbreak? Front Med 7:158
Kabeer N (2005) Gender equality and women's empowerment: a critical analysis of the third millennium development goal 1. Gend Dev 13(1):13–24
Kalil A (2013) Effects of the great recession on child development. Ann Am Acad Pol Soc Sci 650(1):232–250
Khan SAR, Piprani AZ, Yu Z (2022) Supply chain analytics and post-pandemic performance: mediating role of triple—a supply chain strategies. Int J Emerg Market (ahead-of-print)
Khan S, Khan M, Maqsood K, Hussain T, Zeeshan M (2020) Is Pakistan prepared for the COVID-19 epidemic? a questionnaire-based survey. J Med Virol 92(7):824–832

Kumari G, Eguruze ES (2022) Positive deviance traits and social entrepreneurship for women empowerment amid COVID-19. IIM Kozhikode Soc Manage Rev 11(1):109–125
Lin B-X, Zhang YY (2020) Impact of the COVID-19 pandemic on agricultural exports. J Integr Agric 19(12):2937–2945
Loxton M, Truskett R, Scarf B, Sindone L, Baldry G, Zhao Y (2020) Consumer behaviour during crises: Preliminary research on how coronavirus has manifested consumer panic buying, herd mentality, changing discretionary spending and the role of the media in influencing behaviour. J Risk Financ Manage 13(8):166
Mamun MA, Ullah I (2020) COVID-19 suicides in Pakistan, dying off not COVID-19 fear but poverty?—the forthcoming economic challenges for a developing country. Brain Behav Immun 87:163–166
McKernan SM, Mills G, Ratcliffe C, Congdon WJ, Pergamit M, Braga B, Martinchek K (2020) Building savings, ownership, and financial well-being. OPRE Rep 30
McKibbin W, Fernando R (2021) The global macroeconomic impacts of COVID-19: seven scenarios. Asian Econ Pap 20(2):1–30
Meyer BD, Sullivan JX (2003) Measuring the well-being of the poor using income and consumption
Mian A, Rao K, Sufi A (2013) Household balance sheets, consumption, and the economic slump. Q J Econ 128(4):1687–1726
Nawaz A, Su X, Barkat MQ, Asghar S, Asad A, Basit F, Raheel Shah SA (2020) Epidemic spread and its management through governance and leadership response influencing the arising challenges around COVID-19 in Pakistan—a lesson learnt for low-income countries with limited resource. Front Public Health 8:573431
O'Donnell m, Buvinić M, Kenny C, Bourgault S, Yang G (2021) Promoting women's economic empowerment in the COVID-19 context. Center for Global Development, Washington, DC
Ozili PK (2021) COVID-19 pandemic and economic crisis: the Nigerian experience and structural causes. J Econ Adm Sci 37(4):401–418
Pasha HA, Jafarey S, Lohano HR (2000) Evaluation of social safety nets in Pakistan. Social Policy and Development Centre Karachi
Roll S, Chun Y, Kondratjeva O, Despard M, Schwartz-Tayri TM, Grinstein-Weiss M (2022) Household spending patterns and hardships during COVID-19: a comparative study of the US and Israel. J Fam Econ Issues 43(2):261–281
Rothstein J (2021) The lost generation? labor market outcomes for post great recession entrants. J Hum Resour 0920–11206R11201
Samad HA, Zhang F (2018) Electrification and household welfare: evidence from Pakistan. World Bank Policy Res Work Pap (8582)
Shaefer HL, Edin K, Talbert E (2015) Understanding the dynamics of $2-a-day poverty in the United States. RSF: Russell Sage Found J Soc Sci 1(1):120–138
Shahzad MA, Qing P, Rizwan M, Razzaq A, Faisal M (2021) COVID-19 pandemic, determinants of food insecurity, and household mitigation measures: a case study of Punjab, Pakistan. Paper presented at the Healthcare
Siddiqui F, Raghuvanshi A, Anant S, Kumar S (2022) Impact of the COVID-19 on the spending pattern and investment behaviour of retail investors. Ind J Financ Bank 10(1):31–40
Soedarwo VSD, Sulistyowati T, Winarjo W, Bustami MR, Salahudin S, Sihidi IT (2023) Women empowerment during COVID-19: a systematic literature review. WSEAS Trans Environ Dev 19:1117–1131
Tirachini A, Cats O (2020) COVID-19 and public transportation: current assessment, prospects, and research needs. J Public Transp 22(1):1–21
Tsao Y-C, Raj PVRP, Yu V (2019) Product substitution in different weights and brands considering customer segmentation and panic buying behavior. Ind Mark Manage 77:209–220
Wilson JM, Lee J, Fitzgerald HN, Oosterhoff B, Sevi B, Shook NJ (2020) Job insecurity and financial concern during the COVID-19 pandemic are associated with worse mental health. J Occup Environ Med 62(9):686–691

Yousaf M, Zahir S, Riaz M, Hussain SM, Shah K (2020) Statistical analysis of forecasting COVID-19 for upcoming month in Pakistan. Chaos, Solitons Fractals 138:109926
Yuen KF, Wang X, Ma F, Li KX (2020) The psychological causes of panic buying following a health crisis. Int J Environ Res Public Health 17(10):3513
Zippel C, Sherman A (2021) Bolstering family income is essential to helping children emerge successfully from the current crisis. CBPP 25

Sustainable Business Practices: Reducing Carbon Emissions for a Greener Future

Harwinder Kaur and Preet Arora

Abstract In the discussion of global environmental problems, reducing carbon dioxide emissions is a key proposal to curb climate change and maintain the earth's ecological balance. However, this demand is intertwined with the achievement of sustainable development goals and the promotion of a green economy. Carbon dioxide emissions, mostly due to industries and traffic as well as energy production, pose a grave threat to the sustainability of the environment, whose symptoms are climate disturbances, air pollution, and deterioration of ecologies. The accomplishment of such demands much concerted effort toward the consumption of low-carbon alternatives and adoption of sustainable practices across business sectors. The principles of justice, social progress, and environmental protection form the basis for sustainable development to provide a holistic framework integrating environmental considerations into broader development programs. Economic growth should be accomplished in such a way that nobody would have been left behind and economic development would not at all adversely affect future generations. In this concept of the green economy presents a reconcilable means for exploring economic prosperity and environmental sustainability more harmoniously. The use of renewable energy sources, sustainable business practices, and green technologies brings chances to innovate, create employment opportunities, and induce inclusive growth in the green economy. Thus, at this nexus of carbon emissions, sustainable development, and the green economy, an emphasis must lie on coordinated action towards mitigating climate change, in support of the sustainable development of a more sustainable global economy.

Keywords Carbon emission · Sustainable development · Green economy

H. Kaur (✉) · P. Arora (✉)
Chandigarh University, Mohali, Punjab, India
e-mail: harwinderkaurrajput@yahoo.in

P. Arora
e-mail: Preetarora8961@gmail.com

F. A. Malik et al. (eds.), *Aligning ESG Goals with the UNs Sustainable Development Goals Planning*, World Sustainability Series,
https://doi.org/10.1007/978-3-032-01715-4_16

1 Introduction

China's fantastic commercial growth has been the overall reason for its brilliant three-year economic growth. Nevertheless, a financial boom with an industrial orientation would eventually lead to environmental degradation. The International Energy Agency (2008) reports that China is among the biggest emitters of carbon dioxide (CO_2). Between 2004 and 2009, it increased the yearly rise in its carbon dioxide emissions by about 11% (Auffhammer and Carson 2008). China's 11th Five-Year Plan has measurable goals to reduce carbon dioxide emissions to one percent for the first time.

One of the biggest efforts to reduce carbon emissions in the world currently is to reduce carbon emissions by 0.5 billion tons by 2010 (Lin et al. 2008). It's becoming more pressing to know how economic booms affect environmental quality, which will become even more publicized as media attention to environmental issues increases. A number of researchers have empirically modeled the relation between economic growth and environmental quality from emission and income perspectives. Given the results of most of these studies, the results of most of the studies collectively have been estimated to lie within the so-called Environmental Kuznets Curve, or EKC hypothesis. The EKC hypothesis proposes a U-shaped relationship, or an inverted U-shaped correlation, between environmental quality and per capita income and between environmental pollution and per capita profits. Such as Doctor of Economics Anatoly Ivanovich Bazhan, Head of the Economic Research Department of the European Institute of the Russian Academy of Sciences Sergey Anatolyevich Rodinko, Bachelor of Economics, Director of the Ecology and Development Center of the European Institute of the Russian Academy of Sciences, Professor of the Financial University under the Government of the Russian Federation, Kuzmińh Julia Valerievna, Sergej ROGINKO, and VARNAVSKY Vladimir Gavrilovich had set the target for their furious fight against the headwind in the green economy and the cross-border tax on carbon dioxide. The legality of the cross border control of carbon dioxide raises severe issues, according to Varnavski's paper "Cross- border regulation of carbon dioxide in one new instrument of global governance is the European Union.

The new EU system that is being proposed violates international law and threatens the sovereignty of third-country exporting enterprises by monitoring and controlling their greenhouse gas production operations. "He comes to the conclusion that SVAM exhibits every characteristic of protectionism, discrimination and limitations on competition "environment and production sector towards export companies" (Yulia Valerievna Kuzminykh 2024) concludes the following in her work "Impact of a cross-border carbon tax on Russian pulp and paper's economic performance companies": "The cross-border carbon tax has a negative effect on Russia's economic results." Exporters: due to the implementation of the CAR, there was a certain decrease in sales income or a rise in trade costs, this resulted in a decrease of profit margins for Russian exporting enterprises. Consequently, inner systems of obtaining emission trading rights must be implemented in order to upgrade the measures regulating gas emissions and to integrate them into the system of emissions trading (Nath et al.

2013). Therefore, the possible negative effects of the introduction of a cross-border carbon tax cannot be ignored in the "acquisition" of carbon neutrality, To administer this tax in the most efficient manner, it is necessary to have a previous grasp of all the potential benefits and drawbacks.

On the other hand, the good challenges of many nations open possibilities for long-term progress. The road to transition toward a green economic system is too important to skip any stages in it, at least not before the acknowledged existing transition road is fully accomplished (UNEP 2011, Sharif et al. 2023). The Rio +20 Conference is where the idea of "experienced financial system" and, above all, the leading idea of "green growth shaping" first was introduced to the general public (Barbier 2012). That is, it is mainly a marriage between dominating concerns, sustenance of economic boom, and maintenance of the environment (Bina 2013, Kshitij et al. 2022). This way, he pointed out some of the key issues on his path toward a green economy.

The key issues include the emission cuts of carbon, which most countries acknowledged through the Rio +20 Conference (Barbier 2012) and, as such, became the core principle of the Paris Agreement. Sustainable development" is a term that came to be known after "The Brandt land Report" (WCED 1987) exactly explains how to protect today's needs without putting at risk the ability of future generations to satisfy their own needs. Since the green economy aims at green growth, there are areas of overlap between the two pillars of sustainable development: environmental protection n, economic expansion, and protection (Lélé 1991). Most of the countries' childhood transition processes are hindered by the inability to integrate those components with environmental and economic development objectives (Khan and Hou 2021). Economic growth, for its part, is fundamentally attributed to innovation, which is outpacing the growth rate of hydrocarbons (Xuan et al. 2023). The growth in renewable energy usage satisfies neither the cost nor the emissions targets in order for the sustainable developmental vision to be achieved (Sharif et al. 2023). On the other hand, since the new technologies are associated with greater energy needs, then the development of technology can also lead to pollution (Su and Yang 2023).

Furthermore, the economics of emerging markets are driven by the development of financial openness and integration. According to, the result pertains to the dependence of certain industrialized countries' sustainability initiatives on the Kuznetz curve of environmental, ecological footprint, or atmospheric pollution. Put simply, the aforementioned points serve as guiding indications for understanding the connection between green economic development and sustainable development. Nearly every previous study reached this conclusion: The adage "a green economy promotes growth" is frequently used (UNEP 2011; Sharif et al. 2023). However these transitions did not automatically come to be as envisioned for sustainable development (Lorek and Spangenberg 2014). Indeed, the nature of these changes towards sustainability is not well-defined; moreover, it is not at all clear how the momentum of such changes might affect sustainable development given variance in growth rates, income levels, and socio-economic bases so, as earlier empirical studies have shown.

The economy of Vietnam is being closely examined due to its rapid growth and high energy use, which has resulted in an increase in carbon emissions. Hung

(2023). With an emphasis on both economic growth and environmental sustainability, this study looks at how changes affect Vietnam's sustainable development. This research investigates how knowledge could be improved by understanding the between sustainable development and the shift to the green economy, using data from 1996 to 2020 and the autoregressive distributed lag approach. The findings imply that technological innovation drives economic expansion, despite the fact that it tends to impede environmental sustainability.

The UN General Assembly endorsed the SDGs in September 2015. These goals are multiple in number. There are 17 of them, out of 169 total, and they address every aspect of development, encompassing economic, social, and environmental aspects. The Environment Program of the United Nations claims that the green economy actively lessens threats to the environment and ecological shortages while advancing social justice and human well-being.

As far as we know, very few studies address the two main frameworks in Saudi Arabia—sustainable development and the green economy. In order to assess the effectiveness of green economic growth and how it has shown itself in Saudi Arabia's sustainability agenda, this study presents a novel index system approach that is empirically evaluated and compared to previous studies. This manuscript's remaining sections are arranged as follows: The body of research on sustainability, the green economy, and the value of sustainability is reviewed in the second section. The third section presents the study's variables, data sources, and methodology. The study's findings are presented in the fourth section. Concluding the discussion, the final section suggests some implications for policy (Mondal 2023).

2 Review of Literature

The shift towards a green economy has emerged as a necessity in tackling the increasing environmental issues and the world's quest for sustainable development. Researchers from various fields—spanning environmental policy and economics to technology and behavioral science—have enriched the now substantial literature in examining the routes and mechanisms through which this shift is facilitated. Institutional set-ups and governance frameworks feature among the basics raised in the literature. Albanawi (2015) highlights the institutional barriers in Saudi Arabia, including inefficiencies in regulations and minimal public knowledge, that hinder the growth of a dynamic green infrastructure. In light of such institutional constraints, (AlArjan et al. 2021) suggest a tailored policy framework blending the Sustainable Development Goals (SDGs), emphasizing that implementation of global frameworks has to be personalized according to national settings. In a similar context, (Alwakid et al. 2020) highlight the importance of institutional systems and socio-cultural norms in the development of green entrepreneurship, suggesting that formal governance institutions are not enough without linking these to local social and cultural contexts. Around the world, criticisms of the mainstream green growth paradigm have become more visible. Lorek and Spangenberg (2014) and Bina (2013) believe that the mainstream models emphasizing technological efficiency and economic growth are insufficient to deal with ecological constraints and social justice. These authors believe in a paradigm shift towards sustainable consumption and production models that question the growth-driven economic models. This should be more than mere policy adjustments; it demands a reconstruction of societal values, institutional goals, and behavior norms. Financial tools are key facilitators in this transition toward green. Green finance mobilization is a common thread running throughout the literature. Falcone and Sica (2019), in their exploration of Italy's biomass industry, demonstrate the power of financial vehicles to release sustainable potential. Their work showcases how financing specific to the sector, tied to environmental objectives, can facilitate innovation and investment. (Taghizadeh-Hesary and Yoshino 2020) further build on this argument by highlighting the need for public incentives in drawing private sector investment in green technologies, especially in developing nations. Building on these views, (D'Orazio and Popoyan 2019) examine the role of macro prudential regulation in managing climate-related risks to the financial system, placing central banks and financial regulators in a central position in enabling sustainable financial systems. Empirical evidence validates these conceptual models. Kaewsaeng-On and Mehmood (2023), applying quantile regression methodology, establish the beneficial impact of green finance on sustainability performance in newly industrialized economies. Their research confirms the more general hypothesis that environmental policy-lending can produce tangible development benefits. Mondal (2023) also substantiates this connection, illustrating the role played by green entrepreneurship facilitated by focused finance policies in contributing to sustainable economic innovation. Another key research topic is energy policy and greenhouse gas emissions control. Destek and Sinha (2020) carry out a comparative study across OECD nations, showing the complex effects of renewable and non-renewable sources on ecological footprints. Their conclusions promote an equilibrium energy policy that maximizes environmental gain while balancing economic expenses. Khan and Hou (2021) examine interlinks between tourism, energy use, and environmental footprint in IEA nations, emphasizing environmental quality as a moderating factor. Such findings illustrate that energy policy cannot be separated from other social and economic activities. Environmental Kuznets Curve (EKC) examined by Sarkodie and Strezov (2019) presents a theoretical framework for the nexus between economic growth and environmental pollution. Their results imply that environmental damage rises with economic growth but falls once a certain threshold of income is achieved, which would imply decoupling growth from environmental degradation. This model has been contradicted by other research, though. Auffhammer and Carson (2008) point to China's governance issues and how decentralized administrative organizations make controlling emissions difficult. This example highlights the critical role of governance design in environmental policy outcomes. Technological innovation and institutional quality are also highlighted in African settings. Obobisa et al. (2022) maintain that better governance systems and the implementation of green technologies significantly minimize carbon emissions, confirming the importance of policy integration. Sharif et al. (2020) utilize a case study from Nordic nations to prove how environmental taxation, the use of green energy, and technological integration—when implemented under an integrated policy approach—lead to dramatic environmental gains.

Technological devices are also transforming the way in which environmental sustainability is tracked and controlled. Xuan et al. (2022) use satellite imagery to boost agricultural sustainability, demonstrating the significance of geospatial technologies in environmental administration. Liu et al. (2018) give information about ecological management based on biology by categorizing bacterial virulence factors, showing how interdisciplinary science is helping achieve environmental goals. Concurrently, bibliometric analysis software is employed to chart research patterns and measure scholarly impact. Aria and Cuccurullo (2017) present Bibliometrix, a science mapping tool, while Donthu et al. (2021) and Roemer and Borchardt (2015) provide frameworks for measurement-based assessment of research impact via metrics and altmetrics, highlighting the value of academic visibility and evidence-driven policy development. Consumer behavior is also central to the green economy shift. Nath et al. (2013) name environmental concern and perceived consumer effectiveness as key motivation for adopting green products, which indicates that consumer engagement strategy needs to include empowering people. Rejikumar (2016) cautions against the negative impact of "green wash" fear, where consumer mistrust destroys confidence in environmental statements, stunting green market development. Lorek and Spangenberg (2014) return to the imperative of a fundamental change in consumption patterns, bringing individual decisions into line with nature's limits—a view supported by Bina (2013), who insists on broader systemic interventions than market incentives. Social systems, such as healthcare and education, indirectly underpin sustainable development. Appiah et al. (2023) identify systemic shortcomings in Ghana's health infrastructure, with sustainability policies contributing to improved public health. In the context of public trust, Noushad et al. (2021) analyze vaccine uptake in Saudi Arabia, with a focus on institutional trust as a key driver in the effective rollout of public health interventions linked to the SDGs. Educational innovation, such as that referenced by Hung and Chen (2023), uncovers both the potential and the ethical issues involved in mainstreaming artificial intelligence within education systems. While not directly tied to green economy models, these instruments promote education and awarenessboth key to policy adoption and behavioral change.

Methodological and comparative research provides essential information on the diversity of green economy principle implementation. Habiba and Cao (2022) analyze differences in CO_2 emissions and financial growth between developed and developing countries, highlighting the requirements for differential policy intervention.

Zakari et al. (2022) evaluate how innovations in energy efficiency contribute to SDG achievement through context-appropriate strategies. Vieira and Gomes (2009) and Lin et al. (2023) offer methodological advice on the measurement of scholarly productivity and international cooperation to build on the transnational learning to support sustainable development. Not withstanding the abundance of studies, there are still gaps. The literature tends to overplay financial and technological fixes, with too little consideration of socio-political institutions, cultural norms, and equity implications in the longer term. There is a strong need for work that combines behavioral science and economics and policy making, considers justice aspects of green transitions, and broadens empirical analysis to lesser-represented regions, particularly in the Global South. In addition, strengthening education and civic engagement mechanisms will be crucial to the development of resilient and inclusive green economies.

3 Materials and Methods

This method measures, tracks, and evaluates the scientific literature using a collection of quantitative techniques (Bochardt and Chin Roemer 2015). First created by Pritchard in 1969, it quickly became well-liked as a tool for quantitative analysis. It places emphasis on the authors' publications, the leading journal, the techniques employed, and the conclusions drawn.

Bibliometric analysis is arguably one of the most well-known though rather laborious methods for examining a large volumes of scientific data. This allows one to focus on the investigation of the evolution of a particular area in depth and defines its limits. Scholars use bibliometric analysis for many reasons, such as studying the potential intellectual framework for a particular field of interest, or investigating structures and development of collaboration, individual research parts and dynamics of scholarly and journal publishing (Donthu et al. 2021). As for this work, the Scopus database will be used, which is one of the most popular in the majority of bibliometric studies and citation databases (Vieira and Gomes 2009). We explain that the expressions green economy* product sought by us are found in the titles of articles indexed in Scopus This applies to English-language articles, journal source types, and document kinds (Appendix A). This gave us with 481 peer-reviewed research articles from 149 journal sources within the timeframe of 1997 to January 2023 at a 16.33 percent compound annual growth rate. Contributing authors numbered 1069 with a 13.57 references average per document. The whole report on the completeness of the database is as follows: "good": affiliation (C1), abstract (AB), reference (CR), DOI (DI), corresponding author (RP), keyword (DE); "excellent": fields; Author (AU), Document Type (DT), Journal (SO), Language (LA), Publication Year (PY), Title (TI), Total Citations (TC) 0.46.91 is the missing keyword Plus (ID) percentage and no referenced references (NR) or scientific categories (WC). As a result, 481 scientific publications make up our final sample. We use a few bibliometric analysis methodologies that have been proposed, which enable us to perform and probe the scientific maps. To understand the demographic creation of green economy study papers, we use a performance analysis that includes the annual output of the research map as well as the overall research literature gathered by institutions, nations, sponsoring agencies, and national cooperation networks. Furthermore, the quantity of articles that are offered to writers in order to identify the most cited article and the top researchers in the field of the green economy, as well the number of papers published in the journal and the number of citations received by the article.

This research aims at finding out the thematic background path of green economic research and recommending potential topics for further study. For this purpose, we have employed bibliometric analysis techniques-cross-referencing, topic maps, and development maps-to gain an understanding of the thematic backgrounds and development in green economy research. We analyze co-word using the keywords authored to provide current and potential recommendations. Research topics: To assess the bibliometrics of the study, we employ the functionalities of the bibliometric customizable web interface package Biblioshiny for R (Aria and Cuccurullo 2017). We submit the subject map, collaboration templates, Tree Map, topic development map, and cross-reference plan to Biblioshiny. We also make use of the VOS viewer that van Eck and Waltman created in 2009 while building the bibliometric link coverage and keyword analysis.

4 Results

The bibliometric analysis is organized into five major categories. They are classified into five categories: journal level, author level, document level, subject level, and overall generation of scientific literature. The research output includes all of the content prepared utilizing criteria based on nations, stakeholders, and the organization that funded the study. Collaboration with numerous stakeholders and countries throughout the world is another component of the analysis. We utilize fractional articles to pick the principal authors of that body of research and rank journals based on the number of scientific publications to determine the main journal. We looked at the number of local citations the research article received to decide which document was the most important in the field of green economics research. We utilize the scientific.

4.1 Scientific Research Production and Collaborations

Figure 1 the bibliometric analysis indicates high scientific production in the year 2023 compared to 2009 presented in Fig. 1 with a high rise in published articles around the year 2021–2022. This indicates an increase in the interest of researchers in green economy studies. China leads the way by country (Fig. 2) with 410 articles, followed by Pakistan (42), Indonesia (25), Saudi Arabia (17) and UK (15). Geographically, China emerges as a central hub of scientific output, with the darkest shade on the map indicating its leading role in research publications. The extensive international collaborations between China and countries across different continents, including the USA, European nations, and other Asian countries, emphasize China's crucial position in the international research network. This widespread collaboration emphasizes the interconnected nature of contemporary scientific research, highlighting the global efforts in advancing knowledge and innovation

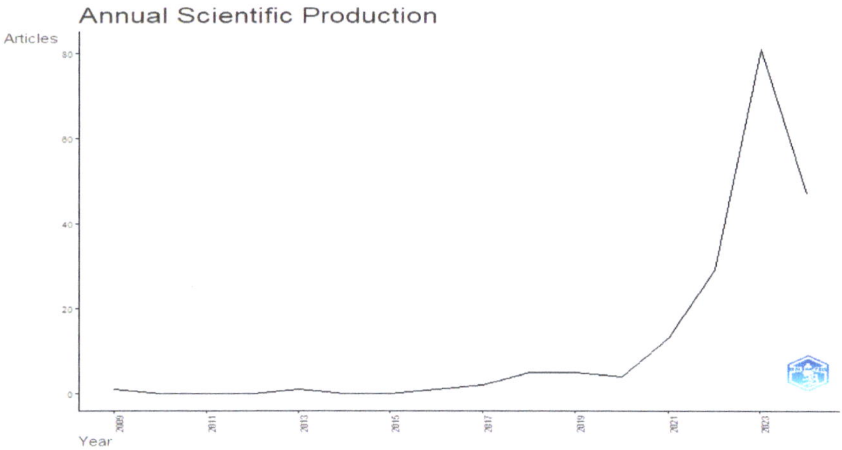

Fig. 1 Publication growth

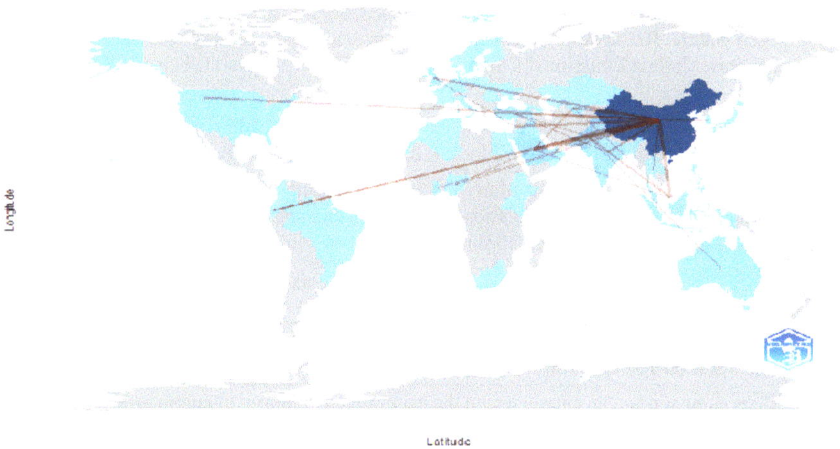

Fig. 2 Country collaboration map

5 Documents

It is important to determine which publications are more important based on citations since this suggests an important research contribution. The ten most significant papers are listed in Table 1 according to the overall number of citations. With 434 citations, the article by Mehmood and Kaewsaeng-on (2024) had the most. This study looks into how energy use, natural resource rent, green investment, and finance development affect carbon emissions (Zhang and others 2020) The study on "impact of technological advancement and industrial structure on the intensity of carbon emissions: Data from 281 Chinese cities," which was the second most referenced paper, was cited 233 times (Razzaq and others 2021). There are 213 citations in the title of the third paper. He is conducting this research on how green innovation affects carbon emissions from consumer and natural sources. A study by Obolisa et al. (2022) uses bibliometric analysis to evaluate the state of research at the moment in the green economy. It generated 176 citations, which naturally vary in topic and sources due to the broad nature of this discipline.

Table 1 Top 10 most cited articles on green technology, carbon emissions, and sustainable investment

No	Article	Total citation
1	Does Natural Resource Rent, Green Investment, and Financial Development Decrease Carbon Emissions? China: A Provincial Panel Analysis (Shen et al. 2021)	434
2	Industrial structure and technological advances' effects on carbon intensity of emissions: Learning from 281 Chinese cities (Zhang et al. 2020)	233
3	Quantile-on-quantile analysis of asymmetric relationships between consumption-based carbon emissions and green technology innovation in the BRICS nations (Razzaq et al. 2021)	213
4	The impact of institutional quality and green technology innovation on CO_2 emissions of African countries (Obobisa et al. 2022)	176
5	The role that many facts of financial growth play in the E7 nations' use of renewable energy: The shift to a sustainable future (Habiba et al. 2022)	158
6	Does using renewable energy lessen the environmental impact of energy use? Evidence from China (Sun et al. 2021)	157
7	Fostering green investments and tackling climate-related financial risks: Which role for macro prudential policies (D'Orazio and Popoyan 2019)	143
8	Do energy policies bring about corporate over investment? Empirical from Chinese listed companies	135
9	Using a statistical dynamic method, the contribution of Australia's energy output to CO_2 emissions and environmental deterioration is evaluated (Sarkodie and Strezov (2018)	119
10	The Clothing and Textile Industry's Circular Economy and Sustainability	116

Source Author's compilation based on citation data retrieved from Google Scholar

Table 2 Most contributing authors

Rank	Authors	A.F	No. of articles
1	Li Z	1.49	4
2	Chen Z	1.41	4
3	Umar M	1.36	5
4	Li J	1.23	5
5	Chen S	1.16	3
6	Gyamfi BA	1.11	4
7	Chen Y	1.7	4
8	Li C	1.03	4
9	Zhang Y	0.65	4
10	Li M	0.7	3

Articles Fractionalized (*AF*)

5.1 Authors

One of the primary objectives of the bibliometric study is to identify writers who have significantly advanced the area. The authors who contributed the most, according to fractionalized counts, are shown in Table 2. However, because fractionalized counting is more reliable than full counting, it is the basis for our author rating. Our results show that Li Z the most contributing author with a fractionalized value of 1.49 followed by Chen Z (1.41), Umar M (1.36), Li J (1.23), Chen S (1.16), Gyamfi BA (1.11), Chen Y (1.7), Li C (1.03), Zhang Y (0.65), and Li M (0.7).

5.2 Thematical Analysis

The main objective of this study is to identify and show the current status of the intellectual landscape of the research field, as well as prospective issues and current research trends. To this end, we employ five different analytic techniques: thematic map (Fig. 4), co-word analysis (Fig. 6), world tree map (Fig. 3), co-citation network (Fig. 7), and theme evolution (Fig. 5). We use the author's keywords as our unit of analysis and conduct a thematic analysis of it. Figure 3: using Biblioshiny to create a word tree map and look for the most relevant keywords for each article The "TreeMap" of words, that came from Author Keywords (DE). The size of the map's box affects the number of repeated keywords, so more terms will show up more frequently. Green finance (311 times), representing 37% of the total occurrence, is the most common keyword. China (48 times, or 6% of the total occurrence), renewable energy (31, 4%), green financing (30, 4%), and sustainable development (26, 3%), complete the top five. Other commonly occurring terms include "green bonds," "green laws," and "green innovation." According to the results, China is leading the way in green financing research, with most of the study concentrating on financing for renewable energy sources and sustainable development. The evolving themes, patterns, and trends that define the research field can be found and recorded using theme maps.

Figure 4 biblioshiny is used for thematic mapping. From their perspective, carbon emission n or energy consumption might be considered to be a niche, while clean energy, green financing policy, or energy transition can be regarded to be a core theme. Among the driving themes are the Sustainable Development Goals, Green Bonds, Climate Change, Carbon Emissions, Green Credit, and Carbon Neutrality. To bring the thematic approach of research studies in integrated area such as green finance research we conduct co-citation network analysis as shown in Fig. 7. Our analysis reveals that three themes have been identified and colour assigned as red and green and blue; it must be emphasized that there are blue clusters wherein red cluster shares one theme which is environmental protection and low carbon emission (Falcone and Sica 2019; Wang and Zhi 2016). The documents in green cluster and blue cluster which have these researchers focus have to some extent wider the scope of research niche the phenomena of green finance. The green cluster includes green finance concerns that are relevant to private participation (Taghizadeh-Hesary and Yoshino 2019), environmental quality concerns (Zhou et al. 2020), and green credit; the blue cluster covers the public spending in the green economic growth initiatives (Zhang et al. 2021), and green finance with productivity and sustainability research.

It has become evident that co-citation analysis is only preliminary and gives a reasonably correct result only for the most frequently cited documents while ruling out documents from. Based only on green finance and sustainability, it appears that our findings marked the beginning of the field of green finance study. But between 2020 and 2022, the field of green finance research saw a significant expansion that allowed for the addition of new areas of study, including industrial structure, environmental regulation, environmental protection, green financing, and policy development.

6 Thematic Evolution Map

6.1 Keywords Analysis

6.2 Co-citation Network

Green bonds are green finance instruments, carbon credit and carbon emissions (Fig. 4, 5 and 6).

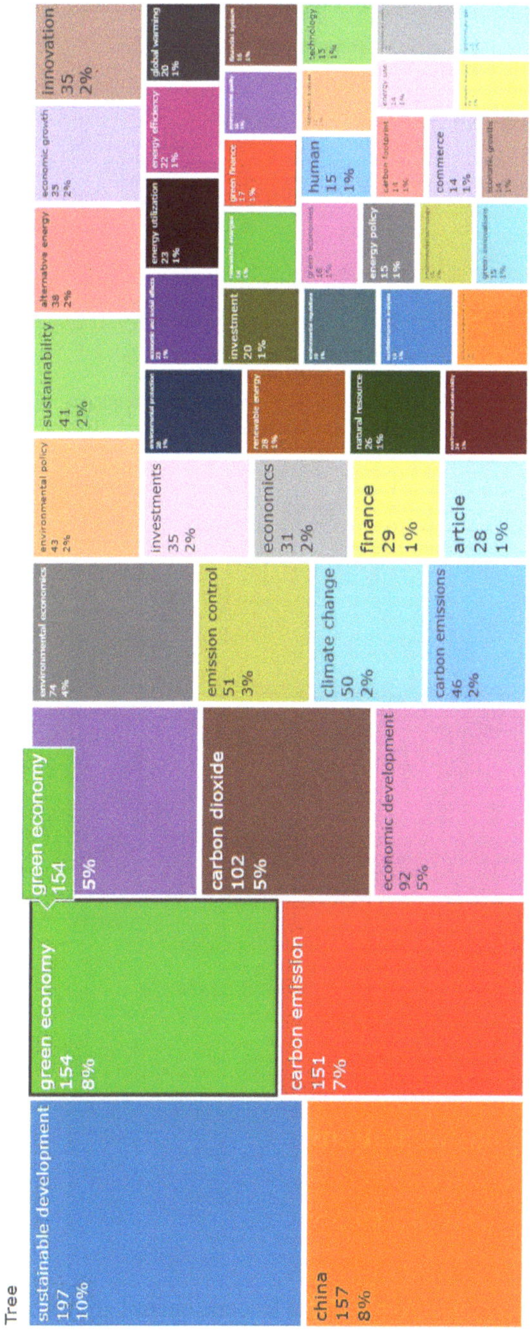

Fig. 3 Word tree map

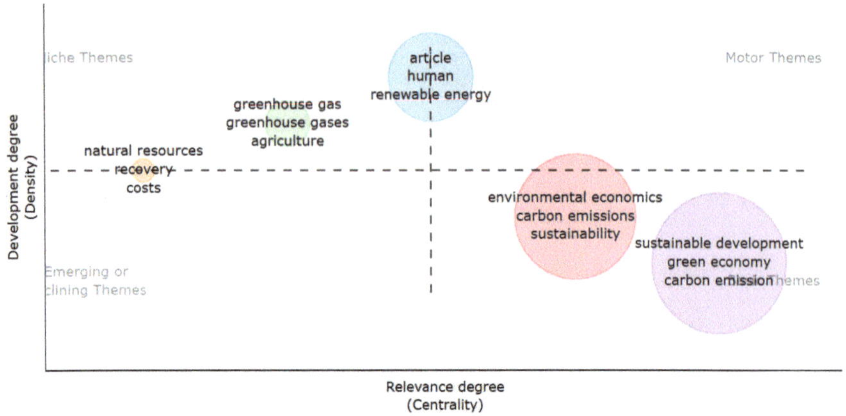

Fig. 4 Thematic map: *Note* Field: Authors' Keyword, Min Cluster Frequency (per thousand docs): 10, Number of Lables: 2, Clustering Algorithm: Walktrap

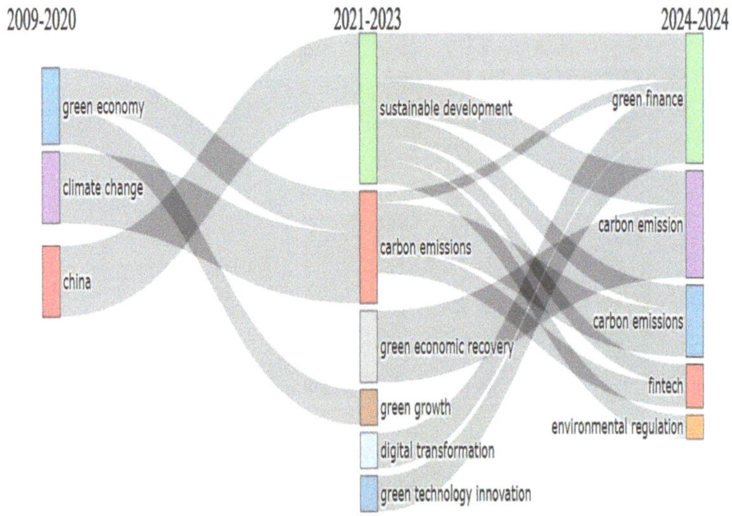

Fig. 5 Thematic evolution map

7 Conclusions and Limitations

7.1 Implementation Challenges

Political Barriers: Resistance from political leaders, interest groups, or stakeholders who are reluctant to enact or support policies that may impact their constituents or vested interests.

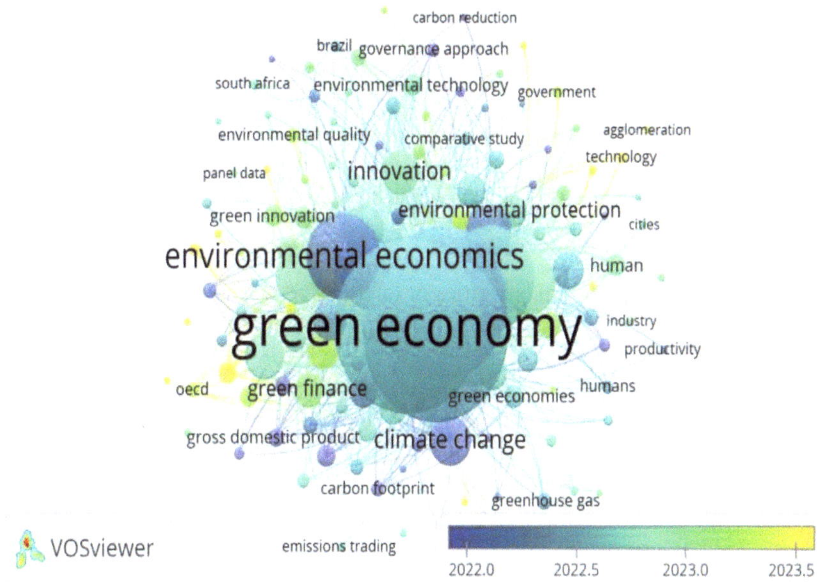

Fig. 6 Co-word analysis

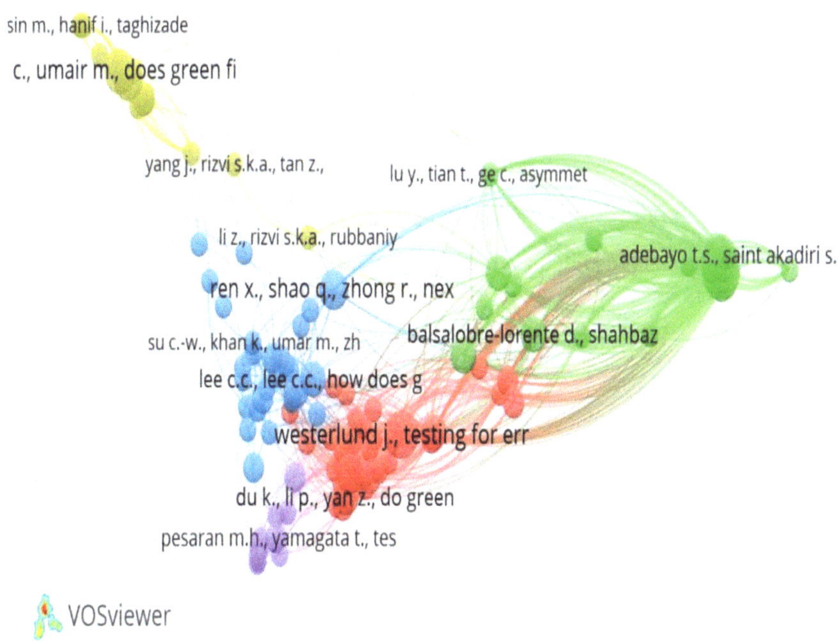

Fig. 7 Co-citation network

Economic Constraints: The high costs associated with transitioning to low-carbon technologies or implementing sustainable practices, which may pose financial burdens on governments, businesses, or individuals.

Technological Limitations: The lack of mature or cost-effective technologies for achieving significant reductions in carbon emissions across all sectors.

Societal Resistance: Public skepticism, lack of awareness, or cultural barriers that hinder acceptance or adoption of sustainable behaviors or policies.

Trade-offs and Conflicting Interests

Sectoral Conflicts: Conflicting priorities or interests between sectors such as energy, agriculture, transportation, and industry, which may hinder the implementation of cohesive and integrated policies.

Trade-offs between the short and long terms: Policies characterized by a trade-off between achieving high short-term economic or political benefits versus longer-term sustainability goals.

Regional Disparities: Disparities in the distribution of costs and benefits of climate policies, particularly between developed and developing regions, exacerbating tensions and hindering global cooperation.

Financial Accessibility: The affordability and accessibility of green technologies, renewable energy sources, and sustainable products/services for low-income or marginalized communities.

Information and Education: Lack of awareness, education, or capacity-building programs to empower communities to adopt sustainable practices or access green solutions.

Institutional Barriers: Structural barriers within institutions or systems that perpetuate inequalities and limit opportunities for marginalized groupsto take part in the process of determining decisions or benefit from sustainable development initiatives.

Technological and Infrastructural Limitations

Resource Constraints: Limited availability of natural resources or materials required for scaling up renewable energy or green technologies.

Infrastructure Challenges: Inadequate infrastructure for supporting the deployment of renewable energy sources, such as energy storage, transmission grids, or charging stations for electric vehicles.

Skills and Workforce Development: Shortages of skilled labor or trained professionals in fields related to renewable energy, sustainable agriculture, or green construction, hindering the transition to a green economy.

7.2 Global Cooperation and Policy Coor

Geopolitical Dynamics: Tensions between nations, competing geopolitical interests, or disagreements over burden-sharing and responsibility for addressing climate change.

Policy Fragmentation: Lack of alignment or coordination among national policies, regulations, or commitments, leading to inefficiencies, inconsistencies, or gaps in addressing global environmental challenges.

Enforcement and Compliance: Weak enforcement mechanisms or lack of accountability measures to ensure adherence to international agreements or commitments related to carbon emissions reduction and sustainable development.

Unforeseen Environmental and Social Impacts

Ecosystem Disruption: Potential negative impacts on ecosystems, biodiversity, or natural habitats due to the expansion of renewable energy infrastructure or land-use changes associated with sustainable development projects.

Social Displacement: Displacement of communities or indigenous peoples from their traditional lands or livelihoods due to large-scale infrastructure projects or resource extraction activities.

Conflict Over Resources: Competition or conflicts over scarce resources, such as water, land, or minerals, exacerbated by climate change impacts or shifts towards sustainable resource management practice.

In conclusion, the complicated interplay between carbon emissions, sustainable development, and the green economy sets the challenge of addressing climate change to promote more inclusive and equitable economic growth through opportunity. Reducing carbon emissions must be done to mitigate adverse impacts on the environment, economy, and society brought about by climate change. However, this must not come at the cost of principles of sustainable development that guides the incorporation of economic, social, and environmental dimensions as pathways to long-term prosperity for both present and future generations.

This means transition towards a green economy: renewable sources of energy, sustainable practices, realization of principles in a circular economy. So much promise to drive innovation, create jobs, and help to achieve resilience in the economy. To be finalized, it poses strong implementation challenges that include overcoming big political, economic, and technological barriers along with access and equity issues over benefits from sustainable development and their equitable distribution among all sectors of society.

These include global cooperation and policy coordination in the cutting of carbon emissions as well as furtherment of goals toward sustainable development. Here, a collective response involving the governments, businesses, civil society organizations, and citizens would work over the geopolitical tensions, trade-offs, and interests that conflict with each other. This also involves institutional frameworks, strengthened through greater transparency, and partnership for resource mobilization and speeding up progress toward a more sustainable, low-carbon future.

Another aspect of achieving a green economy is the imperative to identify and mitigate possible unintended environmental and social effects of the transition to a green economy through appropriate conducting assessments, stakeholder engagement, and adoptable management approaches for reduction and maximum benefit acquisition. This means embracing the principles of sustainability, equity, and resilience to course a better future for everyone-economies that grow without also degrading their natural environment and human well-being placed on the same plane as planetary health (Rejikumar 2016) This nexus of carbon emissions, sustainable development, and the green economy is a multifaceted challenge requiring a holistic and collaborative solution, plugged into science, innovation, and solidarity.

References

AlArjani A, Modibbo UM, Ali I, Sarkar B (2021) A new framework for the sustainable development goals of Saudi Arabia. J King Saud Univ Sci 33(6):101477

Albanawi NI (2015) Saudi Arabian green economy infrastructure: barriers, strategies & opportunity-an analysis. Int J Bus Econ Dev (IJBED) 3(3)

Alwakid W, Aparicio S, Urbano D (2020) Cultural antecedents of green entrepreneurship in Saudi Arabia: an institutional approach. Sustainability 12(9):3673

Appiah EO, Menlah A, Xu J, Susana AA, Agyekum BS, Garti I, Kumah J (2023) Exploring the challenges and roles of nurses in delivering palliative care for cancer patients and co-morbidities in Ghana. BMC Palliat Care 22(1):121

Aria M, Cuccurullo C (2017) Bibliometrix: An R-tool for comprehensive science mapping analysis. J Informet 11(4):959–975

Auffhammer M, Carson RT (2008) Forecasting the path of China's CO_2 emissions using province-level information. J Environ Econ Manag 55(3):229–247

Barbier EB (2012) The green Economy post Rio+20. Science 338(6109):887–888. https://doi.org/10.1126/science.1227360

Bina O (2013) The green economy and sustainable development: an uneasy balance? Eviron Plann C Gov Policy 31(6):1023–1047

Destek MA, Sinha A (2020) Renewable, non-renewable energy consumption, economic growth, trade openness and ecological footprint: evidence from organisation for economic co-operation and development countries. J Clean Prod 242:118537

Donthu N, Kumar S, Mukherjee D, Pandey N, Lim WM (2021) How to conduct a bibliometric analysis: an overview and guidelines. J Bus Res 133:285–296

D'Orazio P, Popoyan L (2019) Fostering green investments and tackling climate-related financial risks: which role for macroprudential policies? Ecol Econ 160:25–37

Eze M (2021) A Comparative Survey Data analysis of declining oil revenue implications on the economic performance of Oil-Exporting countries: Nigeria, Venezuela and Norway. Journal of Resources Development and Management. https://doi.org/10.7176/jrdm/76-03

Falcone PM, Sica E (2019) Assessing the opportunities and challenges of green finance in Italy: an analysis of the biomass production sector. Sustainability 11(2):517

Farjam P, Hekman E, Rouwkema J, Verkerke G (2021) Bone fixation techniques for managing joint disorders and injuries: A review study. J Mech Beh Biomed Mat 126:104982. https://doi.org/10.1016/j.jmbbm.2021.104982

Habiba U, Xinbang C (2022) The impact of financial development on CO_2 emissions: new evidence from developed and emerging countries. Environ Sci Pollut Res 29(21):31453–31466

Hung J, Chen J (2023) The benefits, risks and regulation of using ChatGPT in Chinese academia: a content analysis. Soc Sci 12(7):380

Kaewsaeng-On R, Mehmood S (2023) Quantile modeling for environmental risk: SAARC's journey with green finance, policies, and regulations. J Clean Prod 434:140234. https://doi.org/10.1016/j.jclepro.2023.140234

Khan I, Hou F (2021) The dynamic links among energy consumption, tourism growth, and the ecological footprint: the role of environmental quality in 38 IEA countries. Environ Sci Pollut Res 28(5):5049–5062

Khasanova SS, Saydulgerieva MA (2024) Green economy as a tool for sustainable development and a factor contributing to the formation of a low-carbon economy. In: BIO web of conferences, vol 82. EDP Sciences, p 04003. https://doi.org/10.1051/bioconf/20248204003

Kshitij G, Khanna N, Yıldırım ÇV, Dağlı S, Sarıkaya M (2022) Resource conservation and sustainable development in the metal cutting industry within the framework of the green economy concept: an overview and case study. Sustain Mater Technol 34:e0050

Lélé SM (1991) Sustainable development: A critical review. World Development 19(6):607–621. https://doi.org/10.1016/0305-750x(91)90197-p

Lin FR, Lin SC, Huang TP (2008) Knowledge sharing and creation in a teachers' professional virtual community. Comput Educ 50(3):742–756

Lin Z, Akin H, Rao R, Hie B, Zhu Z, Lu W, Smetanin N, Verkuil R, Kabeli O, Shmueli Y, Costa ADS, Fazel-Zarandi M, Sercu T, Candido S, Rives A (2023) Evolutionary-scale prediction of atomic-level protein structure with a language model. Science 379(6637):1123–1130. https://doi.org/10.1126/science.ade2574

Liu B, Zheng D, Jin Q, Chen L, Yang J (2018) VFDB 2019: a comparative pathogenomic platform with an interactive web interface. Nucleic Acids Research 47(D1):D687–D692. https://doi.org/10.1093/nar/gky1080

Liu L, Iketani S, Guo Y, Chan JF, Wang M, Liu L, Luo Y, Chu H, Huang Y, Nair MS, Yu J, Chik KK, Yuen TT, Yoon C, To KK, Chen H, Yin MT, Sobieszczyk ME, Huang Y, Ho DD (2021) Striking antibody evasion manifested by the Omicron variant of SARS-CoV-2. Nature 602(7898):676–681. https://doi.org/10.1038/s41586-021-04388-0

Liu B, Zheng D, Zhou S, Chen L, Yang J (2022) VFDB 2022: a general classification scheme for bacterial virulence factors. Nucl Acids Res 50(D1):D912–D917

Lorek S, Spangenberg JH (2014) Sustainable consumption within a sustainable economy–beyond green growth and green economies. J Clean Prod 63:33

Marien M (1992) Environmental problems and sustainable futures. Futures 24(8):731–757. https://doi.org/10.1016/0016-3287(92)90104-n

Mehmood S, Kaewsaeng-on R (2024) Charting an economic sustainability path: quantile regression analysis of green finance and financial development in newly industrialized economies. Glob Bus Rev. https://doi.org/10.1177/09721509231224019

Mondal S (2023) The nexus between green entrepreneurship and sustainable development: an econometric study. Glob Bus Rev. https://doi.org/10.1177/09721509231157009

Nath V, Kumar R, Agrawal R, Gautam A, Sharma V (2013) Consumer adoption of green products: modeling the enablers. Glob Bus Rev 14(3):453–470. https://doi.org/10.1177/0972150913496864

Noushad M, Nassani MZ, Koppolu P, Alsalhani AB, Samran A, Alqerban A, Rastam S (2021) Predictors of COVID-19 vaccine intention among the Saudi Arabian population: a cross-sectional survey. Vaccines 9(8):892

Obobisa ES, Chen H, Mensah IA (2022) The impact of green technological innovation and institutional quality on CO_2 emissions in African countries. Technol Forecast Soc Chang 180:121670

Razzaq A, Ajaz T, Li JC, Irfan M, Suksatan W (2021) Investigating the asymmetric linkages between infrastructure development, green innovation, and consumption-based material footprint: novel empirical estimations from highly resource- consuming economies. Resour Policy 74:102302

Rejikumar G (2016) Antecedents of green purchase behaviour: an examination of moderating role of green wash fear. Glob Bus Rev 17(2):332–350. https://doi.org/10.1177/0972150915619812

Roemer RC, Borchardt R (2015) Meaningful metrics: a 21st century librarian's guide to bibliometrics, altmetrics, and research impact. Amer Library Assn

Sarkodie SA, Strezov V (2018) Empirical study of the environmental Kuznets curve and environmental sustainability curve hypothesis for Australia, China, Ghana and USA. J Clean Prod 201:98–110

Sharif A, Kartal MT, Bekun FV, Pata UK, Foon CL, Depren SK (2023) Role of green technology, environmental taxes, and green energy towards sustainable environment: insights from sovereign Nordic countries by CS-ARDL approach. Gondwana Res 117:194–206

Shen M, Zeng Z, Song B, Yi H, Hu T, Zhang Y, Zeng G, Xiao R (2021) Neglected microplastics pollution in global COVID-19: Disposable surgical masks. The Science of the Total Environment 790:148130. https://doi.org/10.1016/j.scitotenv.2021.148130

Shi J, Yu C, Li Y, Wang T (2022) Does green financial policy affect debt-financing cost of heavy-polluting enterprises? An empirical evidence based on Chinese pilot zones for green finance reform and innovations. Technological Forecasting and Social Change 179:121678. https://doi.org/10.1016/j.techfore.2022.121678

Su J, Yang W (2023) Unlocking the power of ChatGPT: a framework for applying generative AI in education. ECNU Review of Education 6(3):355–366. https://doi.org/10.1177/20965311231168423

Sun H, Saeedi P, Karuranga S, Pinkepank M, Ogurtsova K, Duncan BB, Stein C, Basit A, Chan JC, Mbanya JC, Pavkov ME, Ramachandaran A, Wild SH, James S, Herman WH, Zhang P, Bommer C, Kuo S, Boyko EJ, Magliano DJ (2021) IDF Diabetes Atlas: Global, regional and country-level diabetes prevalence estimates for 2021 and projections for 2045. Diabetes Research and Clinical Practice 183:109119. https://doi.org/10.1016/j.diabres.2021.109119

Taghizadeh-Hesary F, Yoshino N (2019) The way to induce private participation in green finance and investment. Financ Res Lett 31:98–103

United Nations Environment Programme. Division of Early Warning, & Assessment. (2011) UNEP year book 2011: emerging issues in our global environment. UNEP/Earthprint

Van De Graaf T, Lesage D (2009) The International Energy Agency after 35 years: Reform needs and institutional adaptability. The Review of International Organizations 4(3):293–317. https://doi.org/10.1007/s11558-009-9063-8

Vieira ES, Gomes JA (2009) A comparison of Scopus and web of science for a typical university. Scientometrics 81:587–600

Xia C, Dong X, Li H, Cao M, Sun D, He S, Chen W (2022a) Cancer statistics in China and United States, 2022: profiles, trends, and determinants. Chin Med J 135(05):584–590

Xia S, Wang L, Zhu Y, Lu L, Jiang S (2022b) Origin, virological features, immune evasion and intervention of SARS-CoV-2 Omicron sublineages. Signal Transduction and Targeted Therapy 7(1). https://doi.org/10.1038/s41392-022-01105-9

Xuan F, Dong Y, Li J, Li X, Su W, Huang X, Zhang Y (2023) Mapping crop type in Northeast China during 2013–2021 using automatic sampling and tile-based image classification. Int J Appl Earth Obs Geoinf 117:103178

Yang Y, Wang ZH, Jin S, Gao D, Liu N, Chen SP, Wang JZ (2016a) Opposite monosynaptic scaling of BLP–vCA1 inputs governs hopefulness-and helplessness-modulated spatial learning and memory. Nat Commun 7(1):11935

Yang B, Wang Y, Qian P (2016b) Sensitivity and correlation of hypervariable regions in 16S rRNA genes in phylogenetic analysis. BMC Bioinformatics 17(1). https://doi.org/10.1186/s12859-016-0992-y

Zakari A, Khan I, Tan D, Alvarado R, Dagar V (2022) Energy efficiency and sustainable development goals (SDGs). Energy 239:122365

Zhang D, Zhang X, Yang Y, Huang X, Jiang J, Li M, Qu J (2021) SARS-CoV-2 spillover into hospital outdoor environments. J Hazard Mater Lett 2:100027

Zhang JJ, Dong X, Cao YY, Yuan YD, Yang YB, Yan YQ, Gao YD (2020) Clinical characteristics of 140 patients infected with SARS-CoV-2 in Wuhan, China. Allergy, 75(7):1730–1741

Zhou S, Wang Y, Zhu T, Xia L (2020) CT features of coronavirus disease 2019 (COVID-19) pneumonia in 62 patients in Wuhan, China. Am J Roentgenol 214(6):1287–1294

Operationalise ESG Practices to Sustain SME Growth: A Malaysian Perspective

Meng Kui Hu and Daisy Mui Hung Kee

Abstract Environmental, social, and governance (ESG) practices have gained significant attention among Malaysia's small and medium enterprises (SMEs). While awareness levels are high, with 80% of SMEs recognising the importance of ESG, the adoption rate remains at 60%, with a stronger emphasis on governance and social aspects rather than environmental initiatives. Of the three ESG pillars, SMEs focus more on governance and social practices than on environmental initiatives. Various stakeholders, specifically government agencies, trade associations, and training providers, have contributed significantly to the remarkable improvement. Despite this, many SMEs need help to effectively integrate ESG practices into their business. Only by doing it correctly can SMEs reap the benefits of ESG implementation. Forging ahead, SMEs must overcome the challenges and operationalise sound ESG practices and strategies into their businesses. In this regard, this chapter aims to identify the primary obstacles hindering ESG adoption and discuss practical strategies to guide SMEs in their ESG journey. More so, this chapter will contribute practical policy recommendations, enabling the regulators and future research directions to drive ESG development for SMEs further.

Keywords Entrepreneurial innovativeness · ESG · Government initiatives · SME · Sustainable growth

1 Introduction

ESG factors have gained significant recognition and drive for adoption among businesses in recent years. Notably, the expectations of businesses, large and small, in embracing ESG are fast gaining in momentum. Although only large companies were

M. K. Hu (✉) · D. M. H. Kee
School of Management, Universiti Sains Malaysia, Penang, Malaysia
e-mail: mengkui.hu@student.usm.my

D. M. H. Kee
e-mail: daisy@usm.my

F. A. Malik et al. (eds.), *Aligning ESG Goals with the UNs Sustainable Development Goals Planning*, World Sustainability Series,
https://doi.org/10.1007/978-3-032-01715-4_17

expected to embrace ESG in the past, nowadays, SMEs are encouraged to integrate ESG factors into their operations. By embracing ESG practices, SMEs can significantly improve their financial health; hence, the need for SMEs to remain updated and compliant (Adenan et al. 2024). Moreover, the green transition that is part of the ESG transformation journey offers significant opportunities for SMEs to enhance efficiency, reduce costs, and unlock new market opportunities. Despite its importance, many SMEs deliberately ignore ESG embracement in their business. They cited various reasons for their actions, the primary ones relate to limited knowledge of environmental regulations, challenges in community engagement, and constraints on resources (Abu Hassan et al. 2023). In other instances, ESG practices remain a low priority to SMEs mainly due to competing demands, limited resources, and the need to focus on other operational issues necessary for business survival. Additionally, many SMEs, especially in rural areas or specialised sectors, face challenges due to inadequate internal resources, a lack of training opportunities, and limited resources that hinder their technical and strategic skills essential for a successful green transition (UNESCAP 2025). Hu and Kee (2022a) emphasise the critical role of global institutions in facilitating ESG integration among SMEs, highlighting how structured support mechanisms, partnerships, and policy alignment can accelerate inclusive development and sustainability. In Malaysia, translating ESG principles into practical, operational strategies for SMEs requires a tailored approach that considers local challenges, policy context, and industry dynamics. This chapter explores how ESG practices can be effectively operationalised to support SME growth in Malaysia.

2 Literature Review

The integration of ESG practices among SMEs in ASEAN is gaining traction, yet significant gaps in readiness and adoption remain. Although approximately 40% of businesses in the region express confidence in implementing sustainability initiatives, nearly 29% remain reluctant or have no plans to embrace ESG frameworks (Confirmis 2025). This disparity highlights the complex interplay between aspiration and operational reality in ESG adoption, especially among SMEs facing structural and financial limitations. Challenges persist while the government supports sustainability through green finance, capacity-building programs, and improved market access. High costs, especially in energy efficiency and waste management, continue to be significant barriers, limiting broader adoption despite growing momentum (Confirmis 2025). Recently, ASEAN has introduced a simplified guide to assist SMEs in supply chains with disclosing ESG practices to stakeholders. The guide is essential because it simplifies the process to enable SMEs across ASEAN to report ESG practices to stakeholders, such as customers, investors, and lenders. This simplified reporting consolidates global and local ESG standards into 38 key disclosures, organised into basic, intermediate, and advanced levels to align with different stages of sustainability readiness (Choy 2025). This initiative aims to encourage more SMEs in ASEAN to embrace ESG practices in their businesses.

In Malaysia, SMEs shared similar predicaments as their peers in other parts of the world. Many perceive ESG as a threat, hence the reluctance to integrate its practices in business operations. By neglecting ESG practices, Malaysian SMEs risk losing RM292 billion (US$68.7 billion) in revenue, according to findings from the Sustainable Finance Institute Asia in 2022. Additionally, SMEs that fail to incorporate ESG practices into their operations may struggle to attract investors, meet customer expectations, qualify for government tax incentives, and secure financing from lending institutions (UNGCNMB 2023). Therefore, SMEs must not remain passive in instituting profound ESG practices to avoid unnecessary business risks. Despite the challenges, studies have demonstrated that implementing ESG practices helps businesses stay competitive, resilient, and responsive to the evolving demands of stakeholders in today's socio-economic landscape. Moreover, SMEs dealing with large companies are encouraged to adopt ESG principles in their business, contributing to a more sustainable and accountable supply chain (Knihová and Jílková 2023).

Considering the rising importance of ESG, SMEs must adopt a more proactive approach in incorporating ESG practices into their operations. The Central Bank of Malaysia emphasises the importance for SMEs to quickly adopt ESG practices into their operations to ensure market competitiveness and sustainability (The Star 2024a). The global enforcement of ESG regulations by advanced nations has impacted the supply chain, hence affecting SME performance. Moreover, the National Blueprint 2021–2025 highlights ESG practices as a crucial enabler for SMEs to remain competitive globally (The Star 2024a). To succeed in their ESG journey, SMEs must develop and implement effective ESG strategies. In this aspect, they need to gain early awareness and proactive planning to adopt ESG practices better, enhancing their overall resilience and performance in the face of emerging sustainability-related risks, expectations, and opportunities. Subsequently, SMEs must work on critical strategies to overcome the primary obstacles hindering their ESG progress and ensure their ESG objectives are attained within the set time frame. It is crucial that SMEs gain support and initiatives from various stakeholders in their ESG journey. Given the background, this chapter discusses the challenges SMEs face in integrating ESG practices for sustainable growth. This chapter also explores practical approaches SMEs can adopt and implement to intensify their ESG journey. Ultimately, this chapter seeks to create a detailed taxonomy of actionable measures that enable SMEs to improve their sustainability and growth through ESG practices. The narratives in this chapter will benefit SME practitioners, policymakers, and academia interested in deepening their knowledge of ESG practices and how the practices can help SMEs to grow their business sustainably.

3 Methodology

The foundation of this chapter is drawn from a discussion of scholarly literature, government policies, and initiatives implemented to promote ESG implementation among SMEs globally, including Malaysia. The discussion is expanded by incorporating a practical perspective from SME practitioners' viewpoints. This is crucial as it provides a solid testimonial on how SMEs implement ESG strategies, enabling them to grow sustainably. Additionally, the methodology includes a compilation of direct feedback from SME leaders on sustainability, offering insights into the benefits and challenges of ESG adoption. Similarly, this chapter discusses best practices for implementing ESG initiatives. Meanwhile, the presentation of this chapter is systematically structured in the following sequence: (1) challenges faced by SMEs, (2) advantages of ESG implementation, and (3) strategy to operationalise ESG practices. Lastly, this chapter also outlines future research directions to further enhance contributions in similar topics. Figure 1 demonstrates the connections between ESG challenges and advantages and how SMEs can operationalise ESG practices to achieve sustainable business growth (SBG).

Fig. 1 Operationalisation of ESG practices for sustainable growth

4 Discussion

4.1 Challenges SME Faced

Although ESG adoption rates among SMEs have progressed over the years, many remain less committed to integrating ESG practices in their business operations. According to Confirmis (2025), about 29% of businesses in ASEAN are hesitant to embrace ESG practices for various reasons. In the Malaysian context, although the level of ESG awareness among SMEs improved significantly over the last two years, from 16% in 2023 to 80% in 2025, the adoption rate of ESG practices grew at a slower rate, from 28 to 60% during the same period (Alliance Bank 2025). Studies have revealed various challenges that SMEs face in embracing ESG practices in their businesses. The broad challenges include limited understanding of the relevance of ESG concepts to their businesses, gaps in the skills needed to incorporate ESG principles in business operations and the reporting procedures of their ESG performance, and obstacles to accessing the tools and funding necessary to start their ESG journey and support their transition to sustainable business practices (UNESCAP 2024). Additionally, other challenges relate to lack of awareness, limited resources, lack of stakeholder support, and ineffective environmental legislation (Journeault et al. 2021). In sum, the following are the primary and common challenges SMEs face in integrating ESG practices into their businesses.

4.1.1 Ambiguous Mindset

Like in executing other business functions, decisiveness in adopting ESG practices is crucial for SMEs. Such leadership is crucial in guiding SMEs to make strategic ESG decisions. Furthermore, dynamic SME leaders can motivate their teams to prioritise sustainable and responsible initiatives. By doing so, leaders who understand and integrate ESG principles in business operations minimise the unfavourable impact on the environment and society, capitalise on sustainable and innovative growth opportunities, and create long-term sustainable profits. However, many SME leaders do not have a clear mindset about the essence of ESG and how its practices can help elevate their business to the next level. According to studies, many SMEs are unaware of the importance of sustainability to their business and the future generation. It is even worse when business owners do not understand what sustainability means. This causes them not to apply sustainability in their business activities, and they often see such practices as irrelevant compared to large companies (Shari et al. 2024). Furthermore, SME leaders are divided on whether ESG adoption enhances their business performance, which is their core business objective. Without certainty, they place less priority on ESG practices. Therefore, many do not prioritise adopting ESG practices (Sheehan et al. 2023).

In Malaysia, the ESG regulatory framework primarily targets publicly listed companies. Additionally, there is a noticeable absence of a reliable ESG model explicitly designed for SMEs to report their ESG data. This gap further deters SME leaders from actively pursuing ESG adoption for their businesses (Adenan et al. 2024). Ironically, many SMEs do not take ESG principles seriously. They believe ESG is only relevant for large corporations and businesses in developed countries (Hu and Kee 2022a, b). In this vein, SME leaders are undecided on implementing ESG practices in their businesses.

4.1.2 Limited Knowledge and Transfer

For SMEs that are receptive to ESG adoption, they have limited understanding of the impacts and benefits of sustainability, along with constraints in time, resources, skills, and expertise. These barriers present substantial obstacles to adopting and integrating sustainable development practices within their business operations (Journeault et al. 2021). Conversely, many SMEs lack knowledge about sustainability issues, which hinders them from adopting sustainability practices. Furthermore, they do not have the expertise to identify and address issues relating to sustainability activities when they want to embark on an ESG journey. The ability to identify the different levels of stakeholders and understand how their ESG strategies align with the overall business interests is crucial to integrate ESG practices effectively (Das et al. 2020). Many SMEs struggle to identify, measure, and report on ESG-related metrics, hampering their ability to effectively implement and communicate their sustainability efforts (Esposito De Falco et al. 2021). This skills gap is further compounded by the fragmented and evolving nature of ESG regulations and standards, making it challenging for SMEs to navigate the complex business sustainability landscape.

Similarly, Malaysian SME leaders face challenges similar to those of their peers in other countries. Many are unaware of how important sustainability is for their businesses and for future generations. The situation worsens when these owners do not fully understand sustainability, leading to confusion and uncertainty about where to start and how to comply with ESG standards. As a result, they are often reluctant to incorporate sustainable practices into their operations, viewing them as irrelevant compared to those of larger companies (Shari et al. 2024). Amid the lack of understanding, SMEs find it complex and challenging to implement ESG practices effectively.

4.1.3 Cost Constraint

As SMEs operate with limited resources, particularly financially, they often prioritise revenue-generating activities. As a result, they tend to place lower importance on adopting ESG practices. This is partly due to the belief that ESG compliance is not mandatory and does not offer direct financial returns. Additionally, the journey toward ESG integration can be complex and overwhelming. The costs associated with planning, monitoring, evaluation, communication, and certification of sustainable practices can significantly impact the profitability of these small businesses, making it a potential obstacle to their growth (Das et al. 2020). On this score, the cost of hiring a consultant to work on ESG adoption, which can be expensive, can discourage SMEs from deciding to embark on the ESG journey.

Financially, most SMEs in Malaysia operate with limited financial resources, making it challenging to allocate funds for sustainability initiatives. The costs associated with adopting new technologies or processes aimed at reducing environmental impact can be prohibitive (Akilah 2024). As a result, SMEs encounter significant financial barriers when trying to implement ESG strategies. Often, they lack access to affordable financing options designed explicitly for sustainability projects (Adenan et al. 2024). Unless ESG adoption is perceived as essential for the business to progress, SMEs will likely avoid implementing it into their business operations.

4.1.4 Regulatory Uncertainty

Another critical challenge hindering SMEs from adopting ESG principles is the lack of clear regulatory frameworks and incentives. While larger corporations may be subject to stricter sustainability requirements, the regulatory landscape for SMEs can be fragmented and inconsistent, making it difficult for these enterprises to navigate the ESG landscape (Mazhorina 2022). Moreover, inconsistent or evolving regulations concerning environmental protection, social responsibility, and corporate governance can confuse SMEs. The absence of clear guidelines and the perception that ESG regulations are constantly changing may discourage SMEs from proactively integrating ESG practices (Mitra and Tam Bui 2024).

Unlike larger companies in Malaysia that can comply with the requirements of the Securities Commission Malaysia and Bursa Malaysia, SMEs are often unable to comply with the regulatory requirements of ESG. Many lack the necessary knowledge of regulations related to ESG practices (Hassan et al. 2023). Furthermore, the uncertainty about potential legal issues and the fear of being accused of greenwashing discourage SMEs from taking their first steps toward ESG adoption. To facilitate their ESG journey, these businesses require support and guidance from the government.

The challenges mentioned above have discouraged and deterred SMEs from exploring the dynamics of ESG and incorporating it into their business operations. If this situation persists among SMEs, it will harm businesses and the nation in the long run. Therefore, SMEs should not let these challenges prevent them from embarking on their ESG journey. They must acknowledge and deepen their understanding of these challenges. They need to find ways to address the issues instead of allowing them to derail their ESG journey.

4.2 Advantages of ESG Adoption

Although SMEs may face challenges in adopting ESG, they can benefit from many advantages if they can successfully implement ESG practices into their business operations. However, not all ESG components exert an equivalent influence on sustainable growth. Specifically, environmental and social performance have a greater effect on sustainable growth, highlighting the unequal impact of different ESG dimensions (Bagh et al. 2024). Ironically, ESG performance promotes sustainable growth by increasing patent applications, lowering agency costs, and reducing corporate risk. Further analysis shows that the environmental score negatively influences ESG, while social and governance scores positively influence it. Ultimately, this research emphasises that strong ESG performance can lead to improved financial performance by effectively driving ESG initiatives (Lin 2024). The following are the primary advantages SMEs can optimise by adopting ESG practices into their business operations.

4.2.1 Enhanced Competitiveness

Sustainability initiatives in businesses, including implementing non-profit components in SMEs through ESG adoption, result in a positive reputation and increased trust among many stakeholders. Additionally, these initiatives create positive images about the company from consumers, communities, and investors, leading to increased profitability and value maximisation (Ikram et al. 2020). This is consistent with the study that found companies implementing ESG strategies often experience a reduction in operational costs by up to 20% due to improved energy efficiency and waste management practices (Finger and Rosenboim 2022). In other words, ESG adoption not only contributes to environmental sustainability but also enhances the financial performance of these enterprises. Additionally, SMEs can adopt innovative approaches to ESG through green technologies and improved corporate governance. This will help meet market demands and boost SME competitiveness (Cahyono et al. 2024). Nevertheless, SMEs prioritising strong ESG practices have a unique opportunity to enhance their global competitiveness. By showcasing their dedication to sustainability and ethical operations, they can elevate their reputation, explore exciting new markets, and attract customers who value responsible business practices (Knihová and Jílková, 2023).

In the Malaysian context, ESG factors play a significant role in mediating the relationship between digital organisational culture and financial performance (Akhtar et al. 2025). Conversely, SMEs that incorporate digital tools and platforms to implement and monitor their ESG activities can enhance their financial performance. Specifically, SMEs can use cloud-based platforms to centralise sustainability data across various dimensions, including energy consumption, water usage, waste management, and emissions monitoring. Ironically, improved financial performance can strengthen the resources of SMEs, allowing them to expand their businesses and invest in crucial areas like research and development, marketing, technology, and talent development. As a result, these activities can boost the competitiveness of SMEs.

4.2.2 Regulatory Compliance

By adopting ESG principles, SMEs must enhance their efforts to ensure regulatory compliance. Achieving compliance helps SMEs fend off legal penalties, attract investment, and promote sustainable growth. Additionally, having a solid compliance record establishes credibility and trust, which are crucial for business growth and resilience. Compliance also fosters a positive corporate culture by promoting ethical behaviour and accountability. This can lead to higher employee morale and retention, as individuals are more inclined to remain with a company that values integrity and fairness (Bello et al. 2024).

As the international regulations on ESG requirements are becoming tighter, Malaysian businesses, including SMEs, must proactively integrate ESG practices into their operations. By doing so, they stand a better chance of exporting their products to ESG-compliant countries. As a trading nation, Malaysia does not have a large domestic market to support itself; hence, domestic businesses need to rely on foreign trade. In this sense, SMEs can incorporate sustainability criteria into procurement practices and leverage larger corporations and government to secure the necessary incentives, resources, and guidance to enhance their ESG efforts (MGTC 2025). Consequently, SMEs can learn and refine their regulatory compliance as they progress in their ESG journey. This is crucial for SMEs, particularly those directly or indirectly exporting their products to foreign companies that are ESG-compliant.

4.2.3 Access to Capital and Finance

In light of the global support for sustainable development, financial institutions actively promote green finance among businesses, including SMEs. They incorporate ESG standards into their credit assessment processes, enabling them to identify and finance projects that positively impact environmental sustainability and social welfare. Financial institutions understand that sustainable companies are more resilient, better managed, and less likely to violate laws or harm the environment (Hakam and Hakam 2024). Furthermore, SMES practising ESG principles can attract a broader range of investors, especially those prioritising sustainable and responsible investments. In recent years, there has been a notable increase in capital directed toward businesses that comply with ESG standards. Therefore, SMEs that effectively integrate ESG practices can discover new funding opportunities (Knihová and Jílková 2023). Against this background, businesses, including SMEs, are encouraged to align their operations with these principles for better financing terms.

Malaysian banks have supported ESG-compliant companies, allocating substantial financing for sustainable projects like renewable energy. For instance, Bank Islam

Malaysia Bhd reached a major milestone by exceeding its initial green financing target of RM4 billion (US$941 million). By September 2024, the bank achieved RM4.8 billion (US$1.13 billion), well ahead of its 2025 goal (The Star 2024b). Separately, Hong Leong Bank unveiled its inaugural sustainable finance framework, which aims to mobilise RM20 billion (US$4.70 billion) over the next five years to support green projects (The Star 2024c). These examples illustrate the proactiveness of Malaysian banks in terms of sustainable development.

4.2.4 Long-Term Value Creation

The adoption of ESG practices offers both short-term and long-term benefits for SMEs. Many SMEs believe that integrating ESG practices can enhance their goodwill and reputation. This enhancement can lead to increased sales, higher demand, and improved employee satisfaction. Furthermore, most SMEs recognise that embracing sustainable business practices can intensify their competitive edge, further boosting their reputation, hence contributing to increased enterprise value and competitiveness (Iqbal et al. 2025). On the other hand, adopting ESG practices improves enterprise value and competitiveness, supported by ethical leadership and stakeholder trust (Arianpoor et al. 2023). Moreover, the younger generation, particularly Gen Z, is concerned about the health of the planet and often influences others to make sustainability-focused purchasing decisions (Wood 2022). They tend to support sustainable brands. Additionally, partnering with institutional investors can benefit SMEs who aim to differentiate their products based on sustainability criteria. These investors typically bring valuable expertise in sustainability, gained from various transactions over time (Drobetz et al. 2024).

A case study on Vitrox, a Malaysian technology company specialising in the design and development of machine vision and electronics, and with a strong emphasis on work-based learning (Kee 2024; Lim et al. 2024), highlights the company's longstanding commitment to ESG principles. Vitrox has integrated ESG values into its company culture, emphasising integrity, trust, and respect. The company prioritises innovation and cost-effective solutions aimed at reducing global e-waste generated during manufacturing processes. This unique culture, grounded in core values of integrity, trust, and respect, has enabled Vitrox to create long-term value, allowing it to remain competitive in both the local semiconductor industry and the global market. The success of Vitrox stresses the importance of incorporating ESG principles into business operations, with a commitment to sustainability evident from top management down to every employee (Tan et al. 2023).

4.3 Operationalise ESG Practices

Given the numerous advantages of ESG adoption, SMEs should optimise their opportunities by taking the necessary steps. They must allocate sufficient resources to create a roadmap for ESG implementation. Rising above challenges, SMEs need to unlock the opportunities of ESG practices to grow sustainably. Despite its importance, implementing ESG strategies can be challenging for most SMEs. Nevertheless, SMEs can implement sound strategies to overcome the primary challenges while optimising available incentives to integrate ESG practices. Research shows that SMEs that excel in ESG practices often take a proactive approach in their outreach efforts. They engage with their internal operations and suppliers, clients, business partners, and regulators. In essence, these SMEs operationalise ESG practices by influencing the larger business ecosystem in which they function, rather than simply managing their internal processes (Stek 2024). By doing so, SMEs can benefit from ESG integration more effectively. Furthermore, government support, along with various agencies and financial institutions, should play a crucial role in educating SMEs about sustainability issues. For instance, a financial institution could require that all SMEs applying for financing participate in a webinar on ESG topics, provided by a relevant agency, before their financing is approved (Shari et al. 2024). Against this background, SMEs must not wait but commence their ESG journey. The following are the critical measures SMEs can take to operationalise ESG into their businesses.

4.3.1 Adopt Dynamic Leadership

As businesses respond to the increasing demand for ESG roles, leaders should focus on fostering a positive culture. This can be achieved by encouraging collaboration, providing educational opportunities, and empowering sustainability champions within the organisation. Additionally, leaders must ensure that their corporate values genuinely reflect a commitment to ESG principles. In this context, SME leaders must lead by example by following best practices and taking proactive steps to protect the environment. By embracing the 3R concept—reduce, reuse, and recycle, SMEs can enhance their resource efficiency and make significant strides in supporting environmental sustainability in their business operations (Hu and Kee 2021). Furthermore, effective leadership in ESG requires a firm grasp of an organisation's external environment and reevaluating existing business models. It involves generating new ideas and forming partnerships to implement them successfully. While ESG management typically focuses on short-term objectives and maintaining control, ESG leadership emphasises long-term vision, coalition-building, and trust. Although managing ESG may appear safer and more prudent, authentic leadership is genuinely necessary (Stek 2024). Effective leaders must have strong planning, organising, and problem-solving skills while fostering relationships and managing team dynamics. Leadership often requires adaptation to change and challenges, while driving innovation and growth in dynamic and uncertain environments (Cahyono et al. 2024).

The positive influence of leadership on ESG development and organisational effectiveness can be attributed to leaders' crucial role in shaping strategies and policies that affect ESG performance in various ways. From an environmental perspective, visionary leaders can initiate projects to reduce environmental impacts, promote sustainable resource use, and guide businesses towards investments in green technologies. From a social standpoint, responsible leaders cultivate a corporate culture that values diversity, supports the community, and prioritises social development (Cheungsirakulvit et al. 2024). Against this background, SME leaders must realign their mindset about ESG adoption and implement proactive leadership in driving their business towards ESG compliance. They need to take charge to ensure success throughout the entire ESG journey.

4.3.2 Instil Innovativeness

SMEs that embrace ESG principles often engage in innovative activities utilising green technologies. These initiatives focus on developing innovative solutions that satisfy market needs and support global sustainability objectives through creative ESG strategies. As a result, SMEs enhance their innovativeness and competitiveness in the process. More so, by focusing on ESG-based innovation, SMEs enhance their competitiveness while providing more sustainable and socially responsible solutions (Cahyono et al. 2024).

In light of the growing importance of innovation in developing solutions that contribute to a sustainable future, Malaysian businesses are encouraged to contribute to a harmonious coexistence between people, planet, and profit (Ibrahim 2025). Besides the government, the private sector, like banks, is doing its part to promote ESG-based innovativeness to promote and accelerate ESG adoption among businesses, including SMEs. For instance, OCBC Bank (Malaysia) Bhd organises the ESG Positive Impact Awards to recognise and celebrate leaders and companies who drive impactful change. The awards are presented to top businesses based on their innovative and impactful ESG campaigns (The Star 2025). This initiative will spur businesses to strive for ESG excellence in elevating their operations, contributing to industry progress, and building stronger communities. Conversely, SMEs must drive innovativeness in their business operations.

4.3.3 Optimise SME Initiatives

As most SMEs are resource-constrained, they can leverage incentives extended by the government to implement their sustainability programmes, such as renewable energy, sustainable supply chains, and innovative technology, to enhance their ESG initiatives (Odayappan 2024). That being said, government support, both financial and non-financial, is essential for promoting sustainability among SMEs (Chong and Kaliappen 2025). Additionally, support from government agencies, financial institutions, and other related organisations is crucial for educating SMEs about sustainability issues. For instance, financial institutions could require SME leaders to gain practical knowledge and skills in ESG criteria before they can apply for financing.

When these leaders understand the importance and benefits of ESG for their businesses, their heightened awareness and interest may encourage them to incorporate ESG principles into their operations (Shari et al. 2024). In ASEAN, various initiatives are implemented to support SMEs in overcoming challenges related to adopting and reporting ESG practices. These initiatives focus on raising awareness and understanding, building skills, providing tools, and facilitating access to finance (UNESCAP 2024). SMEs can significantly benefit from the practical resources and tools offered by ASEAN. Moreover, the United Nations Global Compact Network also provides a platform to support SMEs in their ESG journey. Conversely, SMEs can gain inspiration from case studies featuring successful companies on their ESG journeys. The platform also provides comprehensive guidance on planning, developing, and implementing ESG initiatives. With this, SMEs can learn effective ways to communicate their commitment to ESG practices to others (UNGCMYB 2025).

In the Malaysian context, the government has increased access to financing for SMEs significantly, raising fund allocations from RM9.1 billion (US$2.14 billion) in 2019 to RM31.1 billion (US$7.32 billion) in 2021. Additionally, the Central Bank of Malaysia has introduced two funding facilities as part of the 2022 budget: the Business Recapitalisation Facility and the Low Carbon Transition Facility. Each facility is worth RM1 billion (US$235 million) and aims to assist SMEs in strengthening their capital structure and promoting green financing. In addition to financial support, the government has launched entrepreneurship development programs that cover various aspects, including research and development, training and capacity building, infrastructure improvements, technology transfer, market access, support for social entrepreneurs, and internationalization initiatives. Specifically, SME Corp Malaysia has introduced the ESG Quick Guide to help SMEs grasp the significance of sustainability and ESG practices. This quick guide provides clear steps to assist SMEs in understanding the importance of ESG, setting objectives, identifying relevant ESG indicators, developing a sustainability action plan, and using a straightforward template for ESG reporting (SME Corp 2024).

Ironically, numerous initiatives are offered to SMEs, enabling them to embrace ESG practices. Therefore, SMEs must optimise the targeted initiatives that various government agencies offer to initiate and implement their ESG journey.

4.3.4 Simplify ESG Reporting

ESG reporting is essential for the sustainability of SMEs as it enhances business performance through transparent risk management and the identification of available opportunities. Additionally, ESG reporting provides stakeholders a comprehensive view of a company's performance beyond just financial results. To facilitate ESG reporting and make it more accessible for SMEs, ASEAN has recently launched a simplified guide. This guide aims to assist SMEs throughout the region in effectively disclosing their ESG practices to stakeholders, including customers, investors, and lenders. The document brings together various global ESG reporting frameworks, including the IFRS Sustainability Disclosure Standards and the Global Reporting

Initiative, as well as local guidelines from the ten ASEAN member states. This results in a set of 38 priority disclosures that SMEs can consider tracking and reporting (Choy 2025).

In Malaysia, the Securities Commission Malaysia launched its Simplified ESG Disclosure Guide (SEDG) for SMEs within supply chains. This guide provides practical, structured advice on the disclosures expected of SMEs regarding ESG matters in their supply chains. Moreover, this guide is designed to align with leading global frameworks and reporting standards while referencing local reporting requirements. It includes the disclosures that companies most commonly need to effectively communicate their ESG status. In October 2023, Malaysia became the first country in the world to introduce the Sustainable Enterprise Development Guidelines (SEDG), which provide SMEs in global supply chains with a streamlined and standardised set of guidelines for ESG disclosures. The SEDG features 35 priority disclosures that align with both local and global sustainability guidelines. These disclosures are categorized into basic, intermediate, and advanced levels to accommodate the varying sustainability maturity levels of different SMEs (CMM 2023). This guide is helpful for SMEs as a key reference in their pursuit of ESG implementation.

5 Policy Implications

Taking a cue from the fundamental challenges faced by SMEs in driving ESG adoption, which include the lack of dynamic leadership, it is imperative for policymakers to develop relevant policies to address this primary issue. Development programmes targeting the enhancement of sustainability knowledge and skills among SME leaders can change their mindset on ESG adoption. By participating in these programmes, SME leaders will be more receptive to ESG adoption, hence improving the adoption rate in the long term. Separately, as most SMEs are resource-constrained, policymakers should also develop targeted policies and support programs to bridge the gap in providing functional incentives to SMEs. These initiatives must be able to help SMEs improve their market competitiveness and contribute to their strategic growth (Akhtar et al. 2025). Subsequently, the policy frameworks must be practical and simple for SMEs to understand and apply in driving sustainable growth, hence promoting positive change in the overall SME development (Hakim and Fuad 2024). Simplicity in securing the right incentives is critical in driving sustainable growth among SMEs. In the Malaysian experience, SMEs are often confused by the large number of agencies that hand out various government incentives for various purposes. With over 80 agencies managing more than 240 programmes for SMEs in Malaysia (Barrock 2023), SMEs find it challenging to apply and secure the appropriate support from the government. Ultimately, the creation and implementation of effective and practical policies and incentives can enhance ESG adoption among SMEs, allowing them to achieve sustainable growth.

6 Conclusion

The integration of ESG principles is no longer exclusive to large organisations; it is equally vital for SMEs striving for long-term sustainability. For Malaysian SMEs, operationalising ESG practices offers numerous strategic advantages, from enhancing reputation to increasing organizational value. Yet, this transformation is not without challenges. Limited resources, lack of awareness, and capability gaps often hinder ESG readiness. However, with structured planning, strategic resource allocation, and disciplined execution, SMEs can embed ESG into their core operations to drive meaningful and SBG. As highlighted throughout this chapter, ESG is not a peripheral concern but a strategic imperative in the post-pandemic business context. SME leaders must reframe their mindset, deepen their understanding of ESG principles, and recognise the direct link between ESG disclosures and stakeholder trust. Transparent ESG reporting builds credibility, secures investor confidence, and unlocks new growth opportunities. Therefore, optimising ESG practices is not just a compliance requirement but a catalyst for sustainable business development.

The role of policymakers is pivotal in this journey. To influence behavioural change among SME leaders, policymakers must introduce supportive, practical, and accessible ESG frameworks. Policies should be clearly communicated, easy to implement, and tailored to the unique realities of SMEs. Government initiatives and incentives can act as enablers, facilitating a smoother transition towards ESG-aligned operations. Looking ahead, there is significant potential for further research. Future studies can examine the impact of various internal and external factors on ESG implementation and SME growth. For example, the role of leadership in driving sustainability, or the moderating effect of entrepreneurial innovativeness on the leadership-growth relationship, offers fertile ground for empirical inquiry. These findings will be instrumental for SMEs seeking to develop targeted, effective ESG strategies. In conclusion, ESG adoption is no longer optional, it is a strategic necessity. For Malaysian SMEs, operationalising ESG practices is the pathway to SBG, improved competitiveness, and long-term resilience in the global economy.

References

Abu Hassan MH, Shari W, Abd Wahab N, Asmaa A, Ezanee M, Abdul Wahab NM (2023) Towards sustainable small and medium enterprises (SMEs): awareness and overcoming challenges. Asia-Pac Manage Account J 18(3):1–35. https://www.researchgate.net/publication/377565757

Adenan FA, Ariff A, Adnan SM (2024) ESG in small and medium-sized enterprises: Malaysian perspectives. E3S Web Conf 574. https://doi.org/10.1051/e3sconf/202457403001

Akilah F (2024) ESG implementation in business: Managing the sustainability goals in micro, small, and medium enterprises (MSMEs). SUKUK: Int J Bank 3(I):55–67. https://sukukjournal.org.uk/

Akhtar F, Senadjki A, Vija Kumaran V (2025) Sustainability meets digital culture: the influence of ESG on financial performance in Malaysian manufacturing SMEs. J Innov Digit Transf 2(1):90–108

Alliance Bank (2025) The path to sustainable impact: sectoral insights of Malaysian SMEs. https://www.alliancebank.com.
Arianpoor A, Salehi M, Daroudi F (2023) Nonfinancial sustainability reporting, management legitimate authority and enterprise value. Soc Responsib J 19(10):1900–1916
Bagh T, Zhou B, Alawi SM, Azam RI (2024) ESG resilience: exploring the non-linear effects of ESG performance on firms' sustainable growth. Res Int Bus Financ 70. https://doi.org/10.1016/j.ribaf.2024.102305
Barrock J (2023) Rethink how loans are provided to SMEs, says former MIDF MD Charon. The Edge Malays. https://theedgemalaysia.com/node/674211
Bello HO, Idemudia C, Iyelolu TV (2024) Navigating financial compliance in small and medium-sized enterprises (SMEs): overcoming challenges and implementing effective solutions. World J Adv Res Rev 23(1):042–055
Cahyono P, Ramadhani AD, Rochmawati I, Febriyanty P, Khoirul Mas'ud A (2024) Leadership and ESG-based innovation (environmental, social, governance): implications for SME performance in the start-up environment. J Contemp Admin Manage 2(3):591–599. https://doi.org/10.61100/adman.v2i3.202
Cheungsirakulvit T, Pranee S, Pungnirund B, Techarattanased N (2024) A structural equation model of leadership, technology acceptance model, organisational policies, and ESG score improvement affecting the performance of listed companies in the Thailand Stock Exchange. Int J Dev Adm Res 7(1)
Chong SC, Kaliappen N (2025) Antecedents and consequences for sustainability in Malaysian small and medium-sized enterprises (SMEs). Soc Responsib J. https://doi.org/10.1108/SRJ-01-2024-0009
Choy NY (2025) Asean capital markets forum launches simplified ESG guide to help SMES in supply chains. Edge Malays 1–3. https://theedgemalaysia.com/node/751208
CMM (2023) Simplified ESG disclosure guide. https://sedg.capitalmarketsmalaysia.com/
Confirmis (2025) Southeast Asia SMEs business outlook. www.confirmis.com
Das M, Rangarajan K, Dutta G (2020) Corporate sustainability in SMEs: an Asian perspective. J Asia Bus Stud 14(1):109–138. https://doi.org/10.1108/JABS-10-2017-0176
Drobetz W, El Ghoul S, Guedhami O, Hackmann JP, Momtaz PP (2024) Entrepreneurial finance and sustainability: do institutional investors impact the ESG performance of SMEs? J Bus Ventur Insights 22:1. https://doi.org/10.1016/j.jbvi.2024.e00498
Esposito De Falco S, Scandurra G, Thomas A (2021) How stakeholders affect the pursuit of the environmental, social, and governance. Evidence from innovative small and medium enterprises. Corp Soc Respons Environ Manage 28(5):1528–1539
Finger M, Rosenboim M (2022) Going ESG: the economic value of adopting an ESG policy. Sustainability 14(21):13917
Hakam DF, Hakam LI (2024) Sustainability in small and medium-sized enterprises (SME) financing. Dev Sustain Econ Financ 2–4:100031. https://doi.org/10.1016/j.dsef.2024.100031
Hakim MT, Fuad F (2024) Impact of ESG regulations on sustainable business practices: evidence from Indonesia SMEs. Diponegoro J Account 13(4):1–6
Hassan MHA, Shari W, Wahab NA, Ezanee AAM, Wahab NMA (2023) Towards Sustainable Small and Medium Enterprises (SMEs): awareness and overcoming challenges. Asia-Pac Manage Account J 18(3)
Hu MK, Kee DMH (2021) Advancing SME sustainability: rising above the atrocities of crisis. In Stephens S (ed) Cases on small business economics and development during economic crisis, 1st edn, pp 159–178. IGI Global. https://doi.org/10.4018/978-1-7998-7657-1.ch008
Hu MK, Kee DMH (2022) Global institutions and ESG integration to accelerate SMEs development and sustainability. In: Baporikar N (ed) Handbook of research on global institutional roles for inclusive development. IGI Global, pp 139–156. https://doi.org/10.4018/978-1-6684-2448-3.ch008
Hu MK, Kee DMH (2022b) Recapturing business momentum in accelerating internationalisation for SMEs in the new normal. In Nowinski W, Rymarczyk J, Starzyk K (eds) Covid 19 implications

for the economy. The WSB University in Poznan Press, pp 53–75. https://doi.org/10.58683/9788372053749.03
Ibrahim J (2025) OCBC sponsors ESG positive impact awards. Star Malays 1–5. https://www.thestar.com.my/news/nation/2025/01/15/ocbc-sponsors-esg-positive-impact-awards
Ikram M, Sroufe R, Mohsin M, Solangi YA, Shah SZA, Shahzad F (2020) Does CSR influence firm performance? a longitudinal study of SME sectors of Pakistan. J Glob Responsib 11(1):27–53. https://doi.org/10.1108/JGR-12-2018-0088
Iqbal UP, Nooney LK, Al Ghafri FSS, Daniel TM (2025) Sustainable business practices in SMEs: a retrospective insight on catalysts and hurdles. Cogent Bus Manage 12(1). https://doi.org/10.1080/23311975.2025.2456114
Journeault M, Perron A, Vallières L (2021) The collaborative roles of stakeholders in supporting the adoption of sustainability in SMEs. J Environ Manage 287. https://doi.org/10.1016/j.jenvman.2021.112349
Kee DMH (2024) Enhancing industry-academia collaboration for innovation in the era of Industry 4.0. In: Joshi N, Malik F, Gulati C, Dubey A (eds) Fostering industry-academia partnerships for innovation-driven trade. IGI Global, pp 1–16. https://doi.org/10.4018/979-8-3693-3096-8.ch001
Knihová L, Jílková P (2023) From obstacles to opportunities: enabligh ESG adoption in SMEs. Financne Trhy 1:1–17. https://www.researchgate.net/publication/374368368
Lim SH, Koay LK, Teoh KB, Kee DM (2024) Bridging theory and practice: implementing work-based learning in Malaysian higher learning institutions. In: Joshi N, Malik F, Gulati C, Dubey A (eds) Fostering industry-academia partnerships for innovation-driven trade. IGI Global Scientific Publishing, pp 147–161. https://doi.org/10.4018/979-8-3693-3096-8.ch011
Lin Z (2024) Does ESG performance indicate corporate economic sustainability? evidence based on the sustainable growth rate. Borsa Istanbul Rev 24(3):485–493. https://doi.org/10.1016/j.bir.2024.02.010
Mazhorina MV (2022) ESG principles in international business and sustainable contracts. Actual Prob Russ Law 16(12):185–198
MGTC (2025) Encouraging sustainable adoption among SMEs. Malaysian Green Technology and Climate Change Corporation. https://www.mgtc.gov.my/2025/01/encouraging-sustainable-adoption-among-smes/
Mitra J, Tam Bui H (2024) ESG and SMEs: the state of play
Odayappan M (2024) ESG and the next gen: what's mindful, what works. Star Malays 1. https://www.thestar.com.my/esg/2024/12/06/esg-and-the-next-gen-whats-mindful-what-works
Shari W, Hassan MHA, Ezanee AAM, Wahab NMA, Wahab NA (2024) Ways to promote sustainable practices among Malaysian SMEs from the viewpoints of experts. J Asian Sci Res 14(4):597–611. https://doi.org/10.55493/5003.v14i4.5179
Sheehan NT, Vaidyanathan G, Fox KA, Klassen M (2023) Making the invisible, visible: overcoming barriers to ESG performance with an ESG mindset. Bus Horiz 66(2):265–276. https://doi.org/10.1016/j.bushor.2022.07.003
SME Corp (2024) SME Corp unveils ESG quick guide for MSEMSS. SME Corporation Malaysia. https://smecorp.gov.my/index.php/en/resources/2015-12-21-10-55-22/news/4924-sme-corp-unveils-esg-quick-guide-for-msmes
Stek PE (2024) Are you managing or leading ESG? lessons from ASEAN. Star Malays 1–4. https://www.thestar.com.my/esg/2024/12/06/are-you-managing-or-leading-esg-lessons-from-asean
Tan LP, Choe KL, Ng YK, Lai KF, Khor HC (2023) The Environmental, Social, and Governance (ESG) in Malaysia SME: a case study of Vitrox. In: Proceedings of the 11th international conference on business, accounting, finance and economics (BAFE 2023), vol 272. Springer Nature, p 344
The Star (2024a) Bank Negara: SMEs need to integrate ESG practices into business operations swiftly. Star Malays 1–2. https://www.thestar.com.my/business/business-news/2024/11/26/bank-negara-smes-need-to-swiftly-integrate-esg-practices-into-business-operati

The Star (2024b) Bank Islam surpasses RM4bil green financing target, well ahead of 2025 goal. Star Malays 1. https://www.thestar.com.my/business/business-news/2024/12/26/bank-islam-surpasses-rm4bil-green-financing-target-well-ahead-of-2025-goal
The Star (2024c) HLB launches sustainable finance framework, commits RM20bil to support green projects. Star Malays 1. https://www.thestar.com.my/business/business-news/2024/10/29/hlb-launches-sustainable-finance-framework-commits-rm20bil-to-support-green-
The Star (2025) Inquirer ESG edge impact awards 2025 are now open for entries. Star Malays 1–4. https://www.thestar.com.my/news/nation/2025/02/21/inquirer-esg-edge-impact-awards-2025-now-open-for-entries
UNESCAP (2024) Advancing the green transition of SMEs. https://www.unescap.org/sites/default/d8files/event-documents/%5BEnglish%5DAdvancing%20the%20green%20transition%20of%20SMEs%20-%20Executive%20Summary.pdf
UNESCAP (2025) Unlocking opportunities for the green transition of SMEs. https://www.unescap.org/blog/unlocking-opportunities-green-transition-smes
UNGCMYB (2025) SME ESG hub. UN Global Compact Network Malaysia & Brunei. https://www.esghub.my/
UNGCNMB (2023) Only two options for Malaysian SMEs, comply with ESG practices or lose business opportunities. UN Global Compact Network Malaysia & Brunei. https://www.ungcmyb.org/post/only-two-options-for-malaysian-smes-comply-with-esg-practices-or-lose-business-opportunities
Wood J (2022) Gen Z cares about sustainability more than anyone else—and is starting to make others feel the same. World Economic Forum. https://www.weforum.org/stories/2022/03/generation-z-sustainability-lifestyle-buying-decisions/

Private and Public–Private Energy Investments: Catalysts for Sustainable Development in Asia

Muhammad Ramzan Sheikh, Sana Sultan, and Fareed Shareef

Abstract Private and public–private energy investments play a crucial role in driving sustainable development by enhancing energy infrastructure, promoting innovation, and supporting economic, environmental, and social stability. These investments are key to achieving long-term sustainability goals, particularly in emerging economies. This study explores the impact of private and public–private energy investments on the achievement of sustainable development goals (SDGs) in Asia. Using panel data from 45 Asian countries over the period 1990–2024, the study focuses on three dimensions of SDGs: environmental sustainability (measured by the CO_2 emissions), economic sustainability (measured by GDP growth), and social sustainability (measured by the Human Capital Index). These dimensions serve as dependent variables, with three distinct models employed to analyze the relationship. The study uses the Method of Moment Quantile Regression (MMQR) approach to estimate results, with further robustness checks performed through the Bootstrap Quantile Regression (BS-QR) regressions. The findings suggest that investment in energy significantly boosts all three pillars of sustainabilityenvironmental, economic, and socialthus contributing to the achievement of SDGs in the region. Policymakers should incentivize private and public–private energy investments through favorable regulatory frameworks, tax incentives, and public–private partnerships to accelerate progress toward sustainable development goals.

Keywords Environmental sustainability · Social sustainability · Economic sustainability · Energy investment

M. R. Sheikh (✉)
School of Economics, Bahauddin Zakariya University, Multan, Pakistan
e-mail: ramzansheikh@bzu.edu.pk

S. Sultan
COMSATS University Islamabad, Vehari Campus, Vehari, Pakistan

F. Shareef
Director Public Instructions (Colleges), South Punjab, Pakistan

F. A. Malik et al. (eds.), *Aligning ESG Goals with the UNs Sustainable Development Goals Planning*, World Sustainability Series,
https://doi.org/10.1007/978-3-032-01715-4_18

1 Introduction

As Asian continues to experience rapid economic and population growth, the demand for energy has soared. The region is expected to account for over 60% for global energy demand by 2040. Meeting this demand sustainably is crucial for achieving long-term development goals. The United Nations Sustainable Development Goal 7 (SDG7) aims to ensure access to affordable, reliable, sustainable and modern energy for all by 2030 (Komarulzaman et al. 2023; Ning et al. 2023). To achieve this, substantial investment in energy infrastructure is essential. According to international estimates, Asia must mobilize billions annually in energy investment to address growing consumption needs while minimizing environmental degradation and ensuring inclusive social outcomes (Komarulzaman et al. 2023; Caglar et al. 2022).

Private and public–private energy investments play a vital role in bridging the financing gap and accelerating the energy transition in Asia. While public financing often addresses basic infrastructure needs and energy access in underserved regions, private capital and PPP models bring in innovation, operational efficiency and scalability (Ahmed et al. 2022; Muganyi et al. 2021). These partnerships are particularly valuable in enabling large-scale renewable energy projects, grid modernization and clean technology adoption elements necessary for long-term environmental sustainability (McCauley and Pettigrew 2023; Lup et al. 2023).

Energy investments are not just about infrastructure; they are deeply linked to the three pillars of sustainable development: environmental protection, economic growth and social inclusion. Well-targeted private and PPP investments in energy can stimulate job creation, reduce poverty through improved access to energy and lower emissions through the adoption of cleaner energy technologies (Vercoe 2023). However, imbalances in investment flows and inadequate regulatory framework may hinder the full realization and these benefits across the regions (Banga 2019).

Despite significant efforts, many Asian economies continue to face investment gaps that threaten to derail progress towards SDG7 (Hoa et al. 2024). The disparity between high-performing economies and less developed regions highlights the need for coordinated efforts that heavily rely on fossil fuels, while other lack the institutional and technical capacity to attract and manage sustainable energy investments effectively (Vakulchuk et al. 2023). The transition toward sustainable energy systems requires enabling environments supported by effective policies, regulatory frameworks and access to green financing. Financial innovations such as green bonds, blended finance and risk mitigation instruments are increasingly being adopted to attract both domestic and international capital for clean energy initiatives (McCauley and Pettigrew 2023). PPPs offer a promising route to share risks, reduce public expenditure burdens and expand access to under-financed sectors (Komarulzaman et al. 2023).

This study is important because it looks at how energy investment affects not just the economy and environment, but also social sustainability, which most past studies have ignored. It focuses on Asian countries, where energy needs and social challenges are growing. By using advanced methods, the study gives more accurate results that can help policymakers make better energy decisions that support people, the planet, and the economy.

This study is structured as: Sect. 2 presents the literature review of the existing studies on energy investment and sustainable development goals. Section 3 provides the research questions and objectives. Section 4 is about model specification, data and methodology. Section 5 presents the results and provide discussion. Lastly, Sect. 6 concludes the study and provide recommendations.

2 Literature Review

This section reviews previous studies that have examined the impact of energy investment on environmental, economic and social sustainability. Table 1 summarizes studies examining the impact of energy investment on sustainable development goals. These studies cover various countries, time frames and methodologies.

While some findings indicate that energy investment contributes to higher CO_2 emissions, others suggest it enhances environmental quality by reducing the CO_2 level. However, most of the studies show a negative impact on environmental sustainability as energy investment leads to an increase in CO_2 emissions. The findings also consistently indicate that investing in energy promotes economic growth in both developed and developing countries. After reviewing the existing literature, this study identifies several key research gaps. To the best of our knowledge, this is one of the first studies to comprehensively analyze the impact of energy investment on all three dimensions of the Sustainable Development Goals (economic, social, and environmental). Secondly, there is a noticeable absence of studies that specifically examine this relationship within the context of Asian economies. To the best of our knowledge, no prior study has explicitly investigated the impact of energy investment on social sustainability. Thirdly, prior research has largely relied on the first-generation econometric techniques, which fail to adequately address issues such as slope heterogeneity and cross-sectional dependence critical concerns in panel data analysis involving diverse countries.

3 Research Questions and Objectives

The general objective of the study is to analyze the impact of private and public–private investment in energy on Sustainable Development Goals (SDGs). Based on the specific objectives, the research questions are given as:

- What is the impact of private and public–private investment in energy on environmental sustainability in Asian countries?

Table 1 Summary of the studies on energy investment and environmental sustainability

Author(s)	Country/ region	Time frame	Methodology	Results
Raghutla et al. (2024)	Six global investment countries	1995–2018	Dynamic and Fully Modified OLS	The study revealed that investing in energy led to a decline in environmental degradation by cutting down CO_2 emissions
Ali et al. (2024)	Pakistan	1992–2019	Linear and Non-linear ARDL	According to the study, energy investments contributed to environmental improvement through a reduction in carbon dioxide levels
Udeagha and Ngepah (2023)	South Africa	1960–2020	ARDL	Findings suggested that increased energy investment mitigated environmental harm by decreasing CO_2 emissions
Wu (2023)	OECD	2001–2019	GMM	The study revealed that higher energy investment was associated with greater economic growth
Zhang (2022)	OECD	2011–2020	GMM	According to the findings, expanding energy investment contributed to economic development
Zahoor et al. (2022)	China	1970–2016	OLS	The results showed that boosting energy investment stimulated economic growth
Akinsola et al. (2022)	Brazil	1983–2017	Dynamic OLS and ARDL	The results showed that energy investment played a key role in reducing environmental degradation by limiting carbon emissions
Cheng et al. (2021)	China	1991–2017	FMOLS, OLS and CCR	The research demonstrated that enhanced energy investment was linked to lower CO_2 emissions and improved environmental outcomes
Shahbaz et al. (2020)	China	1990–2017	Bootstrap ARDL	Evidence from the study indicated that energy investments were effective in curbing environmental damage by reducing greenhouse gas emissions
Ahmad et al. (2020)	30 Chinese Provinces	30 Chinese Provinces 2005–2014	GMM	The research indicated that increased spending on energy enhanced overall economic performance

(continued)

Table 1 (continued)

Author(s)	Country/region	Time frame	Methodology	Results
Waqih et al. (2019)	SAARC Countries	1986–2014	ARDL, FMOLS	The analysis confirmed that directing funds into the energy sector helped protect the environment by reducing CO_2 output
Shahbaz et al. (2018)	France	1955–2016	ARDL	The study found that greater investment in energy infrastructure was associated with decreased environmental degradation through reduced carbon emissions
Ahmad and Zhao (2018)	31 Chinese Provinces	2001–2016	CCEMG	It was found that investing more in the energy sector supported economic expansion
Gakuo (2015)	Kenya	1990–2013	OLS	The analysis demonstrated that rising energy investment led to improved economic outcomes

- How does private and public–private investments in energy influence the economic sustainability in Asian countries?
- Is there any relationship between private and public–private investment in energy and social sustainability in Asian countries?

The specific objectives of the study are as follows:

- To analyze the impact of private and public–private investment in energy on environmental sustainability in Asian countries.
- To examine the influence of private and public–private investment in energy on economic sustainability in Asian countries.
- To probe the relationship between private and public–private investment in energy and social sustainability in Asian countries.

4 Model Specification, Data and Research Methodology

Based on the research objectives, the study comprises three modes. The first model analyzes the impact of private and public–private investment in energy on environmental sustainability (measured by the environmental performance index). In this model, the dependent variable is the CO_2 emissions. The independent variables include GDP growth and the square of GDP growth, which are used as determinants of environmental sustainability based on the Environmental Kuznets Curve (EKC) hypothesis. We examined both private energy investment and public–private partnerships in energy investment, as these are the core variables of interest in this study.

Additionally, based on the Pollution Haven Hypothesis, trade openness is included as a determinant of environmental sustainability. The functional and econometric form of the model is given as:

$$CO_2 = f(GDP, GDP^2, PPIE, PPPIE, TRADE) \tag{1}$$

$$\begin{aligned} CO_{2i,t} = {} & \alpha_0 + \alpha_1 GDP_{i,t} + \alpha_2 GDP^2_{i,t} + \alpha_3 PPIE_{i,t} \\ & + \alpha_4 PPPIE_{i,t} + \alpha_5 TRADE_{i,t} + \varepsilon_{i,t} \end{aligned} \tag{2}$$

The second model examines the effect of private and public–private investment in energy on economic sustainability (using GDP growth as the indicator). This model uses annual GDP growth as a dependent variable. The determinants of economic sustainability include labor force participation rate and gross fixed capital formation based on the production function approach. The model also examines both private energy investment and public–private partnerships in energy investment, as these are the core variables of interest. Additionally, trade openness is included as a determinant of GDP growth, supported by endogenous growth theory. The functional and econometric specification of the model is presented as follows:

$$GDP = f(LFPR, GFCF, PPIE, PPPIE, TRADE) \tag{3}$$

$$\begin{aligned} GDP_{i,t} = {} & \alpha_0 + \alpha_1 LFPR_{i,t} + \alpha_2 GFCF_{i,t} + \alpha_3 PPIE_{i,t} + \alpha_4 PPPIE_{i,t} \\ & + \alpha_5 TRADE_{i,t} + \varepsilon_{i,t} \end{aligned} \tag{4}$$

The third model evaluates the impact of private and public–private investment in energy on social sustainability (measured by the Human Capital Index). In this model, the Human Capital Index is taken as a dependent variable. GDP growth is included as an independent variable based on development theories. The model also examines both private energy investment and public–private partnerships in energy investment, as these are the core variables of interest.

Public spending on education and health is considered independent variables, drawing from endogenous growth theory. Additionally, the Gender Inequality Index is included as a determinant, based on Amartya Sen's capability approach. The functional and econometric structure of the model is as follows:

$$HCI = f(GDP, PPIE, PPPIE, HE, EE, GII) \tag{5}$$

$$\begin{aligned} HCI_{i,t} = {} & \alpha_0 + \alpha_1 GDP_{i,t} + \alpha_2 PPIE_{i,t} + \alpha_3 PPPIE_{i,t} + \alpha_4 HE_{i,t} \\ & + \alpha_5 EE_{i,t} + \alpha_6 GII_{i,t} + \varepsilon_{i,t} \end{aligned} \tag{6}$$

The study utilizes panel data from 45 Asian countries covering the period from 1990 to 2024. To estimate the models, the Method of Moment Quantile Regression

Fig. 1 Flow chart of econometric modeling approach

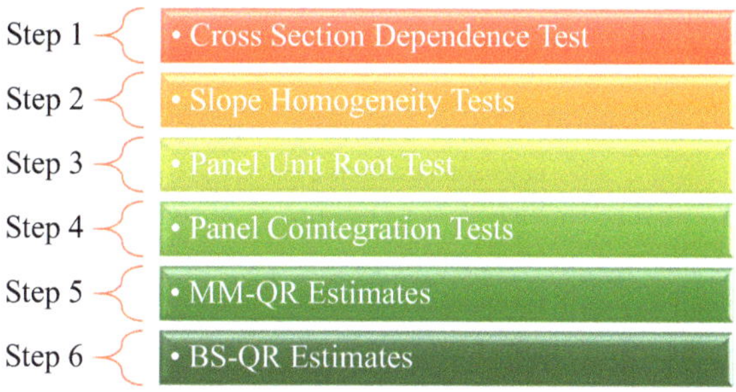

(MM-QR) is employed. Additionally, to ensure the robustness of MM-QR estimates, the study also applies the Bootstrap Quantile Regression (BS-QR) approach. The general equation for Quantile Regression (QR) is:

$$Q_\tau(Y_t|X_t) = X_t\beta_\tau + \varepsilon_t(\tau) \tag{7}$$

where $Q_\tau(Y_t|X_t)$ is the conditional quantile of Y_t given explanatory variables X_t at quantile τ.

β_τ is the quantile regression coefficient vector at τ.
$\varepsilon_t(\tau)$ represents the quantile residuals.

MM-QR extends standard quantile regression by incorporating moment conditions, providing a more robust estimation under certain conditions.

The equation for the Bootstrap Quantile Regression (BB-QR) is the same as that for standard quantile regression:

$$Q_\tau(Y_t|X_t) = X_t\beta_\tau + \varepsilon_t(\tau) \tag{8}$$

However, in the BS-QR, the quantile regression model is estimated multiple times using bootstrap resampling to generate robust coefficient estimates and standard errors. Figure 1 illustrates the econometric modeling approach. Table 2 presents the description of the variables, their measurements and data sources.

5 Results and Discussion

This section presents the results of the study. Table 3 presents the panel descriptive statistics for the study variables. The mean values show that CO_2 emissions average 7.015 metric tons per capita, GDP growth averages 4.434%, and the Human Capital Index (HCI) stands at 0.592. Both private (PPIE) and public–private (PPPIE) energy investments have similar average shares of GDP (0.880% and 0.850%, respectively). Trade openness is high, averaging 88.68% of GDP, while labor force participation rate (LFPR) averages 42.41%. Gross fixed capital formation (GFCF) averages 6.111% of

Table 2 Description of variables, unit of measurement and data sources

Variable (s)	Descriptions	Unit of measurement	Source(s)
CO_2	Carbon dioxide (CO_2) emissions excluding LULUCF	Per capita (t CO_2e/capita)	World Development Indicators, World Bank
GDP	GDP Growth	Annual %	
HCI	Human Capital Index	Scale 0–1	
PPIE	Investment in energy with private participation	% of GDP	
PPPIE	Public–private partnerships investment in energy	% of GDP	
TRADE	Trade openness	% of GDP	
LFPR	Labor force participation rate	For ages 15–24, (modeled ILO estimate)	
GFCF	Gross fixed capital formation	(annual % growth)	
HE	Current health expenditure	(% of GDP)	
EE	Government expenditure on education	Total (% of GDP)	
GII	Gender Inequality Index	Scale 0–1	United Nations Development Programme

GDP, health expenditure (HE) is 4.987%, and education expenditure (EE) is 3.847%. The Gender Inequality Index (GII) has a mean of 0.417.

Substantial variation exists across countries, as indicated by high standard deviations, especially for CO_2, GDP, and GFCF. Skewness and kurtosis values suggest that most variables are non-normally distributed, confirmed by the Jarque–Bera test where all variables (except HCI) have p-values of 0.000, rejecting the null hypothesis of normality at the 1% level. Figure 2 shows the correlation matrix that reveals the relationships among CO_2 emissions, GDP growth, and the Human Capital Index (HCI). CO_2 emissions show a strong positive correlation with HCI, suggesting that higher human capital development is associated with increased carbon emissions, likely due to more industrialization or energy consumption.

However, CO_2 emissions have a weak negative correlation with GDP growth, indicating that economic growth alone does not directly align with rising emissions in the sample. Meanwhile, HCI and GDP growth are very weakly correlated, suggesting that improvements in human capital do not necessarily correspond with short-term changes in economic growth in the observed countries. Table 4 shows the results of the Cross-Sectional Dependence (CSD) test, with all variables displaying significant

Table 3 Panel descriptive statistics

	CO_2	GDP	HCI	PPIE	PPPIE	TRADE	LFPR	GFCF	HE	EE	GII
Mean	7.015	4.434	0.592	0.880	0.850	88.680	42.407	6.111	4.987	3.847	0.417
Maximum	53.598	82.809	0.887	25.440	25.440	437.327	83.317	181.819	23.088	14.199	0.838
Minimum	0.002	−64.047	0.369	0.000	0.000	0.021	17.614	−88.586	1.600	0.000	0.030
Std. Dev	9.352	7.769	0.118	2.020	2.109	56.810	11.980	18.924	2.382	1.569	0.183
Skewness	2.351	−0.100	0.419	7.144	7.143	2.573	0.784	1.492	1.654	1.092	0.166
Kurtosis	9.342	24.362	3.045	70.717	68.200	12.860	3.746	17.467	8.883	6.416	2.384
Jarque–Bera	3973.771	28,637.140	4.167	81,823.980	66,827.100	7128.348	197.840	9355.876	1953.206	632.944	23.742
Probability	0.000	0.000	0.125	0.000	0.000	0.000	0.000	0.000	0.000	0.000	0.000

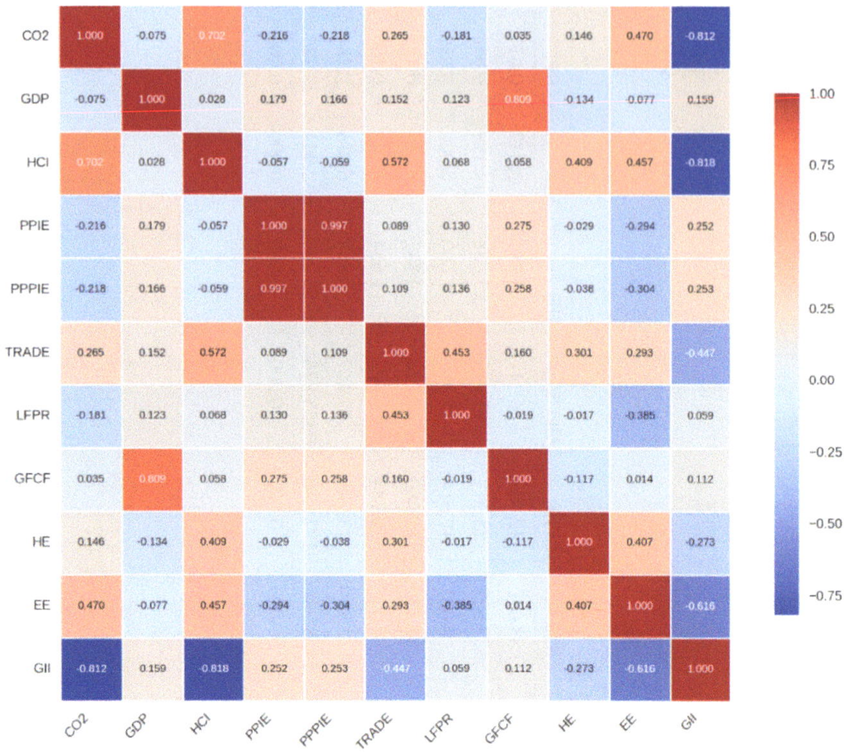

Fig. 2 Correlation analysis

dependence, as their p-values are below 0.05. This indicates that the variables are interrelated across countries, suggesting potential spillover effects. The variable EE is the only exception with a slightly higher p-value but still significant.

Table 5 shows the results of the Slope Homogeneity Test, indicating significant slope heterogeneity across all models. The unadjusted delta test statistics are highly significant, rejecting the null hypothesis of slope homogeneity. The adjusted delta test results also show significant values, but slightly lower than the unadjusted ones.

The results in Table 6 indicate a mix of integration orders across the variables, suggesting that some variables are stationary while others are not, depending on whether a trend is included. For most variables (e.g., CO_2, GDP, PPIE, PPPIE, GFCF, EE), the Zt statistics indicate stationarity when a trend is included, implying they are integrated of order 0 (I(0)) with a trend. However, without a trend, several variables (e.g., GDP, HCI, LFPR, EE) are non-stationary, suggesting they might be integrated of order 1 (I(1)) in the absence of a trend.

Table 7 shows that all three panel cointegration tests (Westerlund, Kao, and Pedroni) indicate significant evidence of cointegration in all three models, suggesting long-term relationships between the variables. The test statistics from each method

Table 4 CIPS Test Results

Variables	CSD	
	t-test	P-value
CO_2	17.417	0.000
GDP	26.935	0.000
HCI	7.790	0.000
PPIE	5.074	0.000
PPPIE	5.190	0.000
TRADE	20.646	0.000
LFPR	69.320	0.000
GFCF	15.349	0.000
HE	32.678	0.000
EE	2.064	0.039
GII	94.823	0.000

Table 5 Slope homogeneity test

Models		Delta test	Hac robust adjusted delta test
1	Adjusted	13.791***	2.148**
	Un-adjusted	13.159***	2.251**
2	Adjusted	10.156***	2.024*
	Un-adjusted	9.659***	2.077*
3	Adjusted	30.765***	1.188*
	Un-adjusted	28.919***	1.176*

*** $p < 0.01$, ** $p < 0.05$, * $p < 0.1$

are consistently significant, supporting the presence of cointegration and confirming that the variables are linked in the long run across the models.

The MM-QR estimates in Table 8 examine the effects of energy investment and economic variables on CO_2 emissions across different quantiles. GDP positively influences CO_2 emission across all quantiles, with the effect strengthening at higher quantiles. This indicates that economic growth contributes more to emissions in Asian countries. GDP^2 has a negative effect, supporting the Environmental Kuznets Curve (EKC), implying that after a certain income threshold, economic growth can help reduce emissions. Figure 3 plots the MM-QR coefficient of energy investment and environmental sustainability. At early stages of economic growth, countries focus on industrialization, increasing production and energy use, often from fossil fuels. This scale effect increases pollution because growth is achieved by expanding economic activity without regard for environmental consequences (Grossman and Krueger 1995; Shahbaz et al. 2015). As income rises, further economies start transitioning

Table 6 Unit root analysis

CSDIPS unit root test				
	Without trend		With trend	
Variables	Lags	Zt statistics	Lags	Zt statistics
CO_2	0	−1.103	0	−1.871*
GDP	0	−16.777***	0	−17.443***
HCI	1	−10.578***	1	11.822
PPIE	1	6.932***	1	7.253
PPPIE	1	6.234***	1	7.443***
TRADE	0	2.723*	0	3.863*
LFPR	1	−3.053***	1	0.133
GFCF	0	−4.876**	0	−3.822*
HE	1	1.386	1	3.098
EE	0	−4.952**	0	−3.221*
GII	1	0.881	1	0.953

*** $p < 0.01$, ** $p < 0.05$, * $p < 0.1$

from heavy industry to cleaner, service-based and high-technology sectors. Investment in energy-efficient technologies and cleaner production methods grows, which reduces emissions despite economic growth (Paramati et al. 2017). With higher income, public demand for a cleaner environment increases. Governments are more likely to implement environmental regulations, carbon emissions as GDP growth beyond a certain point (Zafar et al. 2019; Tenaw and Beyene 2021). This study is in line with the previous empirical findings that support the Environmental Kuznets Curve (EKC) hypothesis, where growth initially leads to increased CO_2 emissions but eventually contributes to environmental improvement as income continues to rise. Notably, studies such as those by Shahbaz et al. (2015), Paramati et al. (2017), Zafar et al. (2019); Tenaw and Beyene (2021).

Private involvement in energy investment greatly raises CO_2 emissions in all quantiles, and the size rises towards the upper quantiles. This implies that energy initiatives headed by the private sector are linked to increasing environmental pressure, particularly in high-emission countries. Though the effect is usually less than that of pure private investment, public–private cooperation in energy investment also has a continuously favourable correlation with CO_2 emissions. This suggests that while PPP-led projects also add to emissions, they could be significantly more ecologically monitored or regulated than private ventures. Due to cheaper costs and quicker return, energy investments in many developing countries are frequently aimed towards traditional energy infrastructure including coal, oil or natural gas. Consequently, more investment increases fossil fuel-based capacity, which increases CO_2 emissions (Shahbaz et al. 2020; Cheng et al. 2021). Often, energy investments seek to increase industrial output and economic expansion. Without robust environmental protections, this results in more energy use and carbon emissions. greater energy

Table 7 Panel cointegration tests

Models	Westerlund	Gt	Ga		Pt	Pa
1	Test	2.666***	3.816***		−2.467***	−3.821***
2		−3.953***	−20.614***		−29.324***	−22.542***
3		−8.894***	−8.408***		−4.653***	−4.211***
		Modified Dickey-Fuller	*Dickey-Fuller*	*Augmented Dickey-Fuller*	*Unadjusted Modified Dickey*	*Unadjusted Dickey-Fuller*
1	Kao test	−2.489***	−3.115***	−3.947***	−3.096***	−3.719***
2		−24.605***	−24.394***	−15.240***	−45.231***	−27.946***
3		3.165***	2.571***	2.535***	3.624***	3.146***
		Modified Phillips-Perron	*Phillips-Perron*		*Augmented Dickey-Fuller*	
1	Pedroni Test	3.162***	−5.626***		4.449***	
2		−17.543***	−23.259***		−25.218***	
3		5.712***	6.532***		5.9999***	

Table 8 MM-QR estimates of energy investment and environmental sustainability

Variables	Location	Scale	Q (0.25)	Q (0.50)	Q (0.75)	Q(0.90)
GDP	0.00271	0.0236	0.0218***	0.0249***	0.0307***	0.0399***
	(0.0556)	(0.0466)	(0.00295)	(0.00394)	(0.00615)	(0.00983)
GDP^2	−0.589***	−0.167	−0.338*	−0.922***	−1.709***	−2.815***
	(0.209)	(0.198)	(0.201)	(0.222)	(0.273)	(0.371)
PPIE	0.0299***	0.0144**	0.0172***	0.0218***	0.0360***	0.0564***
	(0.00576)	(0.00604)	(0.00221)	(0.00295)	(0.00822)	(0.0165)
PPPIE	0.0299***	0.0144**	0.0153***	0.0165***	0.0179***	0.0196***
	(0.00576)	(0.00604)	(0.00234)	(0.00222)	(0.00223)	(0.00243)
TRADE	0.0216***	0.00516	0.0169***	0.0193***	0.0240***	0.0304***
	(0.00434)	(0.00364)	(0.00426)	(0.00396)	(0.00529)	(0.00895)
Constant	1.175***	1.354***	−0.0610	0.591	1.816***	3.497***
	(0.409)	(0.343)	(0.404)	(0.377)	(0.515)	(0.868)

Standard errors in parentheses
*** $p < 0.01$, ** $p < 0.05$, * $p < 0.1$

frequently implies greater manufacturing, but often at the cost of environmental sustainability (Udeagha and Ngepah 2023; Akinsola et al. 2022). Private and PPP energy projects frequently give profit and efficiency first priority; unless they are rigorously controlled or motivated towards renewable energy, they may overlook environmental repercussions. In certain nations, poor environmental governance lets high-emission projects go unregulated (Raghutla et al. 2024; Ali et al. 2024).These results are consistent with the findings of previous studies such as those by Shahbaz et al. (2020), Cheng et al. (2021), Udeagha and Ngepah (2023), Akinsola et al. (2022), Raghutla et al. (2024) and Ali et al. (2024), which also report that increased energy investment, particularly when directed toward fossil fuel infrastructure or lacking green conditionality, tends to raise CO_2 emissions.

Trade openness is positively associated with emissions, especially in higher quantiles, indicating that trade may intensify environmental degradation in already-polluting economies. Trade liberalization boosts economic activity by expanding markets and increasing output. This scale effect leads to greater energy use, especially in energy-intensive and export-oriented sectors, resulting in more CO_2 emissions (Shahbaz et al. 2013). According to Pollution Haven Hypothesis, trade can cause pollution-intensive industries to relocate from developed to developing countries with weaker environmental regulations. Countries that turn pollution havens will have their CO_2 emissions rise as a result of this change (Ertugrul et al. 2016). Often via ships, trucks, and aircraft all of which generate CO_2, trade is the transfer of products across boundaries. Higher trade quantities imply higher transport activity, which greatly contributes to world carbon emissions (Dou et al. 2021). The results of earlier investigations, including those by Shahbaz et al. (2013), Ertugrul et al. (2016); and Dou et al. (2021), help to support our outcome.

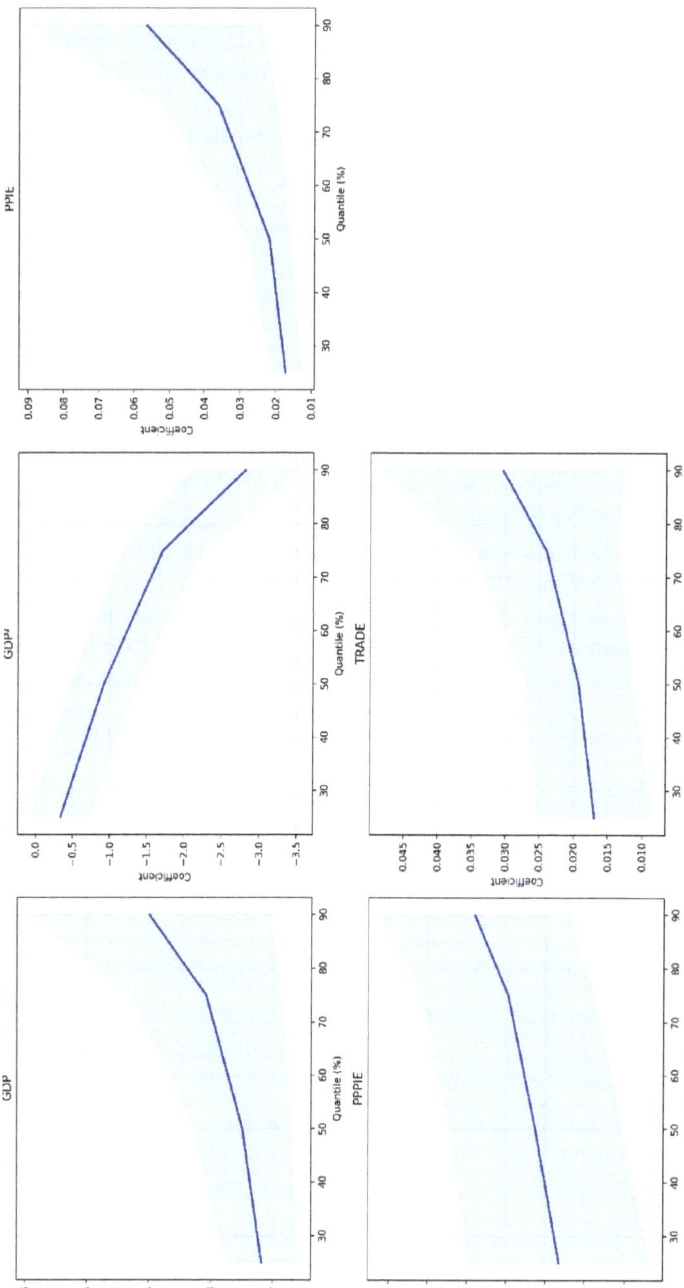

Fig. 3 MM-QR coefficient plot at different quantiles of energy investment and environmental sustainability

Used to confirm the robustness of the MM-QR findings, Table 9 shows the BS-QR projections of environmental sustainability and energy investment. Figure 4 shows the BS-QR coefficient of environmental sustainability and energy investment. The results support the MM-QR conclusions, hence confirming their reliability.

Table 10 shows the MM-QR estimates of energy investment and economic sustainability. Labor force participation rate has a positive and significant effect on GDP growth across all quantiles. This suggests that increasing labor market participation enhances economic growth. Figure 5 plots the MM-QR coefficient of energy investment and economic sustainability. An increase in the labor force participation rate leads to higher economic growth based on the logic of the production function in economics, which states that output is produced using labor and capital. When more people enter the workforce, the total amount of labor available to the economy increases. This directly boosts the economy's capacity to produce goods and services, resulting in higher overall output (Shahid 2014). An increased labor participation makes better use of a country's human resources. Rather than having a large portion of the population inactive or dependent, higher participation ensures that more people are generating income, paying taxes and contributing to economic growth (Amir et al. 2015). As more individuals earn wages, household income rises. This increased income leads to higher consumer spending, which further stimulates economic activity, investment and job creation, creating a positive feedback loop for growth (Cung and Hung 2020) This result is matched with the findings of the previous studies such as Shahid (2014), Amir et al. (2015) and Cung and Hung (2020) which confirm that higher labor force participation contributes positively to economic growth.

Table 9 BS-QR estimates of energy investment and environmental sustainability

Variables	Q (0.25)	Q (0.50)	Q (0.75)	Q(0.90)
GDP	0.00787***	0.0188***	0.0253***	0.0267***
	(0.00104)	(0.00224)	(0.00129)	(0.00126)
GDP^2	−0.00474	−0.00969*	−0.0146***	−0.0135
	(0.00609)	(0.00502)	(0.00469)	(0.0200)
PPIE	0.0253***	0.0219***	0.0125***	0.0779***
	(0.00155)	(0.000989)	(0.00195)	(0.0252)
PPPIE	0.306**	0.140	0.235	1.354***
	(0.122)	(0.0871)	(0.146)	(0.298)
TRADE	0.0107**	0.0249***	0.0300***	0.0173
	(0.00464)	(0.00142)	(0.00575)	(0.0133)
Constant	0.108	0.185	0.553	4.913***
	(0.339)	(0.195)	(0.375)	(1.788)

Standard errors in parentheses
*** $p < 0.01$, ** $p < 0.05$, * $p < 0.1$

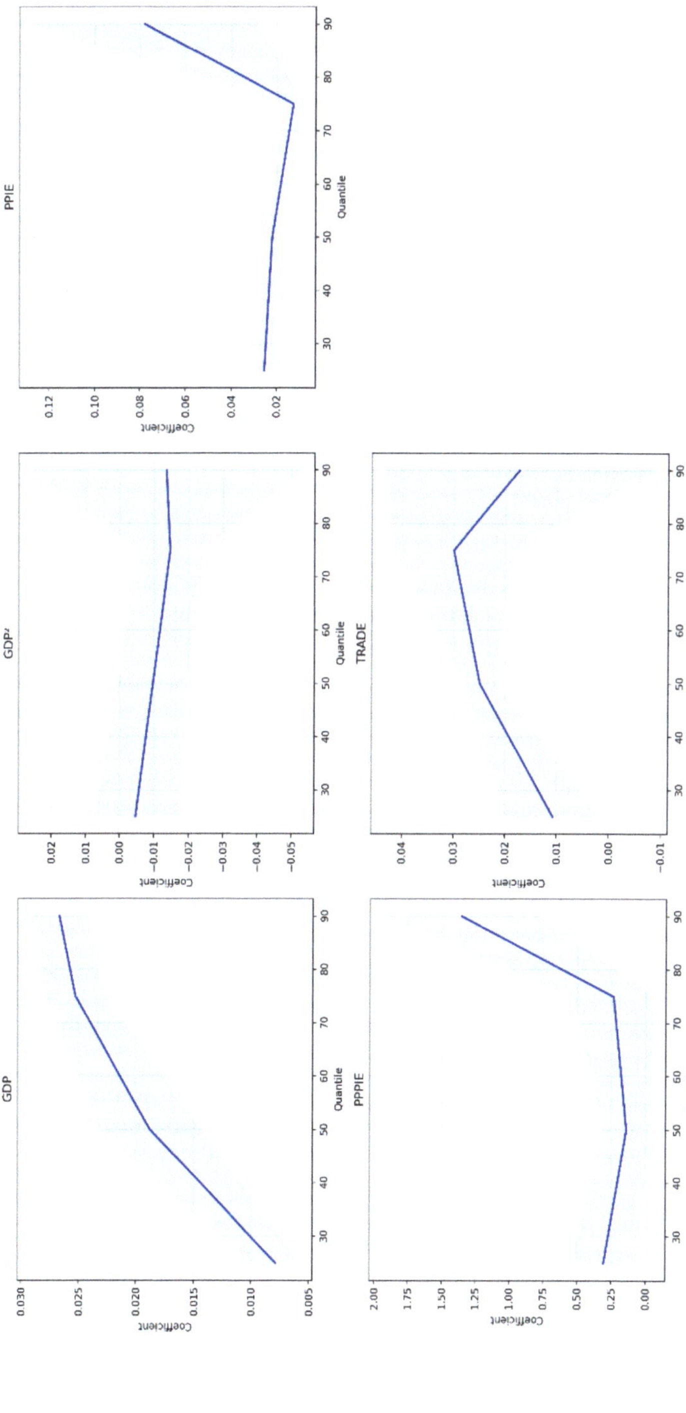

Fig. 4 BS-QR coefficient plot at different quantiles of energy investment and environmental sustainability

Table 10 MM-QR estimates of energy investment and economic sustainability

Variables	Location	Scale	Q (0.25)	Q (0.50)	Q (0.75)	Q(0.90)
LFPR	0.0635***	−0.0162	0.0854***	0.0742***	0.0684***	0.0648***
	(0.0173)	(0.0149)	(0.0297)	(0.0219)	(0.0189)	(0.0176)
GFCF	0.121***	−0.0124	0.128***	0.119***	0.111***	0.103***
	(0.0132)	(0.0117)	(0.0177)	(0.0123)	(0.0101)	(0.0128)
PPIE	0.0745***	0.0637***	0.0616***	0.0583***	0.0547***	0.0507***
	(0.0116)	(0.00963)	(0.0169)	(0.0167)	(0.0171)	(0.0183)
PPPIE	0.0697***	−0.0194	0.0963***	0.0828***	0.0758***	0.0712***
	(0.0158)	(0.0135)	(0.0270)	(0.0200)	(0.0173)	(0.0161)
TRADE	0.000458	0.00600	0.0675***	0.0632***	0.0593***	0.0545***
	(0.00565)	(0.00498)	(0.0155)	(0.0154)	(0.0158)	(0.0169)
Constant	3.290***	2.609***	1.911	3.601***	5.413***	6.982***
	(0.913)	(0.803)	(1.182)	(0.824)	(0.698)	(0.905)

Standard errors in parentheses
*** $p < 0.01$, ** $p < 0.05$, * $p < 0.1$

Gross fixed capital formation consistently and positively contributes to GDP growth across all quantiles, indicating that higher investments in physical assets stimulate economic sustainability. As capital formation increases, the capital stock of the economy grows, which directly enhances the productive capacity. More and better capital allows firms to produce more goods and services, driving economic output (Shuaib and Ndidi 2015). Capital investment usually takes the shape of more sophisticated tools and technology. As workers have access to better equipment and machinery, these investments can help labour to be more productive, therefore boosting the efficiency of production. Higher productivity means higher output from the same level of labour input, hence supporting economic growth (Bal et al. 2016). Usually, more capital creation generates employment possibilities during the building and execution stages of new capital projects. These occupations raise household income, which increases total demand and consumption. This then motivates more economic growth and investment in a self-reinforcing loop (Ajose and Oyedokun 2018). This outcome corresponds to the results of earlier research such Shuaib and Ndidi (2015), Bal et al. (2016) and Ajose and Oyedokun (2018).

Though the impact somewhat decreases at higher quantiles, suggesting that private energy investments are good for economic growth, private collaboration in energy investment greatly increases GDP growth at all quantiles. Investment in energy through public–private partnerships also has a good impact on GDP growth; the biggest impacts are seen in the lower and middle quantiles. This implies that PPP-based energy investments are especially successful in raising low-growth countries. PPP investments in energy result in the creation of more dependable and efficient energy infrastructure. This improved energy supply increases production and helps

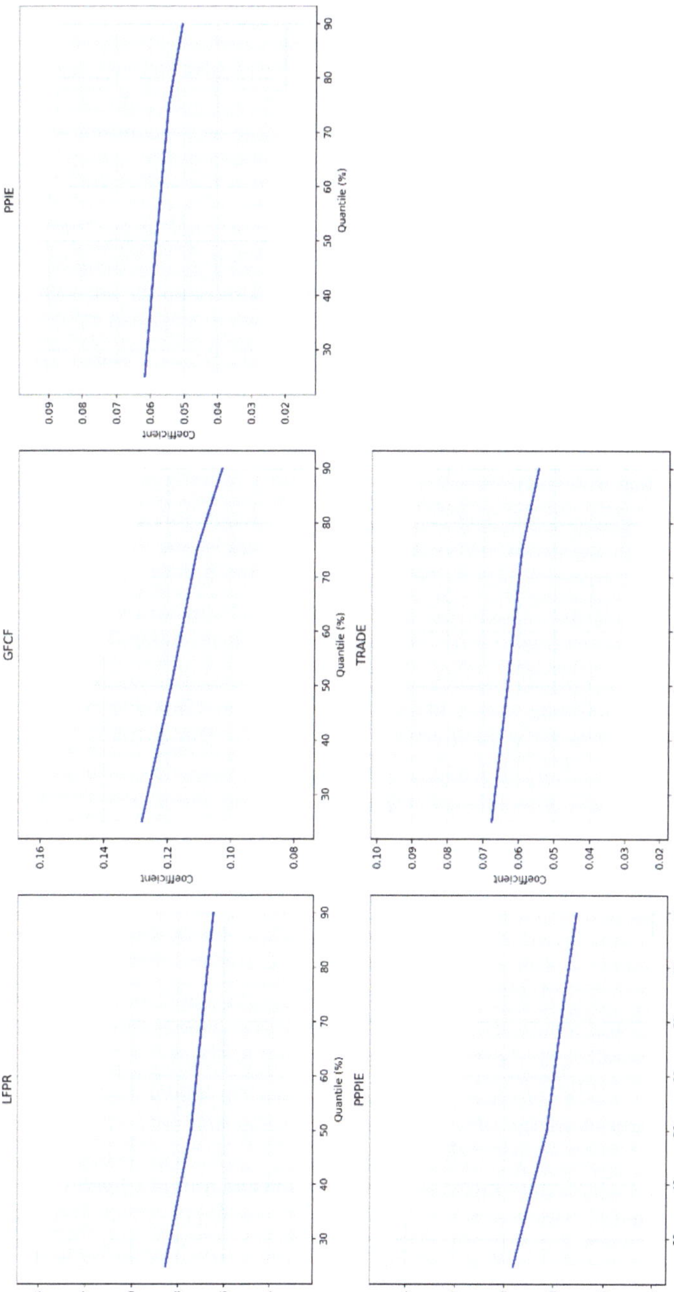

Fig. 5 MM-QR coefficient plot at different quantiles of energy investment and economic sustainability

economic activity by lowering running expenses for companies and sectors. Industries, manufacturing and service sectors all depend on consistent energy availability as they generate GDP growth (Zahoor et al. 2022).

Energy projects, particularly those under PPPs, often involve large-scale construction and operational phases, leading to job creation. Increased employment boosts household incomes, stimulates consumption and generates economic activities through multiplier effects. More jobs and higher income directly contribute to an increase in GDP (Zhang 2022).

PP and PPP investments in energy, especially in renewable energy, contribute to long-term economic sustainability. They help reduce reliance on imported energy, stabilize energy prices and promote energy security. This stability attracts further investment and strengthens economic resilience, fostering sustained GDP growth (Wu 2023). Our findings align with those of Zahoor et al. (2022), Zhang (2022) and Wu (2023), who also demonstrate that investments in energy through private and public–private partnerships contribute to GDP growth.

Trade openness shows a significant positive effect on GDP growth at higher quantiles, indicating that trade contributes more strongly to economic expansion in Asian countries. Trade openness allows countries to access larger international markets, where they can sell goods and services beyond domestic demand constraints. By expanding market opportunities, trade encourages higher output as producers can increase production to meet external demand. This leads to higher overall economic growth (Kleemann and Abdulai 2013). Trade allows countries to focus on producing goods where they have an efficiency edge, boosting productivity (Fitzová and Zídek 2015). Open trade facilitates the exchange of technology and innovation, improving production processes and competitiveness (Keho 2017). This result os consistent with the work of Kleemann and Abdulai (2013), Fitzová and Zídek (2015) and Keho (2017), who argue that trade openness stimulates economic growth.

Table 11 presents the BS-QR estimates of energy investment and economic sustainability, which are employed to verify the robustness of the MM-QR results. Figure 6 plots the BS-QR coefficient of energy investment and economic sustainability. The outcomes are consistent with the MM-QR findings, thereby reinforcing their reliability.

Table 12 presents the MM-QR estimates of energy investment and social sustainability in Asian countries. GDP growth shows a positive and significant impact on HCI across all quantiles, suggesting that higher economic output is associated with better social sustainability through improvements in human capital. Figure 7 plots the MM-QR coefficient of energy investment and social sustainability. As GDP grows, national income increases, providing the government and households with more resources. This enables higher investment in education and training, leading to better better-educated and skilled workforce, which directly contributes to an improvement in the human capital index (Ali et al. 2012). Economic growth often leads to better public spending on healthcare and social services, which enhances overall health and well-being. A healthier population is more productive, contributing to a higher HCI, as human capital is closely linked to physical and mental health (Pelinescu 2015). Economic expansion increases job possibilities and encourages better salaries. This

Table 11 BS-QR estimates of energy investment and economic sustainability

Variables	Q (0.25)	Q (0.50)	Q (0.75)	Q(0.90)
LFPR	0.0688**	0.102***	0.0742***	0.0697***
	(0.0318)	(0.0273)	(0.0179)	(0.0116)
GFCF	0.159***	0.188***	0.190***	0.161***
	(0.0473)	(0.0331)	(0.0305)	(0.0462)
PPIE	0.885**	0.321	0.671*	0.0889
	(0.392)	(0.490)	(0.392)	(0.546)
PPPIE	0.0751***	0.0821***	0.0635***	0.0641***
	(0.0114)	(0.0126)	(0.0141)	(0.0222)
TRADE	0.00800	0.00142	0.00157	0.00655
	(0.00506)	(0.00223)	(0.00186)	(0.00550)
Constant	0.751	2.794**	3.764***	5.543***
	(1.333)	(1.206)	(0.891)	(0.851)

Standard errors in parentheses
*** $p < 0.01$, ** $p < 0.05$, * $p < 0.1$

increases people's capacity to put money into their well-being and education, hence improving human capital. A more competent and capable workforce helps to promote greater development, hence generating a positive feedback loop between GDP and HCI (Sarwar et al. 2021). This finding corresponds with the studies of Ali et al. (2012), Pelinescu (2015) and Sarwar et al. (2021), which show that GDP expansion promotes enhancement in human capital.

In all quantiles, private involvement in energy investment has a negative and considerable impact on HCI. Likewise, public–private partnership investment in energy always demonstrates a strong negative and important impact across all quantiles. Private investors and public–private energy partnerships could give financial profits and infrastructure development first priority above social initiatives like education and healthcare. Consequently, although energy projects could promote economic development, the advantage might not necessarily follow into better human capital as governmental expenditure on social services could be constrained or misdirected. Driven by private sector interests or PPPs, energy projects could emphasise short-term benefits without really considering long-term social or environmental effects. Poorly run energy projects, for instance, might cause environmental damage that could harm public health. A drop in health results in a general fall in well-being and output of the workers, hence harming the human capital.

Health expenditure is positively associated with HCI across most quantiles, reaffirming its role in enhancing human capital. Education expenditure has a significant positive effect on social sustainability in all quantiles. Higher spending on healthcare improves access to medical services, reduces disease prevalence and increases life expectancy. A healthier population is more productive, physically capable and mentally prepared to contribute to economic activities, directly strengthening human

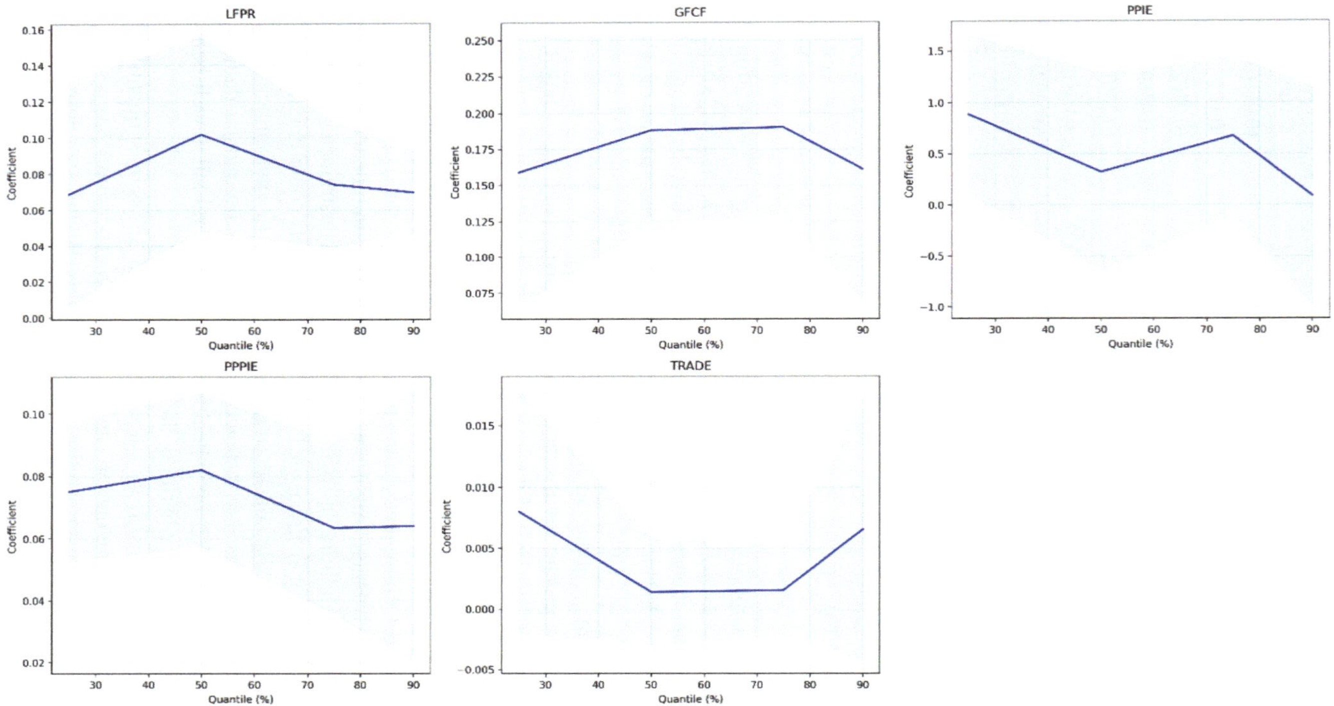

Fig. 6 BS-QR coefficient plot at different quantiles of energy investment and economic sustainability

Table 12 MM-QR estimates of energy investment and social sustainability

Variables	Location	Scale	Q (0.25)	Q (0.50)	Q (0.75)	Q(0.90)
GDP	0.00344**	0.000959	0.00257**	0.00336**	0.00393**	0.00534**
	(0.00136)	(0.000887)	(0.00128)	(0.00134)	(0.00161)	(0.00262)
PPIE	−0.00202*	−0.00251**	−0.00285**	−0.00354**	−0.00479***	−0.00561***
	(0.00116)	(0.00119)	(0.00127)	(0.00142)	(0.00178)	(0.00204)
PPPIE	−0.589***	−0.0271	-0.545***	−0.556***	−0.568***	−0.580***
	(0.0332)	(0.0220)	(0.0397)	(0.0352)	(0.0321)	(0.0317)
HE	0.00940**	0.00187	0.00772**	0.00925**	0.0104**	0.0131*
	(0.00388)	(0.00253)	(0.00367)	(0.00382)	(0.00457)	(0.00745)
EE	0.0124	0.00265	0.587***	0.592***	0.602***	0.621***
	(0.00864)	(0.00563)	(0.0328)	(0.0343)	(0.0384)	(0.0484)
GII	−0.695***	−0.0668	−0.635***	−0.689***	−0.729***	−0.828***
	(0.0685)	(0.0447)	(0.0644)	(0.0679)	(0.0824)	(0.132)
Constant	0.822***	0.0562*	0.772***	0.818***	0.851***	0.934***
	(0.0459)	(0.0299)	(0.0433)	(0.0458)	(0.0557)	(0.0892)

Standard errors in parentheses
*** $p < 0.01$, ** $p < 0.05$, * $p < 0.1$

capital (Dissou et al. 2016). Increased investment in education leads to better school infrastructure, teacher quality and learning outcomes. It equips individuals with the knowledge and skills needed for employment and innovation, which are core components of human capital (Sikayena et al. 2022). This result is consistent with the studies by Dissou et al. (2016) and Sikayena et al. (2022).

Global inequality index exhibits a strong, consistent negative effect on HCI across all quantiles, confirming that greater inequality significantly undermines human capital development. In highly unequal societies, access to quality education and healthcare is often limited to wealthier groups. The marginalized population may lack the resources or opportunities to develop essential skills or maintain good health, which reduces overall human capital formation (Lee and Lee 2018). When inequality is high, it becomes harder for individuals from lower-income backgrounds to improve their socioeconomic status through education or training. This lack of mobility limits the talent pool and undermines long-term productivity and innovation (Castelló-Climent and Doménech 2021). Greater inequality can reduce political and public support for inclusive policies and investments in health and education. Elites may resist the redistribution of spending, resulting in underfunded public services that are essential for building human health (Nga Ndjobo and Ngah Otabela 2023). This finding is in line with the evidence presented in the work of Lee and Lee (2018), Castelló-Climent and Doménech (2021) and Nga Ndjobo and Ngah Otabela (2023), who argue that rising inequality undermines investment in education and health,

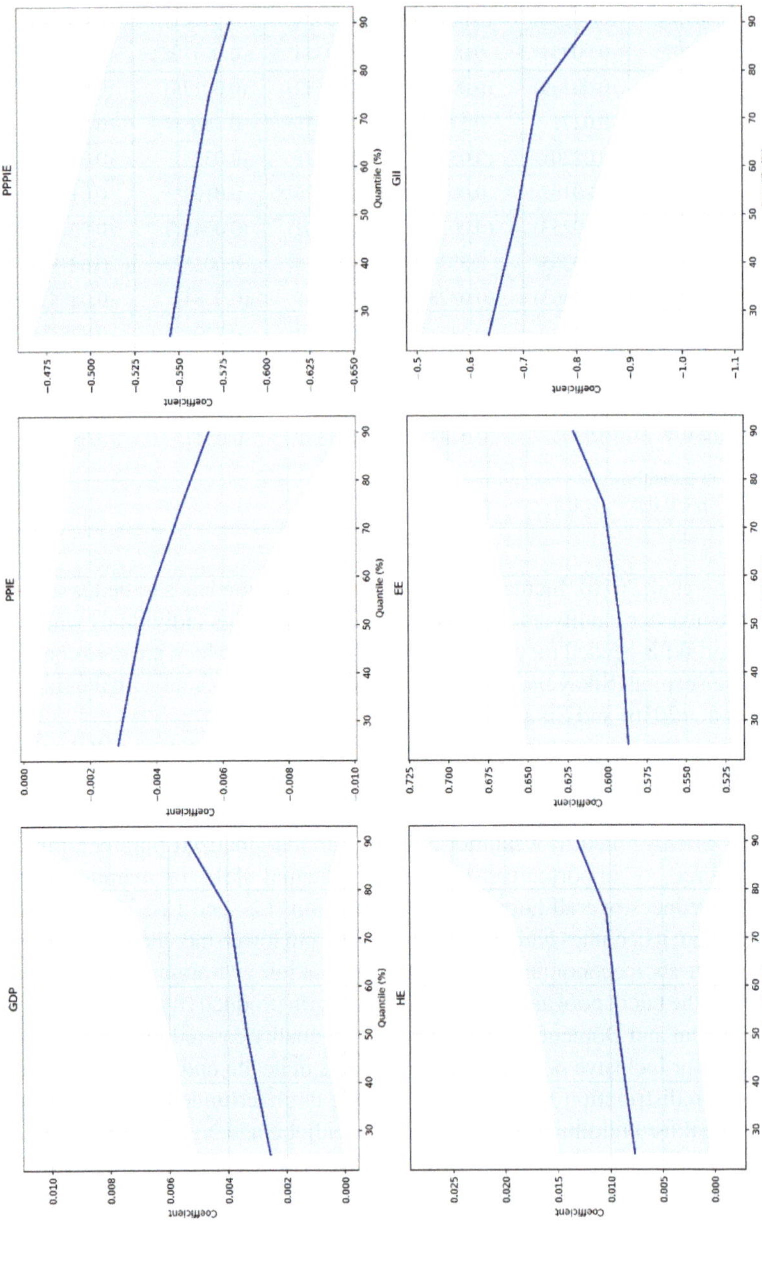

Fig. 7 MM-QR coefficient plot at different quantiles of energy investment and social sustainability

Table 13 BS-QR estimates of energy investment and social sustainability

Variables	Q (0.25)	Q (0.50)	Q (0.75)	Q(0.90)
GDP	0.00246*	0.00273**	0.00277*	0.00408**
	(0.00130)	(0.00119)	(0.00160)	(0.00188)
PPIE	−0.552***	−0.527***	−0.530***	−0.576***
	(0.0349)	(0.0485)	(0.0605)	(0.0635)
PPPIE	−0.608***	−0.610***	−0.614***	−0.616***
	(0.0466)	(0.0546)	(0.0701)	(0.0764)
HE	0.00958*	−0.596***	0.00624	0.0153
	(0.00495)	(0.0641)	(0.00631)	(0.0185)
EE	0.00513***	0.00597**	0.00914*	0.00917**
	(0.00129)	(0.00229)	(0.00479)	(0.00387)
GII	−0.614***	−0.616***	−0.742***	−0.936***
	(0.113)	(0.126)	(0.156)	(0.213)
Constant	0.764***	0.807***	0.861***	0.935***
	(0.0677)	(0.0617)	(0.0935)	(0.141)

Standard errors in parentheses
*** $p < 0.01$, ** $p < 0.05$, * $p < 0.1$

thereby weakening human capital development. It also supports one of the fundamental approaches of economic development known as Amartya Sen's Capability approach of economic development.

Table 13 presents the BS-QR estimates of energy investment and social sustainability, which are employed to verify the robustness of the MM-QR results. Figure 8 plots the BS-QR coefficient of energy investment and Social sustainability. The outcomes are consistent with the MM-QR findings, thereby reinforcing their reliability.

6 Conclusions and Policy Implications

The primary objective of the study is to explore the impact of energy investment on Sustainable Development Goals (SDGs) within Asian economies. To capture the multidimensional nature of sustainable development, the analysis focuses on three key dimensions: environmental, economic and social sustainability. Accordingly, the study develops three separate models. The first model investigates the impact of energy investment on environmental sustainability, the second model explore the influence of energy investment on economic sustainability and the third model aims to examine the relationship between energy investment and social sustainability. The analysis is based on panel data of 45 Asian nations covering the period 1990 to 2024. In the first model, carbon dioxide emission (CO_2) serve as the dependent variable,

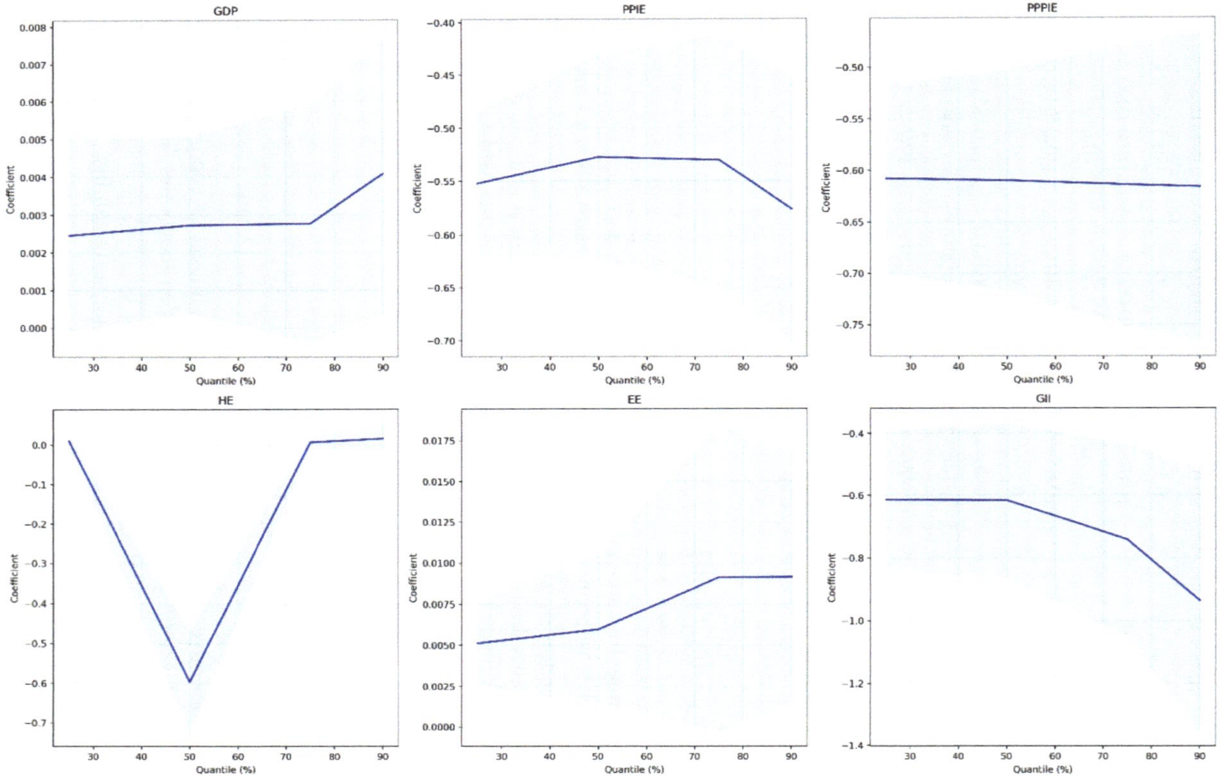

Fig. 8 BS-QR coefficient plot at different quantiles of energy investment and social sustainability

while the independent variables include GDP, GDP squared, private participation in energy investment, public and private partnership in energy investment and trade openness. The second model's dependent variable is GDP growth; the independent variables are labor force participation rate, gross fixed capital formation, private involvement in energy investment, public and private partnership in energy investment, and trade openness. The third model's dependent variable is human capital index; the independent variables are GDP, private involvement in energy investment, public and private partnership in energy investment, health spending, education spending, and gender inequality index. The study employs MM-QR analysis to estimate the results and further applies BS-QR analysis to assess the robustness of the MM-QR estimates. The findings of the study reveal that, in the first model, GDP growth, private participation in energy investment, public and private partnership in energy investment and trade negatively affect environmental sustainability, whereas the squared term of GDP has a favorable impact on environmental sustainability. In the second model, labor force participation rate, gross fixed capital formation, private participation in energy investment, public and private partnership in energy investment and trade openness are positively associated with economic sustainability. In the third model, private participation in energy investment, public and private partnership in energy investment and gender inequality index have a negative impact on social sustainability, whereas GDP growth, expenditure on health and expenditure on education are positively affecting the social sustainability. Based on the results, the study recommends that, given the interconnected nature of SDGs, countries aiming to achieve social and environmental sustainability should promote private participation and public–private partnership investment in the renewable energy sector rather than non-renewable energy. However, to enhance economic sustainability, it is essential to strengthen both private participation and public–private partnership investment in energy.

References

Ahmad M, Zhao ZY (2018) Causal linkages between energy investment and economic growth: a panel data modelling analysis of China. Energy Sources Part B 13(8):363–374

Ahmad M, Jabeen G, Irfan M, Mukeshimana MC, Ahmed N, Jabeen M (2020) Modeling causal interactions between energy investment, pollutant emissions, and economic growth: China study. Biophys Econ Sustain 5:1–12

Ahmed N, Areche FO, Sheikh AA, Lahiani A (2022) Green finance and green energy nexus in ASEAN countries: a bootstrap panel causality test. Energies 15(14):5068

Ajose K, Oyedokun GE (2018) Capital formation and economic growth in Nigeria. Account Tax Rev 2(2):131–142

Akinsola GD, Awosusi AA, Kirikkaleli D, Umarbeyli S, Adeshola I, Adebayo TS (2022) Ecological footprint, public-private partnership investment in energy, and financial development in Brazil: a gradual shift causality approach. Environ Sci Pollut Res 29(7):10077–10090

Ali M, Hashmi SH, Habib Y, Kirikkaleli D (2024) The asymmetric impact of public–private partnership investment in energy on CO_2 emissions in Pakistan. Energy Environ 35(4):2131–2150

Ali S, Chaudhry IS, Farooq F (2012) Human capital formation and economic growth in Pakistan. Pak J Soc Sci 32(1):229–240

Amir H, Khan M, Bilal K (2015) Impact of educated labor force on economic growth of Pakistan: a human capital perspective. Eur Online J Nat Soc Sci 4(4):814–831

Bal DP, Dash DP, Subhasish B (2016) The effects of capital formation on economic growth in India: evidence from ARDL-bound testing approach. Glob Bus Rev 17(6):1388–1400

Banga J (2019) The green bond market: a potential source of climate finance for developing countries. J Sustain Financ Invest 9(1):17–32

Caglar AE, Zafar MW, Bekun FV, Mert M (2022) Determinants of CO_2 emissions in the BRICS economies: The role of partnerships investment in energy and economic complexity. Sustain Energy Technol Assess 51:101907

Castelló-Climent A, Doménech R (2021) Human capital and income inequality revisited. Educ Econ 29(2):194–212

Cheng G, Zhao C, Iqbal N, Gülmez Ö, Işik H, Kirikkaleli D (2021) Does energy productivity and public-private investment in energy achieve carbon neutrality target of China? J Environ Manage 298:113464

Cung NH, Hung DH (2020) The impact of the labor force on economic growth in Vietnam. Int Bus Manag 14(10):79–90

Dissou Y, Didic S, Yakautsava T (2016) Government spending on education, human capital accumulation, and growth. Econ Model 58:9–21

Dou Y, Zhao J, Malik MN, Dong K (2021) Assessing the impact of trade openness on CO_2 emissions: evidence from China-Japan-ROK FTA countries. J Environ Manage 296:113241

Ertugrul HM, Cetin M, Seker F, Dogan E (2016) The impact of trade openness on global carbon dioxide emissions: evidence from the top ten emitters among developing countries. Ecol Ind 67:543–555

Fitzová H, Zídek L (2015) Impact of trade on economic growth in the Czech and Slovak Republics. Econ Sociol 8(2):36

Gakuo EW (2015) The relationship between government investment in energy infrastructure and economic growth in Kenya Doctoral dissertation, University of Nairobi

Grossman GM, Krueger AB (1995) Economic growth and the environment. Q J Econ 110(2):353–377

Hoa PX, Xuan VN, Thu NTP (2024) Determinants of renewable energy consumption in the fifth technology revolutions: evidence from ASEAN countries. J Open Innov Technol Market Complex 10(1):100190

Keho Y (2017) The impact of trade openness on economic growth: the case of Cote d'Ivoire. Cogent Econ Financ 5(1):1332820

Kleemann L, Abdulai A (2013) The impact of trade and economic growth on the environment: revisiting the cross-country evidence. J Int Dev 25(2):180–205

Komarulzaman A, Karunia MR, Andriani A, Siyaranamual MD, Haryani I (2023) Global landscape of renewable energy certificate (REC) systems: a scoping review of the literature. Martin Daniel and Haryani, Indri, Glob Landsc Renew Energy Certificate (Rec) Syst A Scoping Rev Literature

Lee JW, Lee H (2018) Human capital and income inequality. J Asia Pac Econ 23(4):554–583

Lup ANK, Soni V, Keenan B, Son J, Taghartapeh MR, Morato MM, Montañés RM (2023) Sustainable energy technologies for the Global South: challenges and solutions toward achieving SDG 7. Environ Sci Adv 2(4):570–585

McCauley D, Pettigrew K (2023) Building a just transition in asia-pacific: four strategies for reducing fossil fuel dependence and investing in clean energy. Energy Policy 183:113808

Muganyi T, Yan L, Sun HP (2021) Green finance, fintech and environmental protection: evidence from China. Environ Sci Ecotechnol 7:100107

Nga Ndjobo PM, Ngah Otabela N (2023) Can income inequality be affected by the interaction between ICTs and human capital? the evidence from developing countries. J Quant Econ 21(1):235–264

Ning L, Abbasi KR, Hussain K, Alvarado R, Ramzan M (2023) Analyzing the role of green innovation and public-private partnerships in achieving sustainable development goals: a novel policy framework. Environ Sci Pollut Res 1–17
Paramati SR, Alam MS, Chen CF (2017) The effects of tourism on economic growth and CO_2 emissions: a comparison between developed and developing economies. J Travel Res 56(6):712–724
Pelinescu E (2015) The impact of human capital on economic growth. Procedia Econ Financ 22:184–190
Raghutla C, Malik MN, Hameed A, Chittedi KR (2024) Impact of public-private partnerships investment and FDI on CO_2 emissions: a study of six global investment countries. J Environ Manage 360:121213
Sarwar A, Khan MA, Sarwar Z, Khan W (2021) Financial development, human capital and its impact on economic growth of emerging countries. Asian J Econ Bank 5(1):86–100
Shahbaz M, Hye QMA, Tiwari AK, Leitão NC (2013) Economic growth, energy consumption, financial development, international trade and CO_2 emissions in Indonesia. Renew Sustain Energy Rev 25:109–121
Shahbaz M, Nasir MA, Roubaud D (2018) Environmental degradation in France: the effects of FDI, financial development, and energy innovations. Energy Econ 74:843–857
Shahbaz M, Raghutla C, Song M, Zameer H, Jiao Z (2020) Public-private partnerships investment in energy as new determinant of CO_2 emissions: the role of technological innovations in China. Energy Econ 86:104664
Shahbaz M, Solarin SA, Sbia R, Bibi S (2015) Does energy intensity contribute to CO_2 emissions? a trivariate analysis in selected African countries. Ecol Ind 50:215–224
Shahid M (2014) Impact of labour force participation on economic growth in Pakistan. J Econ Sustain Dev 5(11):89–93
Shuaib IM, Ndidi DE (2015) Capital formation: impact on the economic development of Nigeria 1960–2013. Eur J Bus Econ Account 3(3):23–40
Sikayena I, Bentum-Ennin I, Andoh FK, Asravor R (2022) Efficiency of public spending on human capital in Africa. Cogent Econ Financ 10(1):2140905
Tenaw D, Beyene AD (2021) Environmental sustainability and economic development in sub-Saharan Africa: a modified EKC hypothesis. Renew Sustain Energy Rev 143:110897
Udeagha MC, Ngepah N (2023) Can public–private partnership investment in energy (PPPI) mitigate CO_2 emissions in South Africa? fresh evidence from the novel dynamic ARDL simulations approach. Front Environ Sci 10:1044605
Vakulchuk R, Overland I, Suryadi B (2023) ASEAN's energy transition: how to attract more investment in renewable energy. Energy, Ecol Environ 8(1):1–16
Vercoe J (2023) The implementation of renewable energy in malaysia in the context of global initiatives and targets. INSAF| J Malays Bar 40(1):58–76
Waqih MAU, Bhutto NA, Ghumro NH, Kumar S, Salam MA (2019) Rising environmental degradation and impact of foreign direct investment: an empirical evidence from SAARC region. J Environ Manage 243:472–480
Wu H (2023) Evaluating the role of renewable energy investment resources and green finance on the economic performance: evidence from OECD economies. Resour Policy 80:103149
Zafar MW, Shahbaz M, Hou F, Sinha A (2019) From nonrenewable to renewable energy and its impact on economic growth: the role of research & development expenditures in Asia-Pacific economic cooperation countries. J Clean Prod 212:1166–1178
Zahoor Z, Khan I, Hou F (2022) Clean energy investment and financial development as determinants of environment and sustainable economic growth: evidence from China. Environ Sci Pollut Res 1–11
Zhang X, Chen X, Fang Z, Zhu Y, Liang J (2022) Investment in energy resources, natural resources and environment: Evidence from China. Resour Policy 76:102707

Evaluating the Impact of Economic Prosperity and Financial Globalization on Environmental Sustainability: Is the EKC Hypothesis Valid in Turkey? A Fresh Insight

Orhan Şanli, Laeeq Razzak Janjua, and Kiran Jameel

Abstract The purpose of this research is to reveal how globalization influences CO_2 emissions (CO_2) while considering growth, energy use, and population as control variables in Turkey. This study used indices of economic, trade, and financial globalization represent the globalization index. This research work used the ARDL estimation technique to achieve goals. Findings suggest, the variables have a significant long-term link. Globalization has a significant impact on CO_2. Economic growth, population, trade globalization, and financial globalization worsen environmental conditions in the long run. However, economic globalization and energy consumption improve environmental outcomes. Finally, this study has useful insights about the validity of the EKC hypothesis in Turkey.

Kewords Economic growth · Financial globalization · EKC hypothesis · Carbon emission · Energy

---

The original version of the chapter has been revised. A correction to this chapter can be found at https://doi.org/10.1007/978-3-032-01715-4_23

O. Şanli
Department of Economic, Faculty of Economics and Administrative Sciences, Adnan Menderes University, Aydın, Turkey
e-mail: orhan.sanli@adu.edu.tr

L. R. Janjua (✉)
WSB University Poland, Dąbrowa Górnicza, Poland
e-mail: Laeeq.janjua@wsb.edu.pl

K. Jameel
College of Business Management (CBM) Institute of Business Management (IoBM), Karachi, Pakistan
e-mail: Kiran.jameel@iobm.edu.pk

F. A. Malik et al. (eds.), *Aligning ESG Goals with the UNs Sustainable Development Goals Planning*, World Sustainability Series,
https://doi.org/10.1007/978-3-032-01715-4_19

1 Introduction

Since the last quarter of the twentieth century, the globalization process has accelerated throughout the world. There are two main reasons for this situation. Initially, the least developed countries have abandoned mainly import substitution policies. The second is the collapse of the socialist system, which prevents the least-developed countries from integrating into the global system (Khan 1998, p. 103). This situation has connected all countries like a chain with the contribution of open economic policies, and the connection between countries has been strengthened (Mehmood and Tariq 2020). Thus, transportation costs decreased, diffusion and transfer of information and communication technologies increased, and cross-border trade, capital, and foreign direct investments increased significantly (Acs and Preston 1997, p. 1; Khan 1998, p. 103; Lee and Vivarelli 2006).

Globalization, which means that economies, cultures, and people are becoming more connected through commerce, investment, technology, and information, has changed the world we live in today. Globalization has contributed to economic progress, new ideas, and cultural interaction, but it has also had a big impact on the environment, especially when it comes to carbon emissions. There is a complicated link between globalization and carbon emissions, having both direct and indirect effects on climate change around the world.

Globalization makes carbon emissions worse by making international trade and travel bigger. As countries focus on making things and trading with each other over great distances, shipping and flying that uses a lot of carbon has gone up a lot. Cristea et al. (2013) say that transportation related to international trade is responsible for a large and growing amount of carbon dioxide emissions around the world. The more things are moved across borders, the more fossil fuels are used and the more pollution is released.

Globalization has also made it easier for companies to move their factories from industrialized countries with rigorous environmental rules to developing countries with less strict rules. This "pollution haven" hypothesis posits that multinational businesses may leverage less stringent regulatory frameworks to reduce expenses, resulting in heightened emissions in the respective host nations (Cole 2004). Because of this, developing countries typically have to deal with the environmental costs of global manufacturing chains, even though they have smaller historical obligations, which means they contribute more to carbon emissions. Globalization is not bad for the environment in and of itself, though. It can also have good effects on the environment by sharing technology, working together on environmental issues across borders, and spreading new green ideas. For example, cleaner production technologies and energy-efficient practices can be moved from developed to developing countries through foreign direct investment (FDI) and global supply chains (Zhu and Fu 2013). If environmental rules are followed and green expenditures are given priority, this spread of technology can help cut down on emissions. Globalization also makes people more aware of environmental problems around the world, which helps people work together to manage climate change. Global diplomatic processes that are made possible by interconnected institutions and communication networks lead to multilateral agreements like the Paris Agreement. These collaborative frameworks facilitate collective action towards carbon reduction objectives and climate resilience strategies. Still, the overall effect of globalization on carbon emissions relies a lot on how economic integration is handled. Without strict environmental rules and ways to charge for carbon, the environmental costs of globalization may be higher than the advantages. A well-governed globalization framework, on the other hand, can reduce carbon emissions while keeping the benefits of free markets. This framework should encourage low-carbon development, hold businesses accountable for their environmental impact, and support sustainable practices.

In this study, the effect of economic, financial, and trade globalization on carbon emissions is examined in the case of Turkey. At the same time, this study also investigates whether the EKC hypothesis is valid for Turkey. This study differs from many studies in the literature due to the variables used. Because while most of the studies in the literature deal with the variables of political, economic, and social globalization, it is clear that the financial and commercial globalization indices are not adequately examined. The most important reason is that the trade and financial globalization indices were introduced extensively and in detail by Gygli et al. (2019) in 2019 (Kihombo et al. 2022, p. 3984). However, Dreher (2006), the original study on which the globalization index is based, did not divide economic globalization into sub-indices. This situation made it difficult to reveal the impact of trade and financial globalization on the environment with the original globalization dataset. Therefore, there are very few studies in the literature in which these two variables are modeled. In the previous literature, trade and financial openness rates and alternative methods have been used instead of trade and financial globalization. In addition, no study has been found that examines the relationship between trade and financial globalization indices and carbon emissions for Turkey. It is one of the most important reasons for the emergence of this study. In addition to this situation, Turkey's financial and commercial relations with the outside world are pretty strong. At the same time, Turkey is a developing country.

For this reason, economic, financial, and trade freedom and globalization must be vital for Turkey to reach its development goal. In this direction, Turkey has become one of the countries that benefit most from free trade in the world (OECD 2022). While the growth average of Turkey in the last 50 years has been 4.6%, the world growth average has been 3.1%. However, Turkey's rapid growth potential has also brought environmental problems. For example, in the last 50 years, total CO_2 emissions in Turkey have increased approximately eight times and reached 400 million tons. Turkey's total carbon emissions cover 1.1% of global emissions. As seen in Fig. 1, while Turkey's globalization process has increased rapidly, radiated carbon emissions have increased more. So much so that in the last 50 years, while trade and financial globalization has increased approximately two times, CO_2 emissions per capita have increased more than 3.5 times. From this point of view, this study examines the impact of Turkey's economic, trade, and financial globalization on environmental quality. In addition to economic globalization, including financial and trade globalization indices in the model has produced essential and remarkable

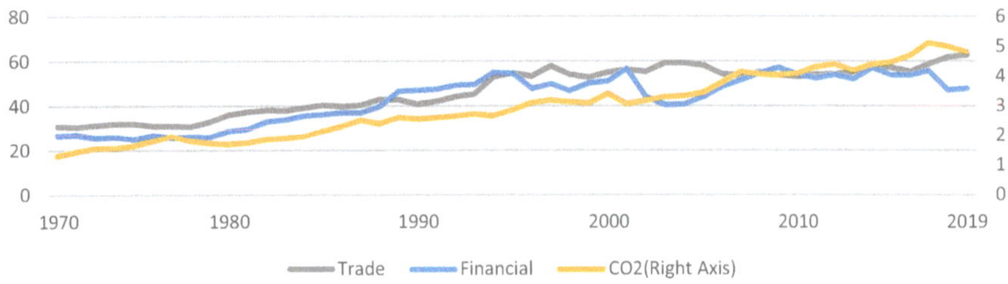

Fig. 1 Trade globalization (overall), financial globalization ındices (overall) and CO_2 emissions (metric tons per capita) in Turkey. Reference: Gygli et al. (2019)

results. Finally, the EKC hypothesis has been tested to predict how the high growth rate of Turkey, which has a growth rate above the world average, will impact the environmental quality in the long run. The study proceeds in the following order; literature review, data, model and methodology, conclusion and discussion, and policy recommendations.

2 Literature Review

A literature review has been conducted on the primary focus of this study: economic, financial, and trade globalization. Consequently, the research exhibits a significant correlation between social, political, and economic globalization and the environment. Nonetheless, additional research employing financial and trade globalization indicators is necessary.

Akdiri and Adebayo (2021) studied the impact of financial globalization on carbon emissions, using India as a case study. The study's NARDL approach results indicate that positive shocks in financial globalization result in reduced carbon emissions. Negative shocks in financial globalization are worsening the state of the environment.

Using the ARDL limits test and the Toda-Yamamoto Granger causality test, Akadiri et al. (2020a) looked at how electricity use, globalization, and real income affect carbon emissions. The study on Turkey indicates that globalization does not influence carbon emissions. But there is a one-way relationship of causality between the two variables.

Yurtkuran (2021) investigated the impact of globalization, agriculture, and renewable energy utilization on carbon emissions in Turkey. The globalization index was made up of political, economic, and social globalization indexes. The ARDL limit test results show that agriculture, energy consumption, and globalization all raise carbon emissions over time.

Ahmed and Le (2021) investigated the influence of ICT and trade globalization on CO_2 emissions in the ASEAN-6 nations. The study utilized Westerlund, Padroni CUP-FM long-term methodology, and panel DH causality methodologies. The study's findings indicate that travel globalization enhances environmental quality. Trade globalization has a direct effect on CO_2 levels.

Rehman et al. (2019) utilized the ARDL method to analyze the effects of globalization, trade, economic growth, oil imports, and energy consumption on Pakistan's ecological footprint. The study results indicate that all variables, excluding oil imports, exert a beneficial influence on the ecological footprint.

Pata and Çağlar (2021) utilized the ARDL method to analyze the impacts of globalization, trade openness, and numerous indicators on CO_2 emissions in China. The study results indicate that globalization and trade openness exacerbate environmental damage. Simultaneously, it was determined that the EKC theory is invalid.

Bilgili et al. (2020) utilized Markov regime change models to analyze the effects of economic, social, trade, financial, and political globalization on Turkey's ecological footprint. The study findings indicate that economic and social globalization augment the ecological footprint, whereas other globalization indices lead to its reduction.

Yilanci and Gorus (2020) analyzed the effects of economic, commercial, and financial globalization on the ecological footprint in MENA nations utilizing the Fourier-Toda-Yamamoto methodology. The study's findings indicate that the ecological footprint serves as the Granger cause of globalization indicators. They also showed that financial globalization can hurt the environment.

Godil et al. (2020) utilized the asymmetric Q-ARDL approach to analyze the impacts of globalization, tourism, and financial development on EFP in Turkey. The study's findings indicate that globalization, tourism, and financial development enhance EFP. The EKC hypothesis is also true.

3 Data, Model, and Methodology

The primary aim of this study is to examine the possible effects of economic globalization, trade globalization, financial globalization, economic growth, the square of economic growth, population, and energy consumption on CO_2 emissions in Turkey. While achieving this goal, the EKC hypothesis was also tested. Globalization indices published by KOF Swiss Economic Institute were used in this study. The data are based on Dreher (2006) and Gygli et al. (2019). CO_2, population, and GDP data are from the World Bank. Energy consumption is taken from Our Word in Data (2022). The time dimension of the study covers the period 1977–2024.

In this direction, the general model of the study was established as follows;

$$\Delta CO_2 = f\,(Gt,\ G2t,\ ECt,\ Pt,\ EGt,\ TGt,\ FGt) \tag{1}$$

As suggested by Soytaş and Sarı (2006), Shahbaz et al. (2012), and Öztürk and Acaravcı (2013), the mathematical view of the log-linear model created for this study is as follows;

$$\ln CO_2t = \beta 0 + \beta 1\ln Gt + \beta 2\ln G2t + \beta 3\ln ECt + \beta 4\ln Pt + \beta 5\ln EGt + \beta 6\ln TGt + \beta 7\ln FGt + \varepsilon t \tag{2}$$

CO_2 represents carbon emissions (metric tons per capita). G is the GDP indicator (per capita, constant 2015 US$). G2 is the square of G. The EC symbol represents the total energy consumption (kWh-person). P represents the total population. The symbols EG, TG, and FG represent economic, trade, and financial globalization. Economic globalization consists of trade and financial globalization. Trade globalization consists of total exports and imports of goods and services. The EKC hypothesis is formulated as follows;

$$E = f(+G, -G2) \tag{3}$$

E is an indicator of environmental degradation. Grossman and Krugger (1991) stated that environmental degradation increases in the early stages of growth and decreases in the following periods.

The long-term model used in this study to examine the long-term relationship is as follows;

$$\begin{aligned}\ln CO_{2t} &= \beta 0 + \sum_{t=1}^{m} \gamma_{1i}\ln CO_{2t-1} + \sum_{t=1}^{n} \gamma_{3i}\ln G_{t-1} + \sum_{t=1}^{o} \gamma_{3i}\ln G_{t-1}^2 \\ &+ \sum_{t=1}^{p} \gamma_{4i}\ln EC_{t-1} + \sum_{t=1}^{q} \gamma_{5i}\ln P_{t-1} + \sum_{t=1}^{r} \gamma_{6i}\ln EG_{t-1} \\ &+ \sum_{t=1}^{s} \gamma_{7i}\ln TG_{t-1} + \sum_{t=1}^{u} \gamma_{8i}\ln FG_{t-1} + \varepsilon_t\end{aligned} \tag{4}$$

To examine the short-term analysis, The mathematical transformation of the model based on the vector error correction model (VECM) is as follows;

$$\begin{aligned}\Delta\ln CO_{2t} &= \beta 0 + \sum_{t=1}^{m} \Delta\beta_{1i}\ln CO_{2t-1} + \sum_{t=1}^{n} \Delta\beta_{3i}\ln G_{t-1} + \sum_{t=1}^{o} \Delta\beta_{3i}\ln G_{t-1}^2 \\ &+ \sum_{t=1}^{p} \Delta\beta_{4i}\ln EC_{t-1} + \sum_{t=1}^{q} \Delta\beta_{5i}\ln P_{t-1} + \sum_{t=1}^{r} \Delta\beta_{6i}\ln EG_{t-1} \\ &+ \sum_{t=1}^{s} \Delta\beta_{7i}\ln TG_{t-1} + \sum_{t=1}^{u} \Delta\beta_{8i}\ln FG_{t-1} + \chi ECT_{t-1} + \varepsilon_t\end{aligned} \tag{5}$$

Here, the symbols m, n, Q, p, θ, ý, ϑ, and q are added to represent the optimum lag lengths determined according to the Schwarz (SC) information criterion. Δ Represents the difference operator. ECT stands for error correction coefficient.

3.1 Unit Root Tests

A unit root is a feature of a time series that shows it is not stationary, which means that its statistical features, such as mean and variance, change over time. A unit root series has a random trend and can give wrong findings when you do regression analysis. Finding a unit root is very important in time series analysis since it helps you choose the right modeling methods, like differencing the data to make it stationary. The Augmented Dickey-Fuller (ADF) test and the Phillips-Perron test are two common ways to find unit roots. If a unit root exists, standard statistical inferences may be rendered erroneous, requiring specialist methodologies such as cointegration analysis.

3.2 ARDL Cointegration Method

Cointegration strategies find correlations between variables that last a long time. Conventional approaches, such as Johansen (1988, 1996), necessitate that variables be stationary at an identical order. Pesaran et al. (2001) created the ARDL approach to get around this problem. It works with variables that are integrated at I(0) or I(1). The ARDL bounds test yields both short-term and long-term outcomes and is suitable for small sample sizes (Pesaran 2015; Narayan 2004). It also figures out the error correction term for both short- and long-term convergence (Shahbaz et al. 2013a, b). The F-statistic must be higher than critical boundaries for cointegration to be proven (Khan et al. 2019; Aslam et al. 2021). Accordingly, the unrestricted error correction model is set up as follows:

$$\begin{aligned}\Delta \text{lnCO}_{2t} &= \beta 0 + \sum_{t=1}^{m} \Delta\beta_{1i}\text{lnCO}_{2t-1} + \sum_{t=1}^{n} \Delta\beta_{3i}\text{lnG}_{t-1} + \sum_{t=1}^{o} \Delta\beta_{3i}\text{ln}G_{t-1}^{2} \\ &+ \sum_{t=1}^{p} \Delta\beta_{4i}\text{lnEC}_{t-1} + \sum_{t=1}^{q} \Delta\beta_{5i}\text{lnP}_{t-1} + \sum_{t=1}^{r} \Delta\beta_{6i}\text{lnEG}_{t-1} \\ &+ \sum_{t=1}^{s} \Delta\beta_{7i}\text{lnTG}_{t-1} + \sum_{t=1}^{u} \Delta\beta_{8i}\text{lnFG}_{t-1} + \chi ECT_{t-1} + \varepsilon_t \end{aligned} \quad (6)$$

In this model, β is the short-run coefficient, and γ is the long-run coefficient. β0 is the constant term. χ indicates the adjustment coefficient of the ECM. The hypothesis for the F test is established as follows;

$$\begin{aligned}&\text{H0: } \beta 1 = \beta 2 = \beta 3 = \beta 4 = \beta 5 = \beta 6 = \beta 7 = \beta 8 = 0 \\ &\text{(There is no cointegration relationship)}\end{aligned} \quad (7)$$

$$\begin{array}{l} H1\!: \gamma 1 \neq \gamma 2 \neq \gamma 3 \neq \gamma 4 \neq \gamma 5 \neq \gamma 6 \neq \gamma 7 \neq \gamma 8 \neq 0 \\ \text{(There is a cointegration relationship)} \end{array} \tag{8}$$

If there is a cointegration relationship here, if H0 is rejected, short-term analysis is started. It is seen that most of the studies Shahbaz et al. (2016, 2019), Solarin et al. (2017), Xu et al. (2018), Zafar et al. (2019a, b), Ahmed et al. (2019), Khan et al. (2019), Salahuddin et al. (2019), Sabir and Gorus (2019), Akadiri et al. (2020b), Pata (2021), Ahmed et al. (2021), Ahmed and Le (2021), Islam et al. (2021), Gaies et al. (2022), Akadiri and Adebayo (2021) examining the effect of globalization and other macro variables on carbon emissions have been conducted with the ARDL bound test method. For this reason, as stated before, the ARDL method was used in this study due to its advantages of the ARDL method and great interest in the literature.

Descriptive statistics values and correlation matrix results are presented in Table 1. Accordingly, the variable with the smallest mean, median, maximum, and minimum values is CO_2. The EC variable has the most effective mean, median, maximum, standard error, skewness, kurtosis, and Jargue-bera values. When the Jargue-Bera probability values are examined, it is seen that the other variables are typically distributed except for the EC variable. However, the Jargue-Bera probability value of 0.99% in Table 5 indicates that the model is usually distributed.

Table 1 Descriptive statistics

	CO_2	G	G^2	EC	P	EG	TG	FG
Mean	1.03	8.71	16.17	30.93	17.85	3.77	3.72	3.82
Median	1.05	8.66	16.04	31.44	17.88	3.90	3.85	3.96
Maximum	1.63	9.39	18.88	34.29	18.23	4.05	4.04	4.13
Minimum	0.28	8.14	12.24	16.63	17.39	3.34	3.22	3.41
Std. dev.	0.37	0.36	1.85	3.031	0.24	0.24	0.27	0.23
Skewness	−0.17	0.34	−0.06	−2.53	−0.26	−0.68	−0.66	−0.48
Kurtosis	1.88	1.97	1.95	12.08	1.94	1.91	1.94	1.76
Jarque-Bera	2.76	3.12	2.26	220.63	2.84	6.27	5.83	5.06
Probability	0.25*	0.20*	0.32*	0.00	0.24*	0.043**	0.051*	0.079*
Sum	50.49	426.81	792.59	1516.05	875.03	185.20	182.70	187.40
Sum Sq. dev.	6.70	6.55	165.30	440.97	2.89	2.98	3.67	2.67
Observations	49	49	49	49	49	49	49	49

Note * Symbol denotes normal distribution according to a 5% significance level. The ** symbol denotes the normal distribution at a 1% significance level

3.3 Results and Discussion

For the ARDL test, the stationarity level of the series should not be I(2). This method is applied if the stationarity level is I(0) or I(1). Therefore, unit root test results of Augmented Dickey-Fuller (ADF), Dickey-Fuller (DF), and Phillips-Perron (PP) are reported in Table 2. Accordingly, some variables are stationary at the I(0), and others at the I(1). None of the series is stationary at the I(2). Accordingly, CO_2, G, G2, EG, TG, and FG are stationary at the first difference according to the three methods. EC and PP are stationary both in level value and difference. The fact that the results are the same in all three methods shows the consistency of the unit root results. The unit root test with Perron break also shows the break years. The fact that the breaking years coincide with similar periods shows the effect of economic and political developments. For example, the periods of 1977–1979, 1992–1996, 2002, and 2009, in which the ruptures occurred, show the impact of events such as crises on a national or global basis. The unit root results show that the ARDL method can be applied.

Table 3 shows the SC and other test results for determining the lag length using the VAR method.

Table 2 Unit root test results

Variable	ADF		PP		Perron break point		
	t stat	p-value	t stat	p-value	t stat	p-value	Break date
CO_2	−1.50(0)	0.52	−1.69(6)	0.42	−2.21(0)	0.96	2002
G	0.32(0)	0.97	0.35(3)	0.97	−1.80(0)	>0.99	2002
G^2	0.32(0)	0.97	0.35(3)	0.97	−1.80(0)	>0.99	2002
EC*	−6.26(0)***	0.00	−6.26(2)***	0.00	−9.56(0)***	<0.01	2009
P*	−5.91(2)***	0.00	−5.92(10)***	0.00	−10.82(0)***	<0.01	1974
EG	−1.45(0)	0.55	−1.48(3)	0.53	−3.30(0)	0.44	1979
TG	−1.07(0)	0.71	−1.06(2)	0.72	−2.43(0)	0.91	1978
FG	−1.55(0)	0.49	−1.54(4)	0.50	−3.25(0)	0.53	1979
ΔCO_2	−6.85(0)***	0.00	−8.08(8)***	0.00	−7.16(0)***	0.00	1977
ΔG	−0.63(0)***	0.00	−6.63(2)***	0.00	−7.28(0)***	<0.01	2009
ΔG^2	−6.63(0)***	0.00	−6.63(2)***	0.00	−7.28(0)***	<0.01	2009
ΔEC	−11.53(0)***	0.00	−14.13(4)***	0.00	−12.11(0)***	<0.01	1977
ΔP	−3.09(10)**	0.03	1.78(7)*	0.07	−5.93(10)***	<0.01	1992
ΔEG	−7.67(0)***	0.00	−7.68(2)***	0.00	−8.73(0)***	<0.01	1996
ΔTG	−6.34(0)***	0.00*	−6.33(2)***	0.00	−6.95(1)***	<0.01	1994
ΔFG	−6.77(0)***	0.00*	−6.78(4)***	0.00	−7.33(0)***	<0.01	1996

Note ***, **, and * indicate 1%, 5%, and 10% severity, respectively

Table 3 Lag selection

Lag	LogL	LR	FPE	AIC	SC	HQ
0	309.8220	NA	2.06e-16	−13.41431	−13.09313	−13.29458
1	722.6936	660.5945	3.98e-23	−28.91971	−26.02905	−27.84211
2	842.8537	149.5326	4.30e-24	−31.41572	−25.95559	−29.38024
3	955.9139	100.4980	1.10e-24	−33.59618	−25.56656	−30.60282
4	1124.174	89.73863*	8.81e-26*	−38.22995*	−27.63086*	−34.27872*

Table 4 ARDL bounds test

Test statistic	Value	k
F-statistic	*10.37271*	7
Bounds critical value		
Significance	Lower bound	Upper bound
10%	2.03	3.13
5%	2.32	3.5
2.5%	2.6	3.84
1%	2.96	4.26

Boundary test results are given in Table 4. The value of the F statistic was determined as 10.37. The F statistic is more significant than the lower and upper critical values (1, 5, and 10% significance). According to this result, there is a strong long-run and cointegration relationship between the variables in this model. In addition, in this study, ARDL (1, 2, 0, 1, 1, 3, 1, 1) model was chosen as the model with the best lags in line with the SC information criterion. Accordingly, the appropriate lag lengths for Eq. 4 are as follows; $m = 1$, $n = 2$, $\alpha = 0$, $p = 1$, $\theta = 1$, $\sqrt{} = 3$, $\vartheta = 1$, $q = 1$.

Table 5 shows the long-time ARDL results, indicating that economic growth (G) elevates CO_2 levels, whilst its square (G2) diminishes them, implying the validity of the EKC hypothesis for Türkiye in the long term. This is in line with research by Bulut (2021), Genç et al. (2022), and others, but it goes against research that says EKC is not valid (for example, Lise 2006; Akbostancı et al. 2009). Notably, emissions are higher during early-stage expansion than during later-stage recovery. This shows that we need more environmental controls. There is a small negative effect of energy use per person on CO_2, which suggests that the environment is getting better but doesn't have strong evidence. Some studies find a negative link (e.g., Ma et al. 2011; Khan et al. 2019), while others find a favorable one (e.g., Cetin et al. 2018; Mohsin et al. 2019). Population growth is strongly correlated with CO_2 levels, mostly due to heightened energy consumption in sectors such as manufacturing and housing. The vast number of visitors and refugees that come to Türkiye each year (up to 140 million) makes this effect even stronger, showing how important it is to have tailored environmental initiatives. The benefits of globalization are mixed: economic globalization makes the environment better, yet trade and financial globalization make CO_2 emissions go up. This difference implies that the composite index might hide the influence of the sub-indices. The results are consistent with Lee and Min (2014) but are at odds with Ahmed and Le (2021) and Miao et al. (2022), who identified beneficial environmental effects stemming from trade and financial globalization. Kalaycı and Hayaloğlu (2019), Koengkan et al. (2020), and Sadiq et al. (2022) are other studies that back this one up.

Table 5 also includes diagnostic test results thus all are stable.

Cusum and Cusumsq test results in Fig. 2 also show that the long-term model is stable and balanced. Diagnostic test results increase the consistency of the model's long-term relationship.

The short-run coefficients are shown in Table 6. G raises CO_2, whereas G2 lowers it, which supports the EKC hypothesis for Türkiye in the medium run. Energy use and population both lower CO_2 levels, but population is the most important influence. In the near term, both economic and trade globalization lower CO_2 levels, while financial globalization has no real effect. This shows that globalization helps the

Table 5 Long-term analysis

Dependent variable CO_2	Coefficient	Std. error	t-statistic	p-values
G	0.485***	0.0595	8.151	0.000
G^2	−0.016*	0.008	−1.984	0.057
EC	−0.006***	0.001	−3.835	0.000
P	0.572***	0.144	3.967	0.000
EG	−7.559***	1.688	−4.477	0.000
TG	3.800***	0.871	4.361	0.000
FG	3.773***	0.794	4.751	0.000
Constant	−15.866***	2.995	−5.297	0.000
R^2	0.99			
Adjusted R^2	0.99			
F-statistic	547.425			
Prob-(F-statistic)	0.000			
Durbin-Watson-stat	2.09			
Diagnostic test	F-stat	p-values		
BP-LM	0.16	0.68		
Ramsey reset	1.00	0.37		
Jargua-Bera	0.01	0.99		
BP-G	0.43	0.96		
ARCH	2.01	0.96		
White	0.37	0.73		
Cusum	Stable			
Cusumsq	Stable			

Note The symbols *** and * represent the significance level relative to 1% and 10%, respectively

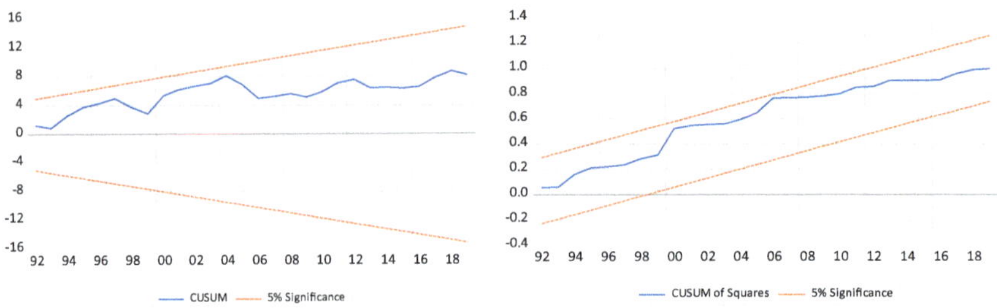

Fig. 2 Cusum ve Cusumsq

environment get better in the near term, but it could make things worse in the long term. The error correction term (ECM-1) is −1.231, which is important at 1%. Since it is between −1 and −2, convergence to equilibrium is quick but not steady (Narayan and Smyth 2008).

Table 6 Short-term analysis (1, 2, 0, 4, 1, 3, 2, 1)

Dependent variable CO_2	Coefficient	Std. error	t-statistic	p-values
ΔG	0.893***	0.093	9.554	0.000
ΔG(−1)	0.232**	0.098	2.353	0.025
ΔG^2	−0.020*	0.010	−1.982	0.057
ΔE	−0.004***	0.000	−4.804	0.000
ΔP	−21.989***	2.415	−9.101	0.000
ΔEG	−0.312**	0.091	−3.414	0.002
ΔEG(−1)	−0.330***	0.092	−3.593	0.001
ΔTG	−0.330***	0.092	−3.593	0.001
ΔFG	−0.840	0.867	−0.968	0.341
Constant	−15.866***	1.556	−10.192	0.000
ECT(−1)	−1.2510***	0.122	−10.184	0.000
R^2	0.85			
Adjusted R^2	0.81			
F-statistic	21.06			
Prob-(F-statistic)	0.000			
Durbin-Watson-stat	2.09			

Note The symbols ***, **, and * represent 1%, 5%, and 10% significance levels, respectively

4 Conclusion and Discussion

The purpose of this research is to analyze the long-term effects of economic globalization, commercial globalization, financial globalization, economic growth, energy consumption, and population on CO_2 in Türkiye. For this purpose, the EKC hypothesis was also tested. The study covering 1971–2019 was conducted with the ARDL bound test method. For this reason, firstly, a unit root test was performed with ADF, PP, and Perron's (1997) methods. It has been determined that the stationarity level is at the maximum I(1).

For this reason, the ARDL method was deemed appropriate, and cointegration, long-term, and short-term analyses were performed. According to the study's long-term results, economic growth and the square of economic growth affect CO_2 at a rate of 0.485 and −0.016, respectively. This result shows that the EKC hypothesis is valid for Türkiye in the long run and that the EKC is an inverted U. Short-term results also show that the EKC is valid. On the other hand, energy consumption has an insignificant but negative effect on CO_2 in both the long and short run. Although this result is insignificant, it shows that energy consumption improves environmental quality. Total population growth increases CO_2 by 0.57 in the long run. However, in the short run, the most influential variable on CO_2 is the total population, which reduces CO_2.

Economic globalization is the variable that most affects CO_2. Increasing economic globalization causes a decrease in CO_2 and, thus, an increase in environmental quality. Accordingly, when economic globalization increases by 1%, CO_2 decreases by approximately 7.55%. Contrary to economic globalization, trade and financial globalization affects CO_2 positively in the long run. Accordingly, trade and financial globalization increase CO_2 by 3.8 and 3.7 percent, respectively. However, according to the short-term results, economic globalization and trade globalization have a reducing effect on CO_2. Financial globalization is not meaningful in the short run. At the same time, the diagnostic tests in Table 5 show the robustness of the model. According to these results, there is no autocorrelation problem, changing variance problem, or model-building error problem in the model, and the model is typically distributed. Cusum and Cusumsq tests in Fig. 2 show that the ARDL model is stable in the long run.

5 Policy Recommendations

One of the factors that increase environmental degradation the most is CO_2 emissions. Therefore, controlling the factors that increase CO_2 emissions is vital for increasing environmental quality. The country examined in this study is Türkiye. Although the results of this study show the presence of EKC in Türkiye, environmental recovery in the later stages of growth is relatively slow. The results show that clean energy should be expanded in areas where energy is used intensively, such as industry, agriculture, and construction, which constitute the central dynamics of growth. The insignificant negative impact of per capita energy consumption on CO_2 supports the need for clean energy. Because approximately 80% of Türkiye's total energy consumption comprises fossil fuels, the results suggest the need for clean energy policies.

Turkey is also one of the vital tourism and refugee centers. So much so that Türkiye, which hosts nearly 4 million refugees, hosts more than 50 million tourists annually. This situation causes the total population of the year to reach 140 million. Many of the refugees reside in areas where fossil fuels are used. In addition, the energy demand required to meet the needs of millions of tourists is also increasing. Local people also intensively use fossil fuels (heating, cooking, electricity). Thus, the amount of energy used to meet the needs of the total population also increases. For this reason, it is essential to consider the energy resources needed by the local people, refugees, and foreign tourists and to support using renewable energy in these areas.

The Turkish economy is a country that benefits significantly from external resources. Türkiye's economic growth increases as foreign economic relations develop, showing that 60% of the total economic size is foreign trade. Therefore, Türkiye's greater integration into the globalizing world, both trade and financially, leads to increased environmental degradation. For this reason, environmental quality needs to increase the use of alternative clean energy sources, especially instead of fossil fuel imports. However, the government must highlight supportive policies in this area due to the cost of clean energy resources. Therefore, energy security in Türkiye is an important area that should be emphasized for the general economic situation and environmental quality.

As a result, new policies are needed in growth, energy consumption, population, and globalization to increase environmental recovery. However, production and consumption costs are among Türkiye's most critical challenges. For example, the increase in exchange rate shocks in Türkiye in recent years has increased production and consumption costs. It may strengthen the priority given to clean energy sources in production and consumption. On the contrary, it may increase the demand for cheap and dirty energy sources. In this case, policies aimed at increasing environmental quality may lead to problems such as high prices, unemployment, and loss of market.

For this reason, it is an essential requirement for the government to announce new and supportive economic policies against such possible problems. For example, Türkiye's signing of the Paris Climate Agreement in 2021 is essential to the clean environment goal. Nevertheless, it turns out that it is not enough. In addition, Türkiye should support national and international clean environment projects and facilitate the use of funds from international projects and governmental projects in the private sector.

References

Acs ZJ, Preston L (1997) Small and medium-sized enterprises, technology; and globalization: ıntroduction to a special ıssue on small and medium-sized enterprises in the global economy. Small Bus Econ 9(1):1–6

Ahmed Z, Wang Z, Mahmood F, Hafeez M, Ali N (2019) Does globalization increase the ecological footprint? Empirical evidence from Malaysia. Environ Sci Pollut Res 26(18):18565–18582

Ahmed Z, Le HP (2021) Linking information communication technology, trade globalization index, and CO_2 emissions: evidence from advanced panel techniques. Environ Sci Pollut Res 28(7):8770–8781

Ahmed Z, Zhang B, Cary M (2021) Linking economic globalization, economic growth, financial development, and ecological footprint: evidence from symmetric and asymmetric ARDL. Ecol Ind 121:107060

Akadiri SS, Lasisi TT, Uzuner G, Akadiri AC (2020a) Examining the causal impacts of tourism, globalization, economic growth and carbon emissions in tourism island territories: bootstrap panel Granger causality analysis. Curr Issue Tour 23(4):470–484

Akadiri S, Alola AA, Olasehinde-Williams G, Etokakpan MU (2020b) The role of electricity consumption, globalization and economic growth in carbon dioxide emissions and its implications for environmental sustainability targets. Sci Total Environ 708:134653

Akadiri SS, Adebayo TS (2021) Asymmetric nexus among financial globalization, non-renewable energy, renewable energy use, economic growth, and carbon emissions: impact on environmental sustainability targets in India. Environ Sci Pollut Res 29:16311–16323

Akbostancı E, Türüt-Aşık S, Tunç Gİ (2009) The relationship between income and environment in Türkiye: is there an environmental Kuznets curve? Energy Policy 37(3):861–867

Aslam B, Hu J, Hafeez M, Ma D, AlGarni TS, Saeed M, Hussain S (2021) Applying environmental Kuznets curve framework to assess the nexus of industry, globalization, and CO_2 emission. Environ Technol Innov 21:101377

Bilgili F, Ulucak R, Koçak E, İlkay SÇ (2020) Does globalization matter for environmental sustainability? Empirical investigation for Türkiye by Markov regime switching models. Environ Sci Pollut Res 27:1087–1100

Bulut U (2021) Environmental sustainability in Türkiye: an environmental Kuznets curve estimation for ecological footprint. Int J Sust Dev World 28(3):227–237

Cetin M, Ecevit E, Yucel AG (2018) Structural breaks, urbanization and CO_2 emissions: evidence from Türkiye. J Appl Econ Bus Res 8(2):122–139

Cole MA (2004) Trade, the pollution haven hypothesis and the environmental Kuznets curve: examining the linkages. Ecol Econ 48(1):71–81

Cristea A, Hummels D, Puzzello L, Avetisyan M (2013) Trade and the greenhouse gas emissions from international freight transport. J Environ Econ Manage 65(1):153–173

Dreher A (2006) Does globalization affect growth? Evidence from a new index of globalization. Appl Econ 38(10):1091–1110

Gaies B, Nakhli MS, Sahut JM (2022) What are the effects of economic globalization on CO_2 emissions in MENA countries? Econ Model 116:106022

Genç C, Ekinci A, Sakarya B (2022) The impact of output volatility on CO_2 emissions in Türkiye: testing EKC hypothesis with Fourier stationarity test. Environ Sci Pollut Res 29:3008–3021

Godil DI, Sharif A, Rafique S, Jermsittiparsert K (2020) The asymmetric effect of tourism, financial development, and globalization on ecological footprint in Türkiye. Environ Sci Pollut Res 27:40109–40120

Grossman GM, Krueger AB (1991) Environmental impacts of a North American free trade agreement. NBER Working Paper No 3914

Gygli S, Haelg F, Potrafke N, Sturm JE (2019) The KOF globalisation ındex–revisited. Rev Int Organ 14(3):543–574

Islam M, Khan MK, Tareque M, Jehan N, Dagar V (2021) Impact of globalization, foreign direct investment, and energy consumption on CO_2 emissions in Bangladesh: does institutional quality matter? Environ Sci Pollut Res 28(35):48851–48871

Johansen S (1988) Statistical analysis of cointegration vectors. J Econ Dyn Control 12(2-3):231–254

Johansen TA (1996) Identification of non-linear systems using empirical data and prior knowledge—an optimization approach. Automatica 32(3):337–356

Kalaycı C, Hayaloğlu P (2019) The impact of economic globalization on CO2 emissions: the case of NAFTA countries. Int J Energy Econ Policy 9(1):356–360

Khan AR (1998) The ımpact of globalization on South Asia. In: Bhalla AS (eds) Globalization, growth and marginalization. Palgrave Macmillan, London

Khan MK, Teng JZ, Khan MI, Khan MO (2019) Impact of globalization, economic factors and energy consumption on CO_2 emissions in Pakistan. Sci Total Environ 688:424–436

Kihombo S, Vaseer AI, Ahmed Z, Chen S, Kirikkaleli D, Adebayo TS (2022) Is there a tradeoff between financial globalization, economic growth, and environmental sustainability? An advanced panel analysis. Environ Sci Pollut Res 29(3):3983–3993

Koengkan M, Fuinhas JA, Santiago R (2020) Asymmetric impacts of globalisation on CO_2 emissions of countries in Latin America and the Caribbean. Environ Syst Decis 40(1):135–147

Lee E, Vivarelli M (2006) The social impact of globalization in the developing countries. Int Lab Rev 145(167)

Lee KH, Min B (2014) Globalization and carbon constrained global economy: a fad or a trend? J Asia-Pacific Bus 15(2):105–121

Lise W (2006) Decomposition of CO_2 emissions over 1980–2003 in Türkiye. Energy Policy 34(14):1841–1852

Ma C, Ju MT, Zhang XC, Li HY (2011) Energy consumption and carbon emissions in a coastal city in China. Procedia Environ Sci 4(2011):1–9

Mehmood U, Tariq S (2020) Globalization and CO_2 emissions nexus: evidence from the EKC hypothesis in South Asian countries. Environ Sci Pollut Res 27:1–13

Miao Y, Razzaq A, Adebayo TS, Awosusi AA (2022) Do renewable energy consumption and financial globalisation contribute to ecological sustainability in newly industrialized countries? Renew Energy 187:688–697

Mohsin M, Abbas Q, Zhang J, Ikram M, Iqbal N (2019) Integrated effect of energy consumption, economic development, and population growth on CO 2 based environmental degradation: a case of transport sector. Environ Sci Pollut Res 26:32824–32835

Narayan P (2004) Reformulating critical values for the bounds F-statistics approach to cointegration: an application to the tourism demand model for Fiji, vol 2, no 04. Monash University, Australia

OECD (2022) Why open markets matter?

Our World in Data (2022) CO_2 emissions. https://ourworldindata.org/co2-emissions

Ozturk I, Acaravci A (2013) The long-run and causal analysis of energy, growth, openness and financial development on carbon emissions in Türkiye. Energy Econ 36:262–267

Pata UK (2021) Linking renewable energy, globalization, agriculture, CO_2 emissions and ecological footprint in BRIC countries: a sustainability perspective. Renew Energy 173:197–208

Pata UK, Caglar AE (2021) Investigating the EKC hypothesis with renewable energy consumption, human capital, globalization and trade openness for China: evidence from augmented ARDL approach with a structural break. Energy 216:119220

Pesaran MH (2015) Time series and panel data econometrics. Oxford University Press, Oxford

Pesaran MH, Shin Y, Smith RJ (2001) Bounds testing approaches to the analysis of level relationships. J Appl Econom 16(3):289–326

Rehman A, Rauf A, Ahmad M, Chandio AA, Deyuan Z (2019) The effect of carbon dioxide emission and the consumption of electrical energy, fossil fuel energy, and renewable energy, on economic performance: evidence from Pakistan. Environ Sci Pollut Res 26:21760–21773

Sabir S, Gorus MS (2019) The impact of globalization on ecological footprint: empirical evidence from the South Asian countries. Environ Sci Pollut Res 26:33387–33398

Sadiq M, Shinwari R, Usman M, Ozturk I, Maghyereh AI (2022) Linking nuclear energy, human development and carbon emission in BRICS region: do external debt and financial globalization protect the environment? Nucl Eng Technol 54(9):3299–3309
Salahuddin M, Ali MI, Vink N, Gow J (2019) The effects of urbanization and globalization on CO_2 emissions: evidence from the Sub-Saharan Africa (SSA) countries. Environ Sci Pollut Res 26:2699–2709
Shahbaz M, Lean HH, Shabbir MS (2012) Environmental Kuznets curve hypothesis in Pakistan: cointegration and Granger causality. Renew Sustain Energy Rev 16(5):2947–2953
Shahbaz M, Hye QMA, Tiwari AK, Leitão NC (2013a) Economic growth, energy consumption, financial development, international trade and CO_2 emissions in Indonesia. Renew Sust Energ Rev 25:109–121
Shahbaz M, Ozturk I, Afza T, Ali A (2013b) Revisiting the environmental Kuznets curve in a global economy. Renew Sustain Energy Rev 25:494–502
Shahbaz M, Solarin SA, Ozturk I (2016) Environmental Kuznets curve hypothesis and the role of globalization in selected African countries. Ecol Indic 67:623–636
Shahbaz M, Gozgor G, Adom PK, Hammoudeh S (2019) The technical decomposition of carbon emissions and the concerns about FDI and trade openness effects in the United States. Int Econ 159:56–73
Solarin SA, Al-mulali U, Sahu PK (2017) Globalisation and its effect on pollution in Malaysia: the role of trans-Pacific partnership (TPP) agreement. Environ Sci Pollut Res 24:23096–23113
Soytas U, Sari R (2006) Energy consumption and income in G-7 countries. J Policy Model 28(7):739–750
Xu Z, Baloch MA, Meng F, Zhang J, Mahmood Z (2018) Nexus between financial development and CO_2 emissions in Saudi Arabia: analyzing the role of globalization. Environ Sci Pollut Res 25(28):28378–28390
Yilanci V, Gorus MS (2020) Does economic globalization have predictive power for ecological footprint in MENA counties? A panel causality test with a Fourier function. Environ Sci Pollut Res 27(32):40552–40562
Yurtkuran S (2021) The effect of agriculture, renewable energy production, and globalization on CO_2 emissions in Türkiye: a bootstrap ARDL approach. Renew Energy 171:1236–1245
Zafar MW, Saud S, Hou F (2019a) The impact of globalization and financial development on environmental quality: evidence from selected countries in the Organization for Economic Co-operation and Development (OECD). Environ Sci Pollut Res 26(13):13246–13262
Zafar MW, Zaidi SAH, Sinha A, Gedikli A, Hou F (2019b) The role of stock market and banking sector development, and renewable energy consumption in carbon emissions: insights from G-7 and N-11 countries. Resour Policy 62:427–436
Zhu S, Fu X (2013) Drivers of export upgrading. World Dev 51:221–233

Achieving Energy Transition Through Economic Prosperity and Financial Development: An OECD Countries Perspective

Iza Gigauri, Mirela Panait, Laeeq Razzak Janjua, and Simona Andreea Apostu

Abstract Renewable energy has gained increasing attention from practitioners and scholars as globalization, industrialization, and urbanization raised the demand for energy on the one hand, and on the other hand, traditional energy sources have obviously affected environmental damage, pollution, and climate change. In addition, sustainable development implies sustainable and efficient use of energy while ensuring energy supply to all citizens around the world. In this regard, renewable energy transition become prevalent for both developed and emerging countries. Therefore, factors influencing the renewable energy transition need to be studied. This research examines the role of economic prosperity and financial development in energy transformation towards zero-carbon energy sources.Data was gathered from OECD countries between 2003 and 2023. The research analyzes 546 observations from various data sources to test variables. Driscoll-Kraay nonparametric technique was used to estimate standard errors. This research demonstrates that the increase in economic growth leads to a renewable energy transition, while an increase in carbon emission causes a decrease in renewable energy consumption in OECD countries. Moreover, the increase in environmental patents gives rise to the renewable energy transition. Furthermore, the expansion of ATMs per one thousand adults measuring micro-financial development, and the increase in bank loans by

I. Gigauri
School of Business and Administrative Studies, The University of Georgia, Tbilisi, Georgia
e-mail: i.gigauri@ug.edu.ge

M. Panait
Petroleum-Gas University of Ploieşti, Romania & Institute of National Economy, Bucharest, Romania
e-mail: mirela.matei@upg-ploiesti.ro

L. R. Janjua (✉)
WSB University, Dąbrowa Górnicza, Poland
e-mail: laeeq.janjua@wsb.edu.pl

S. A. Apostu
Bucharest University of Economic Studies, Bucharest, Romania
e-mail: simona.apostu@csie.ase.ro

F. A. Malik et al. (eds.), *Aligning ESG Goals with the UNs Sustainable Development Goals Planning*, World Sustainability Series,
https://doi.org/10.1007/978-3-032-01715-4_20

governments and public enterprises as well as GDP growth reflecting macro-financial development positively correlates with renewable energy consumption. The study demonstrated the importance of a reliable financial system for energy transition. Since micro-financial and macro-financial development increase renewable energy consumption, policies and strategies should be directed towards the development in order to facilitate clean energy transition.

Keywords Renewable energy · Financial development · OECD · Sustainability · Economic growth

1 Introduction

Recent pandemics, economic, and war crises have changed the political, economic, and social scenes causing transformation and adaptation needs. In this sense, energy becomes the focus of the attention of researchers and professionals. Ever since energy has become available, demand for it has been growing as the lifestyle of humanity changes and the population increases (Armeanu et al. 2023; Tudorică et al. 2024). Urbanization and industrialization caused the use of unsustainable energy leading to environmental degradation (Dilanchiev et al. 2023; Zhang et al. 2023; Nuță et al. 2024). Energy production and consumption impact environmental pollution and negative externalities were registered for the analyzed countries and regions (Tiwari et al. 2013; Panait et al. 2019; Shahbaz and Sinha 2019; Balsalobre-Lorente et al. 2022). Considering the challenges generated by the use of fossil fuels, humanity is in a new energy transition process which, unlike previous transitions, is a politically driven process, the involvement of public authorities being essential considering the complexity of economic, social and technical changes (Khan et al. 2022; Dilanchiev et al. 2023; Andrei et al. 2023). Therefore, developed countries especially show an increasing interest in renewable energy. A modern society living during technological progress consumes more energy. Improved life standards require energy to be used in all economic activities transforming one sort of energy into another (Smil 2017; Puime et al. 2021). Energy is a foundation for economic development, and the production and consumption of energy are strongly related to sustainable economic growth (Østergaard et al. 2021; Andrei et al. 2017; Saqib et al. 2024). The sustainability agenda emphasizes that energy should be produced and consumed in a sustainable way to reduce environmental damage (Sharma and Kar 2015; Andrei et al. 2023).

The global challenges related to poverty, climate change, environmental protection, peace, and prosperity are addressed through the United Nation's Sustainable Development Goals (SDGs). Moreover, the United Nation's Sustainable Development Goals (SDGs) suggest sustainable and efficient use of energy. SDG 7 aspires to provide affordable and clean energy to all citizens around the world, while SDG 13 aims to tackle climate change by monitoring energy-related emissions. Accessible energy can reduce poverty (Neacsa et al. 2020), while renewable energy can address urgent global challenges concerning climate change, environmental pollution, and depletion of resources (Kutscher et al. 2018; Panait et al. 2022a). Previous studies connect energy, emissions, economic growth, and increased urbanization (Armeanu et al. 2021; Akbar et al. 2020). However, unsustainable sources still do not generate a sufficient amount of energy to meet the demand (Lyman 2016; Popescu et al. 2022). Though renewable energy sources have the potential to increase the supply without damaging the ecosystem if their share in the energy sector raise (Sharma and Kar 2015; Gigauri and Vasilev 2022; Khan et al. 2022). Against this background, the European Commission invokes energy production from renewable sources (Brătianu 2016).

Economic growth measured by the gross domestic product (GDP) and GDP per capita, involves the employment level in a country, human development, and social well-being, as well as the number of citizens above the poverty line (Apostu et al. 2022). Recent studies demonstrate the relationship between sustainable economic growth, renewable energy, and economic growth (Panait et al. 2022b). Empirical research in the OECD nations confirmed that financial development determines energy demand and environmental technologies facilitate energy efficiency while decreasing energy consumption (Paramati et al. 2022). It should be noted that various aspects of economic development can be considered when investigating the relationship between renewable energy and economic development (Zhe et al. 2021).

While some authors found a positive relationship between renewable energy consumption and economic growth (Fotourehchi 2017; Khobai and Roux 2018; Temiz Dinç and Akdoğan 2019; Zeren and Akkuş 2020), others could not detect any significant correlation between them (Chen et al. 2020; Can and Korkmaz 2019; Marinaş et al. 2018). However, some studies have revealed that renewable energy consumption positively impacts financial development (Anton and Nucu 2020; Ji and Zhang 2019; Wang et al. 2020).

Thus, the studies seem controversial. In this context, it is important to research the relationship between the three variables in OECD countries. In this context, factors influencing the renewable energy transition need to be explored. Thus, the aim of this research is to investigate the role of economic prosperity and financial development in energy transformation towards zero-carbon energy sources. Data for this study was collected from the OECD countries between 2000 and 2020, available on open sources. This research examines the OECD countries as they share similar economic parameters and they "are not energy-rich" (Gorus and Karagol 2022). The study analyzes 546 observations to test variables. The Driscoll-Kraay nonparametric technique was used to estimate standard errors.

The conducted research reveals that the increase in economic growth leads to a renewable energy transition in the OECD countries, and an increase in carbon emission triggers a decrease in renewable energy consumption. In addition, the increase of environmental patents gives rise to the renewable energy transition. Furthermore, the expansion of ATMs per one thousand adults measuring micro-financial development, and an increase in bank loans by governments and public enterprises as well as GDP growth reflecting macro-financial development positively correlates with renewable energy consumption.

After the Introduction, the paper is structured as follows: Literature Review provides the summary of the related literature, followed by the explanation of the methodology and presenting research results. The subsequent section discusses the findings and derives implications for theory and practice. Finally, the conclusion section includes the limitations of this research and suggests further study directions.

2 Theoretical and Literature Review

Recent studies confirm that financial development contributes to economic growth along with sustainable development goals (Chauhan et al. 2022; Bashir and Qureshi 2023). The energy transition is a complex process that requires important financial funds in order to finance different projects to use renewable energy as a new resources or to change energy mix used by different stakeholders companies, public authorities or households. Because of that, the researchers also focused their attention on the relationship between financial development and energy transition, being studies that focus on the impact that education and financial inclusion have on attracting the necessary funds in the financial circuit for the implementation of specific projects for the production of renewable energy. Another direction of research is the impact that financial development has on the energy transition, the increase in energy consumption generating the intensification of concerns for the production of energy from renewable sources.

The study using data from 1990 to 2015 and VAR analysis found that renewable energy has no significant impact on economic growth but positively influences the financial development in Turkey (Zhe et al. 2021). However, the authors established that economic development and renewable energy consumption together impact the financial development of a country while both financial development and renewable energy consumption have no important effect on economic growth (Zhe et al. 2021). This can be explained by the fact that renewable energy induces the interest of investors and hence, the financial sector can grow (Zhe et al. 2021). In turn, financial development facilitates renewable energy consumption (Zhe et al. 2021).

Anton and Nucu (2020), using panel data from twenty-eight EU member countries between 1990 and 2015, confirmed the positive impact of financial development on renewable energy consumption but reported no effect of the financial market development on renewable energy usage. They concluded that financial development can facilitate renewable energy development encouraging the investments shift from traditional to renewable technologies (Anton and Nucu 2020). Simionescu et al. (2022) investigated panel data from 23 EU states from 1990 to 2020 to conclude that economic development leads to renewable energy consumption and clean energy consumption especially in transport increases economic growth.

The research using data from 72 developing and developed economies confirmed the correlation between renewable energy usage, economic development, capital stock, and trade (Amri 2016). In particular, the listed three factors—renewable energy usage, capital stock, and trade can increase economic growth by 1% (Amri 2016).

Studies analyzing the effect of financial development on energy consumption have produced mixed results. While Chang (2015), Sadorsky (2010, 2011), Ozturk and Acaravci (2013), Çoban and Topcu (2013), Islam et al. 2013), and Paramati et al. (2017) have found a positive impact of financial development of energy consumption, other studies (e.g. Shahbaz et al. 2016a; Topcu and Payne 2017; Ouyang and Li 2018; Destek 2018) have identified a negative influence. In addition, unidirectional or bidirectional relationship between the two variables has been found by Islam et al. (2013) and Furuoka (2015), whereas neutral effect confirmed by another scholars (e.g. Sadorsky 2011; Altay and Topcu 2015; Shahbaz et al. 2017).

2.1 OECD Country Contexts

OECD countries strive to meet the Paris Climate Agreement and reduce emissions by transitioning to renewable energy sources. A systematic literature review revealed the growing potential of the renewable energy market and the need for policy incentives (Li et al. 2020). Prior studies show that around 83% of electricity was produced from unrenewable sources in Australia (OECD country) in 2016 and only 6%—from renewable energy sources (Li et al. 2020). Accordingly, Australia was responsible for the greatest amount of greenhouse emissions worldwide (Hua et al. 2016) in spite of its renewable resources (Hemer et al. 2018). To change this tendency, the country is switching toward more renewable energy closing coal power plants (Li et al. 2020).

Research results based on the expert interview methods found that the adoption of renewable energy in Poland (an OECD member country) can benefit the low-carbon economy, and create new jobs and favorable ecological conditions (Pietrzak et al. 2021). Erdoğan et al. (2023) examined the links between economic development, renewable energy consumption, and non-renewable energy usage on carbon emissions in Canada between 1965 and 2017. Their results confirmed an asymmetric effect of the variables on carbon emissions and suggested that the increase in renewables reduces emissions and enhances economic growth (Erdoğan et al. 2023).

Studying the effect of different energy sources on economic growth through the panel data of 34 OECD countries between 1995 and 2019, Wang et al. (2022) have concluded that all sources—renewables, oil, and natural gas influence carbon emission without contributing to economic development, yet renewable energy still contributes to economic growth.

It is noteworthy that studies carried out in OECD countries demonstrated mixed results. Some early studies confirmed the unidirectional causality between economic development and renewable energy adoption in OECD countries (Salim et al. 2014), while others found a bidirectional relationship between renewable energy and GDP in the OECD region (Ohler and Fetters 2014).

The research performed in the United States (OECD country) using data between 1949 and 2006 has not identified any connection between economic development (measured with GDP) and renewable energy use (Payne 2009). Likewise, further studies conducted in the European countries (some of which are members of the

OECD) established no significant correlation between the two variables (Menegaki 2011; Śmiech and Papież 2014; Omri et al. 2015).

Magazzino (2017) analyzed the relationship between renewable energy adoption and economic growth in Italy (OECD country) using data between 1970–2007, and reported that "if renewable energy consumption increases by 1%, real gross domestic product (GDP) decreases by 0.23%".

Furthermore, studies analyzed the impact of the stock market and FDI inflows on renewable energy consumption through the data between 1993 and 2012 from the EU, G20, and OECD countries concluded that FDI and stock market developments positively affect clean energy usage (Paramati et al. 2017). In addition, unidirectional causality was also found between energy consumption and financial development through the study examining data from 1980 to 2012 (Furuoka 2015). Likewise, the positive impact of clean energy consumption on economic development was maintained by Magnani and Vaona (2013) analyzing data from Italy, by Chang et al. (2015) examining the contexts of Germany and Japan, by Inglesi-Lotz (2016) investigating the OECD countries, and by Bhattacharya et al. (2016) and Bhattacharya et al. (2017) exploring data from multiple countries.

However, some more studies proved the negative effect between the variables. For example, Marques and Fuinhas (2012) examining data from 24 European states from 1990 to 2004 and Bhattacharya et al. (2016) considering 38 countries including OECD countries from 1991 to 2012 suggest the negative consequences of renewable energy consumption on economic development.

A further negative effect of renewable energy consumption on financial development has been shown by Topcu and Payne (2017), who analyzed 32 high-income countries including OECD members through the data between 1990–2014, and stated that the increase in the stock market index causes a decrease in energy consumption. Accordingly, they have found no significant correlation between financial development and energy consumption. These results have been confirmed by more research revealing no statistical impact of the financial development index on energy consumption (Çoban and Topcu 2013; Lee 2013; Ozturk and Acaravci 2013; Shahbaz et al. 2016b; Abbasi and Riaz 2016).

Moreover, Chen et al. (2020) investigating data from 103 countries over the period of 1995–2015 illustrated the linear and positive impact of renewable energy on economic growth for OECD countries, no significant influence of renewable energy on economic development for developed countries, and negative effect for developing economies.

Thus, it can be concluded that the literature review presents varied results on the correlation between renewable energy, economic growth, and financial development for different periods and regions/countries. This study is individualized by the indicators used, the analyzed period and the selected countries, considering the commitment of the OECD countries for the energy transition, they being world leaders regarding the implementation of the necessary measures to achieve the specific objectives of the low carbon economy.

3 Methods and Data

In order to analyze the factors influencing the renewable energy transition, we considered the following variables: renewable energy consumption, CO_2 emissions, GDP per capita, patent applications for environmental development, urban population, ATMs, bank branches, domestic credit to the private sector, and bank credit to government and public enterprises. The data explanation is presented in Table 1.

Data for this study was collected from the OECD countries between 2003 and 2023, available on open sources. This research examines the OECD countries as they share similar economic parameters and they "are not energy-rich", the sample comprising 546 observations.

In order to examine the sample characteristics descriptive analyses of the data were conducted. Summary of descriptive statistics for all variables used in this study can be seen in full on Table 2.

As it can be seen in Table 2, all variables are normally distributed. In case of renewable energy, the values are oscillating between 0.85 and 80.12 with a mean of 20.90 and a standard deviation of 17.31. For CO_2 emissions, we registered a minimum of 2.68 and a maximum of 24.82 with a mean of 7.96 and standard deviation of 3.41. The values for GDP are varying between 6420.84 and 112372.7, the standard deviation being of 23285.9 and the mean of 10.78. The smallest value for PATE for our sample is 0.92 and the biggest is 28.64, the mean being 10.78 and standard deviation being 4.24. The URB values oscillate between 50.75 and 98.44, the mean being of

Table 1 Source of variables and descriptions (Period 2003–2023)

Abbreviation	Variable	Source
RE	Renewable energy consumption (% of total final energy consumption)	IEA
CO	CO_2 emissions (metric tons per capita)	WDI
GDP	GDP per capita (Constant 2015)	WDI
PATE	Percentage of patent applications for environmental development	OECD
URB	Urban population (% of total population)	WDI
FD*ATM	ATMs per 100,000 adults (micro financial development)	The global economy
FD*BB	Bank branches per 100,000 people (micro financial development)	The global economy
FD*DC	Domestic credit to the private sector, percent of GDP (macro financial development	The global economy
FD*BCGP	Bank credit to government and public enterprises, percent of GDP (macro financial development)	The global economy

Table 2 Research results

Variable	Observations	Mean	Std. Dev	Min	Max
RE	546	20.8996	17.31134	0.8528	80.12135
CO	546	7.957189	3.407395	2.682623	24.82465
GDP	546	37042.22	23285.9	6420.838	112372.7
PATE	546	10.77923	4.242394	0.92	28.64
URB	546	74.16937	11.83527	50.754	98.441
FD*ATM	546	80.21309	35.5785	20.25	194.63
FD*BB	546	33.92274	22.29105	4.44	110.94
FD*DC	546	91.40454	42.39621	12.87	304.58
FD*BCGP	546	12.85367	8.109394	0.04	42.22

74.17 and standard deviation being of 11.84. The minimum values for FD*ATM is 20.25 and the maximum values is 194.63, standard deviation is 35.58 and the mean 80.21. The FD*BB values are varying between 4.44 and 110.94, the mean is 33.92 and standard deviation is 22.29. The mean of FD*DC is 91.41, standard deviations is 42.40, the minimum is 12.87 and maximum 304.58. In case of FD*BCGP the minimum value registered is 0.04 and the maximum value is 42.22, the mean is 12.85, and standard deviation is 8.11.

4 Empirical Results

In order to investigate the role of economic prosperity and financial development in energy transformation towards zero-carbon energy sources, the correlation analysis is employed. According to Table 3, the variables are correlated, there is an inverse link between renewable energy and CO_2 emissions, between renewable energy and FD*ATM, FD*BB, and FD*BCGP. Also, a negative link is encountered between CO_2 emissions and FD*DC and FD*BCGP, between GDP and PATE, and FD*BCGP. The PATE has a negative link with FD*ATM and FD*BB, between URB and FD*ATM and FD*BCGP, and between FD*DC and FD*BCGP.

Using the variables we considered 6 models, the dependent variable being renewable energy. Using Generalized Least Squares Estimator the six models are presented in Table 4. According to model 1 and model 3, renewable energy is significantly influenced by GDP, CO_2 emissions, and patent applications. According to model 2, renewable energy is significantly influenced by GDP, CO_2 emissions, patent applications and FD*ATM. Model 4 reflects that renewable energy is significantly influenced by GDP, CO_2 emissions, patent applications, and FD*DC. The results from model 5 indicate that renewable energy is significantly influenced by GDP, CO_2 emissions, patent applications, and FD*BCGP. In case of model 6 renewable energy is significantly influenced by GDP, CO_2 emissions, patent applications, FD*BCGP, FD*DC,

Table 3 Correlation analysis

Variable	RE	CO	GDP	PATE	URB	FD*ATM	FD*BB	FD*DC	FD*BCGP
RE	1								
CO	−0.275	1							
GDP	0.193	0.456	1						
PATE	0.025	0.012	−0.017	1					
URB	0.216	0.276	0.509	0.035	1				
FD*ATM	−0.139	0.021	0.164	−0.002	−0.103	1			
FD*BB	−0.175	0.257	0.289	−0.101	0.204	0.501	1		
FD*DC	0.251	−0.064	0.509	0.117	0.395	0.385	0.308	1	
FD*BCGP	−0.257	−0.132	−0.220	0.027	−0.023	0.162	0.087	−0.140	1

and FD*ATM. In case of all models GDP, and patent applications have a positive impact on renewable energy, and CO_2 emissions have a negative impact on renewable energy. FD*BCGP, FD*DC, FD*BB, and FD*ATM have a positive impact on renewable energy (Table 4).

Using fixed effects, the six models are presented in Table 5. The results in model 1, model 3 and model 4 indicate that renewable energy is significantly influenced by GDP, CO_2 emissions, urbanization, and patent applications. According to model 2 renewable energy is significantly influenced by GDP, CO_2 emissions, FD*ATM, and patent applications. Model 5 reflects that renewable energy is significantly influenced by GDP, CO_2 emissions, urbanization, FD*ATM, patent applications, and FD*BCGP. Model 6 reflects that renewable energy is significantly influenced by GDP, CO_2 emissions, urbanization, FD*ATM, patent applications, FD*ATM, and FD*BCGP. In case of all models GDP, urbanization and patent applications have a positive impact on renewable energy, and CO_2 emissions have a negative impact on renewable energy. FD*BCGP, FD*DC, FD*BB, and FD*ATM have a positive impact on renewable energy.

Table 4 POLS

OLS (driscoll-Kraay standard errors)						
Variable	Model 1	Model 2	Model 3	Model 4	Model 5	Model 6
GDP	0.300 (0.026)***	0.454 (0.036)***	0.344 (0.032)***	0.309 (0.053)***	0.290 (0.025)***	0.265 (0.053)***
CO	−1.236 (0.132)***	−1.264 (0.134)***	−1.251 (0.120)***	−1.242 (0.142)***	−1.245 (0.124)***	−1.174 (0.112)***
URB	0.073 (0.064)	0.350 (0.072)	0.183 (0.032)	0.077 (0.074)	0.066 (0.063)	0.074 (0.119)
PATE	0.019 (0.081)**	0.194 (0.079)**	0.100 (0.069)**	0.197 (0.084)**	0.226 (0.081)**	0.098 (0.073)**
FD*ATM	–	0.433 (0.036)***	–	–	–	0.182 (0.097)**
FD*BB	–		0.443 (0.018)	–	–	0.408 (0.028)
FD*DC	–			0.015 (0.057)**	–	0.249 (0.100)**
FD*BCGP	–				0.097 (0.053)*	0.048 (0.038)*
Constant	1.2839 (0.445)***	3.428 (0.611)***	2.090 (0.372)***	1.255 (0.538)**	1.595 (0.444)***	3.358 (0.836)***
Number of groups	26	26	26	26	26	26
Observations	546	546	546	546	546	546

*** indicates level of significance at 1%, ** indicates level of significance at 5%, and * indicates level of significance at 10%

Table 5 FE (D-K)

Fixed effect (driscoll-Kraay standard errors)						
Variable	Model 1	Model 2	Model 3	Model 4	Model 5	Model 6
GDP	1.012 (0.039)***	0.884 (0.084)***	1.019 (0.045)***	1.009 (0.043)***	1.037 (0.036)***	0.807 (0.146)***
CO	−1.785 (0.107)***	−1.776 (0.107)***	−1.794 (0.104)***	−1.782 (0.116)***	−1.746 (0.105)***	−1.693 (0.130)***
URB	3.531 (0.437)*	3.863 (0.478)*	3.591 (0.462)**	3.544 (0.479)*	3.416 (0.450)*	3.591 (0.509)*
PATE	0.084 (0.023)***	0.077 (0.021)***	0.083 (0.043)***	0.082 (0.028)***	0.084 (0.024)***	0.080 (0.032)**
FD*ATM	–	0.211 (0.061)***	–	–	–	0.306 (0.147)*
FD*BB	–	–	0.017 (0.043)	–	–	0.101 (0.082)
FD*DC	–	–	–	0.014 (0.065)	–	0.024 (0.075)
FD*BCGP	–	–	–	–	0.031 (0.009)***	0.023 (0.008)**
Constant	−1.552 (0.066)***	−0.560 (0.043)***	0.923 (0.281)***	−1.646 (0.361)***	−1.472 (0.132)***	−1.664 (0.164)***
Number of groups	26	26	26	26	26	26
Observations	546	546	546	546	546	546

*** indicates level of significance at 1%, ** indicates level of significance at 5%, and * indicates level of significance at 10%

Table 6 presents the results using Feasible Generalized Least Squares Estimator. Model 1 and model 4 indicate that renewable energy is significantly influenced by GDP, CO_2 emissions, and patent applications. The results in model 2 reflects that renewable energy is significantly influenced by GDP, CO_2 emissions, patent applications, and FD*ATM. According to model 3 renewable energy is significantly influenced by GDP, CO_2 emissions, patent applications, and FD*BB. The results in model 5 indicate that renewable energy is significantly influenced by GDP, CO_2 emissions, patent applications, and FD* BCGP. The results in model 6 indicate that renewable energy is significantly influenced by GDP, CO_2 emissions, patent applications, FD*BB, and FD* BCGP.

In case of all models GDP, and patent applications have a positive impact on renewable energy, and CO_2 emissions have a negative impact on renewable energy. FD*BCGP, FD*DC, FD*BB, and FD*ATM have a positive impact on renewable energy (Table 6).

Considering all the results above, we can conclude that increase in economic growth helpful lead to increase renewable energy transition. Increase in carbon emission decreases renewable energy consumption. Environmental patents increase

Table 6 FGLS

FGLS						
Variable	Model 1	Model 2	Model 3	Model 4	Model 5	Model 6
GDP	0.300 (0.065)***	0.454 (0.069)***	0.344 (0.061)***	0.309 (0.085)***	0.290 (0.065)***	0.265 (0.080)***
CO	−1.23 (0.096)***	−1.264 (0.094)***	−1.251 (0.090)***	−1.242 (0.103)***	−1.245 (0.095)***	−1.174 (0.097)***
URB	0.073 (0.253)	0.350 (0.259)	0.183 (0.283)	0.077 (0.254)	0.066 (0.251)	0.074 (0.262)
PATE	0.196 (0.077)***	0.194 (0.075)**	0.100 (0.073)**	0.197 (0.077)**	0.226 (0.074)**	0.098 (0.075)**
FD*ATM	–	0.433 (0.081)**	–	–	–	0.184 (0.097)**
FD*BB	–	–	0.443 (0.052)**	–	–	0.408 (0.061)*
FD*DC	–	–	–	0.015 (0.100)	–	0.249 (0.108)
FD*BCGP	–	–		–	0.097 (0.041)**	0.048 (0.040)**
Constant	1.283 (0.915)***	1.428 (0.978)***	1.090 (0.864)**	1.255 (0.933)**	1.595 (0.920)*	0.538 (0.120)**
Number of groups	26	26	26	26	26	26
Observations	546	546	546	546	546	546

renewable energy transition (consumption). Increase in ATMs per 100,000 adults (micro financial development) and Bank credit to government and public enterprises, percent of GDP (macro financial development) increase renewable energy consumption.

5 Discussion and Conclusion

The presented study results can contribute to the current knowledge of the relationship between renewable energy, economic prosperity, and financial development and raise awareness of how to meet the increasing need for renewable energy. Affordable and accessible energy is the focus of SDG 7. Sustainable and reliable energy is a fundamental source of economic and social progress. The world needs to transit to renewable energy while advancing development in agriculture, education, healthcare, transportation, business, communication, technologies, and other areas.

This research resonates with previous studies which have demonstrated the positive correlation between economic growth, financial development, and renewable energy consumption (Khobai and Roux 2018; Zeren and Akkuş 2020; Wang et al.

2020). Our research results confirmed that clean energy not only reduces emissions but also facilitates the economic development of a country (e.g. Erdoğan et al. 2023; Chen et al. 2020; Paramati et al. 2017; Bhattacharya et al. 2017). In addition, a reliable financial system in a country can foster clean energy consumption. Since micro-financial and macro-financial development increase renewable energy consumption, policies and strategies should be directed towards the development in order to facilitate clean energy transition.

Furthermore, Sustainable Development Goals can be reached by sustainable and efficient use of energy. Renewable energy sources contribute to the improvement of climate change, ecological issues, environmental protection, as well as, ensure peace, prosperity and alleviation of poverty. An increase in economic development leads to the increase of renewable energy transition, which, in turn, gives rise to a decrease in carbon emissions. This result emphasizes the significance of governmental policies and corporate actions towards clean energy.

The research results have theoretical and practical implications. First, it extends the literature by solving the gap previous studies have shown related to the controversial results about the relationship between renewable energy and economic growth. Moreover, the results will benefit scholars and academics to expand their studies based on the research findings as well as to replicate this study methodology. Next, practitioners and managers striving to reduce the negative ecological footprints of their businesses can rationalize their decisions based on the study results. Likewise, climate change requires actions from the business also within the framework of their corporate social responsibility and sustainability efforts.

Climate change is the main focus of SDG 13, which urges economic agents and governmental bodies to tackle environmental destruction and its impacts. Climate change is tightly intertwined with other sustainable development goals, such as, for example, good health (SDG 3), education (SDG 4), clean water and sanitation (SDG 6), affordable energy (SDG 7), decent work and economic prosperity (SDG 8), sustainable cities (SDG 11), and inclusive societies and national security (SDG 16). Therefore, augmented consumption of renewable energy is one of the highest priorities of the current age reflected by the sustainable development agenda. Inclusive and sustainable societies can be built through policy support for renewable energy supply. On the one hand, accelerated development causes the amplified demand for energy consumption, and on the other hand, environmental pollution, carbon emissions, and climate change need urgent actions to be taken. In this regard, technological advancement and appropriate policy decisions to support clean energy can solve the tension between the two. Decision-makers should encourage all sectors to contribute to the transformation of the energy industry. In this context, the empirical results backing policymakers in incentivizing clean energy consumption can lead to an efficient and sustainable energy sector. It is also significant to promote renewable energy in order to tackle the grand challenges modern society is facing. Adopting green energy would hinder further degradation of the ecosystem and contribute to sustainable goals.

Being aware of the limits of this research, the authors have in mind the realization of analyzes that focus on the countries of the European Union, considering the harmonization of legislation and policies in the field. Despite the efforts, at the EU level we cannot speak of a convergence regarding the stage of the energy transition, which is why studies can be done for groups of countries, namely the old member countries and the new member countries, there being certain differences regarding the level of economic development, the endowment with energy resources, the attitude of public authorities and consumers towards green energy, the challenges generated by the armed conflict in Ukraine which caused concerns regarding energy security and ensuring a fair transition, which affects consumers as little as possible.

References

Abbasi F, Riaz K (2016) CO2 emissions and financial development in an emerging economy: an augmented VAR approach. Energy Policy 90:102–114. https://doi.org/10.1016/j.enpol.2015.12.017

Akbar A, Alam Rehman IU, Zeeshan M, Afridi FEA (2020) Unraveling the dynamic nexus between trade liberalization, energy consumption, CO2 emissions, and health expenditure in Southeast Asian countries. Risk Manag Healthc Policy 13:1915–1927. https://doi.org/10.2147/RMHP.S272801

Altay B, Topcu M (2015) Relationship between financial development and energy consumption: the case of Turkey. Bull Energy Econ (BEE) 3(1):18–24

Amri F (2016) The relationship amongst energy consumption, foreign direct investment and output in developed and developing countries. Renew Sustain Energy Rev 64:694–702. https://doi.org/10.1016/j.rser.2016.06.065

Andrei JV, Mieila M, Panait M (2017) The impact and determinants of the energy paradigm on economic growth in European Union. Plos One 12(3). https://doi.org/10.1371/journal.pone.0173282

Andrei JV, Zaharia A, Graci G, Chivu L (2023) Energy transition or energy diversification? Assessing the complexity of energy ecosystem towards transiting a climate neutral society. Environ Sci Pollut Res 30(50):108477–108511

Anton SG, Nucu AEA (2020) The effect of financial development on renewable energy consumption. A panel data approach. Renew Energy 147:330–338. https://doi.org/10.1016/j.renene.2019.09.005

Apostu SA, Mukli L, Panait M, Gigauri I, Hysa E (2022) Economic growth through the lenses of education, entrepreneurship, and innovation. Adm Sci 12(3):74. https://doi.org/10.3390/admsci12030074

Armeanu DS, Joldes CC, Gherghina SC, Andrei JV (2021) Understanding the multidimensional linkages among renewable energy, pollution, economic growth and urbanization in contemporary economies: quantitative assessments across different income countries' groups. Renew Sustain Energy Rev 142:110818. https://doi.org/10.1016/j.rser.2021.110818

Armeanu DS, Gherghina SC, Andrei JV, Joldes CC (2023) Evidence from the nonlinear autoregressive distributed lag model on the asymmetric influence of the first wave of the COVID-19 pandemic on energy markets. Energy Environ 34(5):1433–1470

Balsalobre-Lorente D, Ibáñez-Luzón L, Usman M, Shahbaz M (2022) The environmental Kuznets curve, based on the economic complexity, and the pollution haven hypothesis in PIIGS countries. Renew Energy 185:1441–1455

Bashir I, Qureshi IH (2023) A systematic literature review on personal financial well-being: the link to key sustainable development goals 2030. FIIB Bus Rev 12(1):31–48. https://doi.org/10.1177/23197145221106862

Bhattacharya M, Paramati SR, Ozturk I, Bhattacharya S (2016) The effect of renewable energy consumption on economic growth: evidence from top 38 countries. Appl Energy 162:733–741. https://doi.org/10.1016/j.apenergy.2015.10.104
Bhattacharya M, Churchill SA, Paramati SR (2017) The dynamic impact of renewable energy and institutions on economic output and CO2 emissions across regions. Renew Energy 111:157–167. https://doi.org/10.1016/j.renene.2017.03.102
Brătianu C (2016) European energy strategies. In: Tanţău, Dima A, Hadad AM (eds) Energy, climate change and sustainability: proceedings of the 10th international conference on business excellence. Business Excellence, Bucharest
Can H, Korkmaz Ö (2019) The relationship between renewable energy consumption and economic growth: the case of Bulgaria. Int J Energy Sect Manag 13(3):573–589. https://doi.org/10.1108/IJESM-11-2017-0005
Chang SC (2015) Effects of financial developments and income on energy consumption. Int Rev Econ Financ 35:28–44. https://doi.org/10.1016/j.iref.2014.08.011
Chang T, Gupta R, Inglesi-Lotz R, Simo-Kengne B, Smithers D, Trembling A (2015) Renewable energy and growth: evidence from heterogeneous panel of G7 countries using Granger causality. Renew Sustain Energy Rev 52:1405–1412. https://doi.org/10.1016/j.rser.2015.08.022
Chauhan S, Verma A, Kumar CV (2022) Effect of capital structure on the financial and social performance of Indian microfinance institutions. FIIB Bus Rev. https://doi.org/10.1177/23197145221099677
Chen C, Pinar M, Stengos T (2020) Renewable energy consumption and economic growth nexus: evidence from a threshold model. Energy Policy 139:111295. https://doi.org/10.1016/j.enpol.2020.111295
Çoban S, Topcu M (2013) The nexus between financial development and energy consumption in the EU: a dynamic panel data analysis. Energy Econ 39:81–88. https://doi.org/10.1016/j.eneco.2013.04.001
Destek MA (2018) Financial development and energy consumption nexus in emerging economies. Energy Sour Part B 13(1):76–81. https://doi.org/10.1080/15567249.2017.1405106
Dilanchiev A, Nuta F, Khan I, Khan H (2023) Urbanization, renewable energy production, and carbon dioxide emission in BSEC member states: implications for climate change mitigation and energy markets. Environ Sci Pollut Res 30(25):67338–67350
Erdoğan E, Serin Oktay D, Manga M, Bal H, Algan N (2023) Examining the effects of renewable energy and economic growth on carbon emission in Canada: evidence from the nonlinear ARDL approaches. Eval Rev. https://doi.org/10.1177/0193841x231166973
Fotourehchi Z (2017) Clean energy consumption and economic growth: a case study for developing countries. Int J Energy Econ Policy 7(2):61–64. Accessed 15 June 2023. https://dergipark.org.tr/en/pub/ijeeep/issue/31921/351179
Furuoka F (2015) Financial development and energy consumption: evidence from a heterogeneous panel of Asian countries. Renew Sustain Energy Rev 52:430–444. https://doi.org/10.1016/j.rser.2015.07.120
Gigauri I, Vasilev V (2022) Corporate social responsibility in the energy sector: towards sustainability. In: Khan SAR, Panait M, Puime Guillen F, Raimi L (eds) Energy transition. Industrial ecology. Springer, Singapore. https://doi.org/10.1007/978-981-19-3540-4_10.
Gorus MS, Karagol ET (2022) Reactions of energy intensity, energy efficiency, and activity indexes to income and energy price changes: the panel data evidence from OECD countries. Energy 254:124281. https://doi.org/10.1016/j.energy.2022.12428
Hemer MA, Manasseh R, McInnes KL, Penesis I, Pitman T (2018) Perspectives on a way forward for ocean renewable energy in Australia. Renew Energy 127:733–745. https://doi.org/10.1016/j.renene.2018.05.036
Hua Y, Oliphant M, Hu EJ (2016) Development of renewable energy in Australia and China: a comparison of policies and status. Renew Energy 85:1044–1051. https://doi.org/10.1016/j.renene.2015.07.060

Inglesi-Lotz R (2016) The impact of renewable energy consumption to economic growth: a panel data application. Energy Econ 53:58–63. https://doi.org/10.1016/j.eneco.2015.01.003
Islam F, Shahbaz M, Ahmed AU, Alam MM (2013) Financial development and energy consumption nexus in Malaysia: a multivariate time series analysis. Econ Model 30:435–441. https://doi.org/10.1016/j.econmod.2012.09.033
Ji Q, Zhang D (2019) How much does financial development contribute to renewable energy growth and upgrading of energy structure in China? Energy Policy 128:114–124. https://doi.org/10.1016/j.enpol.2018.12.047
Khan SAR, Yu Z, Umar M, Zia-ul-haq HM, Tanveer M, Janjua LR (2022) Renewable energy and advanced logistical infrastructure: carbon-free economic development. Sustain Dev 30(4):693–702. https://doi.org/10.1002/sd.2266
Khobai H, Le Roux P (2018) Does renewable energy consumption drive economic growth: evidence from granger-causality technique. Int J Energy Econ Policy 8(2):205–212. Accessed 14 June 2023. https://mpra.ub.uni-muenchen.de/82464/
Kutscher CF, Milford JB, Kreith F (2018) Principles of sustainable energy systems, 3rd edn. CRC Press
Lee JW (2013) The contribution of foreign direct investment to clean energy use, carbon emissions and economic growth. Energy Policy 55:483–489. https://doi.org/10.1016/j.enpol.2012.12.039
Li HX, Edwards DJ, Hosseini MR, Costin GP (2020) A review on renewable energy transition in Australia: an updated depiction. J Clean Prod 242:118475. https://doi.org/10.1016/j.jclepro.2019.118475
Lyman R (2016) Why renewable energy cannot replace fossil fuels by 2050. Energy Econ. Accessed 4 Nov 2022. https://friendsofscience.org/assets/documents/Renewable-energy-cannot-replace-FF_Lyman.pdf
Magazzino C (2017) Renewable energy consumption-economic growth nexus in Italy. Int J Energy Econ Policy 7(6):119–127
Magnani N, Vaona A (2013) Regional spillover effects of renewable energy generation in Italy. Energy Policy 56:663–671. https://doi.org/10.1016/j.enpol.2013.01.032
Marinaş MC, Dinu M, Socol AG, Socol C (2018) Renewable energy consumption and economic growth. Causality relationship in Central and Eastern European countries. PloS One 13(10):e0202951. https://doi.org/10.1371/journal.pone.0202951
Marques AC, Fuinhas JA (2012) Is renewable energy effective in promoting growth? Energy Policy 46:434–442. https://doi.org/10.1016/j.enpol.2012.04.006
Menegaki AN (2011) Growth and renewable energy in Europe: a random effect model with evidence for neutrality hypothesis. Energy Econ 33(2):257–263. https://doi.org/10.1016/j.eneco.2010.10.004
Neacsa A, Panait M, Muresan JD, Voica MC (2020) Energy poverty in European Union: assessment difficulties, effects on the quality of life, mitigation measures. Some evidences from Romania. Sustainability 12(10):4036. https://doi.org/10.3390/su12104036
Nuţă FM, Sharafat A, Abban OJ, Khan I, Irfan M, Nuţă AC et al (2024) The relationship among urbanization, economic growth, renewable energy consumption, and environmental degradation: a comparative view of European and Asian emerging economies. Gondwana Res 128:325–339
Ohler A, Fetters I (2014) The causal relationship between renewable electricity generation and GDP growth: a study of energy sources. Energy Econ 43:125–139. https://doi.org/10.1016/j.eneco.2014.02.009
Omri A, Mabrouk NB, Sassi-Tmar A (2015) Modeling the causal linkages between nuclear energy, renewable energy and economic growth in developed and developing countries. Renew Sustain Energy Rev 42:1012–1022. https://doi.org/10.1016/j.rser.2014.10.046
Ouyang Y, Li P (2018) On the nexus of financial development, economic growth, and energy consumption in China: new perspective from a GMM panel VAR approach. Energy Econ 71:238–252. https://doi.org/10.1016/j.eneco.2018.02.015

Ozturk I, Acaravci A (2013) The long-run and causal analysis of energy, growth, openness and financial development on carbon emissions in Turkey. Energy Econ 36:262–267. https://doi.org/10.1016/j.eneco.2012.08.025

Østergaard PA, Duic N, Noorollahi Y, Kalogirou SA (2021) Recent advances in renewable energy technology for the energy transition. Renew Energy 179:877–884. https://doi.org/10.1016/j.renene.2021.07.111

Panait M, Voica MC, Rădulescu I (2019) Approaches regarding environmental Kuznets curve in the European Union from the perspective of sustainable development. Appl Ecol Environ Res 17(3)

Panait M, Janjua LR, Apostu SA, Mihăescu C (2022a) Impact factors to reduce carbon emissions. Evidences from Latin America. Kybernetes ahead-of-print(ahead-of-print). https://doi.org/10.1108/K-05-2022-0712

Panait M, Apostu SA, Vasile V, Vasile R (2022b) Is energy efficiency a robust driver for the new normal development model? A Granger causality analysis. Energy Policy 169:113162. https://doi.org/10.1016/j.enpol.2022.113162

Paramati SR, Sinha A, Dogan E (2017) The significance of renewable energy use for economic output and environmental protection: evidence from the Next 11 developing economies. Environ Sci Pollut Res 24:13546–13560. https://doi.org/10.1007/s11356-017-8985-6

Paramati SR, Shahzad U, Doğan B (2022) The role of environmental technology for energy demand and energy efficiency: evidence from OECD countries. Renew Sustain Energy Rev 153:111735. https://doi.org/10.1016/j.rser.2021.111735

Payne JE (2009) On the dynamics of energy consumption and output in the US. Appl Energy 86(4):575–577. https://doi.org/10.1016/j.apenergy.2008.07.003

Pietrzak M, Igliński B, Kujawski W, Iwański P (2021) Energy transition in Poland—Assessment of the renewable energy sector. Energies 14(8):2046. https://doi.org/10.3390/en14082046

Popescu C, Panait M, Palazzo M, Siano A (2022) Energy Transition in European Union—Challenges and opportunities. In: Khan SAR, Panait M, Puime Guillen F, Raimi L (eds) Energy transition. Industrial ecology. Springer, Singapore. https://doi.org/10.1007/978-981-19-3540-4_11

Puime FG, Panait M, Andrei J, Gigauri I (2021) Non financial performance of energy companies listed on bucharest stock exchange and relevance for stakeholders. In: Dima AM, Kelemen M (eds) Digitalization and big data for resilience and economic intelligence. Springer Proceedings in Business and Economics. Springer, Cham. https://doi.org/10.1007/978-3-030-93286-2_14

Sadorsky P (2010) The impact of financial development on energy consumption in emerging economies. Energy Policy 38(5):2528–2535. https://doi.org/10.1016/j.enpol.2009.12.048

Sadorsky P (2011) Financial development and energy consumption in Central and Eastern European frontier economies. Energy Policy 39(2):999–1006. https://doi.org/10.1016/j.enpol.2010.11.034

Salim RA, Hassan K, Shafiei S (2014) Renewable and non-renewable energy consumption and economic activities: further evidence from OECD countries. Energy Econ 44:350–360. https://doi.org/10.1016/j.eneco.2014.05.001

Saqib N, Usman M, Ozturk I, Sharif A (2024) Harnessing the synergistic impacts of environmental innovations, financial development, green growth, and ecological footprint through the lens of SDGs policies for countries exhibiting high ecological footprints. Energy Policy 184:113863

Shahbaz M, Mallick H, Mahalik MK, Sadorsky P (2016a) The role of globalization on the recent evolution of energy demand in India: implications for sustainable development. Energy Econ 55:52–68. https://doi.org/10.1016/j.eneco.2016.01.013

Shahbaz M, Shahzad SJH, Ahmad N, Alam S (2016b) Financial development and environmental quality: the way forward. Energy Policy 98:353–364. https://doi.org/10.1016/j.enpol.2016.09.002

Shahbaz M, Van Hoang TH, Mahalik MK, Roubaud D (2017) Energy consumption, financial development and economic growth in India: new evidence from a nonlinear and asymmetric analysis. Energy Econ 63:199–212. https://doi.org/10.1016/j.eneco.2017.01.023

Shahbaz M, Sinha A (2019) Environmental Kuznets curve for CO2 emissions: a literature survey. J Econ Stud 46(1):106–168
Sharma A, Kar SK (2015) Energy sustainability through green energy. Springer
Simionescu M, Rădulescu M, Cifuentes-Faura J (2022) Renewable energy consumption-growth nexus in European countries: a sectoral approach. Eval Rev 47(2):287–319. https://doi.org/10.1177/0193841x221125982
Śmiech S, Papież M (2014) Energy consumption and economic growth in the light of meeting the targets of energy policy in the EU: the bootstrap panel Granger causality approach. Energy Policy 71:118–129. https://doi.org/10.1016/j.enpol.2014.04.005
Smil V (2017) Energy and civilization. The MIT Press
Temiz Dinç D, Akdoğan E (2019) Renewable energy production, energy consumption and sustainable economic growth in Turkey: a VECM approach. Sustainability 11(5):1273. https://doi.org/10.3390/su11051273
Tiwari AK, Shahbaz M, Hye QMA (2013) The environmental Kuznets curve and the role of coal consumption in India: cointegration and causality analysis in an open economy. Renew Sustain Energy Rev 18:519–527
Topcu M, Payne JE (2017) The financial development–energy consumption nexus revisited. Energy Sour Part B 12(9):822–830. https://doi.org/10.1080/15567249.2017.1300959
Tudorică BG, Bucur C, Panait M, Oprea SV, Bâra A (2024) Energetic equilibrium: optimizing renewable and non-renewable energy sources via particle swarm optimization. Utilities Policy 87:101722
Wang R, Mirza N, Vasbieva DG, Abbas Q, Xiong D (2020) The nexus of carbon emissions, financial development, renewable energy consumption, and technological innovation: what should be the priorities in light of COP 21 Agreements? J Environ Manag 271:111027. https://doi.org/10.1016/j.jenvman.2020.111027
Wang Q, Guo J, Li R, Jiang X (2022) Economic growth and carbon emission in the Organization for Economic Cooperation and Development countries: the effects of oil, gas, and renewable energy. Energy Environ. https://doi.org/10.1177/0958305x221138817
Zeren F, Akkuş HT (2020) The relationship between renewable energy consumption and trade openness: new evidence from emerging economies. Renew Energy 147:322–329. https://doi.org/10.1016/j.renene.2019.09.006
Zhang Z, Nuţă FM, Dimen L, Ullah I, Xuanye S, Junchen Y et al (2023) Relationship between FDI inflow, CO2 emissions, renewable energy consumption, and population health quality in China. Front Environ Sci 11:1120970
Zhe L, Yüksel S, Dinçer H, Mukhtarov S, Azizov M (2021) The positive influences of renewable energy consumption on financial development and economic growth. SAGE Open 11(3):215824402110401. https://doi.org/10.1177/21582440211040133

Establishing the Parameters for Effective Methane Mitigation in Dairy Ecosystems

Priya Modi, Manisha, and Chhavi Mathur

Abstract This chapter presents a comprehensive framework for evaluating methane mitigation strategies in dairy farming systems, with particular emphasis on the Indian context. Methane, a potent greenhouse gas primarily produced through enteric fermentation and manure management, represents a significant environmental challenge in dairy production. Through systematic analysis of existing literature and case studies across diverse farming systems, the research identifies six key parameters essential for effective methane reduction: emission reduction effectiveness, economic practicality, environmental co-benefits, scalability, monitoring capability, and social acceptance. The methodology combines theoretical analysis with practical applications, demonstrating how these parameters interact within whole-farm systems. Findings indicate that integrated approaches combining feed optimization, animal husbandry improvements, and manure management yield the most significant and sustainable results. The chapter concludes by outlining policy recommendations that support smallholder farmers in adopting climate-friendly practices while maintaining economic viability. This parameter-based evaluation framework provides a valuable tool for researchers, practitioners, and policymakers working to reduce dairy-related methane emissions while supporting sustainable agricultural development.

Keywords Methane emission · Methane mitigation · Sustainability · Dairy ecosystem

P. Modi (✉) · Manisha · C. Mathur
Vivekananda Global University, Jaipur, India
e-mail: priya.modi@vgu.ac.in

Manisha
e-mail: manisha@vgu.ac.in

C. Mathur
e-mail: chhavi.mathur@vgu.ac.in

F. A. Malik et al. (eds.), *Aligning ESG Goals with the UNs Sustainable Development Goals Planning*, World Sustainability Series,
https://doi.org/10.1007/978-3-032-01715-4_21

1 Introduction

Methane emissions from dairy farming represent a critical challenge in the context of climate change. Although carbon dioxide is more commonly discussed, methane is significantly more potent in terms of its short-term warming potential. Dairy farms release methane mainly through biological processes in cows and from manure management systems. As the demand for dairy continues to grow globally, so does the urgency to address these emissions. Reducing methane is not only essential for meeting environmental targets but also for protecting the future of dairy farming, which is a vital livelihood for millions worldwide.

Mitigation, in this context, refers to the systematic efforts to lower methane emissions through improved techniques, management practices, and technologies. With increasing awareness of climate issues, the importance of methane mitigation in agriculture—especially in the dairy sector—has risen sharply. Effective mitigation offers a dual benefit: it supports environmental sustainability while maintaining or even improving farm productivity. For farmers, this translates into long-term economic resilience; for policymakers, it opens up avenues to meet climate goals without compromising food security.

One of the most important requirements for successful mitigation is the ability to measure it accurately. Without clear and practical indicators, it is difficult to assess whether a strategy is genuinely effective or scalable. Measurement and monitoring systems are essential for identifying what works, where it works best, and how improvements can be made. These systems also play a key role in establishing transparency, building trust among stakeholders, and encouraging wider adoption of mitigation practices.

This chapter sets out to provide a clear and comprehensive understanding of methane mitigation in the context of dairy farming. It explores how methane is produced on farms, what influences its levels, and how we can measure and reduce it effectively. It also introduces a set of core parameters that can guide the evaluation of mitigation strategies. By doing so, the chapter aims to offer practical insights for researchers, farmers, industry experts, and policymakers working together toward a low-emission, sustainable dairy future.

2 Literature Review

Understanding Methane in Dairy Farming: Methane is a highly potent greenhouse gas that significantly influences the Earth's climate system. Though it remains in the atmosphere for a shorter duration than carbon dioxide, it traps far more heat during that period—making it a priority target in climate change mitigation efforts (IPCC 2019). One of the key sectors responsible for methane emissions is agriculture, with dairy farming emerging as a prominent contributor. A clear understanding of how methane is produced in dairy systems and how it contributes to climate change is essential for designing targeted and effective mitigation strategies.

At both global and local levels, dairy farms release considerable quantities of methane. Globally, the livestock sector accounts for a substantial share of methane emissions, with dairy cows playing a central role due to their unique digestive systems (FAO 2018). Countries with expansive dairy industries—such as India, the United States, Brazil, and New Zealand—report particularly high emissions from this source (World Resources Institute 2020). Locally, the intensity of emissions depends on farm size, cattle population, feeding practices, and the level of technological adoption in manure management and housing systems. Smallholder farms may emit less per farm but still contribute significantly at the aggregate level, especially in regions with dense farming communities.

There are two main sources of methane emissions on dairy farms. The foremost among these is **enteric fermentation**, a natural digestive process in ruminant animals like cows. As cows break down feed in the rumen—the first chamber of their stomach—methane is produced and released primarily through burping. This process alone accounts for the majority of methane emissions from dairy cattle. The second major source is **manure management**. When manure is stored under anaerobic (oxygen-limited) conditions, such as in lagoons or tanks, it undergoes microbial decomposition, leading to the release of methane (Animal Production Science 2020).

Factors Affecting Methane Emissions: Methane emissions from dairy farms are shaped by a combination of biological, nutritional, and management factors. These variables influence how much methane is generated both through the digestive process in cows and the way manure is handled on farms (Journal of Dairy Science 2021).

Cow-related factors such as breed, size, and milk yield play a significant role in methane emissions. Generally, larger cows or high-yielding breeds consume more feed, which increases fermentation activity in the digestive system and, consequently, methane production. Indigenous breeds like Gir or Sahiwal in India may produce less methane per animal but might also have lower milk yields, creating a complex trade-off between productivity and emissions (ICAR 2020).

The type and quality of feed offered to dairy cows have a direct impact on the quantity of methane produced. Diets high in fibrous materials like dry grasses or crop residues typically result in more methane due to prolonged fermentation in the rumen. In contrast, diets that include more digestible materials, such as grains or oilseed meals, tend to reduce methane production (Agriculture, Ecosystems & Environment 2021).

Farm management, especially manure handling systems, significantly affects methane emissions. When manure is stored in open or anaerobic conditions, such as in uncovered pits or lagoons, it becomes a substantial source of methane. Alternatives such as composting, aerobic drying, or anaerobic digestion for biogas capture can significantly reduce these emissions (Dairy Australia 2022).

Measurement and Monitoring Approaches: Accurate measurement and monitoring of methane emissions are foundational to understanding their sources and implementing effective mitigation strategies. As methane is a potent greenhouse gas with significant implications for climate change, its quantification—especially in dairy systems—is critical for informed action (WWF 2020).

Methane can be directly measured using specialized equipment that captures emissions at their source. Instruments like portable gas analyzers, infrared sensors, and laser-based detectors allow for real-time readings from specific locations such as cattle sheds or manure storage areas. Systems like Greenfeed can measure enteric methane emissions by attracting cows to a feed station and analyzing their breath (Innovation Center for U.S. Dairy 2020).

In situations where direct measurements are not feasible, emissions are commonly estimated using models and calculations. These estimations rely on factors such as herd size, feed composition, milk production, and manure management methods. Global frameworks like the Intergovernmental Panel on Climate Change (IPCC) offer different levels of estimation—from generalized emission factors (Tier I) to detailed, farm-specific models (Tier III) (IPCC 2019).

Recent years have seen a surge in digital tools designed to support methane monitoring on farms. Affordable sensors can now be embedded in farm equipment or even attached to animals to provide continuous emission data. Mobile apps allow farmers to input basic farm data and receive instant estimates or suggestions for mitigation (USDEC 2021).

3 Objectives

This research aims to evaluate the effectiveness of methane mitigation strategies in dairy farming systems, with special emphasis Indian dairy production. The specific objectives of this chapter are:

To identify and define key parameters that can effectively evaluate methane mitigation strategies in dairy farming.

To analyze how these parameters are correlated and responsible for methane mitigation.

To provide a structured approach for selecting appropriate methane reduction strategies as a policy recommendation for sustainable dairy ecosystem.

4 Research Methodology

This research employs a mixed-methods approach to develop and validate a parameter-based framework for evaluating methane mitigation strategies in dairy farming. The methodology consists of four interconnected components:

The parameter framework was developed through comprehensive analysis of peer-reviewed literature, institutional reports, and policy documents from major dairy-producing regions, with particular focus on emerging economies like India. Key sources included publications from the FAO, IPCC, national agricultural research institutions, and academic journals specializing in dairy science, agricultural systems, and climate change.

A systems thinking approach was applied to understand the interconnections between different parameters and how they function within the broader farm ecosystem. This analysis examined how interventions in one area such as feed management affect other aspects of the farm system such as productivity, economics, and environmental impacts.

5 Discussion

The Six Key Parameters for Evaluating Methane Mitigation:

Based on our research methodology, we have identified six essential parameters that provide a comprehensive framework for evaluating methane mitigation strategies in dairy farming. These parameters collectively address the environmental, economic, technical, and social dimensions of sustainable agriculture.

Effectiveness of Emission Reduction: The foremost parameter is the actual reduction of methane emissions. A method is considered effective when it significantly lowers methane output either per animal or per unit of milk produced. This requires establishing accurate baselines and conducting systematic monitoring post-implementation. Moreover, effectiveness must be consistent across different farm typesranging from smallholder farms in Rajasthan to commercial dairies in Punjab.

Our analysis of mitigation strategies reveals that enteric methane reduction through feed modification typically achieves 10–30% emission reductions, while advanced manure management systems can reduce manure-related emissions by 30–80% (Journal of Dairy Science 2021). However, the effectiveness of any intervention is highly dependent on proper implementation and maintenance.

Economic Practicality and Farmer Viability: A methane reduction strategy is unlikely to succeed if it is not economically practical. Indian dairy farmers, many of whom are small and marginal, operate under tight resource constraints. Any solution must be affordable, easy to implement with existing infrastructure, and ideally lead to cost savings or income generation.

Case studies show that solutions with dual benefits, such as biogas systems that both reduce emissions and provide energy, have higher adoption rates. For example, in Gujarat, community biogas plants serving clusters of small farms have proven economically viable by creating economies of scale while reducing implementation costs for individual farmers (NDDB 2021).

Environmental and Animal Welfare Co-benefits: An ideal methane mitigation strategy should offer broader environmental benefits. This includes reducing other pollutants such as ammonia or nitrous oxide, improving soil fertility, conserving water, and promoting biodiversity. Additionally, animal welfare is a crucial considerationhealthier cows are not only more productive but also emit less methane per unit of production.

Strategies such as improved grazing management have demonstrated significant co-benefits by enhancing carbon sequestration in soils, improving water infiltration, and supporting native biodiversity. Similarly, balanced nutrition approaches that reduce methane also improve animal health and longevity (Agriculture, Ecosystems & Environment 2021).

Scalability and Regional Adaptability: The scalability of any strategy determines its real-world impact. Solutions must be flexible enough to function across the wide range of Indian dairy farmsfrom one-cow family holdings in hilly regions to intensive dairies in peri-urban areas. Regional adaptability is equally important, as climatic conditions, feed availability, and cattle breeds differ across states.

Our analysis indicates that technologies requiring minimal infrastructure changes, such as feed supplements or improved feeding practices, tend to scale more easily than those requiring significant capital investment. However, even capital-intensive solutions like biogas can be scaled through innovative models like cooperative ownership or pay-per-use arrangements (Ministry of Agriculture & Farmers Welfare 2020).

Monitoring, Verification, and Accountability: Robust monitoring systems like methane digesters are necessary to track the effectiveness of mitigation measures. These systems should be simple, affordable, and usable by farmers with basic training. The development of mobile applications for emissions estimation like CAMS Methane Hotspot Explorer has proven particularly valuable in regions with good digital connectivity. Verification mechanisms are equally important for establishing trust and ensuring continued adoption. Third-party verification, either through government extension services or certified agencies, provides credibility and can potentially enable farmers to access carbon markets or sustainability premiums (Ministry of Environment, Forest and Climate Change 2021).

Social Acceptance and Ethical Considerations: Any mitigation strategy must also pass the test of social and ethical acceptability. It is vital that rural families, who form the backbone of India's dairy industry, experience net positive outcomes from adoptionsuch as better livelihoods, job creation, or community-level infrastructure development.

Strategies aligned with traditional practices or cultural values tend to face less resistance. For instance, indigenous methods of composting have been successfully enhanced with scientific principles to improve their methane mitigation potential while maintaining cultural continuity (ICAR 2020).

Integration of Mitigation Strategies in Whole-Farm Systems: A key finding from our research is that methane mitigation is most effective when approached holistically, recognizing the interdependence of various farming practices, environmental conditions, and socio-economic factors. A singular intervention, such as modifying feed or installing biogas units, can bring limited benefits if implemented in isolation.

The interconnected nature of emissions in dairy farming means that interventions in one area affect others. For example, improving feed quality may reduce enteric methane, but if the resulting manure is left unmanaged, emissions from decomposition will continue. Especially in Indian mixed-farming systems, where dairy is often integrated with crop cultivation, these interactions are more pronounced (Ministry of Agriculture & Farmers Welfare 2020).

Our analysis of farm case studies reveals that the most successful approaches combine multiple complementary practices. For example, integrating improved animal nutrition with controlled grazing systems and installing anaerobic digesters can together significantly lower emissions while also boosting productivity and resource efficiency. Such integrated systems also create co-benefitssuch as improved soil health, reduced reliance on chemical fertilizers, and the generation of clean energy from biogas.

Comparison of Mitigation Strategies across Farm Types

Different farming systems require tailored approaches to methane mitigation. Our analysis of various farm types reveals distinct patterns in the effectiveness and applicability of different strategies:

Large Commercial Farms: These operations have typically achieved success with capital-intensive technologies like anaerobic digesters and automated feeding systems. While effective at reducing emissions, these approaches require significant investment, technical expertise for maintenance, and often depend on stable electricity supply. Case studies from commercial dairy operations in Maharashtra show emission reductions of 40–60% through such integrated technological approaches (NDDB 2021).

Small Family Farms: For smallholders, low-cost interventions focused on feed quality improvement and basic manure management have proven most successful. Strategies like introducing leguminous fodder crops, improving feed storage to reduce spoilage, and simple composting techniques have shown emission reductions of 15–25% while improving overall farm productivity (ICAR 2020). Community-based approaches, such as shared biogas facilities serving clusters of farms, have helped overcome the scale limitations that individual smallholders face.

Organic Dairy Systems: These farms often achieve methane reductions through holistic management practices that emphasize soil health, biodiversity, and animal welfare. Practices like intensive rotational grazing, diverse pasture composition, and aerobic composting have shown moderate but consistent emission reductions while delivering substantial co-benefits for ecosystem health (Agriculture, Ecosystems & Environment 2021).

The analysis demonstrates that while each farm type faces different constraints and opportunities, all can achieve meaningful methane reductions when strategies are appropriately matched to their context and capabilities.

6 Policy Implications

Addressing methane emissions in the dairy sector requires more than just on-farm innovationsit demands a strong enabling environment shaped by national and international policies, financial mechanisms, and knowledge systems. Our analysis of the parameter framework reveals several key policy implications for supporting effective methane mitigation in dairy farming.

Alignment with Global Climate Frameworks-Global climate agreements, such as the Paris Agreement and the Global Methane Pledge, have brought methane to the forefront of international climate action. These agreements recognize that reducing methaneespecially from agricultureis essential for achieving short-term climate goals (WRI 2020).

National policies should be designed to help the dairy sector contribute to these global commitments while ensuring they remain economically viable. This includes developing country-specific emission reduction targets for the dairy sector, establishing standardized measurement protocols aligned with international standards, and creating mechanisms for reporting progress against national commitments.

National and State Policy Framework Development

Integration across departments: Policies related to livestock, climate change, energy, and rural development should be coordinated to create synergies rather than contradictions. Currently, programs under different ministries sometimes work in isolation.

State-level adaptation: Given India's agricultural diversity, state governments should develop region-specific action plans that consider local cattle breeds, farming systems, and socio-economic conditions. States like Gujarat and Punjab have demonstrated how tailored approaches can accelerate adoption of climate-friendly practices.

Regulatory frameworks: Clear standards for methane emissions from large dairy operations, coupled with incentives for compliance, can drive technology adoption among commercial producers while generating knowledge that benefits smaller farmers.

Research prioritization: Government funding for agricultural research should prioritize developing low-cost, context-appropriate methane mitigation technologies suitable for smallholder dairy farmers, who constitute the majority of producers in India.

Financial Support Mechanisms

Targeted subsidies: Government programs should provide financial assistance for initial investment in methane-reducing technologies, with subsidy rates inversely proportional to farm size to ensure equitable access.

Credit facilities: Specialized credit products with favorable terms should be developed specifically for climate-friendly dairy investments, potentially linked to carbon reduction outcomes.

Carbon market development: Policies that support the development of simplified carbon credit mechanisms for smallholder dairy farmers could create additional revenue streams. This requires investment in low-cost monitoring and verification systems suitable for small-scale operations.

Payment for ecosystem services: Beyond carbon, dairy farmers who adopt sustainable practices often generate other environmental benefits. Payment schemes that recognize and reward these contributions can strengthen the economic case for adoption.

Extension and Capacity Building

Strengthening extension networks: India's extensive network of Krishi Vigyan Kendras (KVKs), agricultural universities, and state extension departments should receive additional resources specifically for climate-smart dairy training.

Digital knowledge platforms: Mobile-based information services can dramatically expand the reach of technical guidance on methane mitigation, especially important in a country as geographically diverse as India.

Demonstration farms: Each district should establish demonstration units showcasing practical methane reduction strategies appropriate to local conditions, serving as practical learning centers for surrounding farmers.

Farmer-to-farmer networks: Programs that enable peer learning among farmers have proven particularly effective for technology adoption. Policies should support the development of farmer learning groups focused on sustainable dairy practices.

Market Development and Consumer Awareness

Certification systems: Government-backed certification for climate-friendly dairy products would allow consumers to make informed choices and potentially reward sustainable producers with price premiums.

Public procurement: Government institutions could prioritize dairy products from farms implementing methane reduction strategies, creating a stable market for early adopters.

Consumer education: Public awareness campaigns about the climate impact of dairy production and the benefits of sustainable practices can help build market demand for low-emission products.

Cross-cutting Policy Recommendations

Inclusive policy development: Smallholder farmers, women dairy producers, and marginalized communities should be actively included in policy design processes to ensure interventions meet their needs and capabilities.

Adaptive policy frameworks: Given rapid technological change and evolving scientific understanding, policies should include mechanisms for regular review and adaptation based on implementation experience and new evidence.

Data systems development: National and state-level data systems tracking dairy emissions, mitigation activities, and outcomes should be strengthened to guide policy refinement and demonstrate progress.

Public–private partnerships: Government policies should actively encourage private sector involvement in developing and scaling methane mitigation solutions, particularly where commercial viability is possible.

By addressing these policy dimensions, governments can create an enabling environment that supports dairy farmers in transitioning to low-emission practices while maintaining economic viability and contributing to national climate goals.

7 Conclusion

The evaluation of methane mitigation strategies in dairy farming hinges on a number of essential parameters, each of which provides a structured approach to assessing their effectiveness and long-term viability. The research presented in this chapter has identified six critical parameters that collectively form a comprehensive framework for evaluating methane reduction efforts: effectiveness of emission reduction, economic practicality, environmental and animal benefits, scalability and regional adaptability, monitoring and accountability, and social and ethical considerations.

These parameters do not operate in isolation but function as an interconnected system. A strategy that excels in one parameter but fails in others is unlikely to achieve sustainable adoption or meaningful impact. For instance, a highly effective methane-reducing feed additive that is prohibitively expensive for smallholder farmers or requires complex monitoring will struggle to achieve widespread implementation. Conversely, approaches that balance moderate gains across multiple parameters often demonstrate greater long-term success and broader adoption.

The research confirms our initial hypothesis that effective methane mitigation requires a multi-parameter evaluation approach rather than single-focus interventions. The case studies examined across different farm typesfrom large commercial operations to smallholder and organic systemsdemonstrate that successful mitigation strategies are those that address multiple parameters simultaneously while adapting to local contexts. This finding has significant implications for how research, policy, and implementation programs should be designed.

Our second hypothesis regarding the varying importance of different parameters across contexts was also supported by the evidence. For smallholder dairy farmers in resource-constrained settings, economic practicality and simplicity of implementation take precedence, while for larger commercial operations, the absolute scale of emission reduction and verifiability may carry more weight. This underscores the need for flexible application of the evaluation framework rather than rigid standardization.

A holistic, farm-wide approach is essential when addressing methane emissions in dairy farming. As methane emissions are intrinsically linked to various aspects of farm operations, such as feed management, manure handling, grazing practices, and overall farm productivity, addressing only one element in isolation is unlikely to yield significant or lasting results. The most successful case studies consistently demonstrated the value of integrated approaches that considered the entire farming system and its surrounding context.

To create a truly climate-friendly dairy sector, collaboration among all stakeholdersfarmers, researchers, policymakers, industry players, and consumersis paramount. This collective effort can drive the adoption of methane mitigation strategies that not only benefit the environment but also support the economic and social well-being of rural communities. The policy recommendations outlined in this chapter provide a roadmap for creating an enabling environment that supports this transition.

Methane mitigation offers dairy farmers the opportunity to not only reduce their environmental footprint but also enhance resource efficiency, improve animal health, and access additional revenue streams through technologies like biogas production. With appropriate support, these benefits can be realized across the spectrum of dairy operations, from resource-rich commercial enterprises to vulnerable smallholder farms.

The Indian dairy sector, with its immense diversity and scale, has significant potential to contribute to global methane reduction goals while addressing critical issues such as food security, rural development, and sustainable livelihoods. The parameter-based framework presented in this chapter offers a structured approach for evaluating and implementing methane mitigation strategies that are not only effective but also economically viable, socially acceptable, and environmentally beneficial across this diverse landscape.

As the sector evolves, the commitment to a climate-resilient and sustainable future must remain steadfast. By working together and applying the insights from this parameter-based evaluation framework, we can create a more sustainable, prosperous, and climate-friendly dairy sector, making meaningful strides toward a healthier planet and stronger, more resilient farming communities.

References

Agriculture, Ecosystems & Environment (2021) Reducing methane emissions in dairy farming: evaluating the impacts of different mitigation strategies. Agricult Ecosyst Environ 308:107–120. https://doi.org/10.1016/j.agee.2020.107120

Animal Production Science (2020) Strategies to mitigate methane emissions from ruminant livestock: a review. Anim Prod Sci 60(4):522–535. https://doi.org/10.1071/AN19045

Dairy Australia (2022) Managing methane emissions from dairy farms: tools and strategies for Australian dairy producers. Dairy Australia. https://www.dairyaustralia.com.au

Food and Agriculture Organization of the United Nations (FAO) (2018) Mitigation of methane emissions in the livestock sector: a review of the options. FAO. http://www.fao.org

Indian Council of Agricultural Research (ICAR) (2020) Methane emissions in Indian dairy farming: a scientific overview. ICAR. https://www.icar.org.in

Innovation Center for U.S. Dairy (2020) Sustainability in U.S. dairy: methane reduction initiatives. Innovation Center for U.S. Dairy. https://www.usdairy.com

Intergovernmental Panel on Climate Change (IPCC) (2019) Special report on climate change and land. IPCC. https://www.ipcc.ch

Journal of Dairy Science (2021) Methane emissions and mitigation strategies in dairy production: a review of recent research. J Dairy Sci 104(3):1575–1589. https://doi.org/10.3168/jds.2021-02041

Ministry of Agriculture & Farmers Welfare, Government of India (2020) National policy on dairy development in India: towards sustainable dairy farming. Government of India. https://www.agricoop.nic.in

Ministry of Environment, Forest and Climate Change, Government of India (2021) India's commitment to climate action: agricultural sector and methane mitigation. Government of India. https://www.moef.gov.in

National Dairy Development Board (NDDB) (2021) Dairy farming in India: current challenges and opportunities. NDDB. https://www.nddb.coop

U.S. Dairy Export Council (USDEC) (2021) Sustainability in U.S. dairy farming: reducing methane emissions for a greener future. USDEC. https://www.usdec.org

World Resources Institute (WRI) (2020) Climate policy and methane: the role of agriculture in climate change mitigation. WRI. https://www.wri.org

World Wildlife Fund (WWF) (2020) Methane and the climate crisis: why it's a priority for action. WWF. https://www.worldwildlife.org

Mapping the Landscape of Healthcare Training Research: A Post COVID Bibliometric Exploration in Challenging Business Environment

Priyanka Yadav, P. V. Rao, and Garima Mathur

Abstract The goal of this study is to undertake an exhaustive evaluation of the scientific literature on training in the healthcare industry. The methodological approach uses bibliometric analysis to demonstrate and assess quantitatively the volume of literature generated in the area over a given time period. The primary source for data is the Scopus database. The collected data covering the four-year period from 2020 and 2023—the identical frame as the COVID-19 pandemic. 549 papers from a well-chosen corpus are analysed. The findings of this study are then interpreted from a number of perspectives, such as identifying prominent authors, assessing popular keywords, looking at the production of institutions over time, identifying the most frequently cited documents, identifying the major contributing nations in research fields, and identifying the key sources that have generated the most articles over time. The study's findings have broad implications since they can improve and broaden our knowledge of healthcare research projects. These results might serve as a catalyst for the creation of fresh research initiatives.

Keywords Bibliometric analysis · Training · Healthcare · COVID-19 · Network visualisation · Biblioshiny

P. Yadav
SOS Management, Jiwaji University, Gwalior, India

P. V. Rao
Maharani Laxmi Bai College, Gwalior, India

G. Mathur (✉)
Prestige Institute of Management & Research, Gwalior, India
e-mail: garimanmathur@gmail.com

F. A. Malik et al. (eds.), *Aligning ESG Goals with the UNs Sustainable Development Goals Planning*, World Sustainability Series,
https://doi.org/10.1007/978-3-032-01715-4_22

1 Introduction

Bibliometric analysis is a powerful scientific method utilised to evaluate the contributions, influence and relationship or interconnections among research papers, publications and articles, providing valuable insights into their distinctive attributes. This methodology comprises two primary components: performance analysis, which focuses on the contribution of data within specific research domains and scientific mapping, which examines the interconnections among these domains. To effectively explore the dynamics of interactions and collaborations among researchers, co-authorship bibliometric analysis integrates network analytical techniques with datasets from the disciplines of Mathematics, Biology and Physics (Newman, 2004).

Additionally, bibliometric analysis effectively examines the collaboration pattern, author citations and article citations. This technique facilitates the precise identification of the most frequently cited authors and publications and enhances the understanding of impactful contributions within the field. In this study, with the help of statistical and network analytics, various approaches have been employed to effectively utilise datasets sourced from published medical research. This analytical approach thoroughly identifies the common characteristics and connections prevalent within these articles (Chen et al. 2018).

A vital component of the healthcare sector is the development of training programs that aim to enhance the capabilities and adaptability of medical professionals in their respective specialised fields. Furthermore, in particular App-based programs, have proven effective in evolving the medical expertise, especially in the area of nutrition. Research indicates that initiatives focused on teaching life skills are positively correlated with emotional intelligence, which contributes to improved resilience to stress and academic performance (Ononye et al. 2022). Training programs designed to develop empathy among physicians have been associated with the enhancement of communication skills. These programs not only enhance the ability and skills of healthcare professionals to listen actively and engage with the patients but also strengthen a deeper and meaningful doctor-patient relationship. Improved empathy allows physicians to better understand the patient's emotions and concerns, which ultimately leads to better patient outcomes (Karimi & Abdollahi 2019). Training programs in microsurgery play a vital role in enhancement of advanced skills and their effective application within clinical settings. As a matter of fact, integrating hands-on practice with theoretical knowledge contributes significantly to the continual advancement of the microsurgery field, enhancing patient outcomes and fostering innovation in surgical practices (Shahrezaei et al. 2024). In the field of dermatology, there has been significant emphasis on a clearly defined intervention that aims to enhance the knowledge and skills of healthcare professionals through training and education. This initiative is designed to address specific challenges faced within the sector, facilitating the overall quality of care provided to patients (Byrne-Davis et al. 2022). In addressing the stigma related to drug use disorders, it has been established that training interventions play a vital role in promoting effective communication and fostering a deeper understanding between instructors and students in health professional education (Barenie et al. 2023).

Medical practitioners are progressively equipped with advanced competencies due to the adoption of innovative curriculum and enhancement of admission procedures (Jernigan & Carbonneau 2025). Similar to this, a training program focused on positive thinking has been recognized for its significance in enhancing optimism and improving adherence among patients undergoing haemodialysis treatment (Sabouri et al. 2023). The implementation of training programmes has significantly revitalized the domain of electronic health records and medical informatics. This initiative has resulted in the establishment of standardised procedures, a reduction in errors and enhanced the capabilities for retrieving clinical research data (Xiong et al. 2019). To enhance the effectiveness of medical education, it is necessary to evolve training strategies by recognizing emerging trends and integrating creative pedagogical techniques. This approach is suggested from a forward-looking perspective (Azim Majumder et al. 2004). Additionally, a thorough evaluation of learning resources plays a crucial role in strategic educational planning. This relates to optimising student selection and faculty duties through various resource utilisation techniques (Neufeld and Barrows 1974). Programmes for faculty development emerge as essential for raising medical education standards and building leadership effectiveness within healthcare organisations (Bansal and Supe 2007).

In addition, the internship period plays a crucial role since active participation at this time greatly improves the quality of medical services provided by the healthcare industry. By introducing a variety of training patterns, higher education institutions may address the training paradigm and promote the best workforce performance in the healthcare industry (Ferguson 1995). The development of competency-based learning, where seminars and clinical instruction combine to strengthen communication skills, knowledge acquisition, and skill development among healthcare professionals, further accelerates this trend (Hemant Kumar et al. 2023). Last but not least, the inclusion of telehealth training highlights the expectation of further developments in the healthcare industry. The ability of medical students to use cutting-edge innovations to meet the requirements of patients is strengthened by providing them with the knowledge and skills to harness developing technology (Camhi et al. 2020). Together, these academic pursuits help us get a comprehensive grasp of training's crucial position within the healthcare environment.

The need to deploy healthcare staff skilled in critical care and infection control in the face of the COVID-19 outbreak drove the deployment of various training programmes. The experiences and insights of healthcare workers have significantly shaped the creation and development of these methods, thereby ensuring the effectiveness of patient treatment (Liu et al. 2020). The educational environment experienced significant transformation as a result of pandemic, particularly with the implementation of online training platforms for medical practitioners and students. This shift incorporated a variety of instructional methodologies, such as lectures, webinars and online classes, which facilitated vital knowledge upgrade or enhancement and skill development (Saeki 2023). To enhance accessibility and functionality in various scenarios, numerous nations have established comprehensive policies and methodologies to effectively deliver training and instruction to medical professionals through online platforms (Saeki 2023). The popularity of telemedicine is the most well-known example of how the new coronavirus disease-19 spurred a surge in providing of medical care through virtual methods. Procedures, staff with greater engagement (Chauhan et al. 2024), knowledge base (Mathur and Chauhan 2021), and technology were combined to optimize medical care delivery, successfully addressing patient expectations during the pandemic (Wosik et al. 2020).

The implementation of virtual training programs designed to assess infection prevention and control measures shows that African context was not an exception from this paradigm change. This initiative was marked by systematic and caring monitoring and iterative adjustments or refinements, which enabled to achieving the optimal outcome (Kessy et al. 2023). The COVID-19 pandemic serves as that critical period which enhances the approach of e-learning methodologies within the medical education, garnering substantial support. A large group of medical students determined that online learning proved to be more effective than the traditional instructional methods (Mathur et al. 2024). The decision to integrate technology has shown to be effective particularly in the environment where favourable conditions, such as easy access to the internet and electronics devices, were presented (Chandrasiri and Weerakoon 2022). The development of specialized software which significantly improves the communication between the faculty and students, that helps to create a more positive learning environment. This development shows the benefits of online learning which was already well-regarded for its convenience and accessibility (Sujarwo et al. 2020; Nathani et al. 2023). Innovative approaches were developed to deal with challenges of online learning that was experienced during the pandemic, highlighting the vital role of practical clinical experience in medical education. Due to the limited opportunities of hands-on clinical practice, online training initiative were complemented with carefully planned virtual clinical experiences to achieve effective learning outcomes among the participants or the workforce (Jyoti et al. 2023).

It is noteworthy that regulations for safe, supervised hands-on training were developed, aiding medical students' entire educational journey (Alsoufi et al. 2020). Training concerns in the field of surgery, where hands-on experience is essential, have been greatly alleviated by the introduction of virtual surgical clerkship programs. According to these programs not only made it easier for instructors and students to get involved, but they also prevented setbacks, which ultimately led to fewer fatalities worldwide (Calhoun et al. 2020). Online learning gained popularity during the COVID-19 pandemic, which had two benefits: it enhanced satisfaction among learners and institutional effectiveness. Online learning was preferred by students because it was cost effective and time-efficient (Abdull Mutalib et al. 2022). Academic attainment and clinical competency both improved as a result. As a result of the COVID-19 pandemic, medical education suffered a significant transformation that brought together traditional and online teaching approaches, ultimately promoting adaptability, resiliency, and advancement in the healthcare industry.

2 Theoretical Approach of the Study

Training is a vital instrument for enhancing self-awareness, knowledge enrichment, competencies development and increasing self-confidence to complete task effectively. According to Jean Piaget's foundational work, constructivism learning is a conceptual framework within educational theory that facilitates the integration of new knowledge with the pre-existing knowledge. This strategy exemplifies the process of increasing the existing knowledge framework and cognitive structures, also promoting continuous intellectual development and growth. The three distinct types of constructivist learning- Cognitive, Radical and Social which plays a crucial role in shaping educational practices. Cognitive Constructivism basically stands out for its effectiveness of enabling individuals to assimilate new information and shaping and refining their existing cognitive schemas. This style of learning ensures that newly acquired knowledge is firmly retained with the cognitive framework, as such adaptations are grounded in sensory perceptions, experience interaction and cognitive processes (Yilmaz 2011). This method effectively aligns with the role of training programmes as mechanism for enhancing participant knowledge through their cognitive processes. In contrast, Social Constructivism powerfully emphasizes the importance of collaborative learning, arguing that knowledge is created through the dynamic interactions among individuals within social context, taking into account their social and cultural connections (Palincsar 1998). This method clearly highlights the vital role that social interaction acts in the acquisition of knowledge, asserting that learning is most effective in group settings. In this scenario, training is conducted in collaborative groups, fostering cooperation and mutual support among participants. Additionally, Radical Constructivism powerfully emphasizes that knowledge is actively formed through the interactions with the environment and personal experiences rather than formed by sensory systems. This method of learning transforms training into a powerful source for generating new experiences and enable individuals to view concepts from multiple perspectives. This concept strongly supports the idea that individuals develop or form knowledge based on their own experiential insights (Von Glasersfeld 1984).

Constructivism applied into training approaches creates a powerful framework for immersive and interactive learning. Participants are motivated by a strong desire to broaden their intellectual horizons, sharpen their skills and abilities and adjust to ever-evolving circumstances. This engagement enables them to take active participation in the integration of knowledge and information, leading to a deeper understanding and impactful learning experiences. This strategy embodies the idea of proactive knowledge creation, in which students actively strive to improve their cognitive frameworks, leading to a powerful synergy between learning and personal development. The constructivist framework essentially emphasises the reciprocal link between learning and a person's cognitive activities, where training programmes offer the necessary channels for students to actively participate in the creation, absorption, and integration of information.

3 Research Questions

To address the gaps, the study seeks to answer the following questions:

1. What is the publication trends and influential sources in healthcare training research since the onset of COVID-19?
2. What are the most citing countries in the post-COVID healthcare training?
3. What are the major thematic clusters and emerging topics identified in healthcare training literature through keyword co-occurrence?
4. How have the collaboration patterns (cited documents, eminent authors) and institutions shaped the development of healthcare training research in the post-pandemic era?

4 Data Source and Search Strategy

The qualitative aspect of this research project, which was created to collect and analyse data from the Scopus database, distinguishes it. Scopus was chosen as the main repository among the variety of accessible database systems because of its extensive collection of reputable research papers complete with citations and abstracts. The current study focused its data retrieval efforts primarily on Scopus, despite the fact that alternative databases like PubMed and Web of Science also provide significant information on databases.

According to the defined approach, a bibliometric study covering the period from 2020 to 2023 was carried out using Scopus data as the basis. The search approach was carefully planned to produce pertinent results, and by applying specified criteria, 2,533 document results were found. The following search keywords guided the search: TITLE-ABS-KEY ("Training & Development" OR "Training and Development" OR "Training design" OR "Motivation To Transfer Training" OR "Trainer Competency" OR "Training evaluation" OR "Training Transfer") AND (LIMIT-TO (PUBYEAR, 2023) OR LIMIT-TO (PUBYEAR, 2022) OR LIMIT-TO (PUBYEAR 2021) OR LIMIT-TO (PUBYEAR 2020)). The corpus was then reduced to 715 documents through the procedure of topic area reduction; 549 documents were then obtained through the use of keyword criteria.

The study used R software by using Biblioshiny for data analysis. This software synergy made it easier to analyse data using network visualisation techniques by giving a clear and graphical depiction of how relationships within the dataset are connected. In essence, this research endeavour follows a logical and academic approach, utilising cutting-edge software tools and the power of qualitative investigation to understand the complexities of the dataset being studied. The use of Scopus as a primary database and search criteria were strategically adjusted that highlights the accuracy and rigour of this investigation.

5 Inclusion and Exclusion Criteria

The inclusion criteria consist of the title, subject area and keywords of the respective field. Data is collected over a four-year period from 2020 to 2023. The type of document that is included in the selection are articles and review papers and those material are considered which is published in English. The exclusion criteria include the document types such as conference papers, books, letters, notes, editorials and book chapters. Additionally, any paper which is not published in English are also excluded from the selection.

6 Results

6.1 Most Cited Countries

The tabular data provides a comprehensive picture of the influence of research in various nations, providing information on their total citations (TC) and average article citations. This information insights a useful lens to analyse the contribution of these countries on the global academic community and also identifying the trends in the way that research is acknowledged. USA's impressive overall citation count of 461, highlights its prominent role in the research landscape. The average number of citations per article (4.00) shows that American research consistently receives attention; each publication receives on an average of 4 citations. This is a reflection of the country's prolific research output and its long-lasting effects in a various sectors. The second most cited country is the United Kingdom, with a total citation count of 249 and an article average citation is 3.80. China follows in this series is on the third most cited country with a total count of 208 articles with an average article citation is 8.30 shows that each publication receives a great deal of attention, that shows a high level of research output and considerable influence. According to the table, Singapore has a total of 59 citations, with an excellent 9.80 citations per article on average. This information demonstrates Singapore's focuses on producing high-calibre research that is widely recognized, strengthening its position in the worldwide research community (Table 1).

Table 1 Showing most citing countries

Country	TC	Average article citations
USA	461	4.00
United Kingdom	249	3.80
China	208	8.30
Australia	139	3.20
Netherlands	64	5.30
Singapore	59	9.80
Germany	56	2.50
Italy	52	6.50
Turkey	51	7.30
Canada	47	2.00

Exhibit 1 displaying country-wise citation

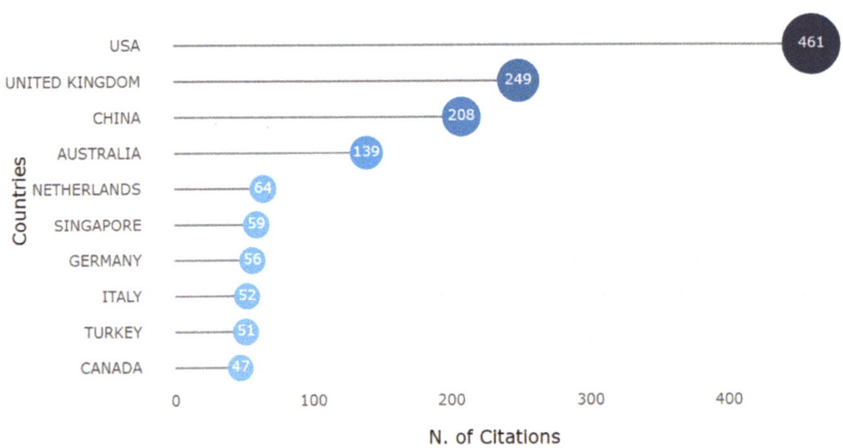

This data offers a window into the research ecosystems of many nations, showing their contributions to research, impact, and global recognition. Total and average article citations together show the depth and breadth of these nations' contributions, illuminating the strength and significance of their research.

7 Most Frequent Keywords

A frequency study of certain phrases within a given corpus of text is presented via the tabular data. Here is a professional analysis of the facts and a final viewpoint of the keywords through the pictorial presentation of the graph.

Exhibit 2 displaying keywords occurrences

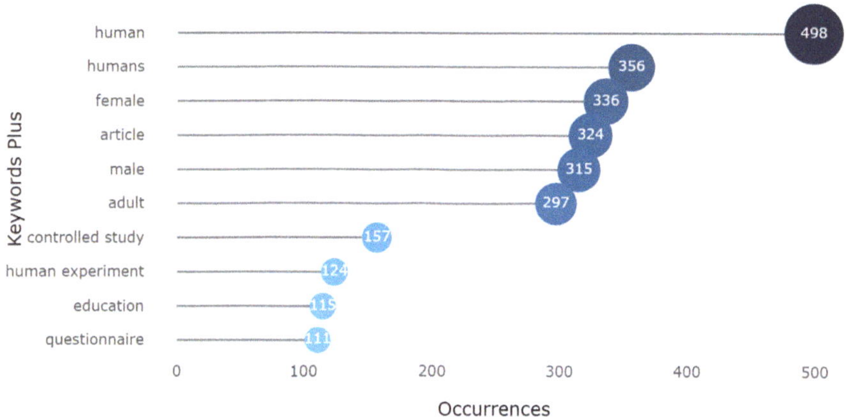

"Human" and "Humans" are often used, with "human" appearing 498 times and "humans" appearing 356 times. This shows that there is a widespread interest in human-related themes. These words denote that people are at the centre of the conversations or material being discussed.

"Female" and "Male": The use of the words "female" (336 times) and "male" (315 times) frequently indicates a focus on gender-related topics. It suggests a knowledge of gender's significance as a variable in numerous situations, maybe including gender studies, health inequalities, and social dynamics.

"Adult": The word's 297 appearances highlight the importance of the adult population. In a time when lifelong learning and the wellbeing of adults have particular value, this may represent discussions and research on adult education, healthcare, and social contributions.

"Controlled Study" and "Human Experiment": The terms "controlled study" (157 occurrences) and "human experiment" (124 occurrences) reveal a strong emphasis on research methods and empirical studies. These phrases emphasise the dedication to meticulous scientific analysis and testing.

"Education": The term "education" is used 115 times, indicating a strong focus on educational issues. This includes conversations on pedagogy, curriculum development, and evolving nature of education in modern culture.

"Questionnaire": This term "questionnaire" appears 111 times and denotes a focus on survey-based data collection techniques. It shows dependence and importance of structured questions to gather knowledge and data across many fields.

Exhibit 3 displaying keywords

As the information shown in the table indicates a textual environment with a significant emphasis on human-centric concerns, gender considerations, adults' discussion, and a dedication to empirical research procedures and educational issues. These themes emphasise align with the current environment and highlight the importance of continue addressing the relevant topics like gender equality, humanism, lifelong learning, academic rigour, and educational progress (Table 2).

As a result, this frequency analysis insight into the dominant ideas and theoretical perspectives in the sample corpus of text. It emphasises the continuous value of researching human-related phenomena and using sound research methods and procedures in today's academic and societal contexts. These results can direct for further investigation and discussion across several academic fields and research areas.

Table 2 Showing keywords occurrences

Words	Occurrences
Human	498
Humans	356
Female	336
Article	324
Male	315
Adult	297
Controlled study	157
Human experiment	124
Education	115
Questionnaire	111

8 Most Cited Documents

The table depicts crucial pointers including the DOI (Digital Object Identifier), total citations (TC), Total Citation (TC) per year, and normalised Total Citation (TC) while shows a thorough overview of a number of research publications. These measures provide information on the intellectual significance and status of each research work among peers (Table 3).

Grøvik E, 2020, J Magn Reson Imaging: There have been 109 citations for this study, which was published in the Journal of Magnetic Resonance Imaging in 2020. It demonstrates constant recognition and impact within its own industry with a remarkable TC per year of 27.25. The normalised TC of 14.50 shows that it still has a substantial influence compared to the year of publication.

Liao Y-Y, 2020, Eur J Phys Rehabil Med: This study report, which was published in the European Journal of Physical and Rehabilitation Medicine in 2020, has received

Table 3 Showing most cited documents

Paper	DOI	Total citations	TC per year	Normalized TC
Grøvik E, 2020, J Magn Reson Imaging	https://doi.org/10.1002/jmri.26766	109	27.25	14.5
Liao Y-Y, 2020, Eur J Phys Rehabil Med	https://doi.org/10.23736/S1973-9087.19.05899-4	75	18.75	9.98
Hawk KF, 2020, JAMA Netw Open	https://doi.org/10.1001/jamanetworkopen.2020.4561	65	16.25	8.65
Nas J, 2020, JAMA Cardiol	https://doi.org/10.1001/jamacardio.2019.4992	44	11	5.85
Felten-Barentsz KM, 2020, Phys Ther	https://doi.org/10.1093/ptj/pzaa114	40	10	5.32
Woon APN, 2021, Nurse Educ Today	https://doi.org/10.1016/j.nedt.2020.104655	37	12.33	9.92
Weintraub D, 2022, Lancet Neurol	https://doi.org/10.1016/S1474-4422(21)00330-6	36	18	25.06
Hussain R, 2020, Int J Occup Saf Ergon	https://doi.org/10.1080/10803548.2018.1465671	28	7	3.73
Sansone P, 2020, J Strength Cond Res	https://doi.org/10.1519/JSC.0000000000002971	26	6.5	3.46
Liu H, 2020, J Adv Nurs	https://doi.org/10.1111/jan.14266	26	6.5	3.46

75 citations. It continues to maintain a commendable TC per year of 18.75, indicating consistent recognition in its field. The normalised TC of 9.98 indicates that its influence has persisted since its publication.

Hawk KF, 2020, JAMA Netw Open: This research study was published in 2020 and has received 65 citations. Its 16.25 TC each year shows a steady level of academic attention and influence, and its 8.65 normalised TC highlights its continued relevance in light of the publishing year.

Nas J, 2020, JAMA Cardiol: There have been 44 citations for this article, which was published in JAMA Cardiology in 2020. With an average TC of 11.0 every year, it continues to be a prominent figure in the academic world. Its continuous importance in relation to its publishing date is highlighted by the normalised TC of 5.85.

Exhibit 4 displaying documents

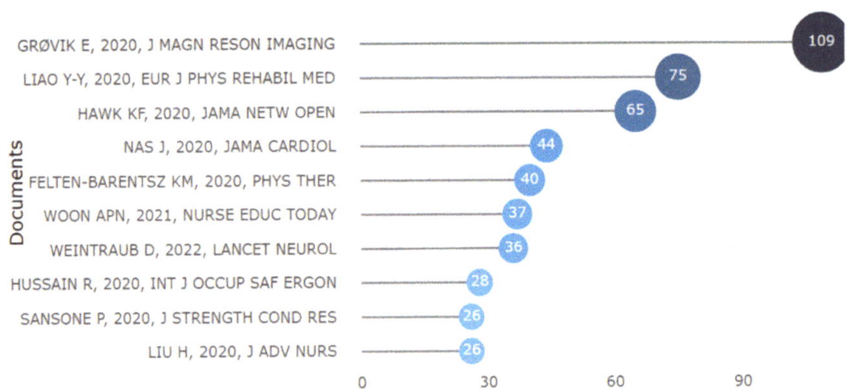

Similar trends can be seen in the other publications in the figure, which have different total citation counts, TC annual rates, and normalised TC. Researcher, scholar, and institution evaluation of the importance and applicability of these contributions within their respective domains is made easier with the use of these metrics, which collectively offer significant insights into the academic impact and recognition of each research work.

9 Most Relevant Sources Production of Time

The table thoroughly overviews the yearly publishing patterns in five different scholarly journals from 2020 to 2023 (Table 4).

The following journals are among them: “BMJ Open,” “Nurse Education Today,” “International Journal of Environmental Research and Public Health,” “Journal of Surgical Education,” and “Nurse Education in Practice.” The table illustrates interpretation of the data:

Table 4 Showing scholarly journals

Year	International Journal of Environmental Research and Public Health	Journal of Surgical Education	BMJ Open	Nurse Education Today	Nurse Education in Practice
2020	3	6	3	3	2
2021	11	11	9	9	6
2022	22	17	14	14	8
2023	22	18	15	14	8

International Journal of Environmental Research and Public Health: This journal released three papers in 2020. By 2023, there were 22 articles in the publication, reflecting a steady growth in productivity. This diary seems to have grown significantly during this time.

Journal of Surgical Education: This journal's publishing rates remained largely consistent, rising from 6 articles in 2020 to 18 articles in 2023. A steady presence in the field of surgical education is indicated by the publications' constancy.

BMJ Open: The journal gradually expanded from its initial 3 articles in 2020 to 15 papers published in 2023. This upward trend suggests that open-access medical research is receiving more support.

Nurse Education Today: Between 2020 and 2023, this journal published 15 articles, a significant increase from the 3 papers published in 2020. The considerable rise in research output points to a growing interest in nursing education-related subjects, which reflects the changing nature of healthcare education.

Nurse Education in Practice: Like Nurse Education Today, this publication showed notable development, going from 2 articles in 2020 to 8 articles in 2023. This shows that nurse education and training are putting more of an emphasis on their practical components.

The value of this information is found in its capacity to provide institutions, researchers, and educators with information about the changing trends and research goals in these publications. It highlights the fields of environmental and public health, surgical education, and nursing education that have shown consistent attention and expansion. This data may be used by academics and decision-makers to find new research fields and allocate funding appropriately. Additionally, educators may keep up with the most recent advancements in their professions to improve curriculum creation and teaching methods.

10 Most Eminent Authors

Specific authors along with their related publication counts and average citation impact are shown as the average number of citations per work. This information provides insights of the output and reputation of these authors in their specific scholarly domain. The table depicts the most prominent authors (Table 5).

The table depicts that Li Y and Neilsen K have the highest articles with 4 articles with average citations 0.52 and 0.70 while Bleich MR has the 3 articles with average citation impact of 2.33. This shows that they have made an important and significant addition to the literature in their specific area.

Exhibit 5 displaying author and documents

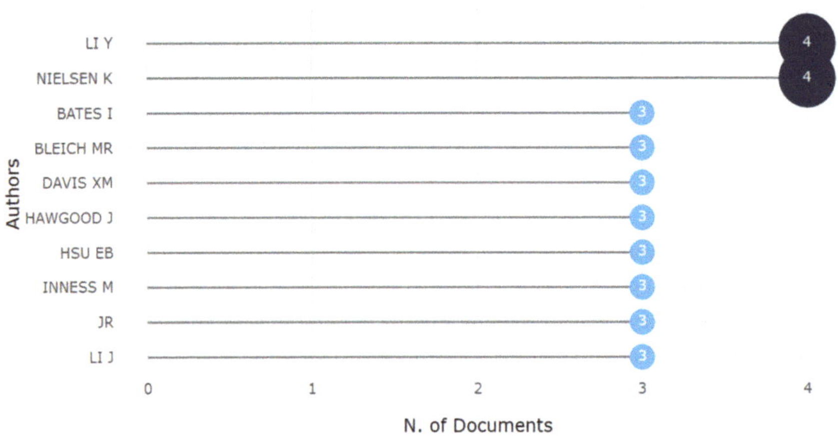

This data is crucial for assessing these author's intellectual contribution and productivity within their specific areas of expertise. It serves to acknowledge for honouring

Table 5 Showing eminent authors

Authors	Articles	Articles fractionalized
Li Y	4	0.52
Nielsen K	4	0.70
Bates I	3	0.47
Bleich MR	3	2.33
Davis XM	3	0.32
Hawgood J	3	0.62
Hsu EB	3	0.32
Inness M	3	0.57
Jr	3	0.24
Li J	3	0.41

authors who have made a significant impact on their fields and for those who have contributing in a more subtle way and still made noteworthy contributions to scholarship. This data may be utilised by institutions, academics and researchers to pinpoint influential thinkers, potential partners and subject-matter experts for a range of academic and research endeavours.

11 Institutions Production Over Time

Data regarding the research articles which is published by various academic institutions over a number of years. The assessment of this data is essential to comprehend the research output and trends of these universities during the specified years. The table demonstrates a detailed analysis of publishing rates of research articles for various prominent academic institutions over a four-year period, from 2020 to 2023. The results shown (Table 6).

King's College London: In 2020, ten research publications were published by this institution. The number of papers published increased steadily over the next

Table 6 Showing institution-wise production of papers

Affiliation	Year	Articles
King's College London	2020	10
King's College London	2021	19
King's College London	2022	25
King's College London	2023	25
Monash University	2020	2
Monash University	2021	6
Monash University	2022	18
Monash University	2023	19
Radboud University Medical Center	2020	14
Radboud University Medical Center	2021	14
Radboud University Medical Center	2022	21
Radboud University Medical Center	2023	21
University of California	2020	0
University of California	2021	11
University of California	2022	14
University of California	2023	18
University of Toronto	2020	0
University of Toronto	2021	13
University of Toronto	2022	13
University of Toronto	2023	19

years, reaching 19 in 2021, 25 in 2022, and another 25 in 2023. This demonstrates a considerable and ongoing dedication to intellectual work.

Monash University: Over the course of four years, Monash University's research output showed a discernible growth. They started off with 2 articles in 2020 and then significantly expanded, publishing 6 papers in 2021, 18 articles in 2022, and 19 articles in 2023. This significant expansion is a result of significant scientific activity.

Radboud University Medical Center: Over the years, Radboud University Medical Centre consistently produced a certain amount of research. In 2020 and 2021, they released 14 articles. In 2022 and 2023, they released 21 articles. This shows that the medical sector has a continuous research focus.

University of California: In 2020, the University of California (UC) had no publications when it first began, but it swiftly developed a research presence.11 papers were published by UC in 2021; this number rose to 14 in 2022 and then to 18 in 2023. This demonstrates a dedication to research and considerable growth over time.

University of Toronto: the University of Toronto started publishing research in 2021.13 papers were published in 2021, and this total stayed largely the same in 2022 and 2023, when they released 13 articles each. This denotes a significant and ongoing presence in research.

Exhibit 6 displaying documents and institutional affiliation

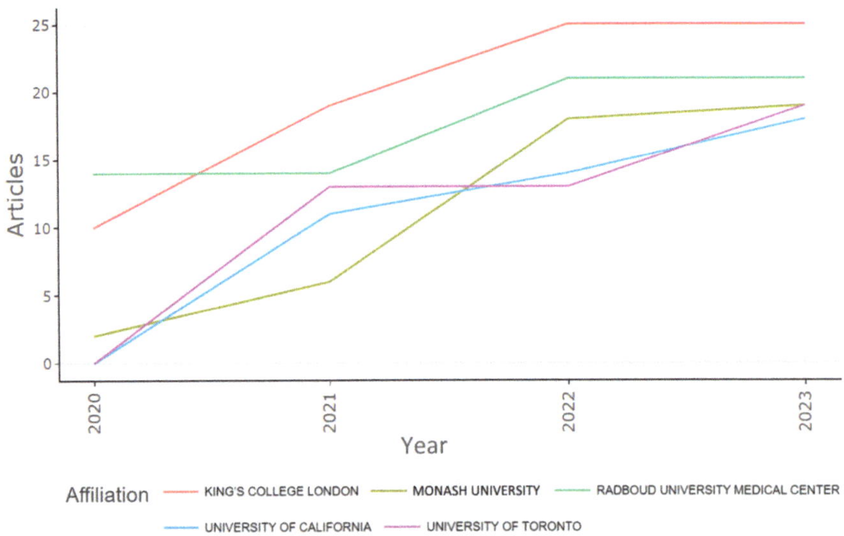

It sheds light on the research output and dedication of various academic institutions. It helps academics, decision-makers, and institutions keep track of research trends, identify organisations that produce high-quality research, and assess the influence and effect of these organisations within the academic and research communities. Making knowledgeable choices about academic alliances, budget allocation, and collaborations is another benefit.

12 Discussion

The analysis of the most frequently used keywords gives a worthwhile understanding of the thematic landscape and research priorities in healthcare training literature in this post-COVID era. Keywords act as incisive indicators of core concepts, reflecting both the scope and the evolution of research directions. In the context of this study, keywords such as "healthcare," "education," training," COVID-19," and "e-learning" emerged eminently, highlighting the academic community's focused interest in digital transformation and efforts for capacity-building in the medical sector.

The word "COVID-19" comes as a central keyword which indicates the pandemic's prominent role in reshaping training models, particularly about infection control, emergency response, and especially various digital learning innovations. Simultaneously, the frequent presence of words like "e-learning," "telemedicine," and "virtual training" emphasize a broader shift toward technology-enhanced learning modalities, which were very essential in overcoming barriers for access and delivery during the pandemic period.

Keywords like "simulation," "competency-based training," and "clinical education" gives insight that various efforts were done to preserve practical skill acquisition despite the limitation of physical interactions. The frequent co-occurrence of these terms' points toward a hybrid training model that integrates theoretical knowledge with simulated practical methods, thereby maintaining the quality of healthcare training.

Moreover, the consistent appearance of terms such as "resilience," "mental health," and "communication skills" reveals an enhancing acknowledgment of the importance of emotional intelligence and well-being in healthcare professional industry. This trend is likely influenced by the psychological impact of the pandemic on healthcare workers and increasing demand for holistic training frameworks.

In conclusion, keyword frequency analysis not only maps prevailing research themes but also reveals emergent priorities, including digital transformation, mental health, and experiential learning. These insights are critical for guiding future research, funding allocation, and curriculum development in the rapidly evolving landscape of healthcare education and training.

13 Conclusion

The enormous scientific literature on training in the healthcare sector has been meticulously and exhaustively reviewed for this study project. The successful completion of this project was made possible by the reliable bibliometric analysis approach, which allowed for the systematic and quantitative evaluation of a sizable body of academic work created during a certain period of time. The study concentrated on a carefully selected corpus of 549 academic publications that were painstakingly obtained from the extensive Scopus database and covered the years 2020–2023, a time frame that was inextricably linked to the global COVID-19 epidemic. The multifaceted analytical lens used in this study included an examination of significant authors, a close examination of popular keywords, a longitudinal analysis of institutional research productivity, the identification of highly cited works, an assessment of the significant contributors in various research domains, and a thorough analysis of key sources that consistently disseminate a significant volume of articles.

This study's implications go well beyond its intended analytical scope. The conclusions reached might both improve and deepen our common understanding of research projects in the complex field of healthcare. These discoveries could also serve as a spark, igniting the development of fresh research projects that are well adapted to meet the constantly changing demands of the healthcare sector. This research project is more than just a dry academic exercise; it makes a significant contribution to the continuous progress of knowledge and practice in the area of healthcare research and training.

14 Implications

This information has important practical, social, research, and management ramifications. It provides strategic guidance for choices about research financing, global collaboration, and research publication options, eventually promoting scientific development, social progress, and economic prosperity. These statistics may be used by institutions, researchers, and politicians to strengthen their positions in the world of research.

15 Policy Implications

The findings of this study offers several implications for national and international frameworks in the area of science, technology and healthcare research. The dominant countries in this area of research should work upon cross border collaborations. The global research policy makers may suggest measures to dominant countries like USA, UK and China for the benefit of developing and developed nations in terms of technology transfer, skill enhancement, funding opportunities etc. The developed nations such as USA and UK can establish an ecosystem for research and may continue to attract diversified investments. Meanwhile, other rapidly growing countries can strategically use funds to enhance visibility.

The regulatory bodies should consider partnerships with top-tier journals to disseminate best practices and research findings among healthcare training institutions. National governments and global health organizations should create structured research exchange programs and joint funding schemes with these leading nations to uplift research volume in the less represented countries.

The major institutions such as Kings college London and Monash University may offer institutionalize network and look for author collaborations to suggest shift towards multidisciplinary and cross-institutional research models.

16 Practical Implication

Because the USA, the UK, and China are so prominent as significant research contributors, institutions and academics all around the world should think about cultivating relationships with researchers from these countries. Such alliances may be advantageous to both parties, advancing international research networks and fostering information sharing.

From a practical perspective, these findings could affect how research money is distributed. While growing nations like Singapore may concentrate on providing high-impact research to improve their worldwide stature, developed nations with large research impact, like the USA, might continue to draw investment in a variety of sectors.

Academic institutions and universities can utilize this information to compare their research output to that of the world's leading institutions. They may make strategic changes and resource allocation decisions by identifying areas where they fail to reach.

In order to guarantee that the social benefits of their research are realised, nations with high average article citation rates should aggressively encourage knowledge transfer methods. Some examples are patents, industrial partnerships, and the commercialization of technology.

This information can support diplomatic efforts in the field of science since nations with significant impacts on the field of research are more likely to be involved in international scientific partnerships that strengthen connections between nations.

17 Social Implication

High-impact research from developed nations like the US and the UK has a significant influence on society by fostering inventions, advancements in science, and a higher standard of living.

Socially, these large citation counts show that the author has power and a global reputation. They boost a country's reputation by luring in talent and capital and, in turn, promoting socioeconomic development.

Scientific literacy is encouraged via high-impact research, which also increases public understanding of the subject. This could motivate more young people to seek jobs in the science sector.

Countries that do considerable research are expected to make large contributions to advancement in public health, technology, and general quality of life, potentially having an impact on the health and well-being of the entire world.

18 Managerial Implication

The data can be used for strategic resource allocation by institutions and government bodies. High-impact research and encouraging international cooperation should be supported by research policy.

Nations looking to expand their international research footprint might utilise these statistics for benchmarking and competitive tactics. It aids in locating research fields where they possess a competitive advantage.

High-impact research frequently results in inventions and has a significant economic impact. This can inform investments and policy choices from a management standpoint, especially in fields where research and innovation are essential.

Research managers and policymakers should coordinate resource allocation with the effect of their work. To stay competitive, high-impact research topics could need more financing.

From a management perspective, this information may direct investments in STEM (science, technology, engineering, and mathematics) education to create a skilled workforce capable of making a significant contribution to research.

Policymakers should encourage strategic international cooperation by utilising the know-how of nations with significant research influence to successfully solve global concerns.

Increasing the sophistication of assessment frameworks for both the impact and quality of research can be a managerial reaction to guarantee both high-impact and high-quality research.

19 Research Implication

To ensure that their work has the greatest possible impact, researchers should carefully select where to publish. Journals connected to nations with high average article citation rates may be selected.

Researchers might proactively look for partnerships with organisations and writers from nations with a solid reputation for research. The reach and effect of their work can be increased through collaboration.

For individual academics, this data highlights the significance of citations and impact indicators in assessing their work and advancing their careers. It emphasises the necessity of pursuing meaningful, high-quality research.

To better address complex global concerns, researchers can conduct interdisciplinary research by merging skills from many domains with an eye towards trends in high-impact nations.

The information can promote debates on research ethics by highlighting the necessity of responsible and ethical research practises, particularly in high-impact fields, to preserve the veracity of scientific knowledge.

20 Limitations and Research Opportunities

Despite the paper's contributions, certain limitations must be addressed because the data was only retrieved from one database, Scopus; however, other databases, such as Web of Science and PubMed, might be utilised to retrieve data for more research. Although we believe Scopus has numerous databases that we utilise for our study, data is also retrieved from other databases to cover a wide range of topics. We also examine materials that have been published in English, therefore the investigation will also be done in other languages. Only Biblioshiny is used to create tables and visualise data as a network, but other software, such as Vos Viewer, Cite Space, and Tableau, may also be used to analyse data for upcoming research projects. Further research can make use of other keywords associated with medical science. Advanced bibliometric analysis, such as co-citation analysis and bibliographic coupling and clustering, can be used by the researchers. The results of our study may aid other researchers in determining patterns and research areas in the medical sciences for future investigations.

References

Abdull Mutalib AA, Md. Akim A, Jaafar MH (2022) A systematic review of health sciences students' online learning during the COVID-19 pandemic. BMC Med Educ 22(1):524. https://doi.org/10.1186/s12909-022-03554-2

Alsoufi A, Alsuyihili A, Msherghi A, Elhadi A, Atiyah H, Ashini A et al (2020) Impact of the COVID-19 pandemic on medical education: Medical students' knowledge, attitudes, and practices regarding electronic learning. PLoS ONE 15(11):e0242905. https://doi.org/10.1371/journal.pone.0242905

Azim Majumder MA, Urban D'Souza UDS, Sayeeda Rahman SR (2004) Trends in medical education: challenges and directions for need-based reforms of medical training in South-East Asia. Indian J Med Sci 58(8):369–380

Bansal P, Supe A (2007) Training of medical teachers in India: need for change. Indian J Med Sci 61(8):472–480

Barenie RE, Cernasev A, Heidel RE, Stewart S, Hohmeier K (2023) Faculty, staff, and student perceptions of substance use disorder stigma in health profession training programs: a quantitative study. Substance Abuse Treat Prevent Policy 18(1):2. https://doi.org/10.1186/s13011-022-00491-6

Byrne-Davis L, Cohen SN, Turner RR (2022) Evaluating dermatology education and training. Clin Exp Dermatol 47(12):2096–2099. https://doi.org/10.1111/ced.15314

Correction to: Aligning ESG Goals with the UNs Sustainable Development Goals Planning

Firdous Ahmad Malik, Silky Vigg Kushwah, Muhammad Shahzad Aslam, Reyed M. Reyed, and A. K. Haghi

Correction to:
F. A. Malik et al. (eds.), *Aligning ESG Goals with the UNs Sustainable Development Goals Planning*, World Sustainability Series,
https://doi.org/10.1007/978-3-032-01715-4

The name of one of the authors in Chapters 14 and 19 was unfortunately published with an error. The originally published version has now been corrected.

The updated versions of these chapters can be found at
https://doi.org/10.1007/978-3-032-01715-4_14
https://doi.org/10.1007/978-3-032-01715-4_19

F. A. Malik et al. (eds.), *Aligning ESG Goals with the UNs Sustainable Development Goals Planning*, World Sustainability Series,
https://doi.org/10.1007/978-3-032-01715-4_23

MIX
Papier aus verantwortungsvollen Quellen
Paper from responsible sources
FSC® C105338

If you have any concerns about our products,
you can contact us on
ProductSafety@springernature.com

In case Publisher is established outside the EU,
the EU authorized representative is:
Springer Nature Customer Service Center GmbH
Europaplatz 3, 69115 Heidelberg, Germany

Printed by Libri Plureos GmbH
in Hamburg, Germany